Chemical Crystallography with
Pulsed Neutrons and Synchrotron X-Rays

NATO ASI Series

Advanced Science Institutes Series

A Series presenting the results of activities sponsored by the NATO Science Committee, which aims at the dissemination of advanced scientific and technological knowledge, with a view to strengthening links between scientific communities.

The series is published by an international board of publishers in conjunction with the NATO Scientific Affairs Division

A	Life Sciences	Plenum Publishing Corporation
B	Physics	London and New York
C	Mathematical and Physical Sciences	D. Reidel Publishing Company Dordrecht, Boston, Lancaster and Tokyo
D	Behavioural and Social Sciences	Martinus Nijhoff Publishers
E	Applied Sciences	Dordrecht, Boston and Lancaster
F	Computer and Systems Sciences	Springer-Verlag
G	Ecological Sciences	Berlin, Heidelberg, New York, London,
H	Cell Biology	Paris, and Tokyo

Series C: Mathematical and Physical Sciences Vol. 221

Chemical Crystallography with Pulsed Neutrons and Synchrotron X-Rays

edited by

Maria Arménia Carrondo

Centro de Quimica Estrutural,
Instituto Superior Técnico, Lisboa, Portugal

and

George A. Jeffrey

Department of Crystallography,
University of Pittsburgh, Pennsylvania, U.S.A.

D. Reidel Publishing Company

Dordrecht / Boston / Lancaster / Tokyo

Published in cooperation with NATO Scientific Affairs Division

Proceedings of the NATO Advanced Study Institute on
Chemical Crystallography with Pulsed Neutrons and Synchrotron X-Rays
Alvor, Algarve, Portugal
March 17-27, 1987

Library of Congress Cataloging in Publication Data

NATO Advanced Study Institute on Chemical Crystallography with Pulsed Neutrons and
Synchrotron X-Rays (1987: Alvor, Portugal)
    Chemical crystallography with pulsed neutrons and synchrotron x-rays / edited by Maria
Arménia Carrondo and George A. Jeffrey.
    p.   cm. — (NATO ASI series. Series C, Mathematical and physical sciences ; vol. 221.)
    "Proceedings of the NATO Advanced Study Institute on Chemical Crystallography with
Pulsed Neutrons and Synchroton X-Rays, Alvor, Algarve, Portugal, March 17–27, 1987"—t.p.
verso.
    "Published in cooperation with NATO Scientific Affairs Division."
    Includes index.
    ISBN-13: 978-94-010-8287-7   e-ISBN-13: 978-94-009-4027-7
    DOI: 10.1007/978-94-009-4027-7

    1. Crystallography—Congresses.   2. Pulsed neutron techniques—Congresses.
3. X-Ray crystallography—Congresses.   I. Carrondo, Maria Arménia, 1948–
II. Jeffrey, George A., 1915–   . III. Title. IV. Series: NATO ASI series. Series C,
Mathematical and physical sciences; no. 221.
QD951.N38   1987
548'.3—dc 19
                                                                              87–26853
                                                                                  CIP

Published by D. Reidel Publishing Company
P.O. Box 17, 3300 AA Dordrecht, Holland

Sold and distributed in the U.S.A. and Canada
by Kluwer Academic Publishers,
101 Philip Drive, Assinippi Park, Norwell, MA 02061, U.S.A.

In all other countries, sold and distributed
by Kluwer Academic Publishers Group,
P.O. Box 322, 3300 AH Dordrecht, Holland

D. Reidel Publishing Company is a member of the Kluwer Academic Publishers Group

Contents

Preface ix

Posters

Organizing Committee

P. Becker   Laboratoire de Cristallographie,
                           C.N.R.S., Grenoble, France

M.A. Carrondo ‡   Centro de Química Estrutural
                           Complexo I, Instituto Superior Técnico
                           Lisboa, Portugal

M.B. Hursthouse   Department of Chemistry, Queen Mary College,
                           University of London, U.K

G.A. Jeffrey ‡   Department of Crystallography
                           University of Pittsburgh, Pittsburgh, U.S.A.

G.H. Lander   European Institut for Transuranian Elements,
                           Karlsruhe

W. Saenger   Institur fur Kristallographie, Freie Universitat
                           Berlin, F.R.G.

‡ Co-Directors
* We deeply regret the absence of Prof. Simonetta from the Università di Milano, Italy, who had agreed to be a member of the Organizing Committee just before his death, in January 1986.

Preface

X-ray and neutron crystallography have played an increasingly important role in the chemical and biochemical sciences over the past fifty years. The principal obstacles in this methodology, the phase problem and computing, have been overcome. The former by the methods developed in the 1960's and just recognised by the 1985 Chemistry Nobel Prize award to Karle and Hauptman, the latter by the dramatic advances that have taken place in computer technology in the past twenty years.

Within the last decade, two new radiation sources have been added to the crystallographer's tools. One is synchrotron X-rays and the other is spallation neutrons. Both have much more powerful fluxes than the previous sources and they are pulsed rather than continuos. New techniques are necessary to fully exploit the intense continuos radiation spectrum and its pulsed property. Both radiations are only available from particular National Laboratories on a guest-user basis for scientists outside these National Laboratories.

Hitherto, the major emphasis on the use of these facilities has been in solid-state physics, and the material, engineering and biological sciences. We believe that there is equivalent potential to applications which are primarily chemical or biochemical.

We have combined synchrotron X-rays and pulsed spallation neutrons in this ASI for two reasons. One is because they have important common properties such that concepts developped by the instrumental scientists using one radiation could be useful to those using the other. The other reason is that both sources have new major facilities which have very recently become operational as for example, the Daresbury and Brookhaven dedicated X-ray synchrotrons and the Rutherford, Argonne and Los Alamos spallation neutron sources. For the near future more sources of both types are planned or in construction in Europe (ESRF), USA (6 GeV synchrotron source), Japan (ALS) and U.K. (ISIS II at RAL).

We have therefore brought to this meeting scientists who know the fundamental properties, advantages and limitations of both the pulsed neutrons and the synchrotron X-rays to interact with chemists, specially chemical crystallographers. The majority of these scientists have had no actual experience in using these national facilities. The portuguese and spanish scientific communities, in particular, shoul benefit not only from the lectures and tutorials but also from the opportunities to make direct contact with some of these instrumental scientists.

Since the participants are mainly chemists, the first part of the course had a strong educational component with emphasis on the basic physics involved in the production and use of these radiations.

The second part was aimed at discussing these techniques, at the frontiers of their applications, such as the exciting potential for real-time structural studies.

The book is organised with the lectures in the same sequence as presented. Each lecture was followed by a discussion, the main points of which are summarized after the lecture.

Prof. Jeffrey has always showed a great enthusiasm for spreading the knowledge of Chemical Crystallography among young researchers providing they are willing to learn the relevant Physics and Mathematics required to be masters of their science. He deplores the crescent trend towards "black-box" science.

This course was designed to overcome this type of deficiency. Prof Jeffrey's long-standing love for Portugal and the Portuguese people, was the "seed" of our association on the organization of this course.The very special climate and scenary of the Algarve, particularly in the early spring, proved to be ideal for the site of the course.

Historically the sea was always an inspiration for creative ideias in the human minds. And so it was in the XV century with Infante D. Henrique, son of the Portuguese King D. João I, who devoted his life to the great enterprise of the Portuguese discoveries departing from the Algarve. The scientific background for these explorations was developped in Sagres, in what is now considered as one of the first Naval Research Laboratories, created by D. Henrique. Appropriately, this was chosen as the site of our excursion.

We wish to thank the sponsors, NATO Scientific Affairs Division in Brussels, and to Instituto Nacional de Investigação Científica, Junta Nacional de Investigação Científica e Tecnológica and Fundação Calouste Gulbenkian in Lisbon, for their generous contributions.

We also thank the Organizing Committee for their help and guidance on the choice of lecturers. We are grateful to Prof. Skyes from Queen Mary

College, University of London, for the organization of the reporters and the preparation of the discussions after the lectures.

Maria Arménia Carrondo
Centro de Química Estrutural, Complexo I
Instituto Superior Técnico, Lisboa, Portugal

CHEMICAL CRYSTALLOGRAPHY: PAST, PRESENT AND FUTURE

G. A. Jeffrey
Department of Crystallography
University of Pittsburgh
Pittsburgh, PA 15260 USA

ABSTRACT. In its early days X-ray crystallography was practiced mainly by physicists, but in recent years the predominant interactions have been with chemists, with the molecular biologist running a close second place. Computer technology and the invention of the direct method for solving the phase problem have made X-ray crystal structure analysis a robotic procedure similar to solution NMR spectroscopy. An important frontier of the science now lies in the use of the special properties of the X-ray and neutron beams which will be discussed at this meeting.

1. PAST

Figure 1 shows the level of interaction between Crystallography and the other sciences as measured by the cross-citations in Acta Crystallographica from 1972-1976. The strongest correlations are with Chemistry.* This was not always so. When I was first introduced to X-ray Crystallography about half a century ago, it was a sub-discipline of Physics. In the hands of the physicists, who were the direct descendants of the founding fathers, von Laue, William and Lawrence Bragg, there had been a period of extraordinary success for X-ray Crystallography in the determination of the atomic structure of the metals, alloys, simple salts, minerals and that branch of inorganic chemistry which provides the structural basis of what is now called Material Sciences. The famous textbook entitled "Structural Inorganic Chemistry" was written when Wells was in a Physics Department. These achievements, which laid the structural basis of inorganic chemistry, physical metallurgy and mineralogy, were all the more extraordinary when one recalls the primitive and frequently home-made X-ray equipment which was then available. The latest in

---

*The fastest growing correlation is with Molecular Biology, but much of the work on the crystallography of biological macromolecules is published in biological journals. A present-day study which took this into account would probably give Chemistry and Biology equal weight.

1

M. A. Carrondo and G. A. Jeffrey (eds.), Chemical Crystallography with Pulsed Neutrons and Synchrotron X-Rays, 1–7.

2

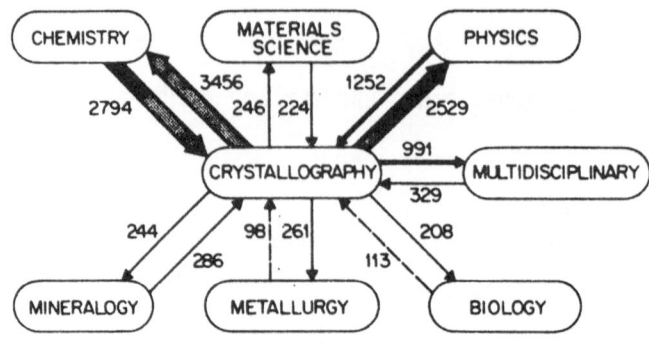

Figure 1

(reproduced with permission, from
Acta Crystallographica, A36, 475, 1980)

commercial equipment was the Unicam oscillation-rotation camera. J.
D. Bernal introduced the reciprocal lattice concept of Paul Ewald to
X-ray Crystallography by writing a paper (1) on how to interpret the
diffraction patterns it provided (2). X-ray Crystallography might
then have become synonomous with solid-state physics, which has since
developed into a quite different discipline, had it not been for the
influence of a few individuals such as J. D. Bernal who was perhaps
the first molecular biologist, Peter Debye, Montieth Robertson and
Linus Pauling who were chemists, and the mineralogist R. C. Evans, who
wrote a classic textbook entitled "Introduction to Crystal Chemistry";
a new Science!. The identity of Crystallography, as distinct from
Physics, was re-asserted by the formation of the International Union
of Crystallography, IUCr, in 1947 (3) which through its Congresses,
publications and commissions, has consistently reinforced its
independent interdiscipinary character. The theme of the scientific
programs of the IUCr Congresses has always been that "Crystallography
is to be developed as a meeting place for the Sciences, not as a
sub-discipline of any one of the major sciences."

By the 1950s, the new frontier for X-ray Crystallography was the
structure of organic molecules. Although the mathematics were known,
the tools for the calculations were hopelessly inadequate;
hand-operated mechanical adding and multiplying machines, the slide
rule, and mathematical tables. Some of these tables were ingeniously
designed, such as the Beevers-Lipson strips, but tables nevertheless.
A more serious deterrent, however, was the lack of interest by the
organic chemists in the three-dimensional shape of molecules. The
concept that a molecule actually had a shape, or, as we now say, a
conformation, seldom appeared in University courses. Until 1960,
chemistry was a configurational science. What was taught was the
properties of various groups, i.e., aldehydes, ketones, alcohols, etc.
and the memorization of syntheses, together with the names of their

designers. The concept of electronic polarization was developed by Ingold and Robertson, but was indicated by pot-hooks, placed in the context of two-dimensional connectivity diagrams. These two-dimensional connectivity diagrams, which adorned the organic text-books, and still do, were quite adequate for the language of chemistry at that time.

A further obstacle was that the physicists, who were the most knowledgeable about X-ray crystallography, and still are, have a natural aversion to organic chemistry. The chemical bond is an anachronism to a physicist, who likes to think of the electronic structure of a molecule as a whole, and express it with a mathematical equation.

The invention of the concept of conformational analysis by Barton and Hassel completely changed the attitude of the chemists. Atomic connectivity formulae ceased to be sufficient. The importance of the shape of the molecule to the organic chemist became recognized. With it came an appreciation of the value of a method capable of determining the shape of molecules, without ambiguity and with exquisite detail and accuracy.

Until about 1965, X-ray crystal structure determination of organic molecules was a very difficult and uncertain method. It was necessary for the investigator to have a good working knowledge of such conceptually difficult subjects as the reciprocal lattice, space group theory, Fourier theory and the Patterson, isomorphous replacement, heavy atom and the direct method, to solve the structure.

2. PRESENT

Now that has changed. The computer that operates the diffractometer is programmed to know about the reciprocal lattice and space group theory. The computer that solves the crystal structure does so using the prepackaged knowledge of a generation of crystallographers, including last year's Chemistry Nobel Laureates, who invented the direct method. The computer also knows how to refine the atomic parameters so as to get the best results from the experimental data. It has also been told how to present them in a way that the chemist can most easily understand them. Consequently, the chemist who operates these machines does not have to know any crystallography; just as the chemist who uses the NMR spectrometer for configurational analysis is not required to know the physics of nuclear magnetic resonance.

This computer packaging of X-ray crystal structure analysis has made the method so attractive to the chemist that it is regarded as an essential analytical tool in many University chemistry departments with major research efforts in organic and organo-metallic chemistry and in the biomedical research laboratories in Medical Institutes and in industrial laboratories. In fact, X-ray crystallography is practiced in almost all major research endeavours where molecular shape at the atomic level of resolution is perceived to be important.

More recent emphasis on the three-dimensional structure of assemblages of molecules, as in the drug industry, has further reinforced the importance of X-ray crystal structure analysis as an analytical tool. Unfortunately, this leads some chemists and molecular biologists to regard the whole of crystallography as an analytical tool. The tail threatens to wag the dog. This development seems to have alarmed some members of the crystallographic community.

This alarm is unwarranted in my opinion and in any case unlikely to change the course of history. Let me mention some advantages from this love-affair between chemistry and crystallography.

1. Some of the non-crystallographic robotic operators may become curious and want to learn about crystallography.

2. Hitherto these robots are not entirely trusted and this has led to a dramatic improvement in the crystallographer's job market, especially in the biomedical applications. The public cynicism concerning the infallibility of computer operators should be encouraged, in this regard.

3. Crystallographers who get permanent positions because of their ability to do service crystallography very efficiently, may become bored with this exercise and become interested in the more exploratory aspects of X-ray crystallography. One such aspect is the experimental measurement of charge-densities and electrostatic potentials. Other examples involve the use of the "big science" machines, which is the topic of this meeting. These are by no means black-box methods, requiring all the crystallographic knowledge and experience available to get significant results.

4. It is a relief to those crystallographers who use crystal structure analysis to support their own research that they are no longer expected to oblige their chemistry colleagues by doing crystal structure determinations on compounds that don't interest them.

Nevertheless, it is disconcerting to observe in the U.S. that the increase in the sale and use of crystal structure analysis equipment in recent years has not been paralleled by an increase in membership of the American Crystallographic Association, which has remained steady at less than 2000 members for the past ten years.* This constant membership suggests that there are an increasing number of chemists and biological scientists who use X-ray crystal structure analysis to get the results they require without the slightest desire to know more about crystallography than they have to.

3. FUTURE

Looking forward to the year 2000, there are two developments which may challenge the present role of X-ray crystal structure analysis in

---

*The ACA does have the distinction of having a higher ratio of both women and Nobel Laureates than any other scientific society (4).

chemistry. One is the rapid development of the so-called super-computers. These make it possible to calculate the three-dimensional atomic structure of molecules. Admittedly, <u>ab-initio</u> methods provide the structure of the isolated molecule at rest, which can, with some molecules, be as far removed from the molecule in the chemist's test-tube as is the molecule in the crystal. But the empirical, or semi-empirical methods, are becoming increasingly sophisticated and adept at calculating the structures of very large molecules. The effects of solvation are being approached by using Monte-Carlo methods. Given a large computer, these methods are easier to use than crystal structure analysis, and faster. Furthermore, they require neither the compound nor the crystals. It is possible to calculate the structure of the molecule and predict its relevant properties before you synthesize the compound, as is done, in fact, in the pharmaceutical industry. Just as NMR spectroscopy has taken over the one-time dominant role of X-ray crystallography for organic configurational analysis, because it is faster and easier, so computers may take over the role of conformational analysis.

The second competitor comes from the Cambridge Crystallographic Data File which contains the organic and organo-metallic crystal structures of more than 50,000 compounds, and seems to be increasing at the rate of about 7000 structures a year. By the year 2000, it should have passed the 100,000 mark. Except for the configurational analysis of a particular compound where NMR fails to give a unique result, I suspect that all the conformational information that the chemist requires about his molecules will be in the Cambridge Crystallographic Data Base, if it is not there already. Relatively few chemists have discovered this data base or succeeded in overcoming the initial hurdle of using it, but this hurdle will disappear as soon as the computer-generation school-boys get to graduate school.

Electrostatic potential surfaces of molecules can also be calculated theoretically, but X-ray crystallography makes it possible to explore experimentally how these potentials respond to the cohesive forces between molecules, providing a chemical crystallography topic of great promise for the future.

I think that it would be as much a mistake to let X-ray crystallography become a sub-discipline of chemistry, as it would have been to regard the structural research of the 1930s as a sub-discipline of physics. This would drop more curtains between the chemists and the physicists, molecular biologists, material scientists and mineralogists, which Crystallography is uniquely suited to keep raised.

With regard to the instruments that we will be discussing in this meeting, they have one property in common. They are very expensive to the taxpayers of the countries that support them. We must expect that the escalating costs of big science, and big medicine, will come under increasingly close scrutiny. These facilities have to prove their worth. Steady-state neutron diffraction is well-established as a chemical structural tool, not only because it gives accurate structural information relating to hydrogen atoms, but also because it

6

reveals the total structure of molecules where hydrogen atoms are chemically very important. But nuclear reactors are not socially popular and the spallation neutron sources are perceived to be less menacing. It remains to be seen whether they can, in fact, replace the nuclear research reactors.

The synchrotron X-ray source is in danger of being perceived to be no more than an inconveniently located X-ray tube by chemists who are accustomed to having X-rays in their own laboratory. The chemists, unlike the nuclear and solid-state physicists, are not attuned to doing their research outside their own laboratories at a large national or international facility. Neither their mores nor their operational budgets seem to be adjusted to spending long periods in teams away from their home.

The use of synchrotron X-rays simply as a high-powered X-ray tube is also being challenged by the scientific instrument makers using a combination of highly reliable rotating-target X-ray tubes with very efficient area detectors. When the X-ray diffraction patterns appear on a television screen in a matter of seconds in ones own laboratory, using a 60 kv, 200 ma rotating-anode X-ray tube, why travel, unless you like travelling. In my view, the major justification for these radiation sources will depend very much upon the unique properties of these radiations, which you will be hearing about during this meeting.

REFERENCES

(1) "The interpretation of rotation photographs is enormously simplified by the use of the mathematical device of the Reciprocal Lattice, first introduced by Ewald." -- from a paper on 'A Universal X-ray Photogoniometer Combining Apparatus for Single Rotation Photographs - Laue Photographs - X-ray Spectrometry - Powder Photogrpahs - Photographs of Crystal Aggregates, Metals, Materials, etc.,' J. D. Bernal, Journal of Scientific Instruments, 4, 273-284 (1927).

(2) Although the principle of the Weissenberg camera was described by K. Weissenberg in Zeits. für Physik. 23, 229-238 (1924), instruments were not manufactured commercially until much later.

(3) 'The Beginning of the Union of Crystallography,' P. P. Ewald, in Crystallography in North America, 1983, pp. 134-135, ACA, New York, NY.

(4) "Crystals and Nobels," G. A. Jeffrey, Physics Today 40, 9-10 (1987).

DISCUSSION

DATA BASE AND ROBOT CRYSTALLOGRAPHY

Dr M B Hursthouse considered that the Cambridge Crystallographic
Data Base would continue to play a positive role as a tool compiled,
and used, by crystallographers, but he shared the worry about the
trend towards 'robot' crystallography.

THE FUTURE OF NEUTRON SOURCES

Dr. J B Forsyth said that Governments saw spallation sources as both
cheaper and safer than nuclear reactors, any larger versions of
which would pose major technical and financial problems.

SCIENTIFIC OPPORTUNITIES WITH NEUTRON SCATTERING

G.H. Lander
European Institute for Transuranium Elements
Postfach 2340, D-7500 Karlsruhe, F.R.G.

and

Intense Pulsed Neutron Division
Argonne National Laboratory, Argonne, Illinois 60439, USA

ABSTRACT   A brief introduction is given to neutron scattering. Emphasis is given to references to textbooks rather than a complete treatise. Examples of vibrational spectroscopy and the determination of spin densities are then given to illustrate new applications of neutron scattering that are of interest to chemical crystallographers.

1.   INTRODUCTION

The neutron was discovered in 1932 by James Chadwick; it has a mass of 1.675×10^{-27} kg, no charge, a spin of 1/2, and a magnetic dipole moment of -1.913 nuclear magneton. Almost from the time of its discovery the unique properties of the neutron in investigating condensed matter were recognised but the fluxes of neutron beams were initially very low. With the building of high-flux reactors in the 1960s the full versatility of neutron scattering came to be appreciated. Neutron beams were initially used by nuclear physicists, then by solid-state physicists, and more recently by material scientists, chemists and biologists. This trend of diversification continues. Figure 1 shows the layout of the beam lines at the Institut Laue-Langevin in Grenoble, France. The ILL started operating in 1972 and is run by the French, German, and UK Governments. With its 36 working instruments and over 750 scheduled experiments per year the ILL is unquestionably the premier neutron-scattering center in the world. There are now some 1500 neutron "users" in Europe, and this number increases steadily as new sources such as the ISIS spallation project at Rutherford Appleton Laboratory are brought on line.

However, there is one major disadvantage of neutron scattering - the available fluxes are low. For example, even at the highest flux sources (ILL, Oak Ridge National Laboratory, and Brookhaven National Laboratory) the "monochromatic" beams ($\Delta E/E_0 \approx 10^{-2}$ where E_0 is the incident energy) have only about 10^8 neutrons cm^{-2}s^{-1} - a figure comparable to an ordinary sealed X-ray tube and therefore at least six orders of magnitude lower than can be expected at either modern synchrotron sources for X-ray fluxes or available from high-powered lasers for light scattering.

How do neutrons therefore compete? The answer is that neutrons are unique and special features of their interaction with matter are exploited. The basic principles of neutron scattering are discussed below and then two examples are chosen to show how unique information can be obtained.

9

M. A. Carrondo and G. A. Jeffrey (eds.), Chemical Crystallography with Pulsed Neutrons and Synchrotron X-Rays, 9–25.
© *1988 by D. Reidel Publishing Company.*

Fig 1 Schematic layout of beam tubes and instruments at the Institut Laue
Langevin's High Flux Reactor in Grenoble, France

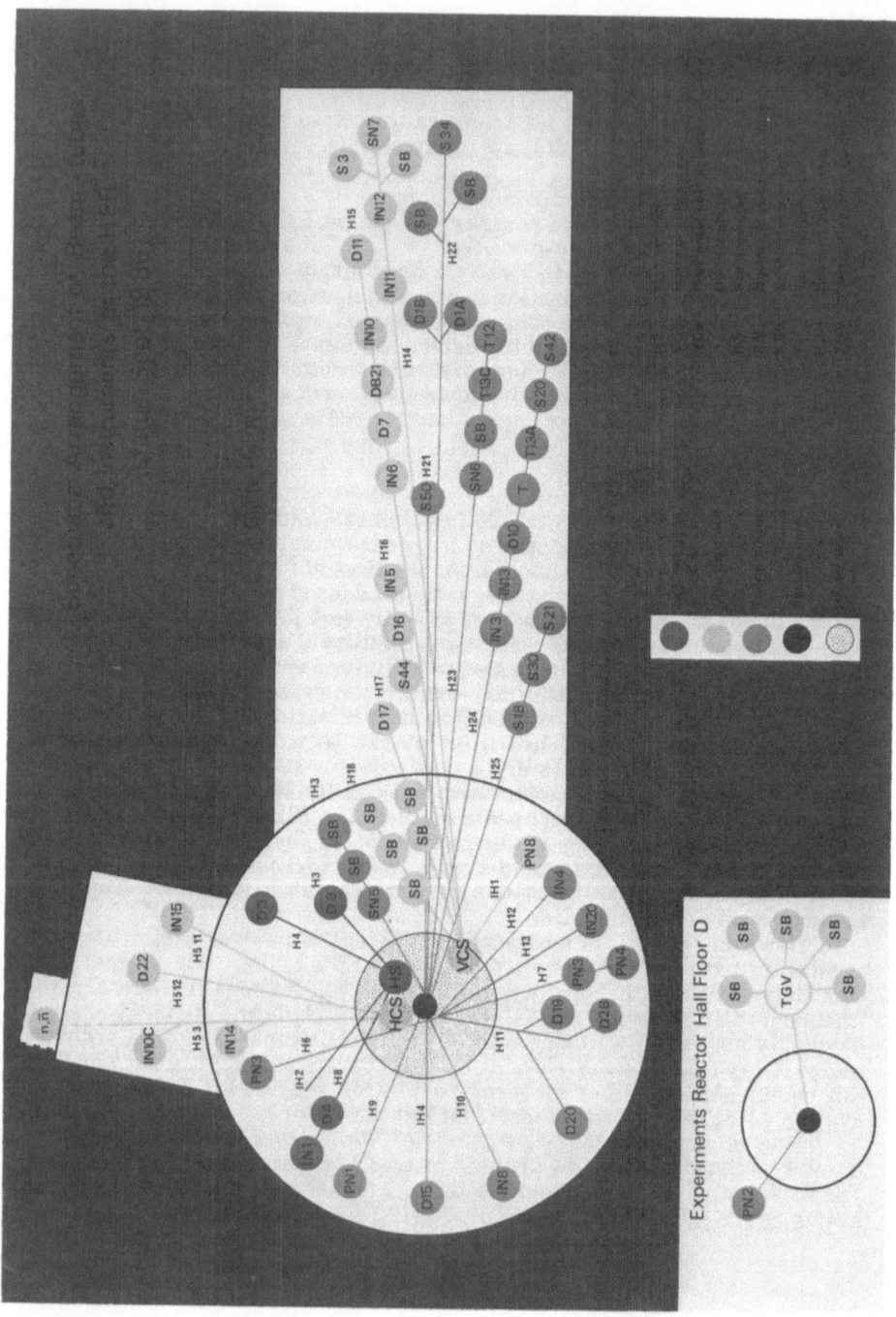

2. FUNDAMENTALS OF NEUTRON INTERACTIONS

References [1-6] are to various books and articles on neutron scattering. In this short article only important concepts will be covered.

2.1. Wave-particle duality

Neutrons are produced either in a fission reactor or by spallation when protons strike a heavy element target. The moderation (slowing down) of neutrons is then accomplished close to where they are produced and neutrons of "thermal" energies, i.e. E = 300K, emerge to be useful. Many energy units are used in neutron scattering, the more common are:

$$E = 0.08617T = 5.227v^2 = 81.81 \, 1/\lambda^2 = 2.072k^2 \qquad (1)$$

where E is in meV, T in °K, v in kms^{-1}, λ in Å and k in 10^{10} m^{-1}. Thus is E = 300 K (25.85 meV) = 1/2 mv^2 and using de Broglie's equation

$$\lambda = h/(mv) \qquad\qquad (2)$$

where h is Planck's constant we find that a neutron of 300K has a wavelength of 1.78 Å.

In fact the energy spectrum of neutron produced in either a reactor or a spallation source is a continuous one from 1 to ~ 1000 meV. By varying the material and temperature of the "moderator" different regions of the spectrum may be favoured, but there is never a characteristic radiation like one finds from X-ray tubes.

The above discussion immediately gives us two uses of neutron scattering
(a)   The wavelength of thermal neutrons is comparable to the interatomic spacing so that we may think of diffraction effects in the same way as we do with X-ray scattering (recall that for CuKα radiation $\lambda = 1.54$Å).

$$\text{Bragg's law is } 2d \sin \theta = \lambda \qquad\qquad (3)$$

where d = d_{hkl} is the interatomic spacing and θ is half the scattering angle.

From Eqs (1) and (2) above we can rewrite Bragg's law in terms of the *time* it takes a neutron to travel a distance L,

$$d_{hkl} = \left(\frac{0.198}{\sin \theta} \right) \frac{t}{L} \, \text{Å} \qquad\qquad (4)$$

where t is in µs and L in cm. Of course since all X-rays have speed c this analogy does not exist with X-rays, even though the phenomena of diffraction (and optics) are similar for the two radiations.
(b)   The energy of neutrons is *small*. Neutrons of 1.8 Å have an energy of 25meV. "Thermal" excitations in solids, e.g. phonons, have energies of between 1 and 15meV. The detection of these processes in neutron scattering is thus relatively simple as a relative large amount of energy can be transferred. In contrast, for CuKα radiation the energy is ~ 8keV so that it is very difficult to detect processes below ~ 0.5eV, which already represents $\Delta E/E_0 = 0.6 \times 10^{-4}$. Using slow neutrons and special techniques [4] energy transfers down to 1µeV (10^{-6}eV) can be detected by neutron scattering. We shall discuss some aspects of inelastic neutron scattering below.

A further advantage of neutrons is that because they have no charge they have very little absorption. (There are a few isotopes, e.g. ^{113}Cd and

^{157}Gd, in which at thermal energies the absorption cross section is very big, $\sigma_{abs} \simeq 20 \times 10^3$ for Cd, because of a neighbouring Breit-Wigner nuclear resonance, but these cases are rare in the periodic table). This means that neutrons penetrate well into materials, they are a bulk technique. The low absorption means that neutrons can penetrate deep into furnaces, cryostats, pressure cells, etc. to view samples under extreme conditions. For example, cryostat walls are made out of aluminium which has a small scattering and absorption cross section. The linear absorption coefficient for neutrons (X-rays) of 1.5 Å is 0.008cm^{-1} (131cm^{-1}) which means that 50% transmission will take place through the thickness of 86cm (5×10^{-3}cm) for neutrons (X-rays). This is quite a difference!

2.2   Nuclear interactions

The fact that the neutron has zero charge means that it penetrates deep into the atom, i.e. it avoids electrostatic repulsion, and will interact with nuclear forces. If we had a complete theory of nuclear forces we could calculate the resulting interaction from first principles, but we do not have such a theory, so the scattering potential is an experimentally determined quantity. The nuclear forces have a range of 10^{-13} to 10^{-12}cm whereas the neutron has a wavelength of $\sim 10^{-8}$cm (1Å). Under these conditions the nucleus acts as a point scatterer, and the nuclear potential may be characterized as a single number.

From scattering theory the total cross section σ is given by the total solid angle (4π) times the square of a scattering potential, which has units of length

$$\sigma_T = 4\pi b^2 \qquad (5)$$

Since the interaction is nuclear it depends on the isotopic state of the nucleus. *Thus different isotopes can have different scattering potentials.* This is an important advantage over scattering of electromagnetic radiation, which except at certain resonance energies, depends linearly on the number of electrons around the scattering nucleus. Furthermore, the neutron has a spin $\pm 1/2$. If the nucleus has spin I then a compound nucleus of spin state $I \pm 1/2$ can be formed and these have different cross sections.

In a normal solid at normal temperatures all these effects are present and it is possible to take average quantities. In fact we can define two cross sections:
(a)   The coherent cross section. Here we take the canonical average over all isotopes and possible spin states

$$b_{coh} = \sum_i f_i b_i \qquad (6)$$

where b_i occurs with relative frequency f_i. If one works through the scattering formulae (see, e.g. Squires p. 22) one can show that the coherent scattering depends on the correlation between the position of the *average* nucleus at different times, and *different* nuclei at different times. In crystallography we are interested in the space correlation of the different average nuclei with respect to one another. All this information comes from *elastic coherent scattering.* Similarly, if we are interested in how assemblies of atoms move (e.g. the wave like motion of a phonon through a solid) then we are interested in *inelastic coherent scattering.*

(b)   The incoherent cross section. Clearly there is an "extra" part of the cross section that arises from the random distribution of the scattering lengths from their mean value

$$b_{inc}^2 = \left(\overline{b_i^2} - b_{coh}^2 \right) \qquad (7)$$

The incoherent cross section gives no interference effects but gives information about the correlation between the position of the *same* nucleus at different times. In the case of elastic scattering (all times equal zero) the incoherent cross section gives no information, other than the presence of such nuclei, e.g. background effects. Thus crystallographers ignore the incoherent cross section. However, if we consider *inelastic incoherent scattering* then the situation is more interesting: here we can observe vibrations of individual nuclei as well as how they move (or diffuse) through the lattice. We shall return to this below.

What are the relative magnitude of σ_{coh} and σ_{inc}? Fortunately, σ_{inc} is almost always small. The magnitude of b_{coh} (recall $\sigma = 4\pi b^2$) is shown for a few elements in Figure 2. Tables of these scattering lengths appear in most of the references 1 to 6. Notice some important differences between the X-ray and neutron values. Some neutron values (e.g.[1]H, Mn, and Ti) are negative; this represents a change of phase because of a nuclear resonance. This situation does not exist in X-rays. The X-ray values of heavy elements are

Fig. 2 The visibility of some atoms and isotopes for X-rays and neutrons. The radii of the circles are proportional to the scattering amplitude b. Negative values of b are indicated by the shading. Taken from Ref. 1. Note that isotopes 1 and 2 are for the element hydrogen, and 58, 60, and 62 are all isotopes of nickel.

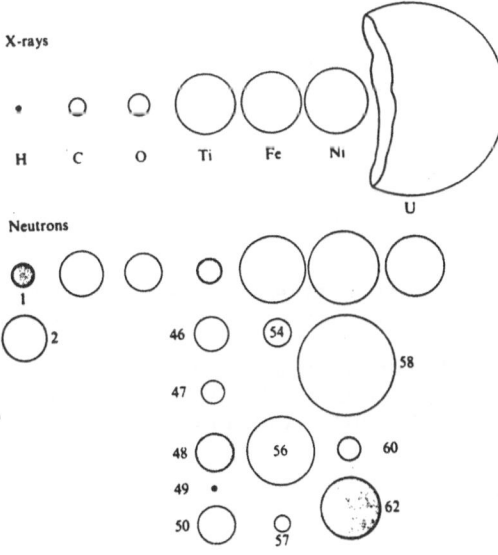

much bigger than neutron values. Thus if one is searching for a light atom in the presence of heavy ones, e.g. experiments on UO_2, neutrons are the preferred technique. ^1H and ^2H (deuterium) have scattering lengths of opposite sign. This suggests that if one substitutes ^2H for ^1H then at a certain value (35,9%) substitution the hydrogen will be invisible to neutrons! This the basis for a whole subfield of neutron scattering, particularly applicable to chemical solutions and biology, and is usually called "contrast variation".

All of the above is about the coherent cross section. What about σ_{inc}? There are only two elements where σ_{inc} is important. One is hydrogen

^1H $\qquad b_{coh} = -0.374 \times 10^{-12} cm$
$\qquad \sigma_{coh} = 1.76 \times 10^{-24} cm^2$ ($10^{-24} cm^2 = 1$ barn)
$\qquad \sigma_{inc} = 80$ barns
$^2H(D)$ $\qquad b_{coh} = +0.667 \times 10^{-12} cm$
$\qquad \sigma_{coh} = 5.59$ barns
$\qquad \sigma_{inc} = 2$ barns

where $\sigma_{inc}^H > 10\sigma_{inc}$ of any other element. The incoherent inelastic scattering is thus completely dominated by the signal of any ^1H atoms present. The second example is V for which $\sigma_{inc}^V = 5.1$ barns. Vanadium has a very small b_{coh} because the two compound nuclei $I \pm 1/2$ have large, but opposite, cross sections. This makes it usual to make sample containers, furnaces, etc. out of V, but the background from σ_{inc} should be kept in mind.

2.3 Magnetic interaction

We noted earlier that the neutron had a magnetic moment. This moment can interact with the moment of the unpaired electrons surrounding a nucleus. For example, the $3d$ shell of Fe has 6 $3d$ electrons that give rise to the magnetic moment in elemental iron and many of its compounds. The interaction of the neutron with these unpaired electrons gives rise to magnetic scattering and enables us to determine unique information about the spatial and dynamical nature of the unpaired electrons in the $3d$, $4d$, $4f$ and $5f$ atomic shells.

Magnetic scattering is more complex than the nuclear interaction we discussed in the previous section.
(a) It can arise from both spin and orbital (i.e. magnetism arising from both the number and motion of the unpaired electrons) contributions.
(b) Because it arises from unpaired electrons that are in orbits away from the nucleus (e.g. the maximum expectation value of a $3d$ electron occurs at $\sim 0.8\text{Å}$ from the nucleus) we cannot expect the simple rule of a "point" scatterer that we found in 2.1 to still be applicable.
(c) Because "spin" is a vector quantity there is a vector relation in magnetic scattering that allows us to determine in which direction the atomic moments are pointing.

We shall return later to magnetic scattering. Although the chemical crystallographer might appear to have little use of this, there are some special applications that demonstrate the unique power of neutrons in chemical matters.

3. SCATTERING THEORY

For details the reader is referred to Ref. [1-6]; here we will just give a few formulae and concepts that are used in the following sections. Again using the de Broglie relationship the neutron momentum **p** may be written

$$p = mv = h/\lambda = \frac{h}{2\pi} \cdot \frac{2\pi}{\lambda} = \hbar k$$

where **k** is the neutron wavevector. For a scattering process (X-rays, neutrons, or electrons) we will have an incident beam of wavevector k_i, a scattered beam of wavevector k_f and the momentum transfer **Q** is defined

$$\mathbf{Q} = \mathbf{k}_i - \mathbf{k}_f \qquad (8)$$

For *elastic* scattering $|k_i| = |k_f| = k$ so that a simple geometrical construction gives

$$Q = 2k\sin\theta = 4\pi\sin\theta/\lambda \qquad (9)$$

so that using Bragg's law $Q = 2\pi/d_{hkl}$. If the neutron gains or loses energy to the system then the energy change ΔE or $h\omega$ is given by

$$\hbar\omega = \Delta E = \frac{\hbar}{2m}(k_i^2 - k_f^2) \qquad (10)$$

3.1 Cross sections

If we consider a neutron beam of incident wavevector k_i incident on a system in a state Ø and interacting via a potential V to leave the system in a state Ø' and the neutron beam with a wavevector k_f then the matrix element we need to calculate the scattering cross section is

$$<k_f\,\text{Ø'}\,|\,V\,|\,k_i\,\text{Ø}> = \int V(r)\,exp\,(i\,\mathbf{Q}.\,r)\,dr$$

where V(r) is a periodic potential and we make use of the fact that the neutron wavefunction may be written as a plane wave.

We can keep the matrix element notation in writing down the differential cross sections, i.e. with respect to a solid angle $d\Omega$ and an energy interval $d\omega$ so that

$$\left(\frac{d^2\sigma}{d\Omega dw}\right)_{coh} = \frac{\sigma_{coh}}{4\pi}\frac{k_f}{k_i}\frac{1}{2\pi\hbar}\sum_{jj'}\int_{-\infty}^{+\infty}<e^{-i\mathbf{Q}.\mathbf{R}_{j'}(0)}\,e^{+i\mathbf{Q}.\mathbf{R}_j(t)}>e^{-i\omega t}\,dt \qquad (11)$$

$$\left(\frac{d^2\sigma}{d\Omega dw}\right)_{inc} = \frac{\sigma_{inc}}{4\pi}\frac{k_f}{k_i}\frac{1}{2\pi\hbar}\sum_{j}\int_{-\infty}^{+\infty}<e^{-i\mathbf{Q}.\mathbf{R}_j(0)}\,e^{+i\mathbf{Q}.\mathbf{R}_j(t)}>e^{-i\omega t}\,dt \qquad (12)$$

where $R_j(t)$ is the position of the j-th nucleus at a time t.

Although Eqs. (11) and (12) are not immediately useful in working out the intensity of any scattering process, they do remind us of the difference between coherent (atom j to atom j') and incoherent (same atom) cross sections, and they also explicitly introduce the time parameter. Of course, for *elastic* scattering as one is interested in crystallography one puts $t=0$ and it is the correlation between atom j and j' at $t=0$. In the case of a periodic function such as a crystal the cross section then becomes directly proportional to the *structure factor*

$$\left(\frac{d\sigma}{d\Omega}\right)_{coh} \alpha \sum_{\tau} \delta(\mathbf{Q}-\tau)|F_N(\mathbf{Q})|^2 \qquad (13)$$

where the nuclear structure factor is

$$F_N(\mathbf{Q}) = \sum_j \overline{b}_j \, exp \, (i\,\mathbf{Q}\cdot\mathbf{R}_j) \, e^{-W_j} \qquad (14)$$

and the delta function indicates that nonzero values are only found at reciprocal lattice vectors such that $\mathbf{Q} = \tau$.

In Eq. (14) the sum is over all atoms in the unit cell and W_j is the Debye-Waller factor. This latter can be thought of as an attenuation factor because the atoms are not stationary but have a small random motion about their equilibrium position $\mathbf{R}_j$. The Debye-Waller factor may be more formally written as

$$W_j = \frac{1}{2} <(\mathbf{Q}\cdot\mathbf{U}_{lj})^2> \qquad (15)$$

where U_{lj} are the normal modes of atom j and this expansion shows that the extent of the Debye-Waller factor is also proportional to Q^2.

4. MAGNETIC SCATTERING AND SPIN DENSITIES

We have discussed briefly above (Sec. 2.3) the general features of magnetic scattering. We now turn to a technique that is of particular interest to chemical crystallographers. Most of the interesting features in the chemistry of the transition and f elements arise from the fact that they have unfilled valence electron shells. In most cases these outer electrons participate in the bonding process and their electron orbits frequently define the spatial character of the packing that occurs in the solid. A determination of the spatial extent of these outer electrons would, of course, be a major step forward in understanding how the solid is formed. Such studies have been completed for light atoms and agree well with modern theory [7], but experiments with X-rays on heavier atoms, e.g. 3d series, are always faced with the problem of the closed shell. Take, for example, the extreme case of uranium with 92 electrons. 86 of these electrons are in the radon core and are of relatively little interest except to fill up space. We need to know the spatial extent of the 6 remaining electrons to a high degree of precision and clearly this is difficult with X-ray scattering, which will always be sensitive to *all* electrons.

In certain cases neutrons offer an alternative approach. This arises because the outer electrons are unpaired and frequently have a large susceptibility. Thus, if we apply a magnetic field the outer electron states are "polarized" and a static magnetic moment exists. The neutron interacts with this moment. Clearly, this technique sees *only* the electrons of interest, but has the disadvantage that it sees the difference distribution between "up" and "down" polarized states.

4.1. Experimental technique of spin density determination

Reviews of this technique and its application to chemical systems have been given by Day [8] and Brown [9]. Our first step is to realize that we measure a difference between two states $\Phi_\uparrow(r)$ and $\Phi_\downarrow(r)$, where these correspond to wavefunctions parallel and antiparallel to the applied magnetic field. If we assume that the spatial extent of these two wavefunctions is the same then

$$\rho_M(\mathbf{r}) = \sum_i (n_\uparrow - n_\downarrow)_i |\Phi_i(\mathbf{r})|^2 = H \sum_i |\Phi_i(\mathbf{r})|^2 \chi_i \qquad (16)$$

where χ_i is the susceptibility of the electron state i. Since χ of the closed shell is usually very small (actually negative), we can in practice correct for this term. This magnetization density is the exact analogy of the charge density in X-ray scattering, and is a periodic function in real space. Thus we may represent it as a Fourier sum of coefficients

$$\rho_M(\mathbf{r}) = \mathbf{M}(\mathbf{r}) = \sum_j \mathbf{F}_M^j(\mathbf{Q}) \, exp(i\mathbf{Q} \cdot \mathbf{r}) \qquad (17)$$

where the sum is over all Fourier coefficients, i.e. all possible Q values. The terms $F_M(\mathbf{Q})$ are the magnetic structure factors and the exact analogue of the nuclear structure factor of Eq. (14). Note that they are vector quantities because the magnetic moment μ_j is a vector. If all the magnetization is associated with a given atom (i.e. there is no appreciable overlap of wavefunction), then

$$\mathbf{F}_M(\mathbf{Q}) = \sum_j \mathbf{\mu}_j f(\mathbf{Q}) \, exp(i\mathbf{Q} \cdot \mathbf{R}_j) e^{-W_j} \qquad (18)$$

where the sum is over the unit cell with coordinates R_j for the different moments μ_j. A term $f(\mathbf{Q})$ has been added here to recognize that the magnetic moment is *not* a point scatterer and has a finite extent in real space. In general $f(\mathbf{Q})$ functions fall monotonically from 1 at $Q=0$ so this indicates that the magnetic intensity [α $f^2(\mathbf{Q})$] will fall steadily as Q increases.

In fact we can show that the magnetic cross section is proportional to

$$|\widehat{\mathbf{Q}} \times \mathbf{F}_M(\mathbf{Q}) \times \widehat{\mathbf{Q}}| = |F_{M\perp}(\mathbf{Q})|^2 \qquad (19)$$

where $\widehat{\mathbf{Q}}$ are unit vectors in the direction of Q, and this means that neutrons are sensitive to the component of *magnetization perpendicular to the Q vector*. Put another way, if $\mathbf{Q} \parallel \mu$ then no magnetic signal is seen - and this is the basis for the investigations of many magnetic structures.

Returning to our problem of a magnetization density we see the vector interaction between a neutron of polarization $\pm P$ leads to a change of sign of the magnetic structure factor, provided μ and P are collinear. This is the case in an induced system in which the neutrons are polarized first parallel and then antiparallel to H, which is also parallel to μ. The neutron intensity will then be proportional to

$$|F_N(\mathbf{Q}) \pm F_M(\mathbf{Q})|^2$$

provided both structure factors are evaluated at the same Q vector. In this case, of course, the induced magnetic structure factor has the same repeat unit as the nuclear structure. The experiment consists of measuring the intensity at each Bragg peak as a function of whether the neutron is polarized parallel or antiparallel to the applied magnetic field.

4.2. Experiment on UCl_4

We shall discuss briefly here the spin density studies that have recently been completed on UCl_4 [10] . These are interesting because they are the first time "covalency" has been seen directly in a $5f$ system. At first glance UCl_4 would seem like a standard ionic system; U^{4+}: $4Cl^-$. Thus the valency is 4. A useful way of thinking about this is that the 4 promoted electrons from the U site, together with the $3p^5$ state on each Cl ion, form four bonding orbitals. This leaves over two $5f$ localized electrons which give rise to the measured susceptibility. The p-d antibonding orbitals are empty. This is a very simple and clean picture and we can calculate exactly the resulting spin density from the $5f^2$ localized electrons. What happens in practice?

Two sections through the difference (observed minus calculated) maps are shown in Fig. 3. There are clearly significant positive perturbations near the U atom, with the atom itself being in a negative hole. The major features of this difference map suggest a simple transfer of spin from the $5f$ Y_0^3 orbital to the $6d(Y_2^2 + Y_{-2}^2)$ orbital. This simple model of the difference plot is shown in Fig. 4.

Our expectation of the magnetization density is that it would be completely $5f$ like, corresponding to a $5f^2$ configuration and tetravalent uranium. This is $\sim 90\%$ correct, but there is a small $6d$ component. The conclusion is that we are seeing a finite susceptibility from the mixing of a small part of the p-d antibonding orbitals with the $5f$ localized state. By implication there must be $5f$ character in the p-d bonding states and we define this as covalency.

The polarized-beam technique we have described here is in fact not a very complex one. There are difficulties of interpretation of the data and special conditions, i.e. obtaining single crystals which do not have phase transitions and have sizeable susceptibilities, but the chemical information obtained [8-10] is of considerable value. It seems likely that more scientists will become interested in these chemical studies as more sophisticated instruments become built. At present only the D3 diffractometer at the ILL has sufficient neutron flux and versatility for performing these studies.

Fig. 3 Two sections of the <u>difference</u> (i.e. observed minus calculated) magnetisation density in UCl4. (a) Section (x 1/4 z) through the uranium and all chlorine atoms: ● uranium Δ chlorine (b) section (x, y, 7/8) through the uranium atom. The chlorine atoms Δ are slightly above or below the plane. Positive and negative contours are shown as continuous and broken lines respectively. The significance level is slightly greater than one contour. All differences are averaged over a cube of side 0.5Å to reduce series termination effects

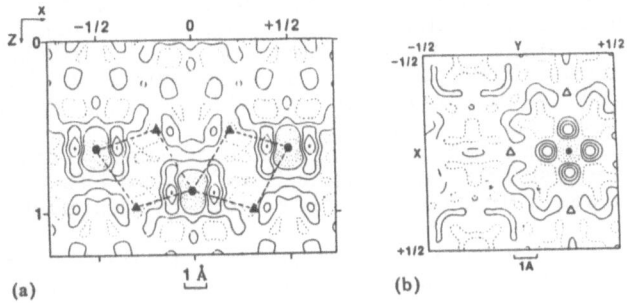

Fig. 4 Modelled density of positive contribution from 6d ($Y_2^2 + Y_{-2}^2$) and negative 5f Y_0^3 at the uranium site taken together and Fourier inverted to give a direct comparison to the observed difference map of Fig. 3

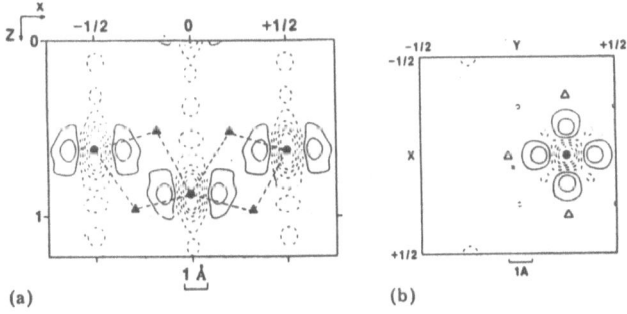

5. INELASTIC NEUTRON SCATTERING

We discussed in Sec. 2.2. some general features of inelastic scattering and we shall return here to some examples of interest to chemical crystallographers. The field of inelastic scattering is a very wide one and unique to neutrons because of their low energy. We recall from Sec. 2 that both the coherent and incoherent cross sections give rise to inelastic scattering. Although the coherent cross section gives us information about phonon modes in the system, the main interest for chemists is in the incoherent cross section from hydrogen atoms. Recall that the incoherent cross section of ^1H is ~ 10 times bigger than that of any other element. Thus, whereas light scattering from vibrational modes is proportional to the polarisability of the atoms, neutron scattering is related directly to simple physical quantitites. Thus Eq. (12) may be reduced to the following form (Ref. 1, p. 300):

$$\left(\frac{d^2\sigma}{d\Omega d\omega}\right)_{inc} = \left(\frac{k_f}{k_i}\right) \sum_j (b_j)^2_{inc} \frac{Q^2 u_j^2}{2M_j} e^{-2W_j} \left\{ \frac{1}{e^{\hbar\omega/kT}-1} + \frac{1}{2}(1\pm1) \right\} x\, N g(\omega) \qquad (20)$$

where μ_j is the atom displacement, M_j the mass, the curly bracket { } gives the two conditions depending whether the neutron gains or loses energy to the system, and $g(\omega)$ is the number of modes of energy $\hbar\omega$ such that $g(\omega)d\omega$ = 3N with N being the number of atoms. The other symbols have been defined previously. Note especially the proportionality to $b_{inc}^2 u^2/M$, which again strongly favours hydrogen with a large b_{inc}, usually large displacements u of the atom, and a small mass.

How is such an experiment performed? A schematic diagram is shown in Fig. 5. This instrument is particularly well suited to a pulsed source, which gives a well-defined pulse of many neutron energies as depicted in A.

Fig. 5 Schematic of neutron inelastic scattering with the so-called constant k_f technique. In A a sharp pulse consisting of many different wavelengths starts from a pulsed source. B is the sample position at which point the λ (or energy) distribution is unchanged, but because of the 12m flight path (L₁) the time distribution is quite wide. C represents the analyzing mirror which selects one energy E_f (a k_f a λ_f) and at D a detector records these neutrons as a function of time

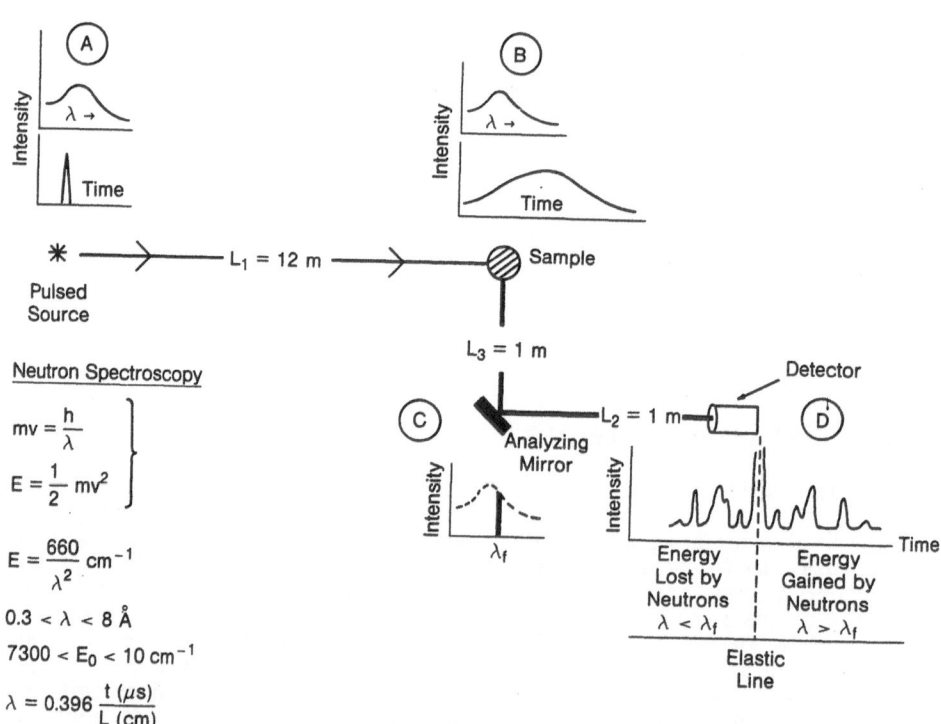

Note that this is quite a simple and compact instrument ($L_2 = L_3 = 1m$) situated at some $\sim 12m(L_1)$ from a pulsed source. Instruments of this type can have a resolution of $\sim 1\%$ of energy transfer over a wide range of energy, say 10 to $5000cm^{-1}$ ($8.065cm^{-1} = 1meV$). They have one disadvantage. Since k_f is fixed and is usually small ($E_f \simeq 4meV \simeq 32cm^{-1}$) in the equation (8)

$$Q = k_i - k_f$$

$|Q| \simeq |k_i|$ if $|k_f|$ is small. Thus we have no control over the momentum transfer Q. As we have seen in Eq. (20), the intensity of the incoherent inelastic scattering basically increases as $Q^2 e^{-2W}$ so moderately large Q's are acceptable. Unfortunately, Eq. (20) represents the probability of a single scattering process and in a real experiment we have to take account of the possibility of double scattering, which is proportional to $Q^4 e^{-2W}$. This puts a restriction on the useful Q range. Other, more complex, neutron instruments [5,6] can fully exploit the Q dependence.

5.1. Examples of inelastic scattering

We show in Fig. 6 an example of how particular vibrational modes can be identified by selective deuteration. Here one can see most clearly that, for example, the mode at $\sim 32meV$ must come predominantly from vibrations of the NH_3 radical. Similarly, the low energy vibrations of the C_6H_5 radical are clearly at $\sim 5meV$. Such studies can give unique information on the potentials in such chemical systems and together with modelling and chemical knowledge can frequently give architectural (structural) information which may not be available from crystallographic techniques [11].

Fig. 6 Example of neutron inelastic scattering illustrating the usefulness of selective deuteration. It should be recalled that D ($= {}^2H$) has a incoherent cross section ~ 40 times less than ^1H

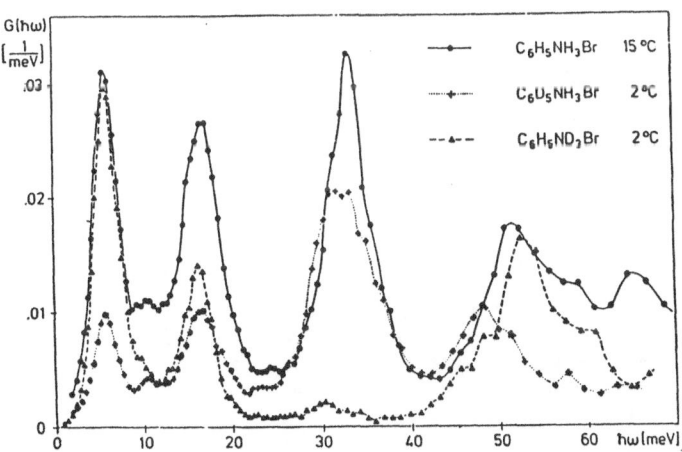

An excellent example of the latter kind of study has recently been performed at the Argonne pulsed source by Brun et al [12] on tetramethyl ammonium cations occluded within zeolites. In synthesizing zeolites, which are used extensively in the catalytic industry, an important technique is so-called "templating" to produce zeolite structures of desired properties. This process consists of letting the zeolite form around an inorganic cation, such as $N(CH_2CH_2CH_3)_4^+$. These large inorganic cations are occluded in the zeolite and when the material is heated they are driven off, leaving large holes that provide the catalytic positions for further molecular reduction. However, the mechanism of this "templating" process remains ill defined. Inelastic neutron scattering has been recorded on $N(CH_2CH_2CH_3)Br$, abbreviated to TMA-Br and TMA occluded in two zeolites Linde type A (LTA) and omega MAZ (Zeolites are basically sodium aluminium silicates). The spectra are shown in Fig. 7.

Fig. 7 Inelastic scattering form TMA-Br (top) and TMA occluded in LTA and MAZ zeolites. Vertical lines represent the values of torsional and bending modes from the calculation for a free ion. See Ref. [11]

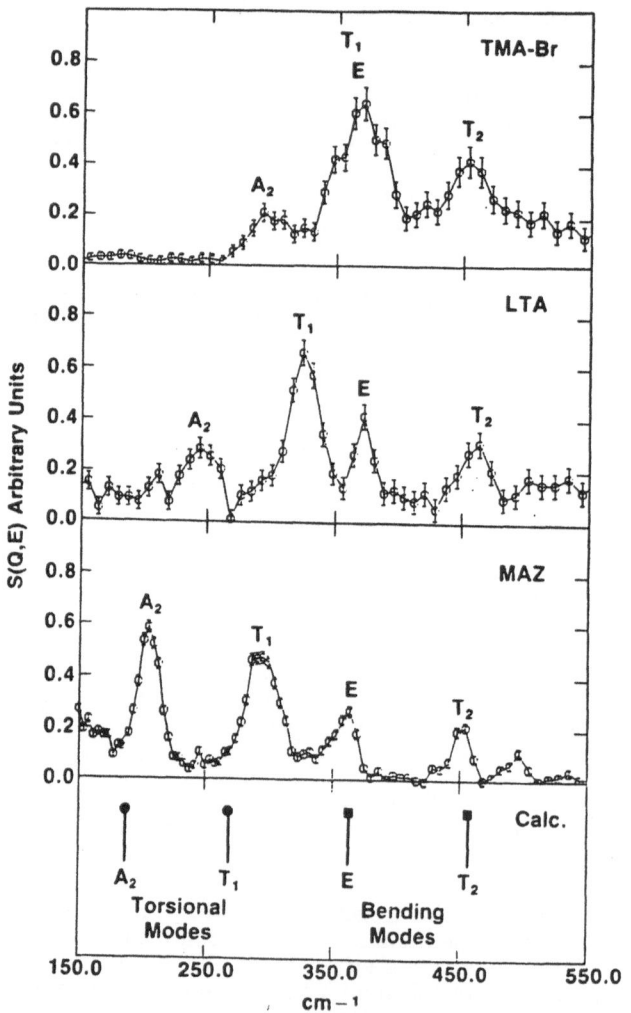

The first point is that it is surprising how free the cations are in the zeolite cages. Notice that the bending modes are basically unaltered from the free-ion to the cation in the zeolite. This was known from Raman scattering since the bending modes are Raman active. However, the torsional modes are not, and they are clearly sensitive to the environment. Since the frequencies are lower than in the halide salt the interaction with the zeolite cage is weaker than with the halide ions. Further the frequencies are even lower for the larger pore MAZ zeolite, and almost represent a free ion as judged by the calculated values.

Observing these torsional modes thus represents a method to observe the interaction between the "template" and the zeolite cage. They give, in a certain sense, structural information which can be exploited in the design of more useful and versatile zeolite catalysts.

6. CONCLUSIONS

We will end by stating again the advantages of neutrons as a microscopic probe
(1) Wavelengths of 1-10Å are ideal for the study of interatomic spatial correlations.
(2) Interaction is with the nuclei. This varies erratically across the periodic table and is often different for different isotopes of the same element. This leads to the possibility of "labelling" and the very productive field of contrast variation, specifically using the big difference between H and D.
(3) Thermal neutrons have an energy of 25meV ($\sim$300K, 200cm-1), whereas X-rays of 1Å have an energy of 12 keV and 1Å electrons an energy of 3.5eV. Neutrons with energies 1-10^3meV allow excitations in the range 1μeV to 0.5eV to be measured as a function of the wavevector Q.
(4) The neutron has a magnetic moment. Both static and dynamic magnetic phenomena can be investigated.
(5) Because of their low absorption neutrons penetrate into materials and are a bulk technique. They can also be used to penetrate furnaces, cryostats, and pressure cells, and so examine materials under extreme conditions.

REFERENCES

[1] For general reading see "Neutron diffraction",G.E. Bacon, Clarendon Press, Oxford, 1975
[2] For emphasis on structural aspects see "Thermal Neutron diffraction", editor B.T.M. Willis, Oxford University Press, Oxford, 1970
[3] For theoretical aspects see "Theory of Thermal Neutron Scattering" W. Marshall and S. Lovesey, Oxford University Press, 1971, revised edition in two volumes by S.W. Lovesey, OUP 1984
See also "Introduction to the Thermal Neutron Scattering" by G.L. Squires, Cambridge University Press, 1971
[4] For an emphasis on Materials Science and techniques such as small-angle scattering see "Treatise on Materials Science and Technology", Vol. 15, ed. G. Kostorz, Academic Press, NY 1970
[5] For pulsed neutrons see "Pulsed Neutron Scattering" C.G. Windsor, Taylor and Francis, London, 1981. Instrumentation at such sources is covered by J.M. Carpenter, G.H. Lander, and C.G. Windsor, Rev. Sci. Inst. 55, 1019 (1984)
[6] A new book on neutron scattering has been published by Academic Press, Methods in Experimental Physics, Volume 23, edited by K. Skold and D.L. Price, 1986
[7] M.L. Cohen, Science 234, 549 (1986)

[8] P.Day, J. Phys. (Paris) 43 C7-341 (1982)

[9] P.J. Brown, J.B. Forsyth, and R. Mason, Phil. Trans. Soc. (London) Ser B290, 481 (1980); P.J. Brown Chemica Scripta 26 433 (1986)

[10] G.H. Lander, P.J. Brown, M.R. Spirlet, J. Rebizant, B. Kanellakopulos, and R. Klenze, J. Chem. Phys. 83 5988 (1985)

[11] Many good examples of chemical information derived from inelastic scattering are given in J.W. White and C.G. Windsor, Rep. Prog. Phys. 47 707 (1984)

[12] T.O. Brun, L.A. Curtiss, L.E. Iton, R. Kleb, J.M. Newsam, R.A. Beyerlein, and D.E.W. Vaughan, J. Amer. Chem. Soc. (to be published)

DISCUSSION

ENERGY ANALYSIS OF X-RAYS AND NEUTRONS

X-ray energies can be determined by energy-dispersive detectors, e.g. Li-drifted Ge. Neutrons are detected by nuclear capture processes which are not significantly energy dependent in the meV range involved, so any energy analysis must be provided before detection takes place.

POLARISATION STUDIES

Polarisation analysis of the scattered beam has been used to distinguish between coherent and incoherent inelastic nuclear scattering, but the further development of polarisation analysis needs high neutron intensities and, for the white beam technique at short wavelengths the provision of expensive polarised neutron filters.

THE SCOPE AND POSSIBILITIES OF CRYSTALLOGRAPHY WITH PULSED NEUTRONS

W.I.F. David
Neutron Division
Rutherford Appleton Laboratory
Chilton, Didcot
Oxon OX11 OQX
U.K.

Abstract

Spallation neutron sources, based on proton synchrotrons, have firmly established pulsed neutron scattering as comparable and complementary to reactor-based neutron scattering. The projected substantial increase in neutron flux and upgrade in instrumentation at spallation neutron sources over the next 10-15 years lead to extrapolated "orders of magnitude" improvements in source performance and resulting science. A selective discussion of several present and future diffraction experiments associated with the present performance and proposed upgrade of the ISIS spallation neutron source is presented.

1. Introduction

The basic ideas of time-of-flight diffraction using pulsed neutron scattering have been known for over 30 years (Lovde 1956; Ringo 1957; Egelstaff 1961). Although the first experiments at pulsed neutron sources were performed as early as 1960 using an electron accelerator at General Atomic, USA (McReynolds and Whittemore 1961) and a pulsed reactor (Blokin and Blokhintser 1961), pulsed neutron sources did not threaten to rival high-flux reactors such as the Institut Laue Langevin, Grenoble, France, until the advent of proton accelerators in the late 1970s. Indeed, the Intense Pulsed Neutron Sources (IPNS) at the Argonne National Laboratory, USA, which began operation in 1981, was the first spallation neutron source that was equipped with a suite of instruments that could be

27

M. A. Carrondo and G. A. Jeffrey (eds.), Chemical Crystallography with Pulsed Neutrons and Synchrotron X-Rays, 27–59.
© 1988 by D. Reidel Publishing Company.

routinely used for condensed matter science. In the 1980s several new spallation neutron sources have been constructed at KENS (Japan), ISIS (UK) and LANSCE (USA). These sources have firmly established pulsed neutron scattering as comparable and complementary to reactor-based neutron scattering. Indeed, the projected substantial increase in neutron flux and upgrade in instrumentation at spallation neutron sources over the next 10-15 years lead to extrapolated "orders of magnitude" improvements in source performance (Table 1). It was concluded, for example, in the proposal for a German spallation neutron source, SNQ, that "it is possible to make great benefit from the enhanced flux during the pulse and to achieve intensities which would be impossible even on a next generation high flux reactor" (Scherm 1984). An accurate assessment of the scope and possibilities of crystallography using pulsed sources is thus difficult and, as a consequence, this article is limited to a selective discussion of present and future science associated with the present performance and proposed upgrade of the ISIS spallation neutron source at the Rutherford Appleton Laboratory.

Table 1   Pulsed spallation neutron sources

Date	Source	Country	Proton Current (μA)	Target	Frequency (Hz)
1972	ZING P	USA	0.04	Pb	10
1976	ZING P'	USA	1→10	Pb	10
1979	KENS-I	Japan	1.5	W	20
1979	WNR	USA	3	W	120
1981	IPNS	USA	10→15	^{238}U	30
1985	KENS-I'	Japan	10	^{238}U	15
1985	LANSCE	USA	30(100)	W	24
1985	ISIS	UK	50(200)	^{238}U	50
1987	IPNS	USA	15	^{235}U	30
1990	ISIS-II	UK	320	W	10
			1280	W	"50"

2. Pulsed Spallation Neutron Sources

Neutrons are produced at spallation sources in a pulsed manner at a repetition rate of typically 24 Hz (LANSCE) to 50 Hz (ISIS) (see Figure 1). A beam of protons, with energies of the order of 500 - 800 MeV, is delivered on to a target such as uranium or tungsten in a burst lasting around 0.4 μs. Copious numbers of neutrons are produced (~ 25 neutrons per incident proton for ^{238}U) principally by being spalled ("chipped") off the uranium nuclei. These neutrons, with energies of the order of MeV, possess wavelengths that are too short for diffraction purposes and must be slowed down to thermal energies using moderators such as water, polyethylene, liquid methane or liquid hydrogen. The precise characteristics of the moderator are governed by the nature of the neutron scattering experiment. The thermal neutron flux may be enhanced by building a moderator in which neutrons undergo a large number of collisions with protons. However, since each collision takes time it follows that enhanced thermal fluxes are accompanied by pulse broadening. High intensity and high resolution instruments tend to view different moderators. The characteristic "white" neutron spectrum thus produced has not only a thermal component but also a pronounced epithermal component ($\lambda < 1$ Å) resulting from the slowing down mechanism prior to thermalization within the moderator (Figure 2). This substantial short wavelength flux is potentially one of the most significant advantages of spallation sources over high-flux reactors for detailed crystallographic studies.

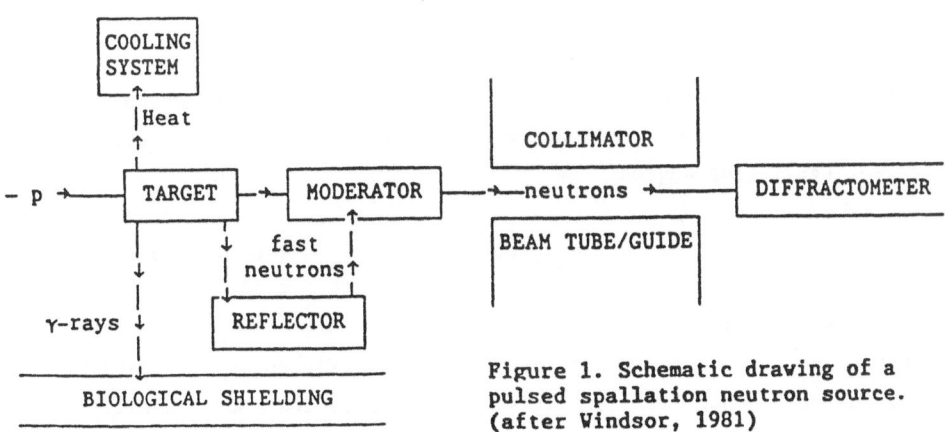

Figure 1. Schematic drawing of a pulsed spallation neutron source. (after Windsor, 1981)

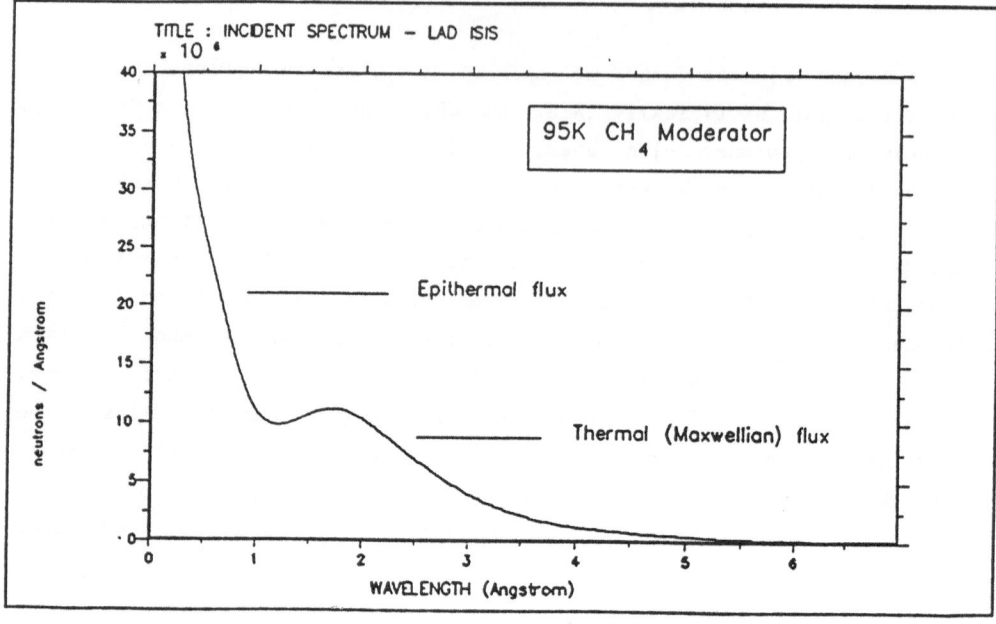

Figure 2.   Neutron flux distribution on a TOF machine

3. Basic Time-of-flight Diffraction

Neutron   diffraction   experiments,   whether   at   a   reactor   or   on   a
spallation neutron source, are governed by Bragg's law,

$$\lambda = 2 \, d \sin \theta$$

In   general,   on   reactor-based instruments a monochromatic neutron beam is
produced   and   different   d   spacings   are measured by moving a detector to
different scattering angles. i.e.

$$\lambda_o = 2 \, d_{hkl} \sin \theta_{hkl} \qquad \text{(monochromatic)}$$

However,   at   spallation   neutron   sources,   because   the   neutron beam is
produced   in   a   pulsed   manner,   neutrons with different wavelengths may be
discriminated by their time of arrival at the detector and thus different d
spacings may be measured at a fixed scattering angle. i.e.

$$\lambda_{hkl} = 2\,d_{hkl}\sin\theta_o \qquad\qquad \text{(polychromatic)}$$

The linear relationship between the wavelength of a neutron and its time of flight may be obtained from de Broglie's hypothesis relating momentum, $p = m_n v$, to wavelength:

$$p = m_n v = m_n(L/t) = h/\lambda$$

where L is the total combined flight path from moderator → sample (primary flight path, L_1) and sample → detector (secondary flight path, L_2). t is the time of flight of the neutron over this distance. Thus

$$t_{hkl} = (m_n/h)L\lambda_{hkl} = 2(m_n/h)Ld_{hkl}\sin\theta_o$$

In convenient units these equations become

$$t_{hkl}(\mu s) = 252.778\ L(m.)\ \lambda_{hkl}(\text{Å}) = 505.555\ L(m.)\ d_{hkl}(\text{Å})\sin\theta_o$$

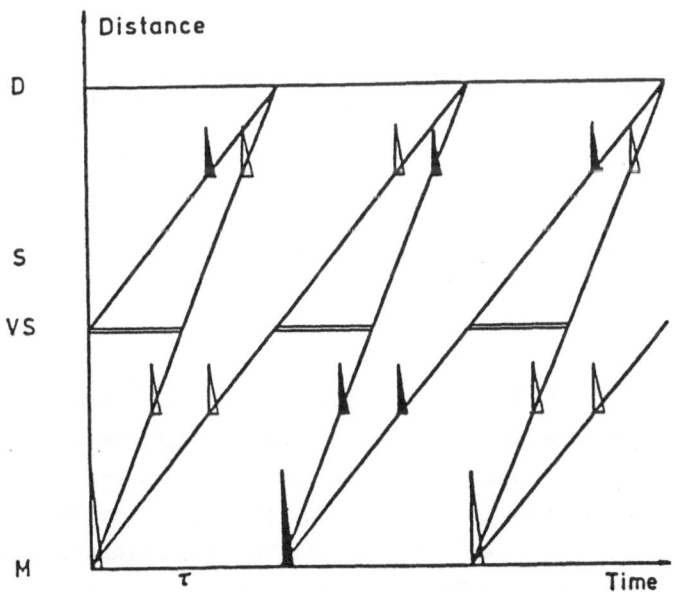

Figure 3. Space / time diagram for pulsed neutron source

Diffraction peaks separate in time (and thus resolution improves) with increasing flight path. Frame overlap becomes a problem with long flight paths but may be eliminated by using velocity selector (VS) choppers.

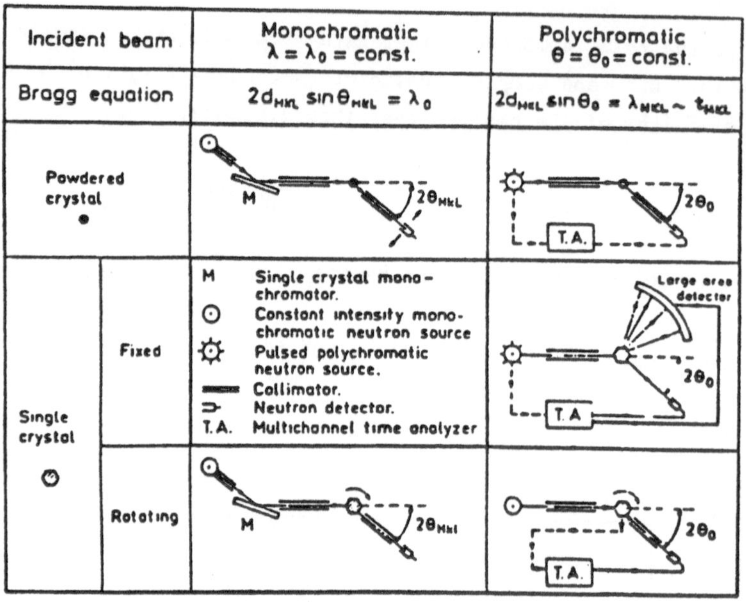

Incident beam	Monochromatic $\lambda = \lambda_0 = $ const.	Polychromatic $\theta = \theta_0 = $ const.
Bragg equation	$2d_{HKL} \sin \theta_{HKL} = \lambda_0$	$2d_{HKL} \sin \theta_0 = \lambda_{HKL} \sim t_{HKL}$
Powdered crystal ●		
Single crystal ◎ — Fixed	M Single crystal monochromator. ⊙ Constant intensity monochromatic neutron source ☼ Pulsed polychromatic neutron source. ▬ Collimator. ⇨ Neutron detector. T.A. Multichannel time analyzer	
Single crystal ◎ — Rotating		

Figure 4. Diagrammatic comparison of TOF and constant wavelength diffractometers

This relationship may be illustrated by a space-time diagram (Figure 3). The linear dependence of time of flight on flight path gives straight lines the slopes of which are proportional to wavelength and inversely proportional to velocity. Neutrons of different wavelengths clearly disperse as a function of time and flight path. Figure 3 also contains another important feature. Given that there is a finite time width to the initial neutron pulse, neutrons of a particular wavelength will propagate non-dispersively (because wavelength and hence velocity are constant) with a pulse structure that is independent of flight path. The principal consequence of this flight-path independent pulse structure is illustrated in Figure 3: resolution improves with increased flight path.

Diffractometers at spallation neutron sources use the time-of-flight technique for measuring diffraction patterns. Figure 4 compares powder and single crystal diffractometers at spallation sources and reactors. In both cases the ability to "time stamp" neutrons at spallation sources leads to a fixed detector configuration and consequent reduction by one of the number of diffractometer axes compared to equivalent reactor machines.

In practice, time-of-flight powder diffractometers generally have banks of detectors at several scattering angles to maximise count rate and to access the largest possible range of d spacings. Large solid-angle area detectors are optimal for TOF single crystal diffractometers allowing volumes of reciprocal space to be scanned simultaneously (see Figure 5). Some scientific consequences of the ability to volume scan reciprocal space will be discussed further in section 5.2.

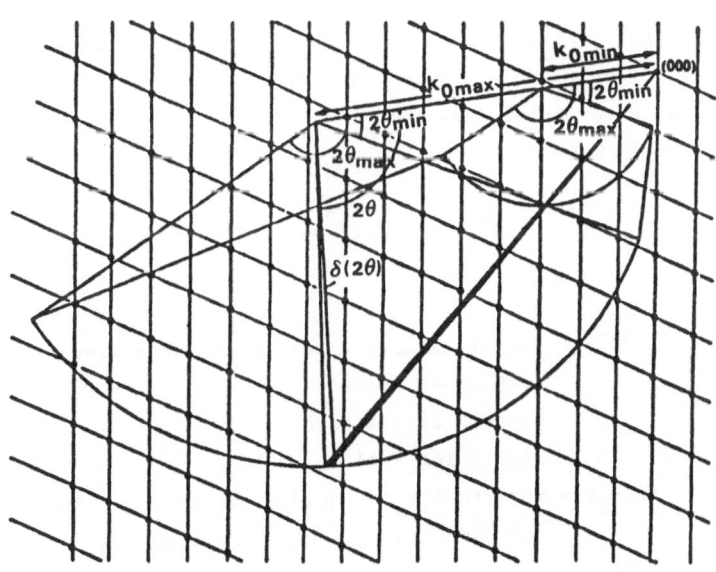

Figure 5. Example of accessible volume of reciprocal space with TOF single crystal diffractometer

4. Resolution considerations

As the neutron production mechanism at a spallation neutron source is rather complicated it is not surprising that the pulse shape is complex. Although a detailed account of pulse shape and instrumental resolution is outwith the scope of this article the basic aspects of overall instrumental resolution (for a TOF powder diffractometer) may be obtained in a simple manner. Consider neutrons of a particular wavelength, λ. They may be "born" at either the front or the back of a moderator, a distance uncertainty , ΔL, typically of the order of 5cm. Because of the slab shape of the moderator this gives an associated full-width-at-half-maximum (FWHM) fractional flight-path uncertainty, $\Delta L/L$, of 0.034/L(metres) which because of the linear relationship between L, t and d gives:

$$(\Delta L/L) = (\Delta t/t) = (\Delta d/d)(\text{moderator}) = 0.034/L(\text{metres})$$

Thus, neglecting other resolution terms (a good approximation in backscattering), it follows that resolution improves linearly with distance. This is graphically illustrated in Figure 6 which compares diffraction patterns of standard Al_2O_3 taken on HRPD (overall flight path 98m.) and LAD (overall flight path 11m.) at ISIS. The resolution gain of HRPD over LAD is of the order of 10 (the ratio of the flight paths).

Other resolution terms arising from geometrical uncertainties associated with sample (S) and detector (D) size may be gathered together in variance form as

$$\sigma^2 = (\Delta l/L)^2 + [(\Delta\theta_1)^2 + (\Delta\theta_2)^2]\cot^2\theta$$

where

$(\Delta l)^2$ = sum of squares of distances travelled in sample and detector

L = overall flight path

2θ = scattering angle

$\Delta(2\theta_1)$ = uncertainty in incident angle from extended sample and moderator/guide/pre-collimation

and $\Delta(2\theta_2)$ = uncertainty in scattering angle from extended sample and detector

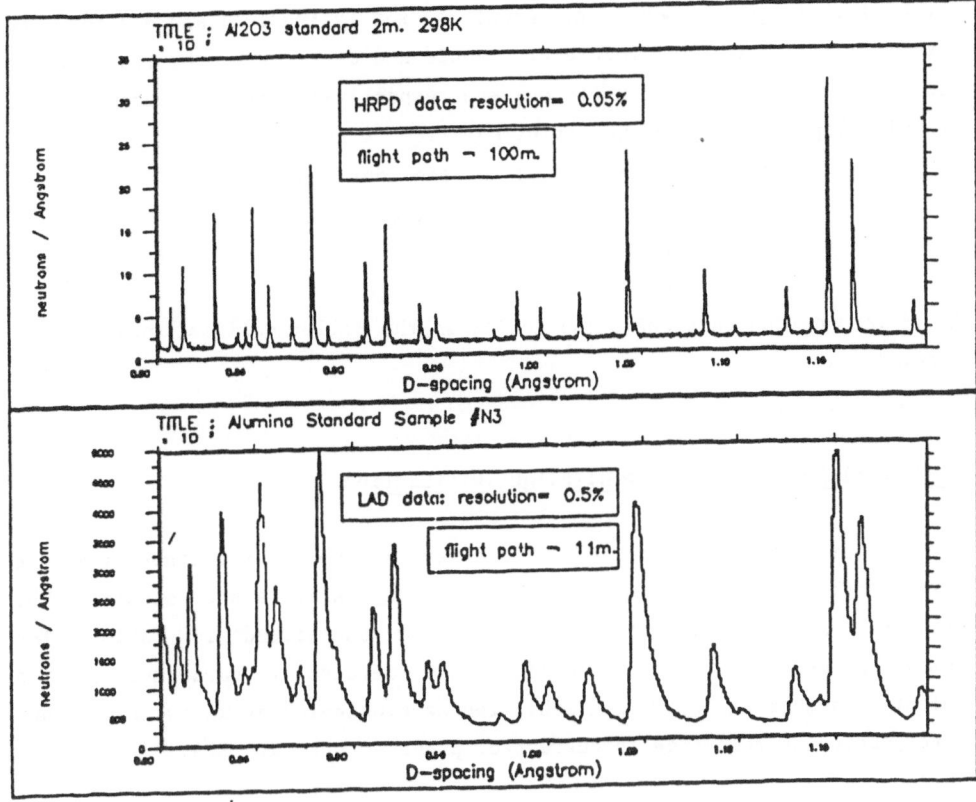

Figure 6. Resolution comparison between LAD and HRPD at ISIS

An empirical measure of TOF powder diffractometer resolution may be
obtained from the full-width at half-maximum expression

$$(\Delta d/d)_{FWHM} = (0.034/L) + 0.0029\sqrt{\left[\ LENSQ + (BEFSQ+AFTSQ)*cot^2\theta)\ \right]}$$

where

$$LENSQ = 4(\eta_i S_i^2 + \eta_o S_o^2 + \eta_D D_D^2)/L^2$$

$$AFTSQ = (\eta_o S_o^2 + \eta_W D_W^2)/L_2^2$$

$$BEFSQ = \begin{cases} (\eta_i S_i^2 + 4M^2/3)/L_1^2 & \text{– beam tube} \\ (0.4\lambda)^2 & \text{– Ni guide} \\ (0.034\Delta\phi)^2 & \text{– Soller collimation} \end{cases}$$

where the shape factors, η, equal 1 for cylindrical and 4/3 for slab geometry: M, S and D are moderator, sample and detector widths respectively (in cm.) – L_1, L_2 and L (in m.) are respectively primary, secondary and overall flight paths. The subscript notation is as follows:

> i = sample shape "seen" by incident beam
>
> o = sample shape "seen" by outgoing beam
>
> D = depth of detector
>
> W = width of detector

λ is the neutron wavelength (Å) and $\Delta\phi$ is the Soller collimation angle (in minutes).

5. Special Features of Time-of-Flight Diffraction

Time-of-flight diffraction offers several important features that are particularly advantageous for various crystallographic studies. The principal aspects of TOF diffraction are summarised in Table 2. The remaining sections of this article are devoted to a discussion of a selected number of crystallographic topics (indicated by an asterisk) that have particular relevance to current or proposed work at ISIS.

5.1 Samples under high pressure – special sample environments

Scattering at fixed angles has a clear advantage in difficult sample environments such as pressure cells, since sample containers need only have fixed windows to allow the passage of incoming and outgoing beams at one (or a few) selected angles. With suitable masking, $2\theta = 90°$ is the optimal scattering angle for a TOF diffractometer. This is dramatically illustrated by comparing diffraction patterns obtained on fixed-wavelength (D1a, ILL, Grenoble by Kuhs, Finney, Vettier and Bliss, 1984) and TOF (SEPD, IPNS, Argonne National Laboratory (ex IPNS Users Bulletin, September 1984)) powder diffractometers from a high pressure form of ice (Figure 7). In the fixed-wavelength case the diffraction pattern is completely dominated by the Al_2O_3 pressure cell. The refinement of the ice structure from the comparatively small sample peaks (shown in block in Figure 7a) represents a tour-de-force in terms of profile refinement. In contrast there are no peaks in the SEPD pattern from the pressure cell (Figure 7b).

The TOF special features important for various studies (after Buras, 1975)

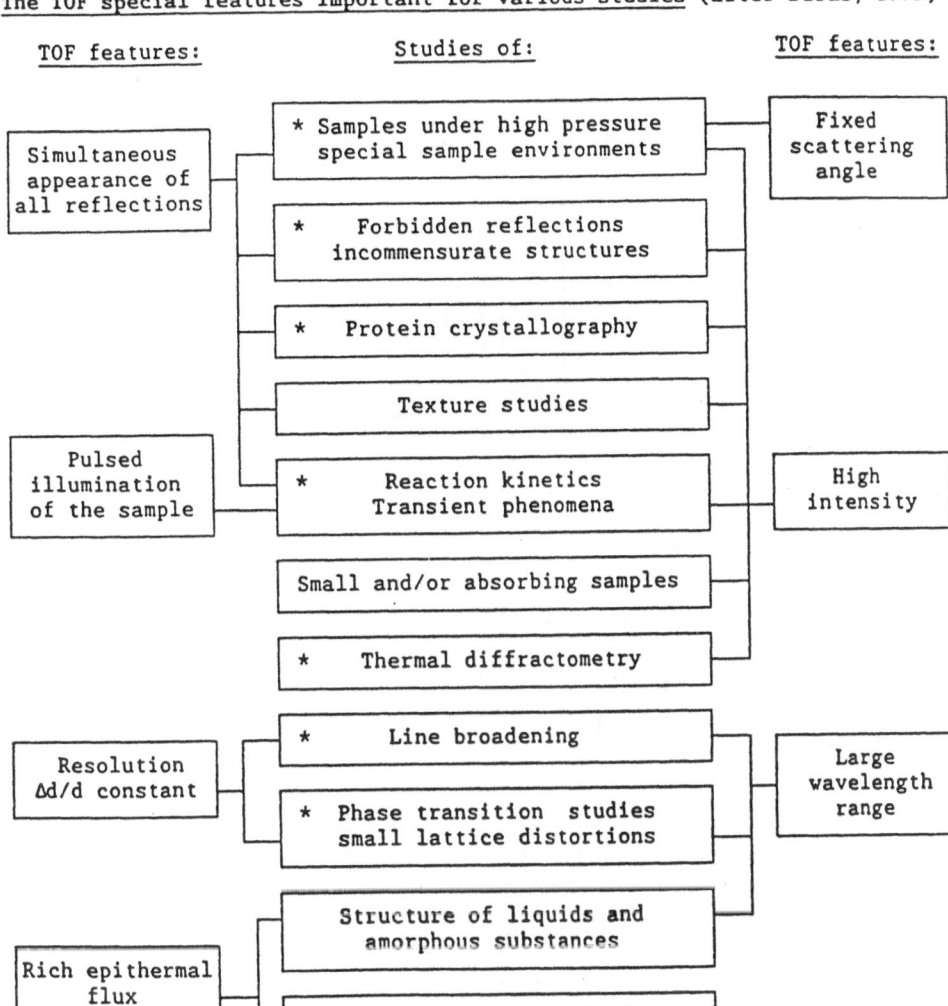

TOF features: Studies of: TOF features:

Simultaneous appearance of all reflections

* Samples under high pressure special sample environments

Fixed scattering angle

* Forbidden reflections incommensurate structures

* Protein crystallography

Texture studies

Pulsed illumination of the sample

* Reaction kinetics Transient phenomena

High intensity

Small and/or absorbing samples

* Thermal diffractometry

* Line broadening

Resolution Δd/d constant

Large wavelength range

* Phase transition studies small lattice distortions

Structure of liquids and amorphous substances

Rich epithermal flux

* High real space resolution studies: anharmonic behaviour order/disorder studies

Table 2

38

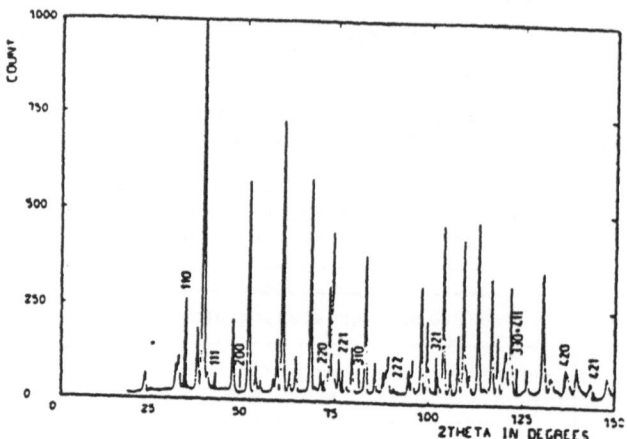

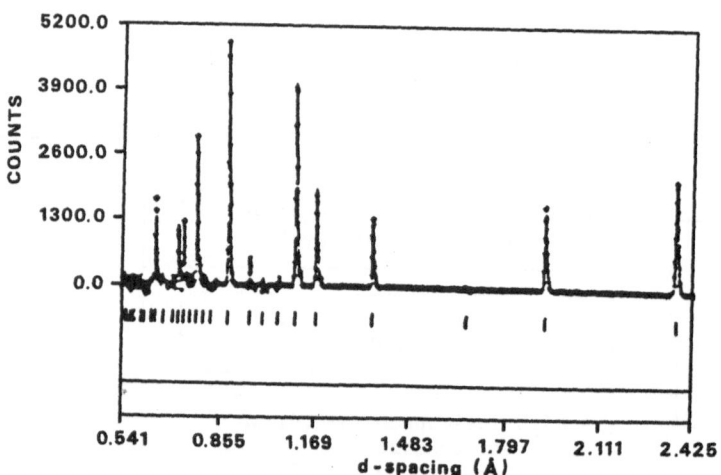

Figure 7. Diffraction patterns for ice VII recorded on D1a (up) and SEPD (90 degrees). D1a pattern: most reflections are from alumina pressure cell – sample peaks are blocked in. SEPD pattern: all peaks are from sample.

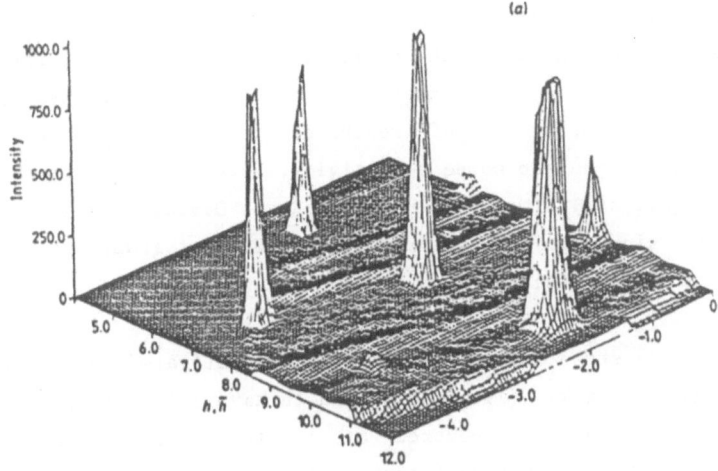

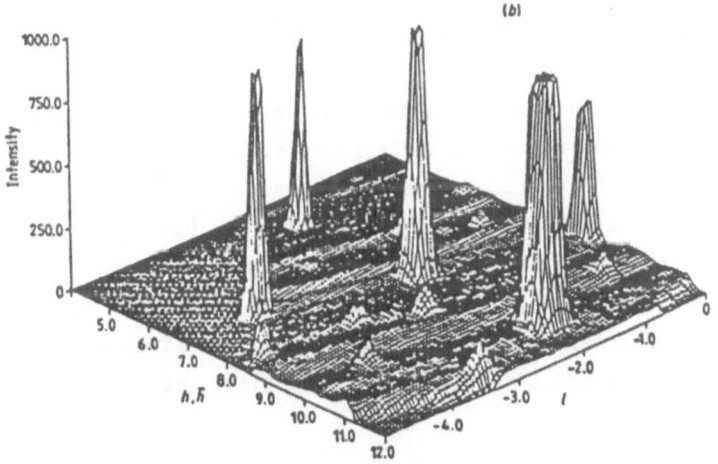

Figure 8.   Recorded neutron-diffraction intensity for proustite at
(a) 61K and (b) 35K

5.2 Forbidden reflections - incommensurate structures

Single crystal diffractometry at pulsed neutron sources is ideally suited to the investigation of weak reflections and incommensurate structures because, with an area detector and time-of-flight analysis, volumes (rather than surfaces) of reciprocal space are surveyed simultaneously with good signal-to-noise sensitivity. Investigations of the material proustite (Ag_3AsS_3) (Nelmes, Howard, Ryan, David, Schultz and Leung 1984) on the single crystal diffractometer (SCD) at IPNS, Argonne National Laboratory highlight this application.

The room-temperature structure of proustite was investigated by Harker (1936) and Engel and Nowacki (1966) and found to have space group R3c. Other phases have since been discovered at low temperatures. Bondar, Vikhnin, Ryabchenko and Yachmenev (1983) found evidence in NQR spectra for two distinct phases between the previously known transitions at ~27 K and ~58 K, confirming the second-order transition at ~58 K but finding another small but distinct first-order change in the spectra at 49 K in addition to the abrupt change at ~27 K. From the character of their spectra they argue that the phase between 49 K and 60 K is incommensurate.

Prior to the neutron diffraction experiment no very clear picture had emerged as to the structural nature of these phase transitions. In the most recent previous structural work, Allen (1984) succeeded in showing the phase below 27 K to be monoclinic, with space group Cc, but was unable to reach any definite conclusion about the phase (or phases) between 27 K and 58 K. The balance of evidence appeared to be in favour of space group R3 in this range (Allen, 1984; Taylor, Ewen and Han, 1984), but Allen was unable to detect any breaking of the c-glide absence conditions (of R3c) in x-ray photographs.

In the TOF single crystal study, attention was concentrated on the reflections that are absent in space group R3c but allowed in R3. Adopting hexagonal indices, these are the (h $\bar{h}$ 0 l) reflections with l $\neq$ 2n. (The R lattice imposes the condition that $-h + k + l = 3n$ for a general (hkil) reflection, and so $h + k = 3n$ in the hh0l plane.) Data were first recorded

at 66 K, above the 58 K transition, and then at 38 K: each run, at one
orientation of the sample crystal, took about six hours. This proved
sufficient to detect additional weak peaks in the hh̄0l plane at 38 K. These
were observed, on warming back towards the transition, in runs at 48 K and
54 K but disappeared at 61 K. Figure 8a shows part of the hh̄0l plane from
one of the runs at 61 K. The ridges are Debye-Scherrer powder-diffraction
rings from the aluminium radiation shield in the cryostat. Figure 8b shows
the same part of the hh̄0l plane from one of the runs made after again
cooling the sample below the transition, to 35 K. The additional peaks
reappeared, and can be seen located in pairs around each c-glide absence
position - most clearly, in this figure, around (9 9̄ 0 3). The very weak
scattering, resulting from multiple scattering, at some of the c-glide
absence positions remains (e.g. at (9 9̄ 0 3)).

A computer peak search through all the 35 K data, covering a region of
reciprocal space around and including the hh̄0l plane, revealed new weak
peaks around many of the Bragg reflections in this region. These satellite
peaks were at one or more of the six positions $(h \pm \delta, k \mp \delta, ., l \pm \delta)$,
$(h \pm \delta, k, ., l \mp \delta)$ and $(h, k \mp \delta, ., l \mp \delta)$ around any (hkil) allowed by the R lattice,
with δ close to 1/3. In the hh̄0l plane, the only satellites strong enough
to be detected around reflections with $l \neq 2n$ (i.e. the c-glide absent
reflections) were at the two positions lying in the plane - namely
$(h \pm \delta, h \mp \delta, 0, l \pm \delta)$ - whilst for reflections with $l = 2n$ the detected
satellites were restricted to the other four, out-of-plane, positions. In
the planes adjacent to hh̄0l (h h̄+I 1 1 and h h̄-I I 1) the reverse was true.
Further investigation of the variation of δ with temperature indicated that
the structure was incommensurate between 58 K and ~ 49 K. Below ~ 49 K
there is a transition to a commensurate phase with $\delta = 1/3$.

5.3 Protein crystallography

The principal advantage of neutron diffraction studies of proteins and
other biological molecules is obvious. Whereas x-ray studies determine the
coordinates of C, N, O and heavier atoms, neutron scattering permits the
localization of hydrogen atoms. The precise location of specific hydrogen
atoms is clearly extremely important for the understanding of many

biological problems. Substantial research into neutron diffraction from biological systems has been undertaken at reactor-based sources, most notably the ILL, Grenoble although no diffraction studies of proteins have to date been undertaken at spallation neutron sources. This does not indicate any inherent shortcomings in the time-of-flight diffraction method but rather that single crystal diffraction at spallation sources is still in its early days. Indeed, the projected performance of protein diffraction at spallation sources is very exciting. Jauch and Dachs (1984), in their performance comparison between the proposed spallation sources, SNQ, and a steady state high flux reactor ($\phi_o = 1.2 \times 10^{15}$ ns^{-1}cm^{-2} i.e. ILL, Grenoble), concluded that SNQ would outperform a high flux reactor by a factor of 10-30 in terms of countrate. Their example, a high-resolution small biological structure determination ($d_{min} \sim 1.2$Å : $V_{cell} = 125000$Å^3 : $V_{sample} = $ 1mm^3 : 50% H/D exchange : 150000 reflections) was calculated to take 10-40 hours (two instruments were described) at SNQ compared with 16 days (400 hours) at a high flux reactor to collect a hemisphere of reciprocal space. The equivalent counting time on a TOF diffractometer viewing a fully coupled moderator (+ proton compressor ring) at the proposed ISIS II source is 30-110 hours, an improvement factor of 4-12 over a high flux reactor. Such counting times from 1mm^3 protein single crystals bode well for protein studies at future spallation sources.

5.4 High real space resolution studies: anharmonic behaviour

In most crystallographic studies, thermal motion is adequately discussed in terms of isotropic (0th rank tensor) or anisotropic (2nd rank tensor) temperature factors. The underlying assumption in this approximation is that the motion of atoms about their mean positions is harmonic in nature (i.e. the atomic potential is harmonic and the atomic wavefunction, Gaussian (the ground state wavefunction of the quantum harmonic oscillator)). This assumption breaks down at high temperatures and in structures where atoms may disorder over a number of crystallographic sites (e.g. superionic conductors). In such structures the deviation from harmonic behaviour, the anharmonicity, may be treated in a perturbative manner. For instance additional higher-order rank (> 2) terms may be added to the formulation of the mean square displacement (and, hence, the temperature factor). Information about anharmonicity, since it may be regarded as a perturbation of harmonic behaviour, is associated with the fine details of the atomic wavefunction / potential. High resolution real-space information, garnered from high $\sin\theta/\lambda$ ($> 1.5\text{Å}^{-1}$) structure factors, is required. Such reflections may be obtained from instruments situated on hot-source beam lines at reactors and, more promisingly in terms of neutron flux, from TOF diffractometers at spallation neutron sources. A recent pilot experiment on SrF_2 (Forsyth, Wilson and Sabine, 1987) using the single crystal diffractometer, SXD, at ISIS highlights the possibilities of crystallographic studies of anharmonicity and disorder.

SrF_2 possesses the face-centred cubic fluorite structure and at elevated temperatures shows high fluorine ionic conductivity. At room temperature this phenomenon manifests itself in a small degree of third order fluorine anharmonicity that has been previously studied by Cooper and Rouse (1971) and Mair, Barnea, Cooper and Rouse (1974). The reason for the anharmonicity may be easily understood. The fluorine ion (site symmetry 43m) is tetrahedrally coordinated by four strontium ions. Fluorine motion along the Sr-F bond is more restricted than that away from the bond. The 43m site symmetry dictates that the simplest description of this behaviour is in terms of a third-order anharmonic component, β. This, in turn, modifies the structure factors such that previously equivalent structure factors in the harmonic approximation become distinct. In particular, for structure factors with the same d spacing (i.e. Σ ($h^2+k^2+l^2$) identical), the anharmonicity manifests itself such that those reflections with $h+k+l = 4n+1$ have higher intensities than those with $h+k+l = 4n-1$.

A total of twelve pairs of (hhk) reflections were measured using a prototype single 20×20 mm. scintillator detector up to a maximum $\sin\theta/\lambda$ of 1.696 Å^{-1} that compares favourably with the previous work of Cooper and Rouse (1971) that stopped at 0.9 Å^{-1}. Of the twelve pairs measured in only one did the F_+/F_- ratio give the wrong sense of the anharmonicity correction. Using Cooper and Rouse's value for β_F of -3.95×10^{-12} erg Å^{-3}, Forsyth, Wilson and Sabine obtained F_+/F_- ratios, calculated from the expression

$$\frac{F_+}{F_-} = 1 - \frac{2b_F}{b_{Sr}} \exp\left[\left(B_{Sr} - B_F\right)\left(\frac{h^2 + k^2 + l^2}{4a^2}\right)\right]\left(\frac{B_F}{4\pi a}\right)^3\left\{\frac{\beta_F}{kT}\right\}\left[|h_1 k_1 l_1| + |h_2 k_2 l_2|\right]$$

that reproduced very well the values derived from the measured intensities. A refinement was performed using the least squares program XFLS3 (Mair and Barnea 1971 and Busing, Martin and Levy 1962), to refine the anharmonicity. Refinement of the measured pairs of reflections resulted in an R-factor of 5.0% for the refinement without anharmonicity and 1.9% when anharmonicity was included. The refined cross component in the third rank tensor, Γ_{123}, was 0.192 ± 0.027, comparing favourably with the best value of 0.189 ± 0.044 obtained from the Cooper and Rouse (1971) refinement. The value of the β_F parameter, refined using the above equation, was $(-4.19 \pm 0.030) \times 10^{-12}$ erg Å^{-3} (c.f. Cooper and Rouse obtained $(-3.95 \pm 0.045) \times 10^{-12}$ erg Å^{-3}).

The pilot study of anharmonicity using TOF single crystal techniques indicates that accurate measurements of anharmonicity may be easily obtained using data of higher $\sin\theta/\lambda$ than previous studies. The data leading to this higher precision result could be collected in less than a day at full ISIS intensity.

5.5 Thermal diffractometry - reaction kinetics - transient phenomena

High intensity neutron powder diffraction at medium resolution has become an increasingly powerful tool over the past five years. Applications have included reaction kinetics and thermal diffractometry (the study of the evolution of a diffraction pattern as a function of temperature). High temporal resolution is required; counting times must realistically be less than ten minutes. This places stringent requirements for high neutron flux that, to date, has been almost solely available on the D1b diffractometer at the ILL, Grenoble. New instruments at both reactors and the new generation of intense spallation neutron sources are being proposed and constructed. These machines will offer order of magnitude intensity improvements over D1b. This opens up the exciting possibility of studying irreversible transformations that occur on the time scale of seconds and reversible stroboscopic measurements on the 10^{-5} second time scale.

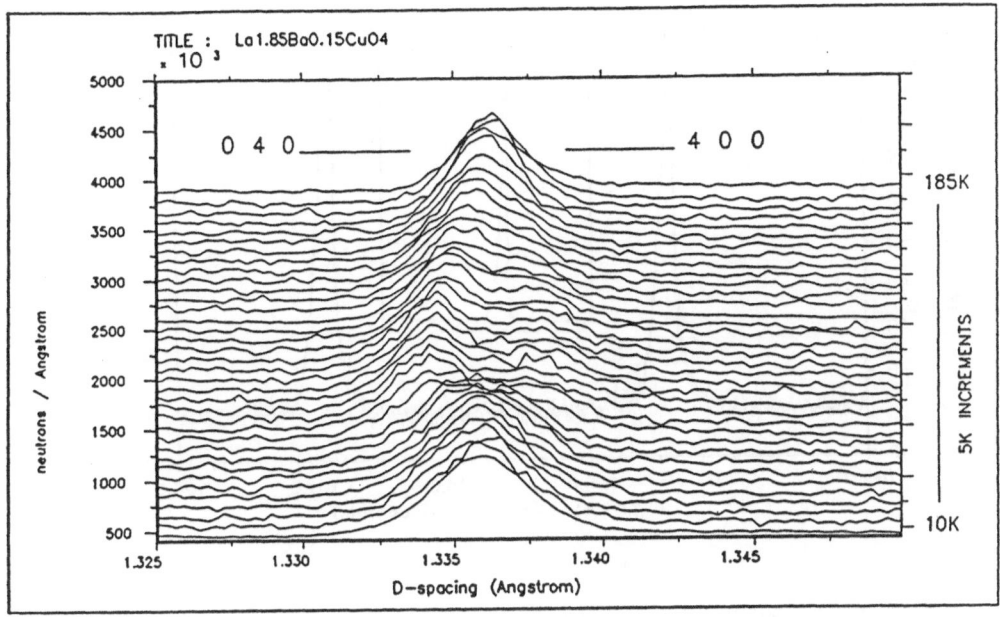

Figure 9. Temperature dependence of the orthorhombic distortion in
(La1.85Ba0.15)CuO4 observed in the 0 4 0 / 4 0 0 splitting.

As an example of high intensity thermal diffractometry recent research
on the new high T_c (T_c=35K) superconductor, $(La_{1.85}Ba_{0.15})CuO_4$, is
discussed (Bednorz and Muller 1986). The experiment was performed on the
high resolution diffractometer, HRPD, at ISIS and highlights the advantage
of combined high intensity and resolution. Neutron diffraction data were
collected for $(La_{1.85}Ba_{0.15})CuO_4$ at 5K intervals from 10K to 200K :
measurements took approximately 30 minutes to collect at each temperature.
Inspection of the diffraction profiles indicated the presence of a small
orthorhombic distortion below ~160K that increased monotonically with
decreasing temperature until ~75K (Figure 9). Below this temperature the
splitting associated with the orthorhombic distortion unexpectedly
decreased, becoming unresolved visually, as a result of peak broadening by
sample effects, at temperatures less than 50K. Both the 75K and 160K

46

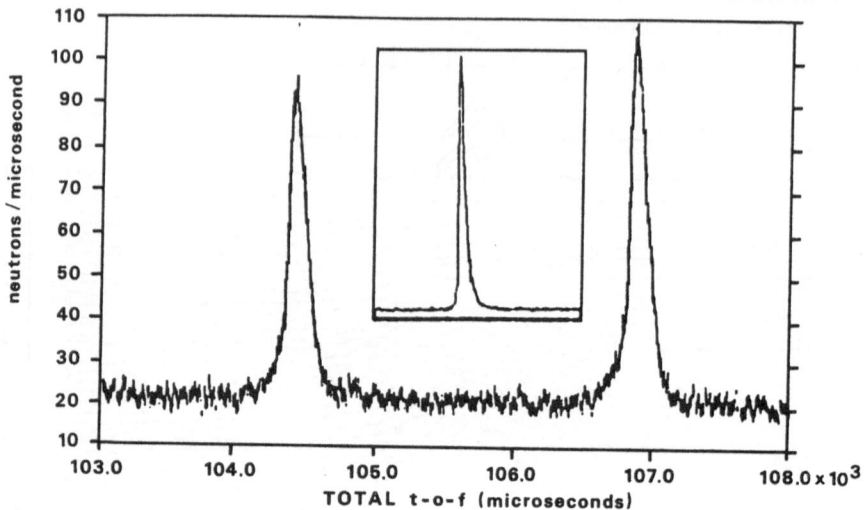

Figure 10.   Anomalous line broadening in langbeinite. Inset indicates
instrumental resolution.

structural   changes   are   accompanied   by   conductivity   anomalies.   Detailed
structural   analysis of the diffraction patterns using the Rietveld profile
analysis   method   confirmed   that $(La_{1.85}Ba_{0.15})CuO_4$ is   isostructural   below
160K   with   the   end-member   compound La_2CuO_4 (space group Abma - Grande,
Muller-Buschbaum and Schvazer 1977). Above 160K the orthorhombic distortion
disappears   and   the   structure   adopts   the   undistorted   tetragonal K_2NiF_4
structure.   Analysis   of the structural parameters in the orthorhombic phase
revealed   an anomalous thermal motion of one oxygen site that is consistent
with   the   presence of a zone-boundary soft optic mode. This has been linked
to   the   possibility   of   enhanced   phonon   coupling   producing   a   high
superconducting transition temperature.

5.6 High resolution powder diffraction - line broadening

The width of a powder diffraction peak is determined by a combination of instrumental and sample effects. Clearly as instrumental resolution the sample contribution becomes more dominant. Recent experience on HRPD at ISIS indicates that sample broadening is generally significant at $\Delta d/d$ resolutions of better than 10^{-3}. Thus, whereas on "previous-generation" instruments such as D1a at ILL, Grenoble sample effects can usually be neglected, analysis of diffraction data on the new generation of high resolution machines such as HRPD must accommodate sample-dependent line-broadening. This leads to a substantially greater complexity in data analysis techniques but, in principle, yields a large amount of information about crystal imperfections such as stacking faults, strain and concentration gradients and crystallite domain sizes. The following example, a commissioning experiment on HRPD, highlights the problems that may be encountered in the study of solid solutions at high resolution.

$Rb_2(FeTi)(PO_4)_3$, a synthetic primitive-cubic langbeinite (a = 10Å: space group $P2_13$), was examined as part of the calibration of the long-wavelength diffraction spectrum on HRPD. In addition to providing suitable well-spaced long d-spacing reflections, $Rb_2(FeTi)(PO_4)_3$ was chosen because of its potential scientific interest as an order-disorder structure. Langbeinite may be regarded as a lower-symmetry variant of garnet (space group Ia3d : e.g. $Ca_3Al_2(SiO_4)_3$) in which two of the three 24c (Ca) sites are occupied in an ordered manner because of the buckling of the corner-linked octahedral/tetrahedral $M_2(XO_4)_3$ framework to accommodate the very large monovalent cation (Rb). This distortion results in two crystallographically distinct but chemically similar octahedral sites over which the Fe and Ti ions may either order or disorder. Although determination of the order/disorder character of $Rb_2(FeTi)(PO_4)_3$ is greatly facilitated by the large difference in Fe (9.54 fm) and Ti (-3.37 fm) scattering lengths, the diffraction data indicated a complex sample-broadened peak shape that was common to all reflections (Figure 10). Rather than the anticipated sharp leading edge and slow exponential tail associated with the instrumental line-shape (see inset, Figure 10), each peak showed a pronounced slow rise indicative of a strain distribution. Initially this was attributed to the presence of a second langbeinite phase at ~5% level because of a successful modelling of a biphasic distribution to individual peaks. This yielded a fractional lattice constant change of $\Delta a/a = 1.55 \times 10^{-3}$. Detailed analysis of this peak description, however, indicated that a two-phase model was inadequate and that a continuous distribution of local ordering/concentration variation of Fe and Ti was more likely. However, as the nature of the sample broadening was uncertain, analytical peak-shape refinement was deemed unsuitable and an alternative procedure involving maximum entropy deconvolution was employed. Initial results confirmed a continuous distribution resulting from strain broadening.

As a result of the problems encountered in peak-shape description the average structure was refined from 64 extracted integrated intensities by grouped-intensities least-squares analysis (GRILS) based on the Mark III version of the Cambridge Crystallography Subroutine Library (Brown and Matthewman, 1987). The resulting structural parameters indicated that the structure was typical of previous langbeinite studies and that, on average, no evidence of Fe/Ti ordering was found.

5.7 Phase transition studies – small lattice distortions

The almost constant $(\Delta d/d)$ resolution of TOF powder diffractometers makes them ideal instruments for the study of structural phase transitions. In the absence of sample broadening, subtle peak splittings associated with lowering of symmetry may be observed with roughly equal precision in multiple orders (nh nk nl) of the same fundamental (h k l) reflection. Thus lattice distortions may, in contrast to conventional instruments, be observed with very high precision in low-index reflections, where the overall density of peaks is very low.

Nickel oxide is an ideal candidate to test resolution. The simple cubic rocksalt structure is distorted to monoclinic or possibly triclinic symmetry by the onset of antiferromagnetic ordering at 525 K. In practice, the structure appears to be rhombohedral, producing a splitting of peaks that requires an instrumental resolution of better than 10^{-3}. Data recorded on HRPD (David, Harrison and Johnson 1986) at ISIS clearly reveal this rhombohedral splitting – Figure 11 shows the (111,11$\bar{1}$), (222,22$\bar{2}$) and (444,44$\bar{4}$) doublets all observed at a resolution of better than 6×10^{-4}. From the fitted peak positions the rhombohedral α angle was refined to high accuracy (α = 90.05953(9)°). In contrast, the (h00) peaks (h = 2n) showed no perceptible splitting, indicating that the monoclinic distortion must be less than 10^{-4} in magnitude. These results have also been confirmed by recent synchrotron measurements on the same sample (Eddy 1986, private communication).

HRPD has been used in a recent refinement of squaric acid (Nelmes, Tun, David and Harrison 1987). Room-temperature single-crystal neutron and X-ray structure determinations of squaric acid ($H_2C_4O_4$) indicate that the material has a planar structure that possesses a small monoclinic distortion from tetragonal symmetry. The small pseudosymmetry ($\Delta d/d$ < 0.001) and hydrogenous content of squaric acid necessitate high resolution for a powder diffraction experiment to be successful. The location of hydrogen using powder diffraction techniques has hitherto been extremely difficult using traditional X-ray and neutron diffractometers. With X-rays, hydrogen scatters so weakly in the presence of other nuclei as to be almost invisible, whereas with neutrons, naturally occurring hydrogen produces very large backgrounds, resulting from a large incoherent scattering cross-section, which masks all but the strongest Bragg peaks. Deuteration obviates the above problems in neutron diffraction studies but is an expensive and often complicating option. The considerably higher resolution of the high resolution powder diffractometer, HRPD, at ISIS over similar machines results in a substantial increase in the peak to background ratio thus allowing problems of significant complexity to be tackled.

The data recorded in 12 hours from a sample 2cm. × 1.5cm. in cross-section and 0.5cm. deep were refined straightforwardly to give an excellent fit (weighted profile R-factor = 4.60% ; expected R-factor = 4.55% ; χ^2 = 1.02) (Figure 12). Atomic positions agree closely with the single-crystal neutron diffraction study; the locations of the hydrogens are substantially different and less well-determined in the single-crystal X-ray study. Lattice parameters obtained from profile refinement indicate the smallness of the monoclinic distortion and the high lattice precision available using the time-of-flight method (a= 6.12890(5)Å; b= 5.26781(5)Å;

50

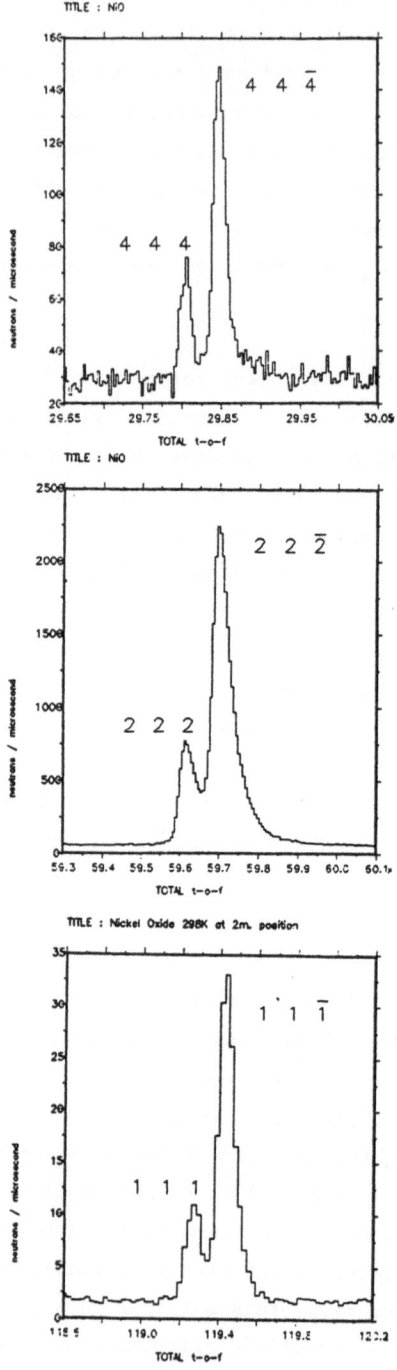

Figure 11. Rhombohedral splitting in NiO measured on HRPD, ISIS.

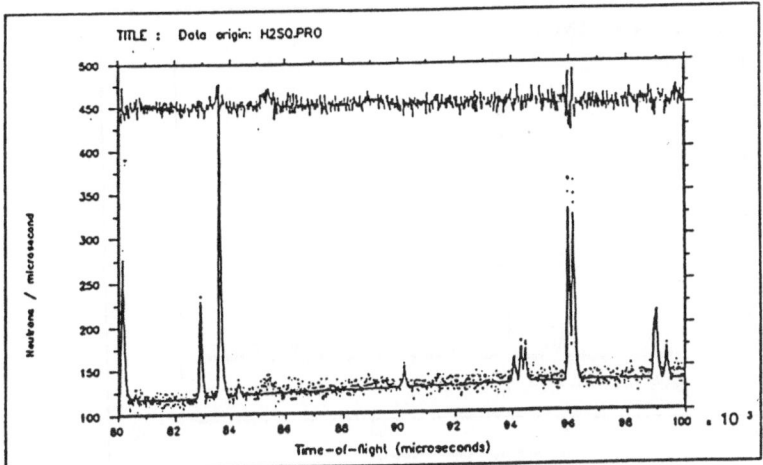

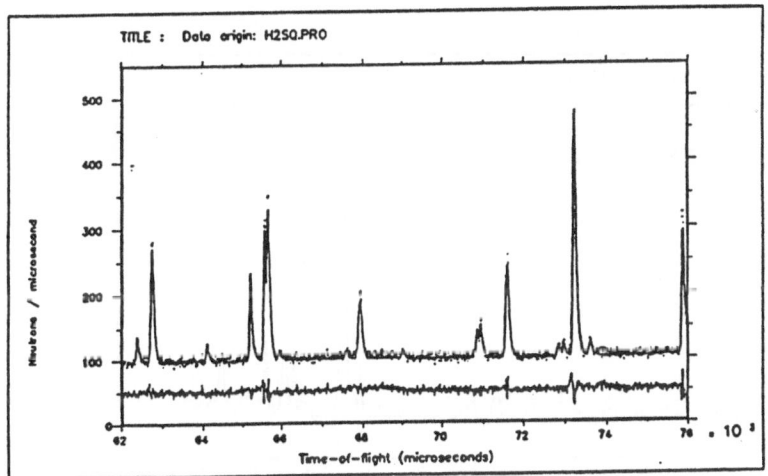

Figure 12.   Observed and calculated HRPD diffraction pattern for squaric
acid. (50000 microseconds ~ 1 angstrom d spacing)

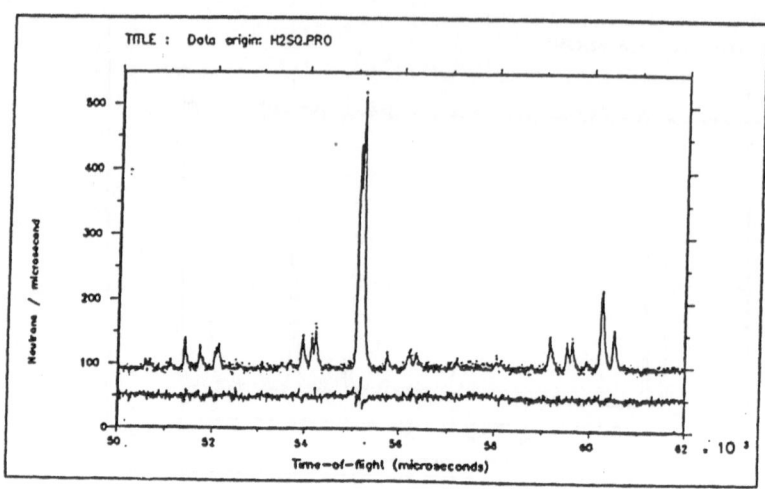

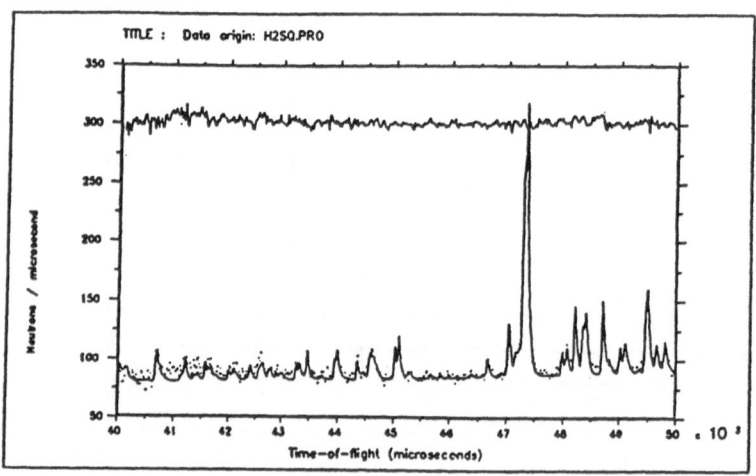

Figure 12.   Observed and calculated HRPD diffraction pattern for squaric
             acid. (50000 microseconds ~ 1 angstrom d spacing)

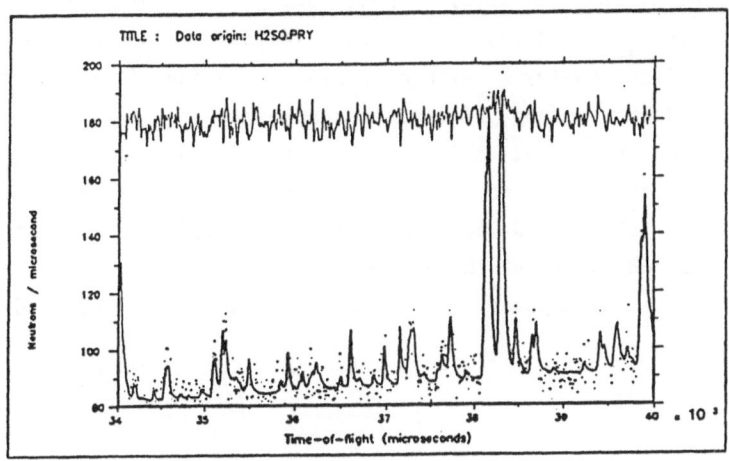

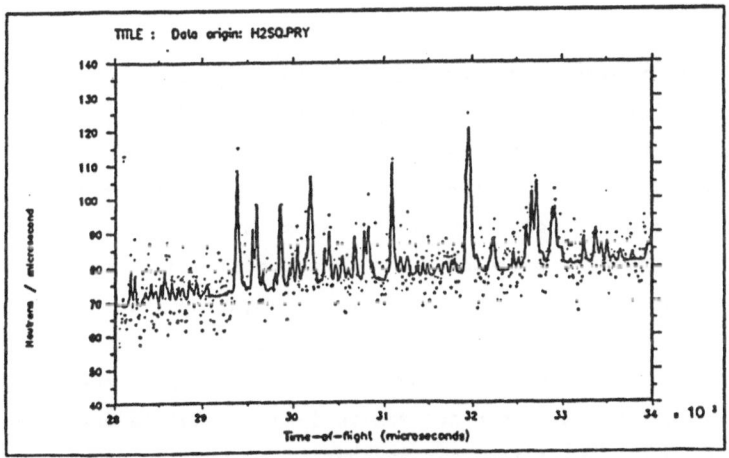

Figure 12. Observed and calculated HRPD diffraction pattern for squaric acid. (50000 microseconds ~ 1 angstrom d spacing)

$c = 6.14025(6)$Å; $\beta = 89.9632(5)°$: spontaneous ferroelastic strains are $\varepsilon_{11} = 9.25(6) \times 10^{-4}$; $\varepsilon_{13} = 3.21(4) \times 10^{-4}$). The initial refinement reported above involved profile analysis with a minimum d spacing of ~1Å. A second data set, collected over a range of shorter d spacings ($d_{min} = 0.54$Å), proved to be even more successful allowing the resolution of the real space scattering distribution (Fourier map – see Figure 13) to be substantially improved. The shorter d spacing data permitted an analysis of the order-disorder character of the O-H--O link between squaric acid molecules. The bond, as anticipated, was found to be fully ordered at room temperature.

6. Conclusions

Spallation neutron sources offer substantial prospects because pulsed neutron beams have intrinsically high resolution and have high fluxes that compare favourably with the best research reactors. Powder diffraction is already established as a technique for both high resolution and high intensity studies. Single crystal studies are only now emerging from a developmental stage because of the complexities of area-detector technology. The prospects are, however, good, particularly for high resolution experiments that exploit the copious epithermal fluxes at wavelengths of less than an angstrom. Some areas of research, such as protein crystallography, are not realistically feasible on current pulsed sources and must await the next generation of spallation neutron sources such as the proposed upgrade to ISIS. Such machines will undoubtedly play a very important role in the future of all areas of crystallography.

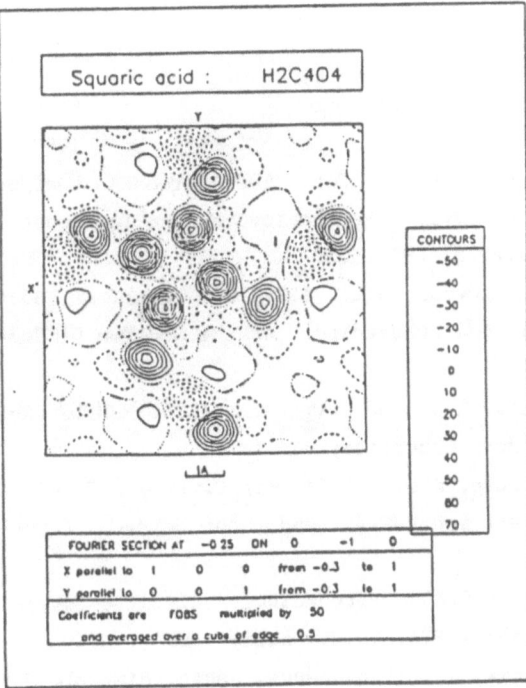

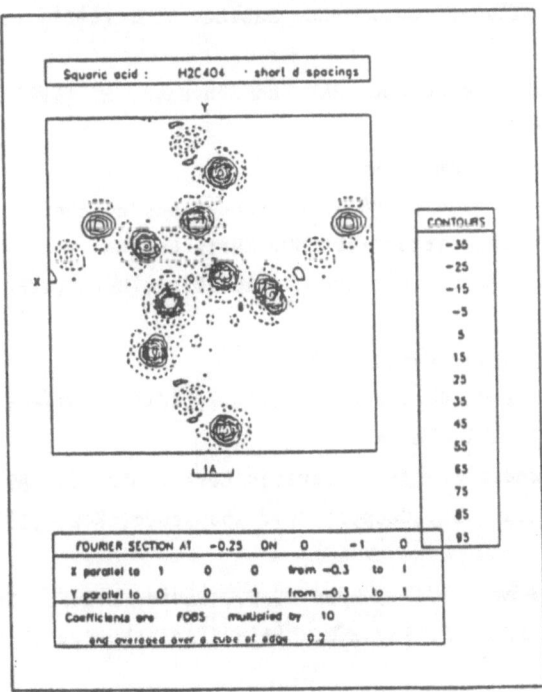

Figure 13. Fourier map (from observed Fourier components) for squaric acid: left – dmin = 1A   right – dmin = 0.54A

References

Allen, S. (1984) Phase Transitions

Bednorz, J.G. and Muller, K.A. (1986) Z. Phys. B64, 189

Bloklin, G.E. and Blokhintsev, D.L. (1961) Atomnaja Energia 10 437

Bondar, A.V., Vikhnin,V.S., Ryabchenko,S.M. and Yachmenev, V.E. (1983) Sov. Phys.-Solid State 25 1497-1501

Brown, P.J. and Matthewman, J.C. (1987) RAL report no. 87-025

Buras, B. (1975) in Proceedings of the Neutron Diffraction Conference, Petten, RCN-234

Busing, W.R., Martin, K.O. and Levy, H.A. (1962) ORFLS. Report ORNL-TM-305. Oak Ridge Nationasl Laboratory

Cooper, M.J. and Rouse, K.D. (1971) Acta Cryst., A27, 622-628

David, W.I.F., Harrison,W.T.A. and Johnson,M.W. (1986) RAL report no. 86-068

Egelstaff, P.A. (1961) Proceedings of the symposium on neutron time-of-flight methods, E.A.N.D.C. Saclay p.69

Engel, P. and Nowacki, W. (1966) Neues. Jahrb. Mineral. Monatsch. 8 181

Forsyth, J.B., Wilson, C.C. and Sabine, T.M. (1987) (submitted to Acta Cryst. B)

Grande, V.B., Muller-Buschbaum, Hk. and Schwazer, M. (1977) Z. Anorg. Allg. Chem. 428, 120

Harker, D. (1936) J. Chem. Phys. 4 381

Jauch, W. and Dachs, H. (1984) in Proceedings of the Workshop on Neutron Scattering Instrumentation for SNQ, Maria Laach

Kuhs, W.F., Finney, J.L., Vettier, C. and Bliss, D.V. (1984) J. Chem. Phys. 81 3612-3623

Lowde, R.D. (1956) Acta Cryst. 9 151

McReynolds, A.W. and Whittemore, W.L. (1961) Inelastic scattering of neutrons, p421, IAEA Vienna

Mair, S.L. and Barnea, Z. (1971) Physics Lett., 35A 286-287

Mair, S.L., Barnea, Z., Cooper, M.J. and Rouse, K.D. (1974) Acta Cryst., A30 806-813

Nelmes, R.J., Howard, C.J., Ryan, T.W., David, W.I.F., Schultz, A.J. and Leung, P.C.W. (1984) J. Phys. C: Solid State Physics 17 L861-L865

Nelmes, R.J., Tun, Z., David, W.I.F. and Harrison, W.T.A (1987) (in preparation)

Ringo, G.R. (1957) Handb. Phys. 33 590

Scherm, R. (1984) in Proceedings of the Workshop on Neutron Scattering Instrumentation for SNQ, Maria Laach

Taylor, V., Ewen, P.J.S. and Han, T. (1984) Ferroelectrics 55 83-86

Windsor, C.G. (1981) Pulsed Neutron Scattering

DISCUSSION

NEUTRON PULSE SHAPE

The shape can be represented by a gaussian of second moment 2-6 µs convoluted with an exponential decay of time constant 10-30 µs. The result is a pulse with a fast rise and a slower decay, the latter being due to the neutrons emerging from the moderator in a manner similar to that of photons from a black body.

TIME RESOLUTION OF PULSED NEUTRON EXPERIMENTS

Extrapolation of experience with ISIS suggests that, with the proposed ISIS II in high resolution mode, a complete data set of intensities and lattice spacings for a reasonably complex material could be collected in seconds. In high intensity mode for a relatively simple material, it might be possible, in principle, to get a complete data set with every pulse, i.e. 50 times a second. For a reversible process, which can be repeated many times to build up statistically significant counts, the ultimate limit for time resolution is the pulse width of about 10 µs.

Sophisticated time-resolved experiments have not yet been done on the present ISIS, but they would certainly be of interest.

THE UMBRELLA EFFECT

Skewed 2θ values are obtained when they are measured appreciably above or below the plane of the incident beam, because the intersection of the diffraction cone with the plane of the detector is curved.

FUTURE DEVELOPMENT OF PULSED NEUTRON POWDER DIFFRACTION INSTRUMENTS

High intensity instruments will probably use extremely large solid angle area detectors placed close to the sample, resolution-focussing, and firmware to map all the data.

For high resolution work, line broadening due to sample structure may limit the improvement which can be obtained by increasing the length of the flight path. Such broadening is shown by 90% of the samples studied in the 100 m flight path of ISIS, but for the 10 Hz source in ISIS II there may be advantages in extending the flight path to 200 m, at least for backscattering experiments. High pressure work also requires 90° scattering where resolution is determined by $d\theta.\cot\theta$, implying the use of sophisticated pre- and post-collimation and long secondary flight paths. For the high resolution powder diffractometer, long secondary flight paths and low angle scattering are planned for the exploration of larger unit cells. Although it would be unrealistic ever to contemplate studying proteins in that way, organic and inorganic unit cells of up to 20-30 Å along one edge should be feasible.

THE EFFECT OF INCOHERENT SCATTERING BY PROTONS IN BIOLOGICAL SAMPLES

In the example of neutron diffraction by the 50% deuteriated protein single crystal discussed above, data accurate to within 10% were required. Much of that error can be attributed to incoherent scattering by the remaining protons, and the problem is the same for steady state and spallation sources.

COMPARISON OF INTERATOMIC DISTANCES OBTAINED FROM EXAFS AND FROM CRYSTALLOGRAPHIC DATA

Metal-fluorine distances in strontium fluoride measured by EXAFS can be as much as 5% larger than those based on unit cell dimensions and equilibrium atomic positions. These differences can be explained by the deviation of the fluorine from its equilibrium position due to the third order anharmonic component of its vibrational motion.

CRYSTAL ORIENTATION IN NEUTRON DIFFRACTION BY POWDERS

The squaric acid sample was taken straight from the bottle, without any grinding to randomise crystal orientations. However, the Debye-Scherrer geometry of the HRPD includes all the ring and so removes that one aspect of preferred orientation.

SOME ASPECTS OF DIFFRACTION PHYSICS WITH PULSED NEUTRONS

B T M Willis

Chemical Crystallography Laboratory
9 Parks Rd

and

Atomic Energy Research Establishment
Harwell
Oxon OX11 0RA
England

ABSTRACT. By means of pulsed neutron diffraction without energy
analysis, it is possible not only to study Bragg scattering but also
to study processes involving inelastic neutron scattering. Thus the
same diffractometer can be employed for examining elementary
excitations as for crystal structure analysis. We discuss the nature
of the thermal diffuse scattering (TDS) from acoustic phonons, and
show that the sound velocity in the crystal can be determined from the
TDS 'edges' on either side of the Bragg reflection. One edge arises
from phonon emission (Stokes process) and the other from phonon
absorption (anti-Stokes), and each is best observed in back scattering
with 2θ close to $180°$. On the other hand, if 2θ is $90°$, TDS may be
forbidden in the neighbourhood of the reflection and the measured
Bragg intensity is then free from the 'TDS error'.

GLOSSARY OF SYMBOLS

$\underline{B}$	reciprocal-lattice vector
c_s	phase velocity of sound
$\hbar$	Planck's constant/2π
j	polarization index of normal mode
$\underline{k}_o, \underline{k}$	initial and final wave vectors of neutron
k_B	$2\pi/\lambda_B$
Δk_o	$k_o - k_B$

M. A. Carrondo and G. A. Jeffrey (eds.), Chemical Crystallography with Pulsed Neutrons and Synchrotron X-Rays, 61–75.
© 1988 by D. Reidel Publishing Company.

Δk	$k - k_B$
m_n	neutron mass
q	wave vector of normal mode
$\underline{Q}$	scattering vector $(= \underline{k} - \underline{k}_o)$
$\underline{Q}_\Theta^{el}$	vector defining locus of elastic scattering through angle 2Θ
t	time-of-flight
t_B	time-of-flight for Bragg scattering
Δt	$t - t_B$
v_n	neutron velocity
β	c_s/v_n
ε	index for phonon absorption (-1) or phonon emission (+1)
ε'	$\varepsilon\Delta\Theta/|\Delta\Theta|$
Θ	half the scattering angle
Θ_B	Bragg angle
$\Delta\Theta$	$\Theta - \Theta_B$
λ_B	Bragg wavelength
ζ	angle between $\underline{B}$ and $\underline{q}$
$w_j(\underline{q})$	frequency of normal mode $(j\underline{q})$

1. INTRODUCTION

Experiments in neutron scattering are divided traditionally into two categories: elastic and inelastic. In an elastic scattering experiment there is no exchange of energy between the incoming neutron and the scattering system, whereas in an inelastic experiment the neutrons are scattered with a gain or a loss of energy. This division has given rise to two types of instrument: a neutron diffractometer for elastic scattering measurements, and a triple-axis neutron spectrometer for inelastic studies.

In this paper we shall show that excitations in a crystal can be measured without carrying out an energy analysis. This means that the same instrument, a diffractometer, can be employed both for structural studies (elastic scattering) and for examining elementary excitations (inelastic scattering).

The diffraction method of studying excitations was first suggested by Elliott and Lowde (1955), who treated the case of

scattering from a fixed-wavelength source; their theory was then
applied to the study of magnons (Samuelson, 1968) and of phonons
(Hohlwein, 1977). We shall show that the replacement of a continuous,
fixed-wavelength source by a pulsed neutron source appreciably
enhances the power of the diffraction method, and we shall illustrate
this with reference to some experimental results on pyrolytic
graphite.

The theory of the method hinges on the properties of the
'scattering surfaces' for the elastic and the inelastic scattering of
thermal neutrons. These surfaces are described in the next section,
where it is shown that their topologies are quite different for a
pulsed and a continuous neutron source.

2.   ELASTIC AND INELASTIC (ONE-PHONON) SCATTERING SURFACES

The one-phonon scattering surface is defined as the locus in
reciprocal space of the end points of the wave vectors $\underline{k}$ of neutrons
which are scattered inelastically in a one-phonon process. To
understand its properties, it is helpful to consider first the simpler
case of the elastic scattering surface.

For elastic scattering - which includes Bragg scattering - there
is no change of energy or wavelength on scattering, i.e.

$$k = k_o \tag{1}$$

where k_o and k are the magnitudes of the wave vectors of the incident
and scattered beams respectively. In a fixed-wavelength experiment,
the scattering angle 2Θ is a continuous variable, and the scattering
surface defined by (1) is the **Ewald sphere** (Fig. 1a). On the
other hand, in time-of-flight diffraction, 2Θ is kept fixed and the
incident wavelength varies; the scattering surface is then a
right-circular cone with its axis along the incident beam and with
semi-angle $(\pi/2) - \Theta$ (Fig. 1b).

The geometry of the one-phonon scattering surface is determined
by the conservation relations for energy and momentum, when the
neutron exchanges one quantum (phonon) of energy with the crystal.
Let $\hbar\omega_j(\underline{q})$ be the phonon energy, where ω is the frequency and j labels
the branch of the dispersion curves. $(\hbar = h/2\pi)$. $\underline{q}$ is the wave
vector of the normal mode of vibration; as $q \to 0$ the three acoustic
modes $(j = 1, 2, 3)$ are identified as the three sound waves with the
same propagation vector $\underline{q}$.

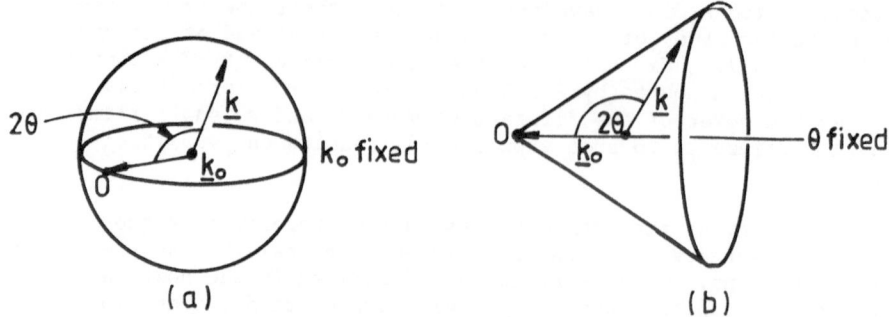

Figure 1.   Scattering surfaces for the elastic scattering of neutrons
X-rays etc.   (a) Fixed wavelength and variable scattering angle, 2Θ:
Ewald sphere.   (b) Fixed angle and variable wavelength:
right-circular cone. (a) corresponds to conventional neutron or
X-ray diffraction, whereas (b) refers to Laue or time-of-flight
diffraction.   O is the origin of reciprocal space.   Bragg scattering
occurs when the vector $\underline{k} - \underline{k}_o$ coincides with a reciprocal-lattice
vector.

The conservation of momentum for a one-phonon process with
neutron-energy gain (phonon absorption) is expressed by

$$\underline{k} - \underline{k}_o = \underline{B} + \underline{Q} \tag{2}$$

where $\underline{B}$ is the reciprocal-lattice vector (i.e. $\hbar\underline{B}$ is the momentum
taken up by the crystal).   The magnitude of $\underline{k}_o$ is related to the
wavelength λ in the incident white beam by

$$k_o = 2\pi/\lambda,$$

and if λ_B is the wavelength giving rise to Bragg scattering at $\Theta = \Theta_B$,
then

$$\lambda_B = 4\pi \sin \Theta_B / |\underline{B}|.$$

Extending (2) to include the energy-loss process, we obtain:

$$\underline{k} - \underline{k}_o = \underline{B} - \varepsilon\underline{q} \tag{3}$$

where ε is either +1 or -1. ε = +1 refers to phonon emission and ε =
-1 to phonon absorption.

The conservation of energy on scattering is expressed as

$$(\hbar^2/2m_n) (k^2 - k_o^2) = -\varepsilon\hbar w(\underline{q})$$

with m_n the neutron mass. This can be rewritten in the form

$$(\hbar/2m_n)\ (\Delta k - \Delta k_o)\ [(\Delta k + \Delta k_o) + 2k_B] = -\varepsilon w(\underline{q}) \tag{4}$$

where

$$\Delta k = k - k_B,\ \Delta k_o = k_o - k_B,\ k_B = 2\pi/\lambda_B.$$

For scattering with small energy transfer $(q \to 0)$, we can neglect the term $(\Delta k + \Delta k_o)$ in (4). This is the same approximation as that adopted by Seeger and Teller (1942) and later workers in discussing the fixed-wavelength case. The neutron velocity, v_n, is given by

$$v_n = \hbar k_B/m_n;$$

and if the acoustic modes propagate at the phase velocity c_s without dispersion, then we have

$$w = c_s q. \tag{5}$$

Hence (4) reduces to

$$k - k_o = -\varepsilon\beta q \tag{6}$$

where $\beta = c_s/v_n$. (6) determines the topology of the one-phonon scattering surfaces.

Let us now consider the nature of the TDS when the detector is offset by $\Delta\Theta$ from the Bragg setting:

$$\Theta = \Theta_B + \Delta\Theta.$$

In Fig. 2 the vector $Q_{\Delta\Theta}^{el}$, which is inclined at an angle $(\pi/2) - \Theta$ to the incident radiation, represents the locus of elastically scattered events received by the detector. The lattice point P does not lie on this vector and so Bragg scattering is not observed.

If Q in Fig. 2 is the end-point of the scattered vector $\underline{k}$ for neutron energy gain $(\varepsilon = -1)$, then QS = βq and QN is $\beta q \cos\Theta$, where N is the foot of the perpendicular from Q to the elastic scattering vector. Thus we find that

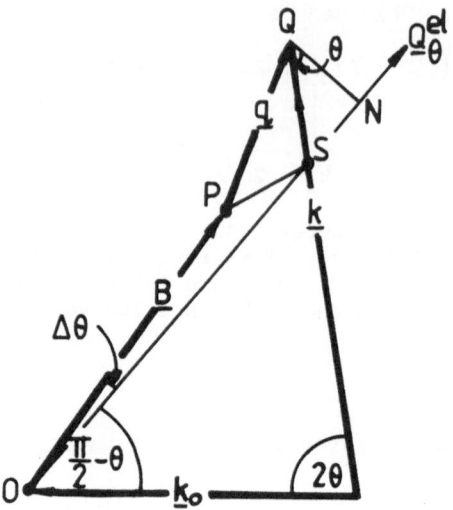

Figure 2. The scattering angle, 2Θ, is equal to 2ΔΘ plus twice the Bragg angle. The vector Q_Θ^{el} defines the locus of all elastic scattering events for the angle 2Θ. Q is the end point of the phonon wave vector $\underline{q}$, and N is the foot of the perpendicular from Q to the elastic scattering line.

$$QN/QP = \beta \cos \Theta$$

and so Q is constrained by the condition that the ratio of its distance from the fixed point P to its distance from the fixed line Q_Θ^{el} is equal to $(\beta \cos \Theta)^{-1}$. If β is constant, i.e. the elastic waves propagate isotropically, the one-phonon scattering surface is a conic section whose eccentricity e is

$$e = (\beta \cos \Theta)^{-1}.$$

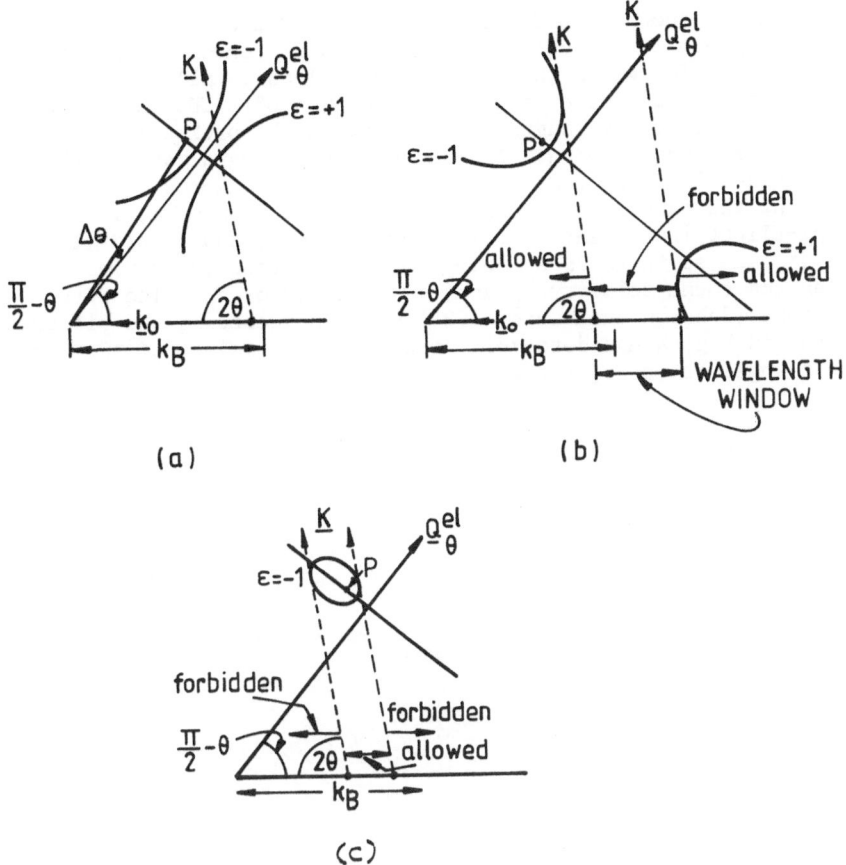

(a)

(b)

(c)

Figure 3. Diagrams illustrating the change in the one-phonon
scattering surface with increasing β, where β is the ratio of the
sound velocity to the neutron velocity; (a) β < 1; (b) 1 < β < sec Θ;
(c) β > sec Θ. P is the reciprocal-lattice point and the diagrams are
drawn for isotropic sound velocities and for Θ = 40°, ΔΘ = +6°. The
broken lines are vectors parallel to the scattered beam; TDS takes
place when these vectors intersect the scattering surfaces, indicated
by heavy lines.

For e > 1 the scattering surface for isotropic crystals is a
hyperboloid of two sheets with the reciprocal-lattice point at one
focus; one sheet corresponds to ε = -1 and the other to ε = +1 (Figs.
3a, b). For e < 1 the surface is an ellipsoid of revolution: only
one scattering process is then possible, either phonon absorption if
ΔΘ > 0 (Fig. 3c), or phonon emission if ΔΘ < 0.

68

It can be shown (Willis, 1986) that, if β lies in the range 1 <
β < sec Θ, there is a wavelength 'window' which encompasses the Bragg
reflection and for which TDS is forbidden. This is illustrated in Fig.
3(b) for the special case of Θ = 40° and ΔΘ = 6°: TDS cannot take
place for incident wavelengths lying between the broken lines and near
the Bragg wavelength $2\pi/k_B$. The range of β giving a wavelength window
is rather small for 2Θ less than 90°, but in a back-scattering
instrument the range can become very large. For example, in the
time-of-flight instrument described by Steichele and Arnold (1973) Θ
is 88.5°, and so there is a window for all neutron velocities lying
between the sound velocity c_s and one-fortieth of c_s. The HRPD
instrument (Johnson and David, 1985) used in recording the diffraction
patterns in Fig. 4 has detectors at even higher values of 2Θ: for the
highest-angle detector, sec Θ is over 50.

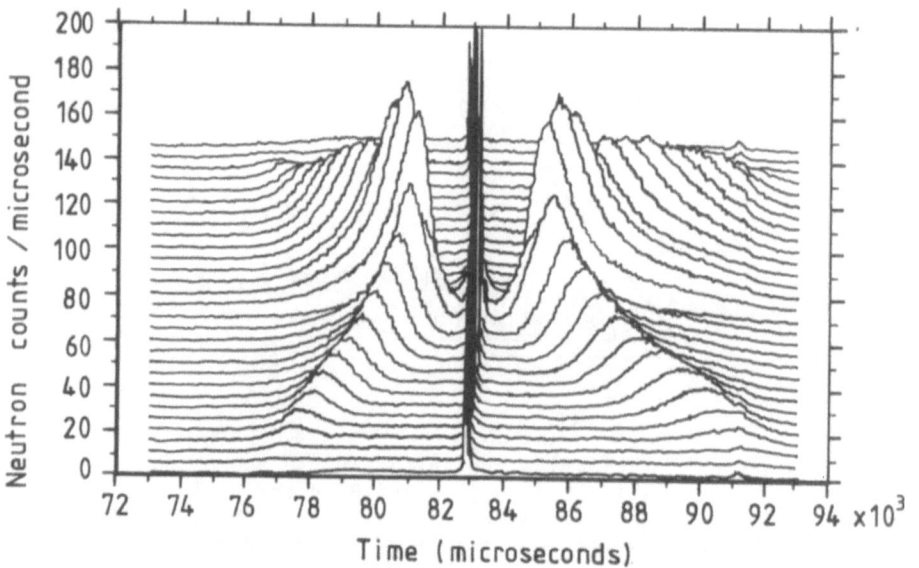

Figure 4. The scattering pattern around the (004) reflection from
pyrolytic graphite, showing the time windows at different off-set
angles 2ΔΘ. There are thirty detectors and Bragg scattering (ΔΘ = 0)
is recorded in detector no. 15.

3.   WIDTH OF WAVELENGTH WINDOW AND MEASUREMENT OF SOUND VELOCITY

The wave number of the incident radiation at the two edges of the window is given by (Willis, 1986)

$$\frac{\Delta k_o}{k_B} = \frac{\Delta\Theta \cosec^2 \Theta}{\varepsilon'(\beta^2 - 1)^{-\frac{1}{2}} - \cot \theta} \tag{8}$$

where the expression has been generalized to cover all combinations of ε and $\Delta\Theta/|\Delta\Theta|$. ($\varepsilon'$ is the product of ε and $\Delta\Theta/|\Delta\Theta|$.) For positive $\Delta\Theta$, Δk_o is positive for phonon emission and negative for absorption; for negative $\Delta\Theta$, the signs of Δk_o are reversed.

The distance from source to sample can be much larger than that from sample to detector: in the HRPD the ratio is 50:1. Under these circumstances, the time-of-flight t depends principally on k_o and so all elastic and inelastic events associated with the same incident wave-length are integrated together into the same time channel. The time window is then given by the same expression (8) as for the wavelength window:

$$\frac{\Delta t}{t_B} = - \frac{\Delta\Theta \cosec^2 \Theta}{\varepsilon'(\beta^2 - 1)^{-\frac{1}{2}} - \cot\Theta} \tag{9}$$

The two edges of the window correspond to phonon absorption ($\varepsilon = -1$) and emission ($\varepsilon = +1$). In Fig. 3(b) the Bragg wavelength is not in the middle of the window, but this asymmetry disappears as Θ approaches 90°. Equations (8) and (9) reduce to

$$\Delta k_o/k_B = - \Delta t/t_B = \varepsilon'(\beta^2 - 1)^{\frac{1}{2}}|\Delta\Theta| \ .$$

when $\Theta = 90°$, and the two edges are now symmetrically displaced in time on either side of $t = t_B$. The total width of the time window is

$$t_{tot} = 2t_B(\beta^2 \ 1)^{\frac{1}{2}} \Delta\Theta \tag{10}$$

Fig. 4 illustrates the appearance of the TDS windows in the diffraction from pyrolytic graphite. The instrument used was HRPD at the spallation neutron source ISIS. Each curve in Fig. 4 is given by a different detector, and, as Θ is nearly 90°, each window is symmetrical about $t = t_B$. The window width increases with $\Delta\Theta$, in accordance with (10), and the sound velocity is obtained by plotting $\Delta t/t_B$ versus $\Delta\Theta$ (see Willis, Carlile, Ward, David and Johnson, 1986).

We have assumed above that the crystal is acoustically isotropic, i.e. that β is independent of the direction of sound propagation. This situation, of course, is rarely found in practice. The extension of the TDS theory to the anisotropic case has been carried out by

Schofield and Willis (1987). They show that the principal change in the theory is to replace the <u>phase velocity</u> of the phonons by their <u>group velocity</u> in the expressions determining the position of the edges of the windows.

4. BRAGG REFLECTIONS FREE FROM TDS

THe 'TDS error' in measuring Bragg intensities arises from the scattering of the radiation by acoustic modes of vibration in the crystal. It is impossible to avoid this error with X-rays, and so it must be calculated using the measured sound velocities. With time-of-flight pulsed neutron diffraction, there are certain conditions under which there is no TDS error at all. We shall discuss these conditions here.

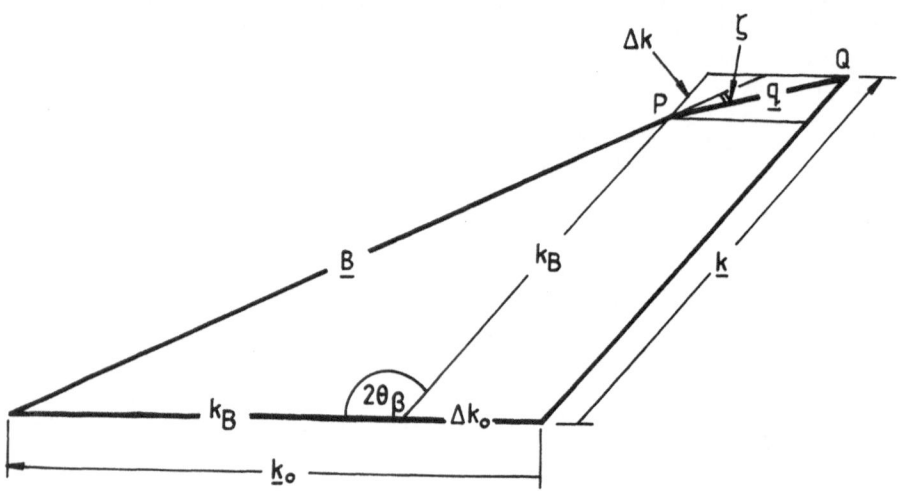

Figure 5. Vector diagram for scattering to a point Q by a phonon of wave vector $\vec{PQ}$. The scattering angle is twice the Bragg angle.

For Bragg scattering, the scattering angle is equal to twice the Bragg angle: $2\Theta = 2\Theta_B$. The vector diagram for this angle is shown in Fig. 5, which is equivalent to Fig. 2 with $\Delta\Theta = 0$. We see from Fig. 5 that

$$q \cos \zeta = (\Delta k_o + \Delta k) \sin\Theta_B \tag{11a}$$

and

$$q \sin \zeta = (\Delta k_o - \Delta k) \cos\Theta_B \tag{11b}$$

where ζ is the angle between the reciprocal-lattice vector and the direction of propagation of the phonon. Inserting (11) into (4), and using (5),yields

$$\frac{\sin\zeta}{\cos\Theta_B}\left[v_B + \frac{\hbar}{2m_n} \cdot \frac{q\cos\zeta}{\sin\Theta_B} \right] = \varepsilon c_s \quad . \tag{12}$$

Bragg scattering occurs at the centre of the Brillouin zone, $q = 0$, so that (12) becomes

$$\sin \zeta = \varepsilon c_s \cos\Theta_B/v_B \quad .$$

There will be no TDS if the magnitude of the right-hand side exceeds unity, i.e. if

$$v_B < c_s \cos \Theta_B \quad . \tag{13}$$

Thus, TDS is forbidden if the incident neutrons are slower than the sound velocity multiplied by the cosine of the Bragg angle. (This situation cannot arise in X-ray diffraction, as X-ray photons are 10^5 times faster than sound velocities.)

We also have the relations

$$k_B = \pi/(d_{hkl} \sin\Theta_B) = (m_n/h)v_B \tag{14}$$

where d_{hkl} is the plane spacing. Combining (13) and (14) we obtain

$$d_{hkl} > \frac{h}{m_n} \cdot \frac{1}{c_s} \cdot \csc 2\Theta_B \quad .$$

The minimum d spacing for which there is no TDS is given by

$$d_{min} = \frac{h}{m_n} \cdot \frac{1}{c_s} \cdot \csc 2\Theta_B, \tag{15}$$

and the range of spacings free from TDS is

$$d_{min} < d < d_{max}$$

where d_{max} is the largest spacing in the crystal.

In a time-of-flight experiment, the Bragg angle is one half the scattering angle. For a detector in the forward scattering position with 2Θ close to zero, or for a detector in back scattering with 2Θ approaching $180°$, $\operatorname{cosec} 2\Theta_B$ in (15) will be very large and it is unlikely that any Bragg reflection will be uncontaminated by TDS. Clearly, the most favourable choice of scattering angle to avoid TDS is $90°$, as $\operatorname{cosec} 2\Theta_B$ is then a minimum (= 1) and d_{min} is given by

$$d_{min} = h/(m_n c_s) . \tag{16}$$

Equation (16) has been used in Table I to calculate d_{min} for three cubic crystals. The sound velocities are given for both longitudinal and shear modes of vibration: the velocities span a range of values, because they vary with the direction of propagation in the crystal. The largest value of d_{min} in (16) corresponds to the slowest transverse mode. In the fourth column of the Table, d_{min} is given for this mode only, as this ensures that lattice planes with spacings exceeding this d_{min} are free from TDS associated with all modes. The final column in Table I gives the maximum d-spacing, as calculated from the cell dimensions.

The Table shows that for silicon, a relatively hard crystal, all the reflections up to 620 are uncontaminated with TDS. On the other hand, for a soft crystal such as lead, d_{min} exceeds d_{max} and so none of the reflections is free from TDS. Yttrium iron garnet, which has a large cell and hence a large d_{max}, has quite a high number of uncontaminated reflections.

TABLE I

Cubic crystals: lattice planes free from TDS at $2\Theta_B = 90°$

crystal	c_s ($\mathrm{km s}^{-1}$) for longitudinal modes	c_s ($\mathrm{km s}^{-1}$) for shear modes	d_{min} (Å)	d_{max} (Å)
lead	2.0 – 2.3	0.5 – 1.1	6.92	2.84
silicon	8.4 – 9.4	4.6 – 5.8	0.84	3.12
yttrium iron garnet	7.2 – 7.3	3.7 – 3.9	1.05	8.74

5. CONCLUSIONS

In a one-phonon neutron scattering experiment, using pulsed neutron
diffraction, gaps or 'windows' appear in the incident and scattered
wavelengths and in the total time-of-flight. These windows are
associated with each Bragg reflection, and their edges are determined
by the sound velocity in the crystal. The sound velocity is readily
obtained, in the absence of energy analysis, from the edges of the
time-of-flight window. This velocity is needed in correcting Bragg
reflections for the 'TDS error', although crystals examined at a
scattering angle of 90° may have no TDS error at all.

There are some interesting analogies between the interaction of
slow neutrons and sound waves in a crystal, and the interaction of
particles and waves in other media. For example, when the sound
velocity in the crystal resolved in the scattering direction is equal
to the neutron velocity, there is a large number of phonon states
contributing to the scattered radiation: similarly, in the case of
the sonic boom, when the aircraft velocity in the direction of the
observer equals the sound velocity, there is quite a stretch in the
flight for which noise arrives simultaneously to produce the bang.
Cerenkov radiation is the optical analogue of the sonic boom; eq (13)
gives the critical velocity below which there is no TDS, but it also
gives the critical particle velocity (with c_s the velocity of light in
the dielectric medium) below which there is no Cerenkov radiation.
Finally, we note that eq (10) contains the Lorentz transform
$(1 - v_n^2/c_s^2)^{1/2}$, familiar in the theory of special relativity.

ACKNOWLEDGEMENTS

Work described in this paper was undertaken as part of the Underlying
Research Programme of the UKAEA. Many of the ideas in it were
discussed with Dr C J Carlile of the Rutherford Appleton Laboratory
and with Dr P Schofield of AERE, Harwell. The experimental work was
performed on the pulsed neutron source ISIS with help from Dr W I F
David and Dr W T A Harrison.

REFERENCES

Elliott, R J and Lowde, R D (1955). Proc Roy Soc London, Ser A, **230**, 46.

Hohlwein, D (1977). Proceedings of a Symposium on Neutron Inelastic Scattering, IAEA Vienna, p 197.

Johnson, M W and David, W I F (1985). Rutherford Appleton Laboratory, Report RAL-85-112.

Samuelson, E J (1968). Phys Lett. A**26**, 160.

Schofield, P and Willis, B T M (1987). Acta Cryst A**43**, 000 .

Seeger, R J and Teller, E (1942). Phys Rev **62**, 37.

Steichele, E and Arnold, P (1973). Phys Lett A, **44**, 165.

Willis, B T M (1986). Acta Cryst. A**42**, 514.

Willis, B T M, Carlile, C J, Ward, R C, David, W I F and Johnson, M W (1986). Europhys Lett. **2** (10), 767.

DISCUSSION

LIMITATIONS OF THE ABOVE APPROACH TO THERMAL DIFFUSE SCATTERING

The method should also be applicable to relatively simple organic molecules, but probably not to large biological molecules for which the conventional treatment of acoustic modes is not appropriate.

Longer counting times, perhaps by one order of magnitude, than those usual for single crystal neutron diffraction may be needed to observe the features due to TDS.

THE EFFECT OF THERMAL DIFFUSE SCATTERING ON STRUCTURE REFINEMENT

The extent to which peaks are affected by TDS depends strongly on the angle of scattering. Hence investigators refining structures from data collected at various angles should not assume that temperature factors are independent of scattering angle. Similar problems will arise in charge density studies and in the measurement of electrostatic potentials of organic molecules.

Provided the elastic constants of the material are known, or can be inferred from sound velocities measured by the above method or in some other way, the problem could be solved either by choosing conditions under which it is absent or by making the appropriate corrections. For large biological molecules, however, there remains the question of how to treat their normal modes of vibration.

COMPARISON OF SOUND VELOCITIES MEASURED BY DIFFERENT METHODS

Measurement of sound velocities by the above method is very convenient. In the few cases where comparison with conventional methods is possible, there is agreement to within 2%.

More interesting, however, would be to explore differences between such methods. For example, the ultrasonic technique uses lower frequencies ($\omega < 10^{10}$ cps) than neutron scattering techniques ($\omega \sim 10^{12}$ cps). Hence phonon-phonon interactions are more likely to occur in ultrasonic measurements, leading to sound velocities which are less than those given by neutrons.

APPLICATIONS OF NEUTRON SCATTERING IN CHEMISTRY. PULSED AND CONTINUOUS SOURCES IN COMPARISON

H. Fuess
Institut für Kristallographie der Universität
Senckenberganlage 30
6000 Frankfurt
Federal Republic of Germany

ABSTRACT. Applications of both elastic and inelastic scattering of neutrons in chemistry are presented. The special requirements for instrumentation at a steady state reactor and at a pulsed source are given and performances for special applications are compared throughout the paper. Both types of sources are equally well suited for most purposes. Gain factors could be realised by proper development of instruments which are conceived to use the peak flux of a pulsed source by time of flight techniques.
  The examples of single crystal diffraction include high precision electron and magnetization density work. It is emphasized that novel applications will be feasible if polarization analysis of scattered neutrons is introduced on a great scale. The contribution of neutron scattering to our understanding of the structure of solutions is demonstrated and some applications in material sciences are reported. It is pointed out throughout the article that neutrons contribute to the knowledge of static structure as well as to dynamical processes. Some examples of inelastic neutron scattering as spectroscopy on molecules, molecular groups and fragments demonstrate the great potential of the neutron to elucidate diffusion, rotation or tunneling motion in molecules and crystals.
  It is concluded that a definite need exists for a neutron source with higher fluxes than those at present available.

1. INTRODUCTION

The essential aim of research in chemistry is the production and identification of new materials. The characterization is achieved by the chemical formula and the molecular and crystal structure. In addition the changement of the structure and the mobility are of importance. Therefore a complete description of a compound is given by static arrangement and by the displacement of atoms due to vibrations. The structure determination is essentially achieved by diffraction of X-rays, the dynamical displacement of atoms is investigated by electromagnetic waves, by measuring energy changes. The neutron is a unique and marvellous

M. A. Carrondo and G. A. Jeffrey (eds.), Chemical Crystallography with Pulsed Neutrons and Synchrotron X-Rays, 77–115.
© *1988 by D. Reidel Publishing Company.*

particle because it allows both types of investigations as it matches
<u>simultaneously</u> typical energy and wavelength scales in condensed matter.
The convenient neutron wavelengths (1 - 10Å) are ideal for the study of
interatomic distances and the available neutron energy range from 1 -
1000 meV (1 meV $\sim$ 8cm^{-1}). The vibrational energy in condensed matter is in
that range and energy transfer from a molecule or crystal to a neutron
is in the same range as that of the incident neutron itself. The energy transfer is therefore easily detectable. An energy transfer occurs of
course during an X-ray diffraction process. But X-rays of 1Å wavelength
have energies of about 12 keV, an energy transfer of several meV is
accordingly very difficult to detect.

The general interaction of a neutron with condensed matter is a
combination of several processes which occur simultaneously. The range
covered by energy and momentum transfer is given in Fig. 1. A comparison with X-ray scattering is presented in Fig. 1a, whereas a summary of
accessible studies is given in Fig. 1b.

It is, however, not feasible, to register at the same time all
scattering processes in a single experiment. We shall therefore first
describe briefly the main scattering processes occuring when a neutron
beam is interacting with condensed matter.

The neutron is scattered by the nuclei and due to its own magnetic
moment by the magnetic moments of atoms and nuclei. The scattering process is normally elastic (momentum transfer) and inelastic. Both interactions have coherent and incoherent components. The variety of interactions together with the relevant applications is given in Table 1.
The complete scattering process is a superposition of all individual
interactions. Highly specialised instruments have been conceived and
constructed in order to separate a particular scattering process out
from the others.

The application of neutrons to chemistry started with diffraction
experiments on hydrogenous materials. A great deal of effort was then
spent on the determination of magnetic structures and on magnetization
experiments. The study of large crystalline molecules like proteins and
polymers became feasible with the advent of high-flux reactors. Almost
all scattering contains information of relevance to chemists and we shall
give examples in the following section. A historical account of neutron scattering is given in the book "Fifty Years of Neutron Diffraction" edited by Bacon (1986). A review on neutron diffraction experiments
in chemistry was published by the author (Fuess 1979).

The instruments used for elastic scattering were conceived along
the successful developments of X-ray diffraction (powder diffraction;
small angle diffraction; four-circle Eulerian cradle) whereas the inelastic spectrometers had to apply highly specialized facilities adapted
for precise ranges of energy and momentum transfer.

Table 1: The interaction of neutrons with matter

| Neutron Scattering | | | |
| elastic (momentum transfer) | | inelastic (energy transfer) | |
coherent	incoherent	coherent	incoherent
three-dimensional order	ordered defects	collective motion	individual motion
crystal structure	lattice defects	phonons	spectroscopy
H-positions	vacancies	phase transitions	rotation of molecules
cation order	interstitials	(soft modes)	
	clusters		plastic crystals
Liquids			
magnetic structure	spin glasses	magnons	
magnetization density			
magnetic form factors			

Table 2: Neutron Scattering:

a) Steady state reactors

Location	Flux $(10^{14}\mathrm{cm}^{-2}\mathrm{sec}^{-1})$	Neutrons available	Number of instruments	Year
Ill Grenoble	12	H, Th, C	40	1972
KFA Jülich	2	Th, C	15	1964
Harwell	2	Th, C	15	1950
LLB Saclay	5	H, Th, C	22	1982
Brookhaven NL	9	Th, C	11	1960
Oak Ridge	10	Th	9	1960

b) Spallation Sources (proton driven), Target U^{238}

	Incid prot. Energy MeV	Pulse freq.(HZ)	Pulse width(sec)	Year
KENS, Japan	500	15	0.5	1980/85
IPNS, Argonne	500	30	0.1	1981
SNS Rutherford,UK	800	50	0.2	1985
SNQ,Jülich,FRG	1100		100	+

+ Project discontinued; H = hot, Th = thermal, C = cold

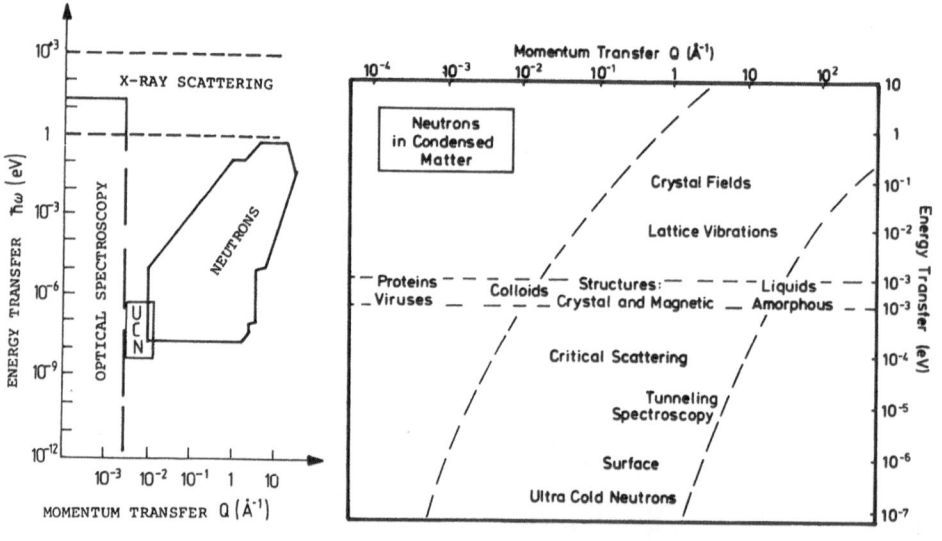

Fig.1: Range of energy and momentum transfer in neutron scattering.
a) Comparison with X-ray scattering and optical spectroscopy (UCN
means Ultra Cold Neutrons). b) Application of momentum and energy
transfer in condensed matter research.

2. STEADY STATE AND PULSED SOURCES

The conventional neutron facilities are nuclear fission reactors which became operational by about 1948. The most important figure for the application of neutrons for scattering experiments is the flux and a continuous flux of $1.5 \cdot 10^{15}$ n sec^{-1} cm^{-2} (at 57 MW) is the maximum at present available. Plans exist to build a reactor at Oak Ridge (US) with a flux of 5.10^{15} n sec^{-1} cm^{-2}. Higher fluxes are not accessible due to cooling problems. The spallation process at proton or electron accelators potentially leads to higher fluxes but they are only available during a short period of time. Peak fluxes of 10^{17} n cm^{-2} sec^{-1} are available during the duration of the incoming pulses. The time average flux of existing spallation sources is still inferior to a steady state reactor. A considerable gain is obtained if all the neutrons in the pulse can be applied to experiments. This means that the scattering experiment should be in phase with the pulse rate. The most suitable method for this type of experiments is the time of flight technique which has successfully been applied for many inelastic scattering experiments. Some instruments have been built recently for neutron diffraction using the time of flight technique at spallation sources.

The fundamental differences between a pulsed source and a reactor may be summarized in two points

(i) The pulsed source (spallation or fission) produces neutrons in bursts of 1 to 50 μsec duration (spaced 20 – 100 millisec apart). During this short time a high intensity of neutrons is available but only for a short period. The pulsed source therefore requires modified techniques which make full use of this intensity. These techniques are essentially based on time of flight (TOF) where the time a neutron needs to travel a certain distance is the basic quantity. The neutron pulse contains neutrons of a wide energy spectrum. Some preselection is therefore required.

(ii) The spectral characteristics of neutrons from pulsed sources differ from those of reactor neutrons. There is a much larger component of high-energy neutrons with energies above 100 meV which may eventually lead to new applications.

In Table 2 some of the neutron research facilities are given. The SNQ-project in Germany which was under discussion for several years was finally abandoned. It is seen from Table 2 that many research reactors are equipped with hot and cold sources in order to extend the available energy of neutrons. Cold sources usually produce low energy (long wavelength) neutrons by moderation with liquid hydrogen at 22K; hot sources produce short wavelengths. The hot source at the ILL, Grenoble, consists on a graphite bloc and moderates neutrons at about 2000°C. Due to the large amount of high energy neutrons in a spallation source no hot moderator is needed. The wavelength distribution of neutrons is given in Fig. 2 together with the distribution of X-rays from synchrotron radiation and from a conventional tube.

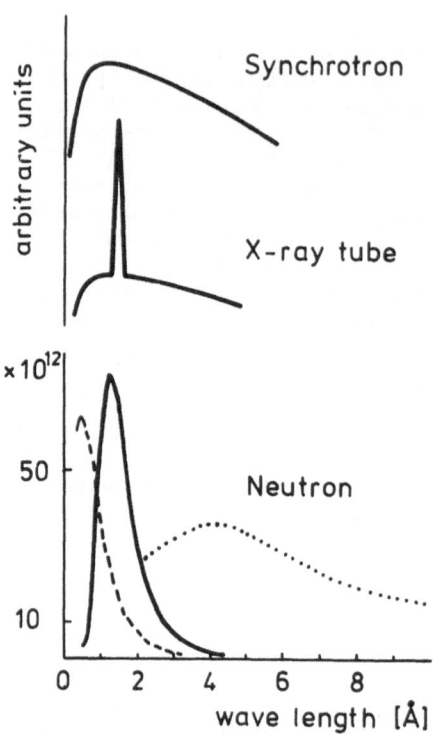

Fig. 2: Schematical wave-length distribution of hot, thermal and cold neutrons together with X-ray spectra from a synchrotron and a conventional X-ray tube.

At present no clear advantage for pulsed or steady sources may be derived from the experiments. The pulsed sources have definitely a potentially greater flux if TOF-techniques for all possible applications can be developed. Fig. 3 gives a summary of the instruments for elastic and inelastic scattering which were proposed for the SNQ at Jülich in Germany. This figure illustrates that all different scattering processes of Table 1 may well be covered by TOF techniques on a pulsed source.

This contribution does not intend to favour one or the other source. We shall try to give examples of applications which are relevant to chemistry. In some cases a comparison of results obtained with both kinds of sources may be possible. If a comparison is made it is based on the proceedings of two workshops held in recent years. One (Scherm and Stiller, 1985) resumes the type of instrumentation planned for the German spallation source SNQ; the other (Lander and Emery, 1985) gives an account of a comparison between proposals made for an advanced neutron source in the United States.

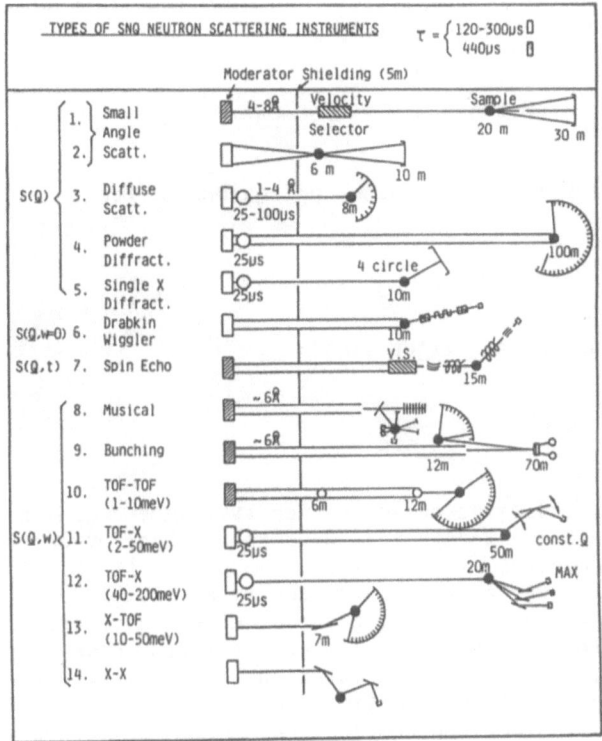

Fig. 3 : This figure is from the contribution of Colin Windsor in the "Proceedings of the Workshop on the SNQ". The symbols mean □ moderator at ambient temperature, ▨ = cold moderator, ── = beam without neutron guide, ══ = with neutron guide, o = chopper, ● = sample.

3. STRUCTURAL INVESTIGATIONS

3.1. Performance of diffractometers at both sources

The basic experiment for structure investigation is the diffraction experiment with no energy transfer ($\Delta E = 0$). The purely coherent elastic part leads to Bragg peaks which allow a determination of the position of the atoms in a crystal. Furthermore the structure of liquids, solutions, glasses and amorphous material is investigated by diffraction methods. As an entire session is devoted to powder diffraction we shall not cover powder work here.

Experiments at a pulsed source use the "energy dispersive" technique in which θ, the Bragg angle is fixed and the time t a neutron travels from the sample to the detector is observed. This time may be related to the neutron wavelength λ by

$$\lambda = h/mv = (0.3955\overset{\circ}{A}cm/\mu sec)\ t/L$$

where L is the total flight path. The application of Bragg's law

$$2d_{hkl}\ \sin\ \theta = \lambda$$

relates the spacing to the flight time

$$d_{hkl} = \frac{0.3955}{2xL}\ \frac{t}{\sin\ \theta}$$

In single crystal studies at pulsed sources the Laue-technique with a white beam of neutrons is applied. Detection is performed by a position sensitive detector. The determination of the flight time t leads to the neutron wavelength diffracted by an individual reflection. The time analysis of the scattered neutrons allows a separation of different orders in the Laue pattern. The area detector gives a picture of the scattered neutrons. This representation not only includes the Bragg peaks but also additional superlattice peaks and diffuse scattering between the peaks. In fact the combination of the TOF-method with a position sensitive detector allows the simultaneous registration of a good part of reciprocal space as shown in Fig. 4. It immediately makes clear that the presence of a large $\Delta\lambda$ in the incident beam gives rise to many details in reciprocal space.Fig. 4 shows the (1kl) plane from a single crystal of the organic conductor $(TMTSF)_2ClO_4$ at 15 K (Leung et al. 1984).The superlattice reflections indicate a phase transition due to shifts in the atomic parameters. Large area detectors are now in use with steady reactors and monochromatic incident beam. In this case a gain is only obtained for large unit cells where many reflections occur simultaneously.

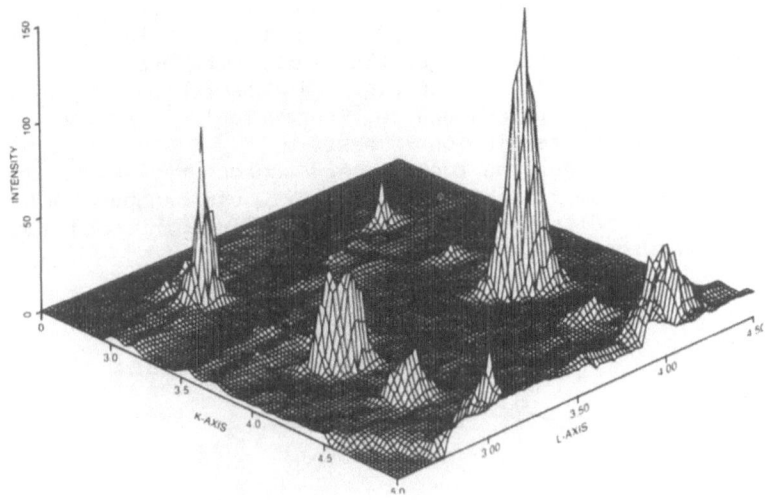

Fig. 4: Two-dimensional detector screen of an area detector
showing the intensity distribution on the (1,k,1)
plane of the organic conductor $(TMTSF)_2ClO_4$ (Leung et
al., 1984)

Table 3: Basic features for single crystal diffractometers

	TOF (pulsed source)	Reactor
Method Mode	Laue θ fixed λ variable	Bragg θ variable λ fixed
Corrections	extinction and multiple scattering are wavelength dependent	better accuracy in intensities
Resolution	constant over range of θ $\Delta S/S$ (S=2sin θ/λ) $2\theta = 90^o$: ~ 0.01	depends on θ $\Delta \lambda/\lambda = 0.01$
Flux at sample (n cm^{-2} sec^{-1})	$\sim 3 . 10^8$ (averaged)	$7.5 . 10^7$

A comparison between typical instrument parameters at both types of sources is given in Table 3 (after Lander and Emery, 1985). This comparison is based on the time-averaged flux integrated over a useful wavelength range. The shaded area in Fig. 5 is the region of reciprocal space viewed with the pulsed source. The optimal diffraction angle is near to 90° where the resolution is best.

The flux estimates for the two proposed sources (pulsed or reactor with $5 - 10^{15}$ n sec^{-1} cm^{-2}) indicate that they would be approximately equivalent for large cells with somewhat higher flux performance for the pulsed source with smaller sized cells.

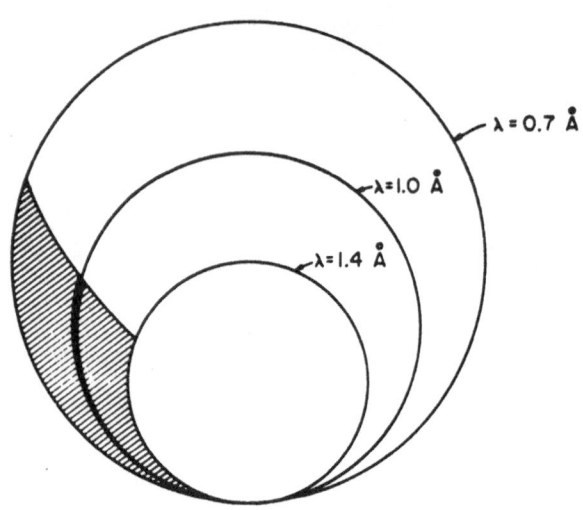

Fig. 5: Comparison of reciprocal space viewed by pulsed source (shaded) and steady state reactor (after Lander and Emery, 1985)

The correction of a number of diffraction effects, especially extinction, which are wavelength dependent, may limit the accuracy of data obtained by a multiwavelengths method at a pulsed source and thus reduce the possible gain factors because smaller crystals have to be used.

3.2. Single crystal studies

The traditional studies of hydrogenous materials continue to be of great interest. Metal hydrides, organometallic compounds (Fig. 6) and zeolites (Fig. 7) are among the more popular crystals studied. These studies are clearly limited by the requirement for crystal size ($\geq 1mm^3$).

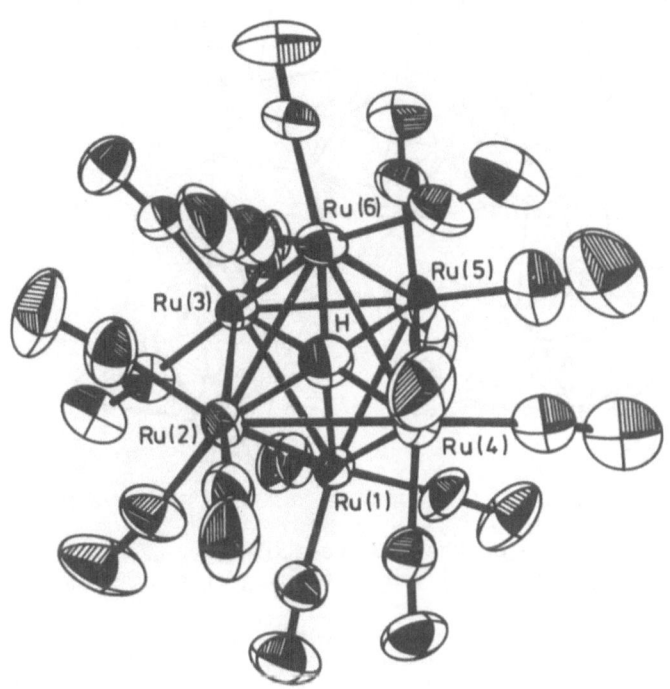

Fig. 6: The structure of the anion $[HRu_6(CO)_{18}]^-$ representing the hydrogen atom in the middle of the Ru_6-cage. (after Jackson et al., 1980)

The tremendous amount of information on hydrogen bonding was summarized in diagrams which correlate geometries in hydrogen bonded systems.

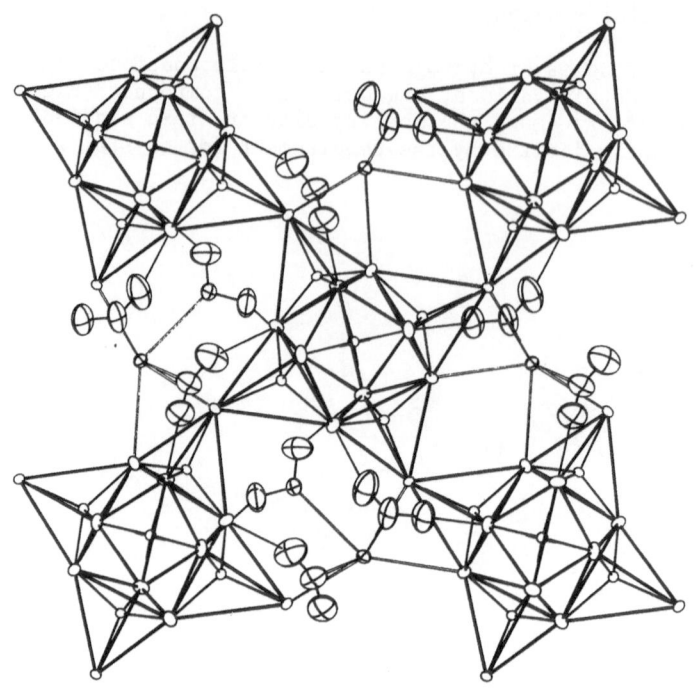

Fig. 7: Crystal structure of the zeolite scolecite
$Ca_8 [(AlO_2)_{16} \cdot (SiO_2)_{24}] \cdot 16H_2O$. (Joswig et al. 1984)

Fig. 8 represents the O–H...O hydrogen bonds. The O–H distance is given as a function of O...O for a number of neutron diffraction studies. (Joswig et al. 1982) The very short hydrogen bridges (with O...O distances below 2.45Å) are of special interest. It seems that several O–H distances are possible for the same O...O distance. Furthermore the existence of symmetrical hydrogen bonds is still an open and controversial question.

Another traditional application of neutron diffraction is the determination of cation distributions. Successful applications were reported for structural investigations of minerals and inorganic solid state chemistry.

Neutron diffraction on protein single crystals made use of the unique ability to measure the degree of H/D exchange of almost all labile protons due to the large difference in scattering power between these two isotopes. Proteins of higher molecular weight have been examined than by NMR studies. New results were recently obtained for solvent regions between molecules and salt bridges using the enormous difference between the scattering power of D_2O and H_2O as a solvent. Further

experiences have elucidated the nature of the interactions between proteins and surface probes such as deuterated alcohols.

Furthermore the difference in scattering length between nitrogen
on one side and carbon and oxygen on the other side was used to determine the correct orientation of side chains.

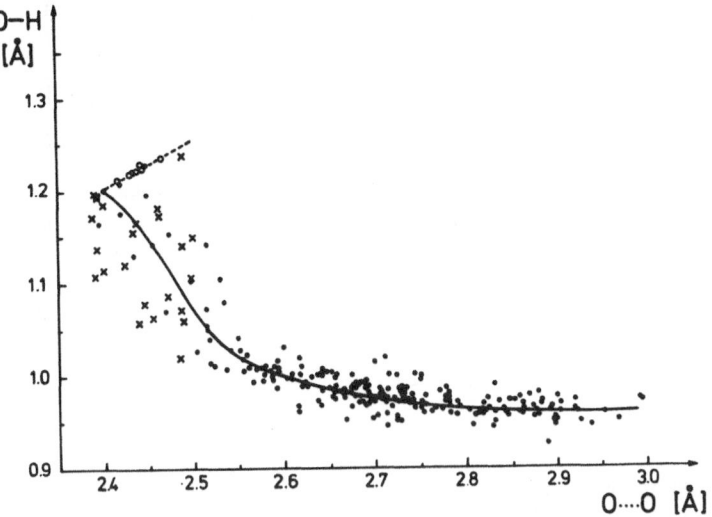

Fig. 8: The O-H-distances as a function of O...O for O-H...O hydrogen
bonds. ----symmetrical bonds; ——calculated curve: + very short
H-bridges (with O...O below 2.45Å)

3.2.1. <u>High precision: electron distribution.</u> Combination of X-ray
and neutron diffraction (X-N-method) provides detailed knowledge of the
distribution of bonding electrons and non-bonded electron pairs("lone-
pairs") . The method is based on the assumption that an unbound atom is
spherical and that any deviation from sphericity is due to chemical bonding. Therefore the density of a neutral chemical atom calculated at
the position determined by neutron diffraction is subtracted from the
total density measured by X-rays. In Fig. 9 the experimental deformation density in 1-(trichlorosilyl)-1,2,3,4-tetrahydro-1,10-phenan-
throline is presented. This compound is obtained by a reaction of the
type Lewis acid-Lewis base and a pentacoordinated silicon is obtained.

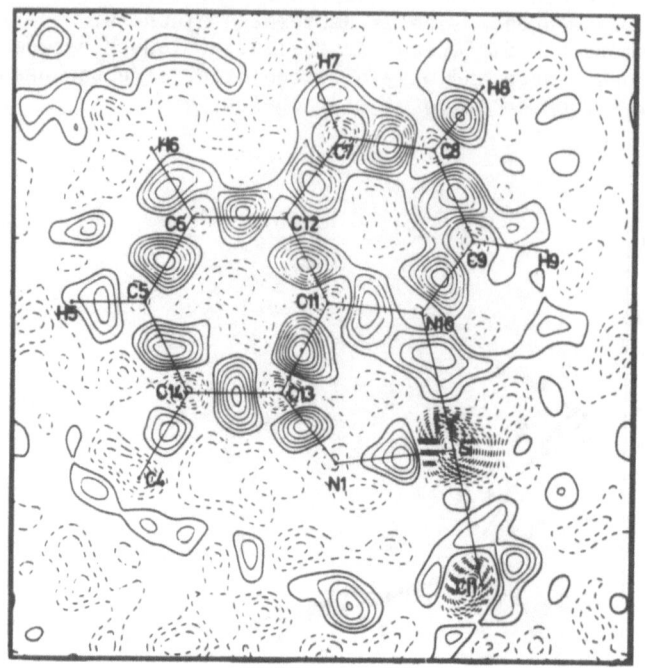

Fig. 9: Deformation electron density in 1-(trichloro)-1,2,3,4,-
        tetrahydro -1,10-phenanthroline (Klebe et al. 1984)

The electron density elucidates the divergent nature of the two sili-
cium-nitrogen bonds, the shorter one (1.74Å) showing the bonding electrons at the center of the connecting lines of both atoms, the longer
one (1.97Å) is interpreted more as an addition of the Si to the electron pair of the N10 nitrogen (Klebe et al. 1984). Deformation electron
densities in the carbon-carbon and carbon-hydrogen bonds is clearly
displayed and correspond to observations in other organic molecules.
The comparison of experimental deformation densities with theoretical
calculations based on molecular orbital theories contributes to the
understanding of chemical bonding. The work on thiosulfates for example
(Fuess et al. 1985) demonstrates clearly the importance of d-electrons
for bonds between sulfur and oxygen.
      Further fields of interest for high precision data include the
distribution between ions and vacancies in non-stoichiometric oxides
or carbides of transition metals. The determination of precise temperature factors is more easily achieved with neutrons than with X-rays because nuclear neutron scattering does not depend on the scattering angle
as the form factor of X-ray diffraction. Therefore a precise evaluation
of possible anharmonic effects in thermal motion is much easier to

analyse from precise neutron diffraction data.

As already mentioned previously data collected at a steady state reactor are expected to be more precise as all necessary corrections are applied for a single wavelength. A pulsed source, however, naturally has a wavelength band and corrections (e.g. extinction) which depend on wavelength are more troublesome. For work which demands high accuracy (electron density and atomic distribution; anharmonicity in temperature factors) a diffractometer at a reactor seems to be preferable, whereas fast data collection rates and complete investigation of the reciprocal space ask for a pulsed source.

3.2.2. Magnetization density studies by polarized neutrons. It has been stated in the introduction that neutrons possess a spin of $s \pm 1/2$ which leads to magnetic scattering. The diffraction process of unpolarized neutrons produces a superposition of nuclear magnetic intensities. The evaluation of the magnetic part of the diffracted intensity established a great number of ferro-, antiferro- and ferrimagnetic structures. They are the base for the formulation of the magnetic space groups and the prediction of possible magnetic symmetries (see e.g. Opechowski and Guccione, 1965). We shall not include magnetic structures here but refer to a compilation of experimental work by Oles et al. (1985).

Magnetization densities may give some insight into the cohesive interatomic interactions, like chemical bonding or covalent contributions to bonding. This is possible because d-electrons contribute to the cohesion of the solid as well as to its magnetism. In order to determine magnetization densities a precise knowledge of the magnetic structure factor is required. Precise measurements of magnetic structure factors can only be achieved with polarized neutrons. The term polarized neutrons means that all neutrons in the incoming beam have a single spin state ($s = +1/2$ or $s = -1/2$).

Traditionally polarized neutrons were produced by monochromator crystals and 50% of all neutrons were lost. Considerable progress in the development of neutron polarizers for steady-state sources has been made in recent years. This was especially possible by the development of polarizing mirrors and supermirrors which consist of a number of bilayers of various thickness and composition. Their polarizing efficiency is dependent on wavelength. High efficiency polarizing devices are now available for neutron wavelengths of 2Å and longer. These devices can be used with equal benefit for both types of sources. For steady state reactors further developments of polarizers with high reflectivities are needed for wavelengths shorter than 0.5Å. For pulsed sources the development of broad-band high energy polarizers with higher transmission is needed but seems to be difficult. More facilities with higher neutron fluxes and with polarization devices are required because the analysis of the polarization state of the neutron after scattering will allow for new applications which include (i) separation of nuclear and magnetic scattering (ii) diffusive motion of polymer molecules in solution (iii) flexibility of macromolecules (iv) diffusion in solids. It is hoped that polarization analysis will become a routine method for future applications.

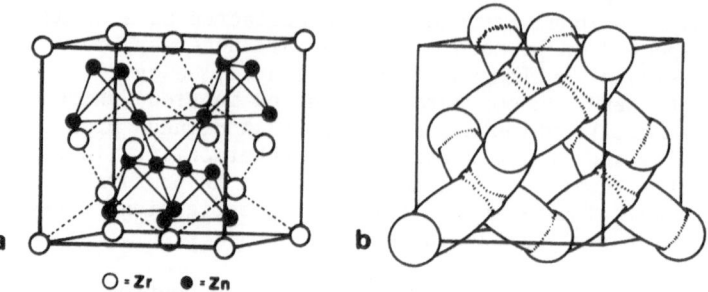

O · Zr   ● · Zn

Fig. 10:   (a) Positions of the zirconium and zinc atoms in ZrZn$_2$
           (b) Schematic representation of the magnetization density

In the past the most common use of polarized neutrons has been the
determination of magnetic moment densities in ferro- and paramagnetic
magnets. In a polarized beam experiment the following ratio is measured.

$$R = \frac{I^+}{I^-} = \frac{(F_N + F_M)^2}{(F_N - F_M)} = \frac{(1+\gamma)^2}{(1-\gamma)^2} \qquad \gamma = \frac{F_M}{F_N}$$

This is the ratio between the intensities with incoming spin up
(s = +1/2, leading to I$^+$) and down (s = -1/2 and I$^-$). These intensities
are both composed by nuclear and magnetic scattering.
     With a magnetic portion of about 1%, this method still yields an
R-value of 1.04, allowing a precise determination of F$_M$, the magnetic
structure factor if F$_N$, the nuclear structure factor, is known. The
magnetic scattering is governed by a magnetic form factor and decreases
with increasing sin θ/λ . An experimental determination of magnetic
structure factors drawn as a function of sin θ/λ may be compared with
theoretical calculations directly. The spatial distribution of the electrons in unfilled inner shells (mainly 3d or 4f-shells) is obtained by
a Fourier inversion of the magnetic structure factors which produces a
magnetization density. A magnetization density for a compound with a 4d-
element is presented in Fig. 10 for the cubic Laves phase ZrZn$_2$ (Brown
1986). This density extends over the tetrahedral framework of the zirconium atoms. This delocalized density cannot solely be due to the
overlap of 4d-electrons of Zr since the overlap between 4d-functions at
a distance of 3.2Å which separate the Zr atoms is relatively small. Some
contribution of either the zinc electrons or the more diffuse 5s and 5p
electrons of Zr is therefore expected. A direct comparison of the experimental density with two densities from model calculations is shown
in Fig. 11 where the model of Fig.11b reproduces better the experimental
density than Fig.11c.
     It has been concluded that the bands which contribute to the weak
ferromagnetism of ZrZn$_2$ are based on the 4d bands with some admixture
of Zr 5p functions. No significant contribution from the Zn electrons is
observed.

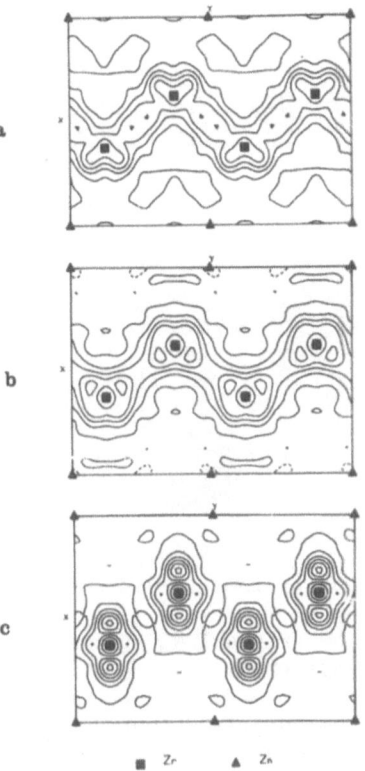

Fig. 11: Observed and calculated magnetization densities in ZrZn$_2$, sections perpendicular to [1$\bar{1}$0], atoms marked.
a) Observed density at 5K, contours at 0.004μ_B/Å^3
b) calculated density with 75% Zr 4d and 25% Zr 5p character.
c) Calculation with 75% Zr 4d and 25% Zn 4p character.

■ Zr   ▲ Zn

Another approach to interprete magnetization densities has been described by Fender and coworkers (1986) who analyzed their data in form of population analysis of different d-functions in complexes of 3d-elements. For the Tutton salts of manganese and nickel (ND$_4$)$_2$ M(SO$_4$)$_2$ · 6D$_2$O with M = Mn, and Ni they refined the spin population. In Mn^{2+} no reduction in the spin population of the d-orbitals (t$_{2g}^{3.1}$ e$_g^{2.0}$) is observed but a small 4s$^{0.2}$ population is present which together with the overlap population of -0.1 spins balances the 0.2 net positive spin found on the water molecules. In the Ni compound the spin population is almost entirely e$_g$ (t$_{2g}$ = -0.06, e$_g$ = 1.64, 4s = 0.38). The Ni-OD$_2$ show more covalent spin transfer than was seen for the similar Mn-OH$_2$system.

Magnetization density studies on more complicated systems will be feasible with higher intensity sources, whether reactors or pulsed sources. To illustrate future applications we present in Fig. 12 the magnetization in the organic free radical α,α-diphenyl-ß-picryl-hydrazyle. The measurement of the magnetic intensities was performed at 4.2K in a field of 4.6 Tesla. Besides an important concentration of magnetization at the central N-N-bond, density is observed distributed over the entire molecule. (Boucherle et al., 1982).

94

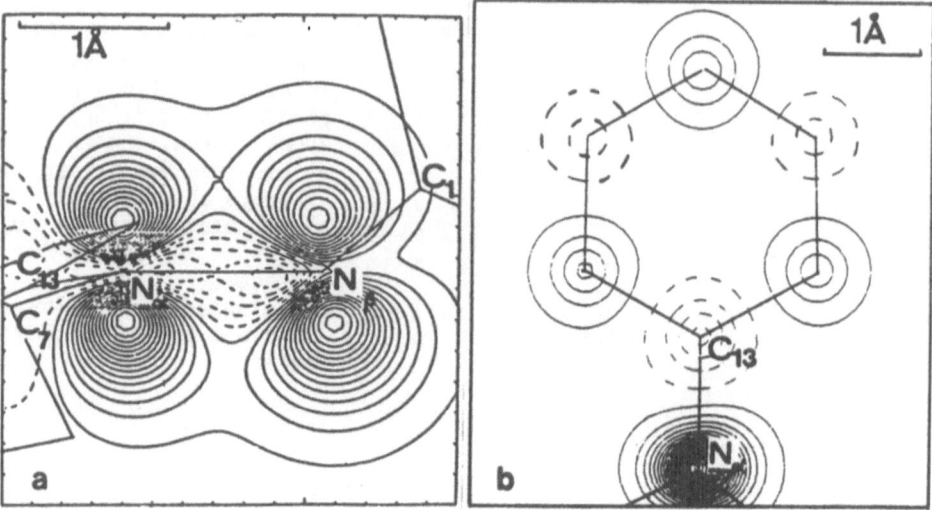

Fig. 12: Projection of the magnetization density in α,α-diphenyl-ß-
    picryl-hydrazyle
    a) along a direction parallel to the trigonal N_α-plane
    b) perpendicular to one of the phenyl rings

Even more information could be extracted from polarized neutron scattering experiments if all types of instrument could be equipped with
polarization analysis techniques. This would allow to explore the full
information which is included in a neutron scattering experiment (momentum and energy transfer and an analysis whether the spin direction has
been inversed during the scattering process).

4. LIQUIDS, SOLUTIONS AND AMORPHOUS SOLIDS

Neutron diffraction has provided a powerful tool for elucidating the
structure of liquids and glasses. Current studies include binary metallic or nonmetallic glasses (Fig. 13), liquid alloys, molten salts and
aqueous solutions.
    The observed scattering function S(Q) in Fig. 13 may immediately be
interpreted in terms of interatomic distances. In a multicomponent system like a solution of a salt in water, several correlations and interatomic distances have to be determined experimentally. In order to obtain a structural model for such a solution several distances have to
be observed (cation-cation; cation-oxygen; cation-hydrogen; anion-anion).
Each of these distances is expressed by a partial structure factor
$S_{ij}(Q)$ whose Fourier transforms, the radial distribution functions,

represent the distribution of j-type atoms about a central i-type atom.

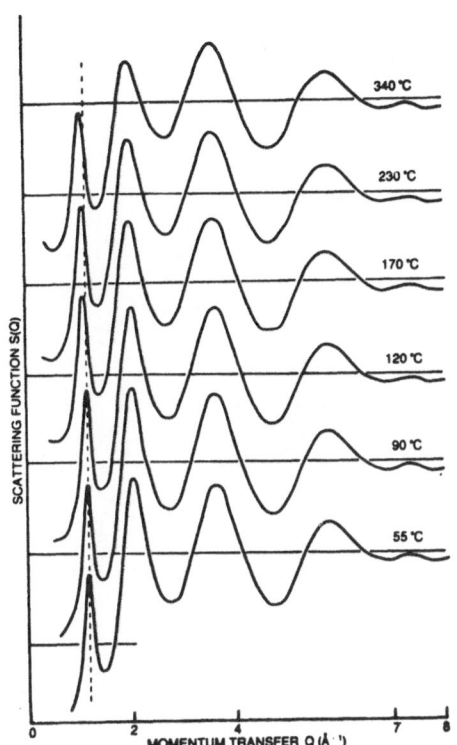

Fig. 13: Temperature dependence of the observed scattering function S(Q) for glassy and liquid P_2Se_3 (Misawa and Watanabe, cited from Lander and Price, 1985)

The great advantage of neutrons over X-rays in this type of study is the different scattering power of isotopes. The isotopic replacement (see Enderby and Neilson, 1981) allows a distinction between atoms which are identical scatterers for X-rays. By isotopic exchange the individual correlation can be separated out of the totality, thereby facilitating the derivation of mean distances for every two particles.

A great deal of work was reported for chloride salts (LiCl, NaCl; $CaCl_2$, $NiCl_2$ and others) in H_2O and D_2O. For a review see Enderby (1983). To exemplify some of the results we present the $NiCl_2$ case in some detail. This particular salt is well suited for the isotopic replacement method as nickel exhibits several isotopes with very diverse scattering lengths. In addition two isotopes of Cl are available. The distribution function for the environment of the Ni^{2+} ion in D_2O is represented in Fig. 14. This function shows maxima at 2.07Å and 2.67Å which correspond to the mean distances between Ni and O and Ni and D, respectively. The maximum at 4.5Å indicates a second coordination shell, which for the first time could be proved in this way. This hydration shell becomes more pronounced for smaller concentrations.

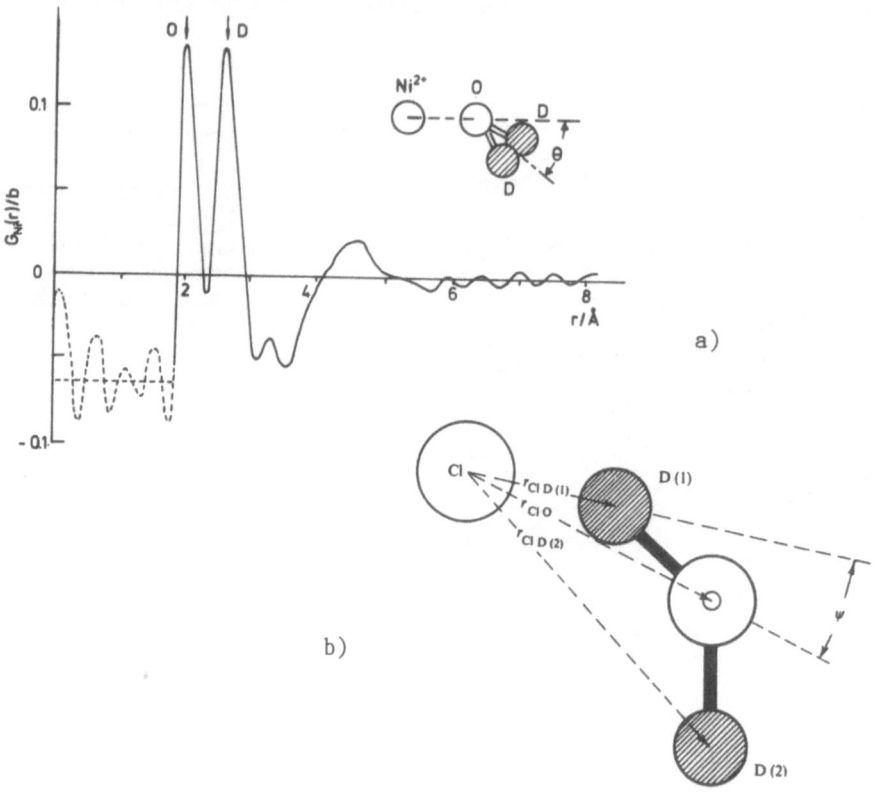

Fig. 14: a) Distribution function of a solution of $NiCl_2$ in D_2O
(4.41 molal). Inlet: the definition of the angle Θ (see
tab. 4). b) The geometry of the $Cl-D_2O$ interaction.

The number of water molecules in the hydrate shell decreases with increasing concentration (Table 4). Besides this general information,
data was also obtained on the exact geometry of the Ni-water interaction. To describe this geometry, an angle Θ, formed between the plane
of the water molecule and the Ni-O connecting line, is defined. For low
concentrations the plane of the water molecule lies in this connecting
line, but above a concentration of 1 molar it deviates from this plane
(see inlet in Fig. 14 and Table 4). The hydration of the Cl^- ion was
determined accordingly from the solution of various chlorides and yielded a mean Cl-D distance of 2.25 + 0.03Å and a Cl-O distance of 3.20(4)
to 3.34(4)Å (Fig. 14b). Thus in the anion-water complex bonding occurs
between Cl^- and deuterium. At the same time it was demonstrated that
the influence of the cation on the geometry of the Cl-water interaction
is small.

Table 4: Hydration of Cations in a solution of D_2O.
Errors given by the authors in parantheses.
(see Enderby 1983)

Solution	Mol.	Distance Ni – O	Ni – D	$\theta(°)$	Hydr. number
$NiCl_2$	0.086	2.07(3)	2.80(3)	0(10)	6.8(8)
	0.46	2.10(2)	2.80(2)	17(5)	6.8(8)
	0.85	2.09(2)	2.76(2)	27(5)	6.6(5)
	1.46	2.07(2)	2.67(2)	42(4)	5.8(2)
	3.05	2.07(2)	2.67(2)	42(4)	5.8(2)
	4.41	2.07(2)	2.67(2)	42(4)	5.8(2)
$Ni(ClO_4)_2$	3.80	2.07(2)	2.67(2)	42(8)	5.8(2)
$CaCl_2$	1.0	2.46(3)	3.07(3)	38(9)	10.0(6)
	2.8	2.39(2)	3.02(3)	34(9)	
	4.5	2.41(3)	3.04(3)	34(9)	6.4(3)
LiCl	3.57	1.95(2)	2.55(2)	40(10)	5.5(3)
	9.95	1.95(2)	2.50(2)	52(15)	3.3(2)

Neutron diffraction on single crystals of $NiCl_2$ hydrates yielded a similar geometry in the crystalline state. Diffraction diagrams of the
melts of the chlorides $CaCl_2$, $BaCl_2$, $SrCl_2$ and $ZnCl_2$ showed that the
arrangement of chlorine atoms around the cation and the mean Cl-Cl
distances were influenced only by the size of the cation and not by its
electronic structure.

It should be mentioned that an enormous amount of work has been
reported on amorphous alloys (structural and magnetic properties); on
ceramics, gels and other materials with no crystalline order. The investigations were performed by diffraction experiments in a suitable
range of momentum transfer mostly with neutrons with short wavelength
to cover a wide range of momentum transfer. Many investigations on solutions and amorphous material were carried out by small angle scattering
(SAS). A detailed comparison of a SAS instrument at a pulsed source and
at a reactor was reported by Seeger and Pynn (1985). Special emphasis
was given to the resolution in both cases. These authors conclude "that
the choice of the instrument should depend on the aims of the experiment
and that neither can claim overwhelming superiority. An advantage of the
pulsed source is that the entire range is measured with similar resolution with only one experimental arrangement". An advantage for a
pulsed source may be seen if high spatial resolution if the radial distribution factor is required as in the case of measurements in solution. In this particular case data to wave vectors Q ($Q = 4\pi \sin\theta / \lambda$)
up to $50Å^{-1}$ are required. This creates a need for short wavelength
neutrons which are naturally available at a pulsed source.

5. SOME APPLICATIONS IN MATERIAL SCIENCES

The investigation of texture in material science constitutes a relatively recent application of neutron scattering, but the neutron as a
probe for material testing has a promising future. This future requires
intense neutron beams because relatively long measuring times are needed
at present neutron facilities. The main advantages a neutron beam offers
as compared to X-rays are: (i) The neutron registers bulk properties due
to its great depth of penetration (ii) neutron scattering is sensitive
to magnetic properties and their variation as a function of external
stress or strain. In addition they are complementary to X-rays in any
case as they produce different contrast. It was again the neutron small
angle scattering technique which produced the first result to materials
research on metals and polymers.

In polymer research the isomorphous replacement method has been
instrumental for structural investigation in the solid state. For polyethylene for instance neutron scattering from polymers which contained
deuterated chains within hydrogenated material helped to clarify a controversy and gave evidence for the so-called solidification model
(Fig. 15).

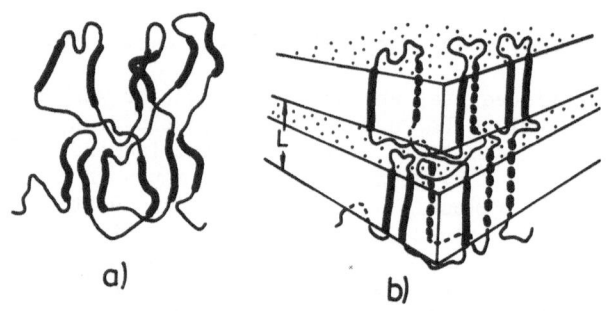

Fig. 15: Structure of polyethylene a) in the melt b) in the solid
state (Stamm et al., 1979)

This model assumes that the structure in the solid occurs by straight-
ening coil sequence which are already present in the melt. Long range
diffusion processes are thus avoided. Most work on polymers was performed by small angle scattering.

Neutron small angle scattering gave furthermore evidence of inhomogeneous regions of clusters and exsolutions in metals and alloys
with sizes between 10Å and $10\ 000\text{Å}$. The small angle scattering techniques

integrate over the sample segment of interest and thus provides information on bulk properties. In addition information on surfaces may be obtained due to differences in scattering power. Fig. 16 gives an example of the annealing process of ZrO_2 dispersed on the surface of an PtRh-alloy. The coarsening of the particles during the annealing process is clearly detected.

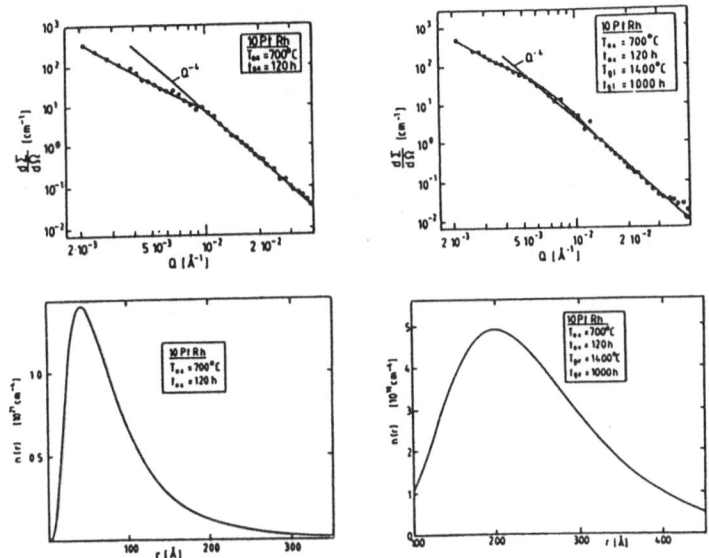

Fig. 16: Neutron scattering of ZrO_2-coarsening as function of annealing. The ZrO_2 particles are dispersed on PtRh-alloy. The annealing conditions are given in each picture. The two pictures above are the scattering curves; below the distribution of particle size is plotted. The mean radius increases from 30Å to 250Å. (After Schmatz, 1984)

A second technique already applied with success in materials research is neutron radiography which gives direct images with a resolution down to 0.05mm. Most metals are penetrated to a great depth by neutrons but the penetration is quite small for hydrogen containing metals. This difference in penetration ability for low energy neutrons offer the possibility to examine the penetration of oil into a metal. By this technique the oil flow in an engine was followed.

The distribution of crystallites in polycrystalline material and especially its changes under external conditions of stretch and strain is of great importance. These texture investigations look at the position, intensity and orientation distributions of Bragg reflections of polycrystalline material and especially its changes under external conditions of stretch and strain is of great importance. These texture investigations look at the position, intensity and orientation distributions of Bragg reflections of polycrystalline materials.

The evolution of Bragg reflections under stretch are demonstrated in Fig. 17 which describes the transformation of an austenitic steel from a face centred cubic lattice (fcc) to a body centred (bcc) lattice under stretch (Jung, 1983). During this transformation an intermediate hexagonal phase (hcp) occurs. Texture, that is preferred orientation of the crystallites, is already present in the original material. Models of the transition mechanism may be tested on a microscopic scale by determining the volume fractions of the three phases involved. The decrease of the fcc reflection is continuous with increasing stretch whereas the bcc reflection increases. The hcp-transition phase has its maximum at about 12% and its intensity finally almost vanishes. It was already mentioned that systematic work in material sciences requires the investigation of many samples. Suitable measuring times can only be obtained with high fluxes. The methods employed for these investigations (small angle scattering, diffraction techniques) are the same as these already discussed previously. The relative advantages and disadvantages of reactors and pulsed sources must not be repeated here. Perhaps the reservation made for precise intensity measurements at a pulsed source are not valid for the type of investigation in material sciences as most measurements are relative and compare the change of material under external conditions.

This chapter was based upon an article of the late Professor Schmatz (1984) from Karlsruhe which was published in a proposal for the German SNQ-project.

6. STRUCTURE OF ADSORBATES

Surfaces as sites of chemical reactions merit the special interest of chemists as many chemical reactions occur on surfaces. Layers of gas molecules on surfaces have therefore extensively been studied by neutron scattering. Many effects of monolayers on substrates can be observed by optical methods. The explanation of infrared and Raman techniques does, nevertheless, require the existence of a gap in the adsorption spectrum of the substrate. The adsorption of the adsorbate has to be in this gap. In all other cases only reflection techniques may be applied. Because neutrons are not subjected to these restrictions, the application of neutron scattering on surfaces has produced a number of results providing data on the structure of adsorbed layers as well as on vibrations and diffusion of adsorbed molecules. The structure of a mono-molecular layer of adsorbed gas molecules on a grafoil surface is shown in Fig. 18. At first a structure is formed in which the adsorbate (e.g. N_2, D_2, O_2 or CH_4) displays an elementary cell in a rational proportion to the structure of the substrate. In many cases the ($\sqrt{3}x \sqrt{3}$ -30°) structure is then formed (Fig. 18a) in which the adsorbate cell is $\sqrt{3}$ longer than the graphite cell and is turned at an angle of 30° in relation to the latter.

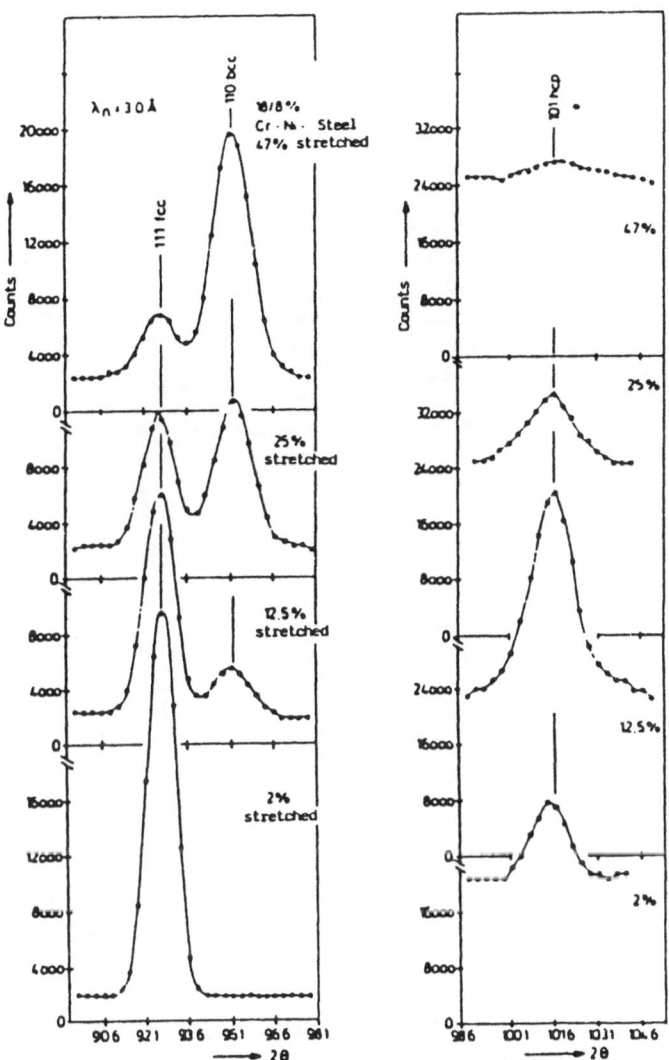

Fig. 17: Shape and intensity of three Bragg reflections measured as a function of stretch in an austenitic steel . The fcc→bcc phase transition is demonstrated by the decrease of the 111 fcc and the increase of the 110 bcc reflection (left). At right the intensity change of the 101 hcp reflection indicates the hexagonal transition phase.

102

With further increase in coverage, the cell dimensions change, hence-
forth appearing in a non-commensurable relationship to the substrate
cell (Fig. 18b).

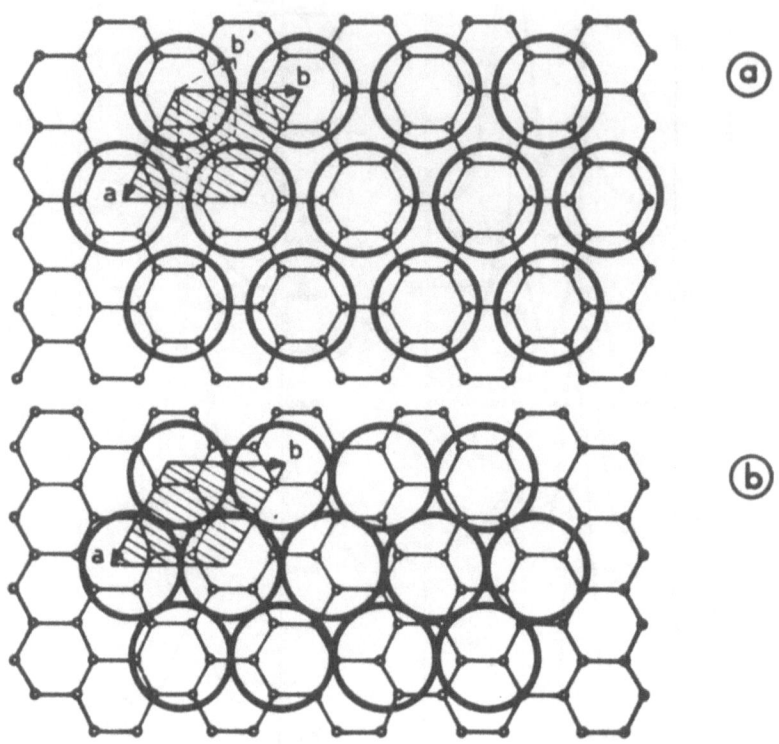

Fig. 18: Schematic drawing of an adsorbate on grafoil a) low coverage
with a commensurate ($\sqrt{3}$- $\sqrt{3}$ -30)-structure b) high coverage
with an incommensurate structure

This is also documented by the shifting of the position of the 10 reflection (Fig. 19) which drastically changes its position as soon as
the coverage exceeds the monolayer. By changing coverage and temperature
complete phase diagrams were mapped out which demonstrate all kinds of
order-disorder, commensurate-incommensurate and solid-liquid transformations which are known from bulk crystalline material.

First inelastic neutron scattering studies were undertaken which
are sensitive to the coverage as demonstrated in Fig. 20 which presents a phonon energy loss spectrum of ethylene molecules adsorbed on
graphite. The drastical change in the observed spectra between the two
different coverages was interpreted as a reorientation of the ethylene
molecules resulting from an increase in coverage.

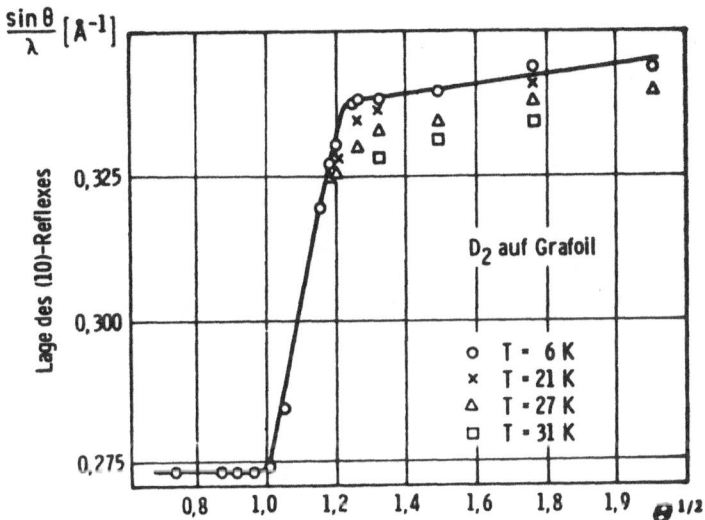

Fig. 19: Change of the position of the O1-reflection as a function
of coverage for the systems D_2 on grafoil

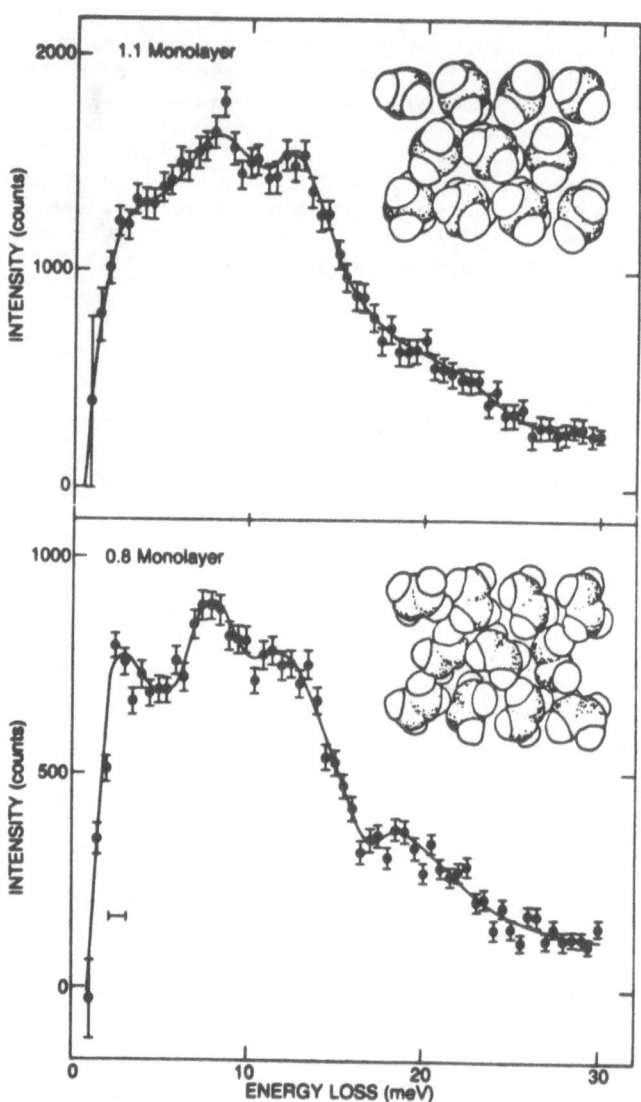

Fig.20: Phonon energy loss spectra for ethylene on graphite. The change in energy transfer as a function of coverage is interpreted by a reorientation of the molecules

7. NEUTRON CHEMICAL SPECTROSCOPY

7.1. Techniques and instrumentation

The aim of spectroscopy with neutrons is to understand fundamental and applied aspects in the dynamics of molecules and fluids. This includes force fields and crystal fields which are responsible for intercalation, physi- or chemisorption and for transport on surfaces. The basic quantity measured in neutron scattering is the energy transfer ΔE. As a consequence specialised instruments which are adapted to a given ΔE range were developed. Low and medium resolution is covered by the beryllium filter technique and by time of flight machines. Three types of spectrometers were conceived for high resolution. (The energy transfer in neutron scattering is normally given in meV; 1 meV $\sim 8 cm^{-1}$)

    (i) time of flight with a resolution of 10-50 μeV resolution

    (ii) backscattering (Birr et al., 1971) with 0.1 - 10 μeV resolution.

    both techniques are essentially used for inealstic incoherent scattering.

    (iii) spin-echo (Mezei, 1972) with an energy resolution of 0.01 - 0.1 eV is mostly used for coherent studies.

In order to obtain higher resolution cold neutrons are particularly suitable (with a given energy resolution of $\Delta E / E_o$, ΔE diminishes as E_o, the incident energy, becomes smaller). The availability of cold neutrons $(\lambda > 4\text{Å})$ in the last decade has led to a significant breakthrough in the absolute energy resolution. Ultra-cold neutrons termed UCN (see Fig. 1) with $\lambda > 400\text{Å}$ may add to that gain, especially for low Q studies.

A second division in neutron inelastic scattering, besides the resolution is the range of energy transfer. Whereas all energy transfers above $\Delta E \sim 2$ meV are said to be inelastic scattering, small energy transfers below that value are termed quasielastic scattering because they introduce a broadening of the elastic line and do not give rise to well separated inelastic lines.

Chemical spectroscopy has found the following areas of application

(A) Use of momentum transfer (in combination with energy transfer) between 0.1Å^{-1} and 10Å^{-1} to define

a) eigenvectors of molecular vibrations not accessible to IR and Raman measurements

b) elastic (EISF) and inelastic (IISF) coherent structure factors for molecular motion like (i) ion diffusion (ii) continuous reorientation or diffusion (iii) jump rotations, (iv) rotational diffusion by small steps.

B) The use of the large incoherent cross section of hydrogen (isotopic replacement)

C) High energy resolution of neutron scattering at low energy transfer which has revealed phenomena like the tunneling diffusion of molecules or molecular fragments at low temperatures.

Polarization analysis which we already mentioned in chapter 3.2.2. added to inelastic spectrometers could considerably improve results on inelastic machines, as it might help to separate incoherent from co-

herent scattering which may both occur during a dynamical process. Such
a separation has been reported in a study of the reorientational motions
of the molecule C_2Cl_6 in its plastic crystalline phase (Gerlach et al.,
1982). Polarization analysis would furthermore benefit to (i) rotational
diffusion processes in plastic and liquid crystals (ii) oscillations and
diffusion on physisorbed and chemisorbed molecules. (iii) measurement
of the density of states in glasses and polymers at low energy transfers
(iv) solvent and solute motions in ionic and other solutions.

Another option which requires high neutron fluxes is time-dependent spectroscopy. Chemical processes which may possibly be investigated
are (i) reactions in zeolites, like the isomerization of cyclopropane to
propene; (ii) competitive adsorption (iii) changes of intercalation and
adsorption. To cover all these needs a series of instruments were proposed for advanced sources. They are given in Table 5 (after Lander and
Emery, 1985) for the high resolution part. It is obvious from Table 5
that backscattering and spin-echo spectrometers would not gain from a
pulsed source as they use only the time averaged flux. Ultra-cold neutrons may be better suited at a steady state reactor but a gain is fore-
seeable for ultra-cold neutrons if high reflectivity, long-wavelength
monochromators can be developed. Time of flight instruments would, however, largely benefit from a pulsed source.

Table 5: Expected total gains (improved machines and new source)
relative to existing spectrometers on a 10^{15} n sec^{-1} cm^{-2}
reactor

Instrument	Δ E[μeV]	Reactor 10^{16}	pulsed source 10^{17}(peak)
TOF	10 – 50	X 30	X 150
Backscatt.	0.1 – 10	X 10 – 50	X 10 – 50
NSE [a]	0.01 – 0.1	X 10 – 50	X 10 – 50
UCN	~ 0.1	X 10	X 1
NSE + 3-axis	5 – 50	X 10	X 1

[a] NSE: Neutron spin-echo

Essential ameliorations whether they concern improved resolution, time
resolved spectroscopic experiments or the application of polarization
analysis on a wide scale need higher fluxes.

7.2. Motion of molecules and molecular fragments

As already mentioned above the isotopic replacement of hydrogens by
deuterium or partial deuteration of hydrogenous substances often helps
in the assignment of individual maxima to particular vibrations.

This phenomenon is illustrated in Fig. 21 for the phonon density
of states of anilinium bromide ($C_6H_5NH_3Br$), and two partially deuterated derivatives. The deuteration of the $NH_3^{\oplus}$ -group leads to a complete disappearance of the band at 33 meV which is therefore attributed to the motion of this group (Schweiss et al., 1983). The aim of

that study was to correlate the structural phase transition (monoclinic to orthorhombic at 25°C) to the motion of the two different groups of the molecule. Additional information was obtained by the analysis of the quasielastic line (Fig. 22) which becomes broader and broader with increasing temperature. This line broadening was interpreted as a rotational diffusion of the $-NH_3^\oplus$group. These jump diffusion processes were observed in the orthorhombic high temperature and – with markedly diminished rate – in the monoclinic phase. From the correlation times at different temperatures the activation energies (24kJ / mol for the monoclinic and 5.5 kJ / mol for the orthorhombic phase) were derived.

Comparison with calculated elastic incoherent structure factors suggests jumps of 60° for a hindered rotation of the $NH_3^\oplus$ group.

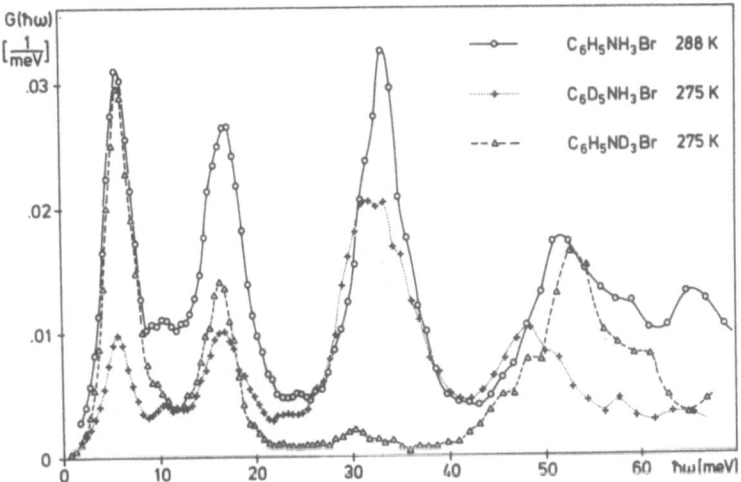

Fig. 21: Comparison of the generalized density of states for anilinum bromide with those for its partially deuterated derivatives (monoclinic phase)

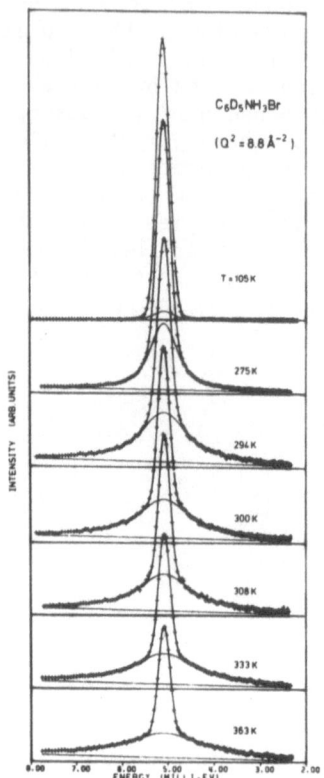

Fig. 22: Temperature dependence of the quasielastic line broadening of $C_6D_5NH_3Br$. The continuous lines are computer
fits, representing phonon bachground, Lorentzian–Gaussian
convolution describing the quasielastic scattering and
sum of purely elastic Gaussian and Lorentzian–Gaussian
convolution

The backscattering technique which we already mentioned several times
owes its existence from the fact that the relative energy resolution
$\Delta E / E$ varies as the cotangent of the Bragg angle Θ if reflection from
a single monochromator is used. Hence, a Bragg angle of $\Theta = \pi/2$ gives
best resolution and corresponds to backscattering. An instrument built
according to these requirements is most useful if the energy width of

the scattering is almost independent on the wavevector. A scattering process which follows these rules is rotational tunneling of the molecules, molecular groups or fragments in crystalline solids. The tunneling occurs in a given field and between states of different angular orientation. In NH_4ClO_4 (Fig. 23) the field is created by $ClO_4^{\ominus}$ anions and it is mainly electrostatic in nature. The large number of lines in the spectrum is explained by the interaction of the crystalline field and the rotating $NH_4^{\oplus}$ - group (Prager et al., 1976).

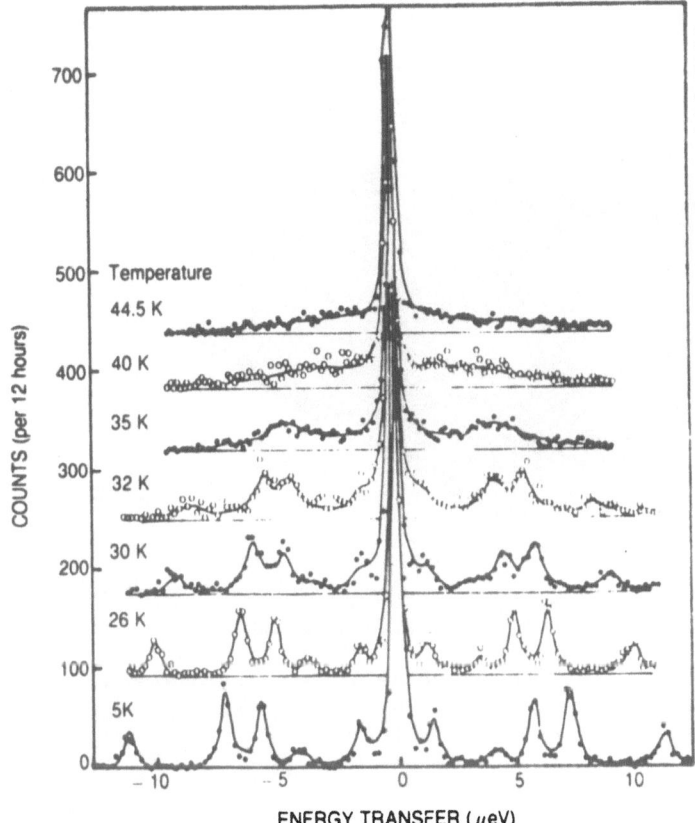

Fig. 23:
High resolution energy spectrum of NH_4ClO_4. The peaks show the various states that the ammonium ion accesses by rotational tunneling. Above 30K the tunneling disappears.

The momentum and energy transfer of water in its supercooled form is shown in Fig. 24. The spectrum represents mainly the scattering function of hydrogen as its large incoherent scattering cross-section dominates the entire scattering process.

The most prominent peak at 420 meV is due to O-H stretch vibrations. The unusual increase of the vibrational frequency with increasing temperature is due to the breaking of the hydrogen bonds which tend to reduce the frequency of the O-H stretch vibrational modes (Chen et al., 1984). In a more recent paper (Teixeira et al., 1985) these findings were confirmed but the additional analysis of the quasielastic scat-

110

tering gave information on the relaxation times of jump diffusion of protons and its temperature dependence.

A review on the reorientation of the methyl group in molecular crystals as studied by inelastic neutron scattering is given by Cavagnat (1985) who compares the results from neutrons with those by different physical methods, especially for toluene and nitromethane.

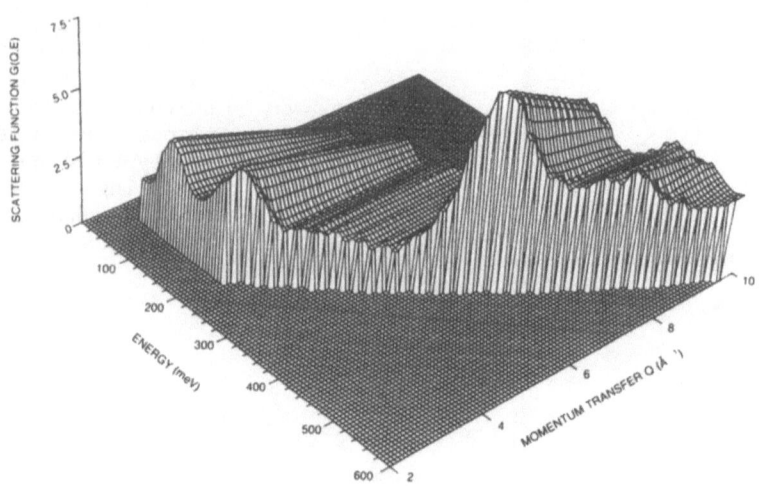

Fig.24: Proton scattering density function in supercooled water at -15ºC.

Other molecular crystals studied in great detail by neutrons include methane (see Press, 1981), carboxylic acids (Stoeckli et al. 1986) and several ammonium and methylammonium salts

8. CONCLUSION

The present contribution has given a certain number of applications of neutron scattering to problems which are relevant to chemists. These examples have proved our initial statement that neutrons are a unique and valuable tool for research in condensed matter sciences. The fields range from pure fundamental physics (moment, charge, lifetime of the neutron) to applied engineering (flow of hydrocarbons on engines). The results presented here are in most cases complementary to findings obtained by other diffraction or spectroscopic methods and represent therefore essentially new findings.

As far as the leitmotiv of our contribution is concerned, a comparison of steady state and pulsed source, we may state that either source is good, provided, it supplies the maximum amount of neutrons over the whole energy range useful for application. A more detailed view is given in Table 6 where the gain estimates of the German SNQ-project are summarized as compared to the existing high-flux reactor (10^{15} n sec^{-1} cm^{-2}) at the ILL. This gain includes machine improvement as well as instrument improvement and is based on an optimistic view.

The estimated gain factors in Table 6 are compared with an existing steady state reactor not with a future new one as in Table 5.

Table 6: Intensity comparison for instruments at the SQN-project and the existing HFR at the ILL, Grenoble.

Type of Instruments		Gain
TOF Powder Diffractometer resolution 2×10^{-4}		35
TOF Single Crystal Diffractometer	cell = 6Å	10 - 35
	cell = 50Å	9 - 30
	cell = 300Å	3
Small Angle Scattering	beam $\emptyset$ 10 mm res. 10%	2
	beam $\emptyset$ 10 mm res. 1%	20
Diffuse Scattering		7
Triple Axis Spectrometer		1
Inelastic TOF	thermal beam	14
	cold beam	17
Inverted TOF with Backscattering	res. 10 - 50 μeV	12 - 15
Multi-Crystal Backscattering	res. 0.1 - 1 μeV	12
Spin echo	res. 20 μeV	3 - 6

The essential advantage of the pulsed source is with TOF techniques but to repeat the Shelter Island Workshop (Lander and Emery, 1985): "To a large extent the future needs of the scientific community could be met with either a $5 - 10^{15}$ n cm^{-2} sec^{-1} steady state source or a 10^{17} n cm^{-2} sec^{-1} peak flux spallation source.". This statement pronounced for the United States is definitely true for Europe as well.

REFERENCES

Axe, J.D. and Nicklow, R.M.; Physics Today p. 27 (1985)

Bacon, G.E. (1986) Fifty years of neutron diffraction, Adam
    Hilger Bristol, UK

Birr, M., Heidemann, A., Alefeld, B.; Nucl. Instr. Methods
    95, 435 (1971)

Boucherle, J.X., Gillon, B., Maruani, J., Schweizer, J.;
    J. de Physique, Coll. 43, C7, 227 (1982)

Brown, P.J.; Chemica Scripta 26, 433 (1986)

Cavagnat, G.; Journ. chim. phys. 82, 239 (1985)

Chen, S.H. Toukan, K. Loong, C.K., Price, D.L. Teixeira,
    J.; Phys. Rev. Lett. 53, 1360 (1984) Cited after Lan-
    der and Price (1985)

Enderby, J.E. and Neilson, G.W.; Rep. Prog. Phys. 44, 38
    (1981)

Enderby, J.E.; Rev. Phys. Chem. 34, 155 (1983)

Fender, B.E.F., Figgis, B.N., Forsyth, J.B.; Proc. R. Soc.
    Lond. A 404, 139 (1986)

Fender B.E.F., Figgis, B.N., Forsyth, J.B., Reynolds, P.A.,
    Stevens, E.; Proc. R. Soc.Lond. A 404, 127 (1986)

Fuess, H.;in "Modern Physics in Chemistry", Vol. 2, (E.
    Fluck and V. Goldanskii, eds.), Academic Press, 1-193
    (1979)

Fuess, H., Bats, J.W., Cruickshank, D.W.J., Eisenstein, M.;
    Angew. Chem. Int. Ed. 24, 509-510 (1985)

Gerlach, P., Schaerpf, O., Prandl, W., Dorner, B.; J. de
    Physique, Coll. 43, C7, 151 (1983)

Jackson, P.F. Johnson, B.F.G. Lewis, J., Raithby, P.R. Ma-
    partlin, M., Nelson, W.J.H., Rouse, K.D., Allibon, J.,
    Mason. S.A.; J. Chem. Soc. Chem. Comm. p 295 (1980)

Joswig, W., Bartl, H., Fuess, H.; Zeitschrift f. Kristal-
    lographie 166, 219-223 (1984)

Joswig, W., Fuess, H., Ferraris, G.; Acta Cryst. B 38,

2798-2801, (1982)

Jung, V.; Arch. Eisen-Hütten-Wesen II, 1 (1983)

Klebe, G. Bats, J.W. Fuess, H.; J. Amer. Chem. Soc. 106, 502 (1984)

Lander, G.H. and Emery, V.J.; Nucl. Instrum. Meth. in Phys. Res. B 12, 525-561 (1985)

Lander, G.H. and Price, D.L.; Physics Today p 38 (1985)

Leung, P.C.W., Schultz, A.J., Wang, H.H., Emge, T.J. Ball, G.A., Cox, D.D. Williams, J.M.;Phys. Rev. B 30, 1615 (1984)

Mezei, F.; Z. Phys. 225, 146 (1972)

Oles, A., Sikora, W., Bombik, A., Konopka, M.; Magnetic Structures Determined by Neutron Diffraction, Akademia Gorniczo-Hutnicza, Krakow (1985)

Opechowski, W. and Guccione, R.; Magnetism Vol. II A; (G.T. Rado and H. Suhl, eds.), Academic Press, New York (1965)

Prager, M., Alefeld, B., Heidemann, A.; in Proc. XIXth Cong. Magn. Resonance, Ampere, Heidelberg, Geneva, p 389 (1976) Cited after Pynn and Fender (1985)

Press, W.; Single Particle Rotation in Molecular Crystals; Springer Tracts in Modern Physics, Vol. 92 (1981)

Pynn, R., Fender, B.E.F.; Physics Today, 47 (1985)

Scherm, R. Stiller, H. (eds.); Proceed. Worksh. Neutron Scatter. Instrument; Ber. der KFA Jülich, Nr. 1954 (1985)

Schmatz, W.; in SNQ-Project Proposal; Report of the KFA Jülich, p. 24 (1984)

Schweiss, B.P., Fuess, H. Fecher, G. Weiss, A.; Z. Natur-forsch. 38a, 350 (1983)

Seeger, P.A. and Pynn, R.; Nucl. Instr. Meth. Phys. Res. A 245, 115 (1986)

Stamm, M. Fischer, E.W., Dettenmaier, M., Convert, P.;
    Farad. Disc. Poy. Soc. Chem. 68, 263 (1979)

Stoeckli, A., Furrer, A., Schoenenberger, Ch. Meier, B.H.,
    Ernst, R.R., Anderson, I.; Physica 136 B, 161 (1986)

Teixeira, J., Bellissent-Funel, M.C., Chen, S.H., Dia-
    noux, A.J.; Phys. Rev. A 31, 1913 (1985)

DISCUSSION

ELASTIC INCOHERENT SCATTERING

It was noted that the use of elastic incoherent scattering to study lattice defects (Prof Fuess's Table 1) extended the statement in Dr Lander's paper (section 2.2 b) that this interaction gave only background effects. The latter referred to a given nuclear species, but change of nuclear species can, in certain circumstances, lead to changes in elastic incoherent scattering which reveal details of momentum transfer in the solid. For chemical crystallography with neutrons, however, the predominant effect of this interaction is the strong incoherent background produced by hydrogen atoms, which in some powder samples may make the Bragg reflections difficult to see.

HYDROGEN BONDS

As it is uncertain how the frequencies observed in Raman and infrared spectroscopy are to be correlated with interatomic distance, further work on hydrogen bonds by elastic and inelastic neutron scattering is desirable. But the large thermal motion of hydrogen atoms, even at low temperatures, will restrict the precision with which their positions can be specified.

Completely symmetrical hydrogen bonds now appear to occur, with perhaps one exception, only where there is a symmetry requirement for such a configuration.

LIMITATIONS OF DEUTERIUM SUBSTITUTION

Since complete deuteriation cannot always be achieved and the hydrogen and deuterium forms are not identical, it may sometimes be necessary to study both forms.

BONDING AND MOTION OF WATER IN ZEOLITES

In the very large cage zeolite mentioned above (figure 7) there are 5 cyrstallographically distinct sites for water molecules. In 3 the molecules are completely mobile and in the other 2 there are only very weak bonds to the framework with 0 . . 0 distances of 2.90 - 2.95 Å. This water should thus be seen as a cluster of molecules whose motion does not involve the breaking and formation of specific hydrogen bonds.

SINGLE CRYSTAL PULSED NEUTRON DIFFRACTION

J. B. Forsyth
Rutherford Appleton Laboratory
Chilton, Didcot
Oxon, OX11 0QX, UK

ABSTRACT.   Diffraction data from single crystals is obtained using the white-beam, stationary crystal, Laue technique. Separation of orders of the same reflection, which therefore fall on the same spot on the detector, is by their different times of arrival. The geometry of diffraction, the resolution of time-of-flight (TOF) diffractometers and their design is discussed. The integrated intensity of reflection is compared to that for the classical, monochromatic beam, moving crystal diffractometer and the overall speed of data collection is deduced. The success of the TOF technique is largely dependent on the provision of a suitable area, position-sensitive detector (PSD): its qualities are specified and a practical detector is described. Finally, some consideration is given to the suite of programs required to determine the crystal orientation matrix and to reduce the data to observed structure factors.

1. INTRODUCTION

The origins of pulsed neuton diffraction date back to the seminal paper by Lowde in 1956 and development, in the period to 1969, is covered by Buras (1969) and by Turberfield (1970) . Although a pulsed beam can be produced from a steady state reactor by using a conventional, a statistical or a Fourier chopper, the development of the subject has been mainly stimulated by the construction of inherently pulsed sources based on a pulsed reactor, an electron linac or a proton synchrotron. Most of the diffraction measurements made at the first generation of pulsed neutron sources were on powdered materials; examples where single crystals were used are a study of $BaTiO_3$ on the Tohoku University linac by Niimura et al (1973), measurements on Mn_5Ge_3 at the Neutron Booster, Harwell by Day and Sinclair (1970) and work on crystals of deuterated naphthalene and lanthanum magnesium nitrate at the IBR-30 by Balagurov et al (1979).

The advent of the second generation of pulsed neutron sources, such as the IBR-2 at Dubna, IPNS at Argonne National Laboratory, WNR at Los Alamos and ISIS at the Rutherford Appleton Laboratory, has produced significant advances in the performance of their associated single

M. A. Carrondo and G. A. Jeffrey (eds.), Chemical Crystallography with Pulsed Neutrons and Synchrotron X-Rays, 117–135.
© *1988 by D. Reidel Publishing Company.*

crystal diffractometers and in the treatment of the data they produce.

The discussion begins with a brief resumé of the aims of single crystal diffractometry and its strengths relative to measurements made on powders. The remaining sections are concerned with the practical aspects of the subject, starting with the Ewald construction appropriate to a white, pulsed incident beam . Particular attention is paid to the conditions which must be fulfilled before accurate results can be expected and the expression which connects the observed intensity to the related structure factor is given. Some space is devoted to a discussion of instrumentation, instrumental resolution and their optimization for various types of study. Lastly, a brief description is given of the software programs which are required to obtain and interpret the observed intensities.

2. THE AIMS OF SINGLE CRYSTAL DIFFRACTOMETRY

The most usual objective of a single crystal diffraction study is the determination of the moduli of a set of structure factors which is as complete as possible out to a choosen limit in reciprocal space. The structure factor F_{hkl} of a Bragg reflection is related to the distribution of scattering material within the crystallographic unit cell by the expression:

$$F_{hkl} = \Sigma \; A_r f_r(\underline{\kappa}) \exp(i\underline{\kappa}.\underline{r}) \exp(-W) \qquad 2.1$$

where $\underline{r}$ is the position of the rth scattering centre with scattering power $\bar{A}_r$ and form factor $f_r(\underline{\kappa})$, $\underline{\kappa}$ is the scattering vector and $\exp(-W)$ is the Debye-Waller temperature factor for the scattering centre.

The distribution of scattering material within the crystalline unit cell may then be calculated, since:

$$\rho_{\underline{r}} = (1/V) \; \Sigma \; F_{hkl} \exp(-i\underline{\kappa}.\underline{r}) \qquad 2.2$$

where V is the volume of the unit cell. Each F_{hkl} has an associated phase angle relative to F_{000} and the determination of these phases constitutes the crystallographic "phase problem". For centrosymmetric crystals, the phase is either 0° or 180° i.e. the sign of F_{hkl} is either positive or negative respectively. Interference between nuclear and magnetic scattering occurs in the polarized neutron scattering from some magnetic materials and this can be used, in the case of centrosymmetric crystals, to determine the amplitude and phase of the magnetic structure factor with respect to its nuclear counterpart.

Single crystal samples generally give much stronger diffraction spectra than powdered material, so the range of structure factors that can be successfully determined is correspondingly greater. The three-dimensional resolution of spectra in reciprocal space, as opposed to the single dimension obtainable from powdered samples, allows better atomic resolution to be obtained in real space and larger unit cells to be studied. Information is not lost due to the overlap of inequivalent reflections with the same or very similar d-spacing. Finally, external forces or fields may be applied along specific directions in a single crystal and this is of the utmost importance in the study of phase changes and magnetic materials.

3. DIFFRACTION IN RECIPROCAL SPACE

As is well known, the consequences of Bragg's law are most easily visualised in reciprocal space by using a geometrical construction due to Ewald (1911). A sphere of radius $1/\lambda$ is drawn about the incident beam direction such that it touches the origin of the appropriately oriented reciprocal lattice. Planes are in a position to reflect the wavelength λ if their corresponding reciprocal lattice point lies on the surface of this so-called Ewald sphere. The direction of the diffracted beam is given by the radius vector from the centre of the sphere to the reciprocal lattice point. Figure 1(a) illustrates the Ewald construction in two dimensions and Figure 1(b) shows that it corresponds to Bragg's law since:

(a) the incident and scattered beams are coplanar with the normal to the planes.
(b) the angle made by the incident beam to the normal is $(90-\theta)$ and the beam is scattered through 2θ.
(c) $d^*/(2/\lambda) = \sin\theta$ or $\lambda = 2d\sin\theta$.

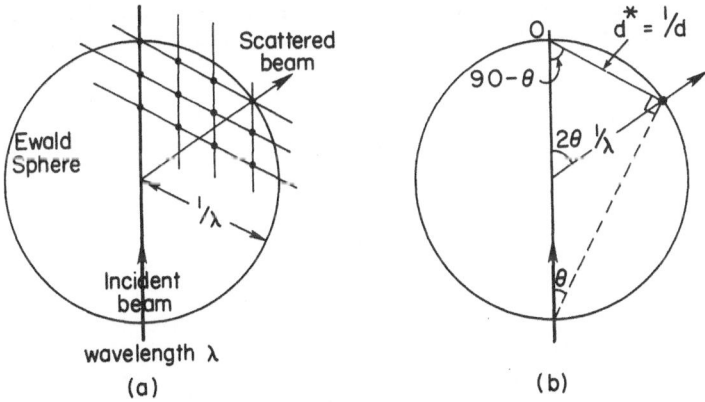

Figure 1. (a) The Ewald construction in two dimensions and (b) its equivalence to Bragg's law.

In classical single crystal diffractometry, the beam is continuous and roughly monochromatic; the beam incident on the sample has an angular divergence and the crystal under examination will usually have a mosaic spread. Under these conditions the Ewald sphere in the plane of scattering is shown in Figure 2. There is a finite volume of reciprocal space included between the spheres representing the extremes of the incident beam divergence and its wavelength spread. Each reciprocal lattice point is also drawn out to cover a finite area, on a sphere centred at the origin, by misalignment in the crystal mosaic.

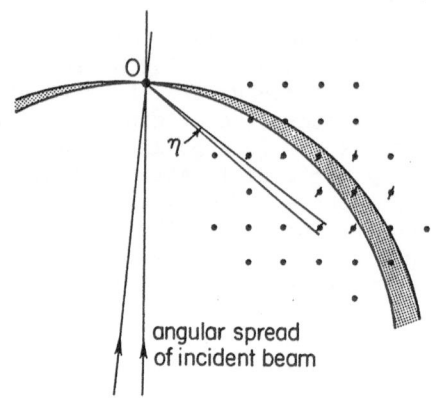

Figure 2 The Ewald construction appropriate to conventional, quasi-monochromatic diffraction. The crystal mosaic spread is η.

In the TOF method of single crystal diffractometry, all the reflections lying completely within the bounds of the Ewald spheres corresponding to the maximum and minimum wavelengths within the incident beam will reflect at a single setting of the crystal, as illustrated in Figure 3.

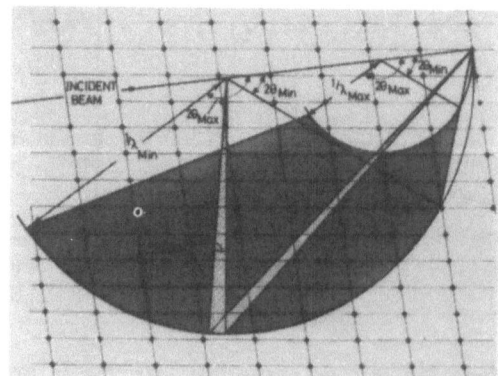

Figure 3. The Ewald construction appropriate to white beam (Laue) diffraction. If the source is pulsed, orders of the same reflection, which therefore lie in the same radial direction from the origin, will arrive at different times in the detector and may therefore be separated by TOF.

4. DIFFRACTED INTENSITIES

Having described the geometry of crystalline diffraction, we now consider the method of measuring a diffracted intensity in such a way that it can be related to the fundamental quantity F_{hkl}. Unfortunately, you will recall that this relationship is not unique, but is strongly dependent on the quality of the actual crystal being investigated. In perfect crystals, such as semiconductor quality silicon, coherence is maintained throughout the whole sample, even if it is several centimetres in linear dimensions. In this case, energy is dynamically exchanged between the incident and diffracted beams, the angular width over which the crystal reflects is very narrow and the first Born approximation, which ignores the energy transferred to the diffracted beam, cannot be invoked. The theoretical treatment of the diffraction by perfect crystals is complicated and beyond the scope of this lecture. In practice, however, most materials crystallize imperfectly and include many dislocations, grain boundaries and other defects which reduce the range of diffraction phase coherence to the order of 10^{-6} m. The crystal may then be considered to be composed of small mosaic blocks, within each of which coherence is maintained. Deviations may also occur in the direction of the normal to a set of crystal planes on passing from one mosaic block to another and the angular width of reflection is consequently broader. In the imperfect or kinematical limit, the first Born approximation applies.

In classical single crystal diffractometry, the crystal is rotated, in an incident beam intensity I_0, through its reflecting position at a uniform angular velocity ω about an axis perpendicular to the normal of the reflecting planes and the diffracted counts N are collected in a suitably placed detector. The normalized quantity $N\omega/I_0$ is termed the **integrated intensity** and it is this quantity which is related to the square of the structure factor for an ideally mosaic crystal in which the kinematical limit is reached:

$$N\omega/I_0 = \lambda^3 V |F_{hkl}|^2 / (V_0)^2 \sin 2\theta \qquad 4.1$$

where V is the volume of the crystal, V_0 is the volume of its unit cell and λ the incident wavelength. The geometrical factor $1/\sin 2\theta$ corrects for the relative times taken by a reciprocal lattice point to traverse the surface of the Ewald sphere. The method of measurement ensures that all parts of the crystal have an equal chance of diffracting the full range of wavelengths and angles of incidence present in the incident beam.

In the TOF technique, integration over the crystal mosaic spread is performed as a function of wavelength, so $i_0(\lambda)$, the incident neutron intensity per unit wavelength range must also be measured. The number of neutrons detected is given by:

$$N = i_0(\lambda)\lambda^4 V |F_{hkl}|^2 / (V_0)^2 2\sin^2\theta \qquad 4.2$$

The requirements for a well-centred crystal, completely bathed in a uniform beam and for a detector with uniform response are the same as those required in classical, continuous source, monochromatic beam, rotating crystal diffractometry.

To make effective use of a pulsed source, as much of the wavelength range must be used as possible and the detector must cover an extensive range of scattering angles in two dimensions. It should be remarked at this point that such two-dimensional, position-sensitive detectors (PSDs) can also be used to advantage on conventional diffractometers, particularly when large unit cells are involved which lead to a high density of reciprocal lattice points, many of which are simultaneously in a position to diffract.

The dependence of the diffracted intensity on crystal perfection is striking; an equivalent measurement made on an ideally perfect crystal would yield integrated intensities proportional to $|F_{hkl}|$, not $|F_{hkl}|^2$. The phenomenon by which diffracted intensity is lost as the degree of perfection increases is termed 'extinction' and it was first treated quantitatively by Darwin (1922), who introduced the idea of a mosaic crystal. In his model, the two parameters required to define perfection are the coherence length or linear dimension of the mosaic blocks and η, defining the angular misorientation of the blocks as a Gaussian distribution:

$$W(\Delta) = (1/\eta \{2\pi\}) \exp(-\Delta^2/2\eta^2) \qquad 4.3$$

The loss of intensity due to increased coherence is termed primary extinction. Insufficient angular misorientation allows one reflecting block to shield another with the same orientation, but which is further into the crystal, giving rise to secondary extinction. The most commonly used procedure for correcting observed intensities for the presence of extinction is due to Becker and Coppens (1974), who extended Darwin's treatment and included an alternative Lorentzian distribution of mosaic spread. Corrections for intensity losses of up to factors of 2-3 in the strongest reflections can usually be made with an accuracy of some 5%, and experimental justification for the model may be sought in measurements made at different wavelengths. It is clear that extinction must tend to zero as λ goes to zero and, with it, the scattered intensity (Equations 4.1 and 4.2) or as the volume of the crystal becomes vanishingly small. The pioneering experiments on the weak pulsed sources were on much larger crystals than are now used on the best reactor sources and, hence, extinction was significant.

5. SINGLE CRYSTAL DIFFRACTOMETERS

The mechanical layout of the single crystal diffractometer SXD on the ISIS pulsed source is illustrated in Figure 4.

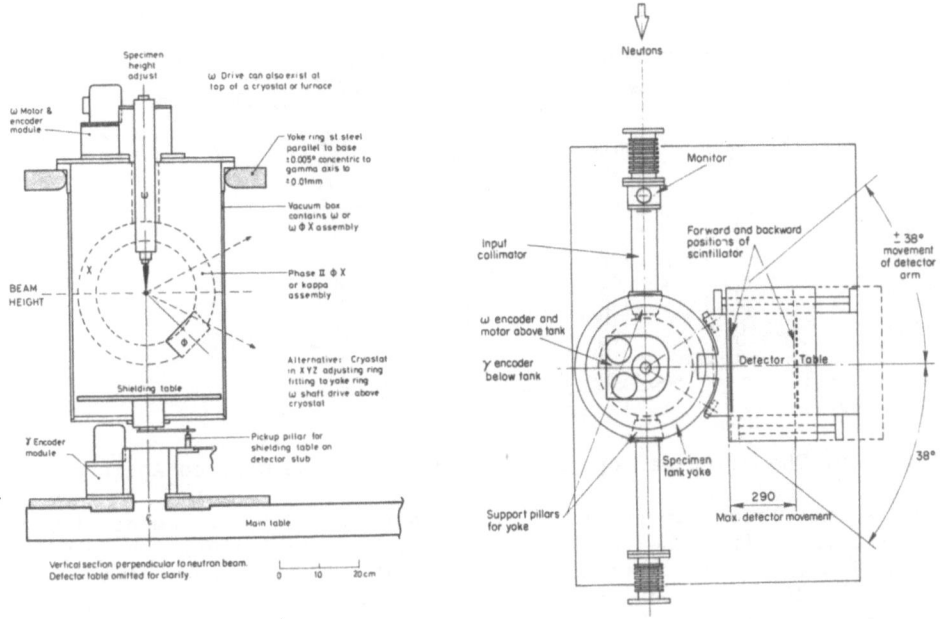

Figure 4. The diffractometer SXD at ISIS. At the left in side elevation and at the right in plan. The sample is 8 m from the moderator.

Its detector is based on the scintillation principle, which gives the following advantages for TOF experiments, when compared to gas detectors:

(a) the inherent speed of response is much higher (50 ns, cf 5 μs).
(b) the efficiency is higher for a given depth and hence the path-length uncertainty can be reduced.

Difficulties in their use stem from the low light output from the fast, ^{6}Li containing glass scintillator and its sensitivity to γ-rays.

In the SXD, a ^{6}Li scintillation detector some 300x300 mm in area can be positioned at 300-600 mm from the sample, which is 8 m from the ambient moderator AP. The normal beam coordinates γ, ν are restricted to 15° to 165° and -30° to 30°, respectively. The detector uses the principle of the Anger gamma-ray camera (Anger 1958), which was first successfully adapted for neutron use at the Argonne National Laboratory by Strauss et al (1981). A continuous sheet of scintillator, some 1-2 mm in thickness, is viewed through a thick (≃40 mm) slab of glass or plastic by a close-packed array of photomultipliers, PMs (Figure 5).

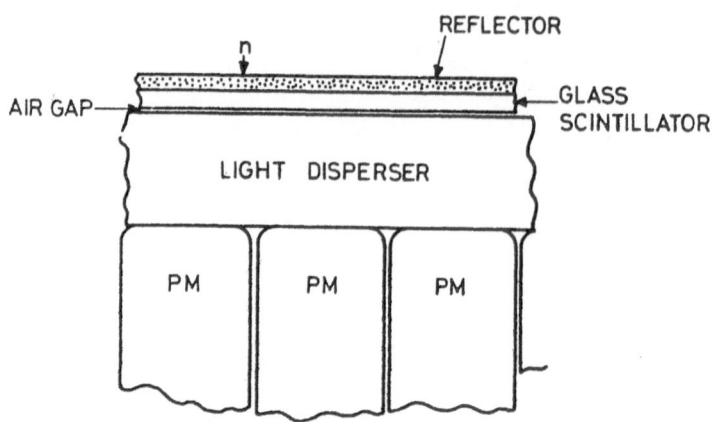

Figure 5. The principal optical components of the neutron Anger camera.

The scintillator is fronted by a white reflecting surface of Al_2O_3 and is separated from the thick disperser by an thin air gap, whose purpose is to limit the light finding its way to the PMs to a cone of semi-angle of some 40°. Position determination is made by comparing the amplitudes of the signals which come from the PMs and correspond to the same neutron event. In the original gamma-camera, this comparison was performed on-line by analogue circuitry based on a resistor network accepting the signals from all PMs (typically 19 in number). Strauss et al (1981) adopted the same principle in their 49 PM neutron detector, but in the Rutherford Appleton detector, which has 45 PMs, the signal processing is limited to the PM nearest the event and its six neighbours. Two benefits arise from this strategy:

(a) the area to be digitized is smaller, so the number of bits is reduced for a given pixel size. This in turn speeds up the process, which can be completed in under 1 μs for 6 bits, so dead-time losses are reduced.

(b) PMs which contribute no positioning information, due to the dimensions of the light cone, are excluded from the signal processing and so do not contribute noise.

SXD commissioning experiments with a single, 20x20 mm scintillation detector have already shown that useful data can be obtained to very high values in $\sin\theta/\lambda$. The thermal anharmonicity parameter in SrF_2 has been determined to better precision than previously obtainable with reactor data. Selected pairs of reflections, which occur at the same value of $\sin\theta/\lambda$ and have intensities which differ solely due to the anharmonicity, were measured out to a $\sin\theta/\lambda$ limit of 1.7 $Å^{-1}$. Figure 6 shows the {hhh} spectra from this material. Although an extensive data set was measured with wavelengths from 0.3 - 4 Å, a single extinction parameter for the mosaic spread in the Becker-Coppens formalism was sufficient to obtain an overall R-factor of 0.041.

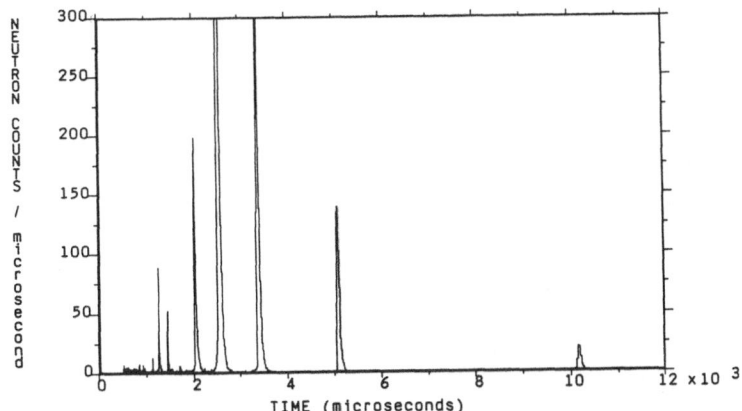

Figure 6. The {hhl} spectra from a single crystal of SrF_2 on SXD.

It is clear that the wide coverage of reciprocal space, which is obtained through the use of an area PSD and the white-beam TOF technique, is excellent for the location of satellite reflections and the study of phase changes in general. It is therefore necessary to provide for non-ambient specimen environments, of which low and high temperatures and high pressures are the most frequently needed. Joule-Thompson refrigerators with special crystal orienters have been developed for the instruments working at ANL and Los Alamos; these diffractometers are illustrated in Figures 7 and 8, respectively.

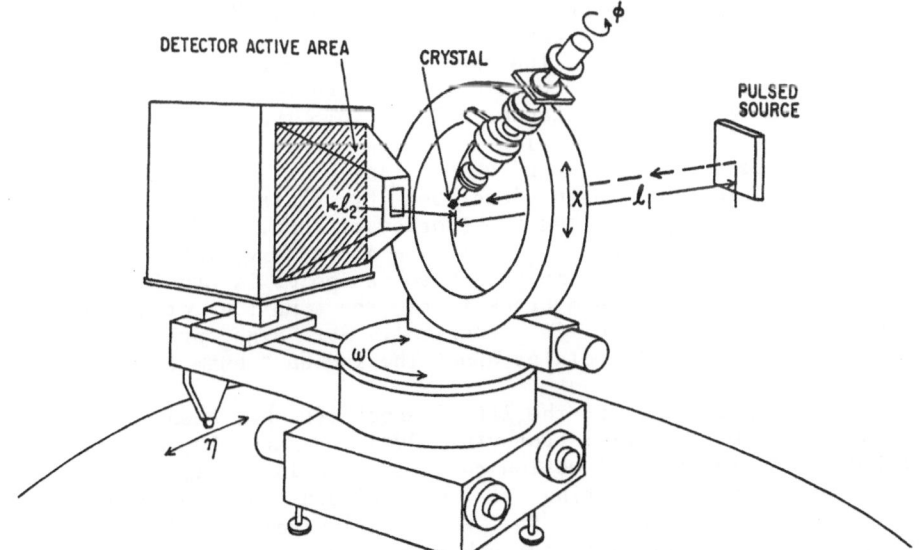

Figure 7. The SCD single crystal TOF diffractometer at ANL.

126

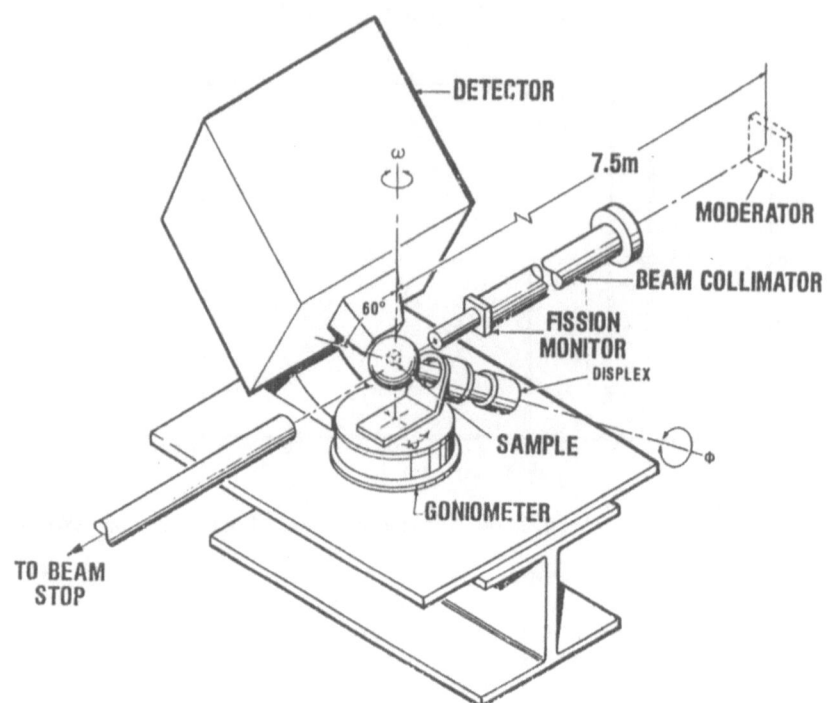

Figure 8. The single crystal diffractometer at the pulsed neutron facility, Los Alamos.

The ANL instrument, designed by Schultz et al (1984), uses a Displex cooler mounted on a conventional phi-chi circle with the latter held at a fixed ω angle of 45°. The specimen enclosure is elongated in the directions of the main incident and transmitted beams, so that the main vacuum windows to the cooler cannot scatter into the detector centred at γ = 90°, ν = 0°. The χ rotation is limited to 90°, so some 35 crystal orientations are needed to cover 2π solid angle in reciprocal space. The Los Alamos design has been described by Alkire et al (1985). The detector is a fixed, gas-filled PSD (Borkowski-Kopp, 1975), 250 x 250 mm in sensitive area, centred at γ = 90°, ν = 45° at a variable distance of 0.25 - 0.5 m from the sample. A Displex cooler is provided with a φ rotation, which is held on a bracket from the ω motion at χ = 60°. The specimen vacuum enclosure is spherical. For high pressure work a cylindrical pressure cell rotates about the incident beam (Figure 9).

A liquid helium cryostat of the ILL 'Orange' type has been modified for use on the SXD at ISIS (Forsyth 1987). Figure 10 shows how the restriction of the crystal motion to a single axis, ω, allows the major sources of window scattering to be excluded from the detector. The provision of temperatures down to 1.2 K is principally of interest in the study of magnetism and, as such, the volume of data to be measured is usually much less than in a full nuclear structure refinement of a

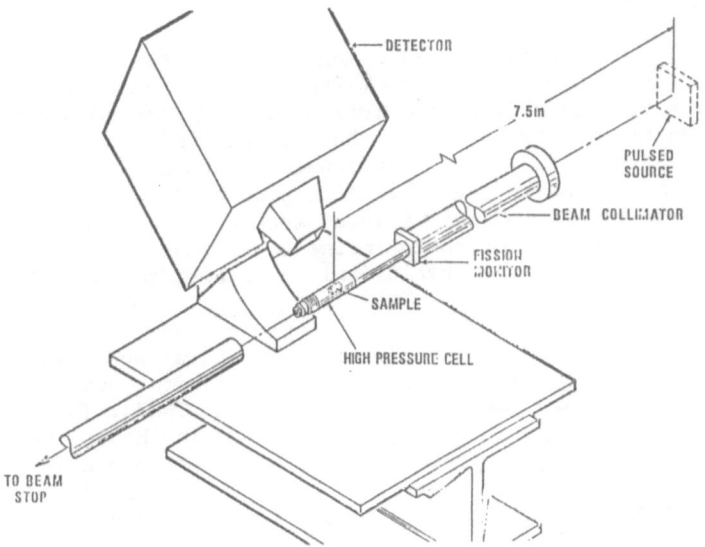

Figure 9. The high pressure cell used for single crystal diffractometry
at Los Alamos.

fairly complex crystal. The necessity of using several different
mountings of the specimen is therefore considered to be preferable to
the acceptance of a highly structured background.

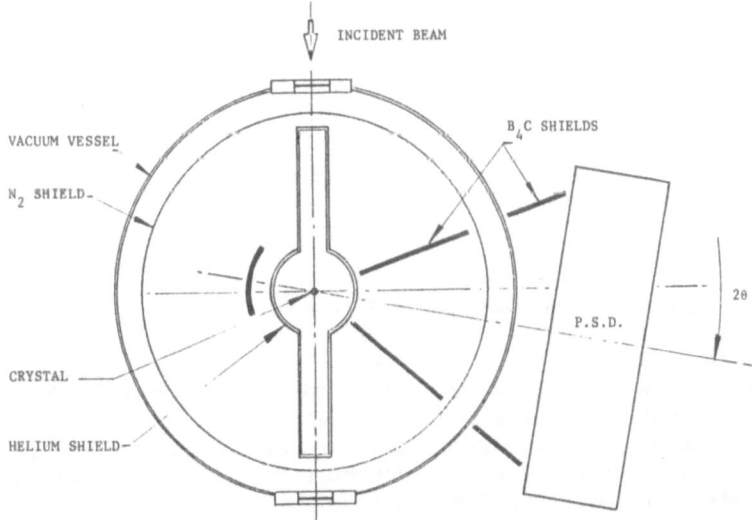

Figure 10. The liquid helium cryostat for use on the SXD at RAL. The
crystal is rotated from above, via a long support tube. B_4C shields
defining the limits of the scattered beams are coupled to the detector
motion and rotate within the cryostat between the N_2 and He bins.

6. INSTRUMENTAL RESOLUTION

The resolution characteristics of a monochromatic beam diffractometer have been given by Cooper and Nathans (1968). Similar calculations for a TOF instrument have been made by a number of authors including Buras and Holas (1968) and Stoica (1975). Since there is always a conflict between resolution and scattered intensity, instruments should, in general, have their resolution matched to the scientific problem under investigation. An instrument designed to study simple structures to high $\sin\theta/\lambda$ will have very different resolution characteristics to those of an instrument optimized for the study of biological crystals to modest atomic resolution.

A full treatment of the resolution function, including scattering into a two-dimensional detector and the effect of finite sample size and anisotropic mosaicity has been given by Wilkinson (1987). The effect of an incident beam divergence of $\pm\alpha_i$ and $\pm\beta_i$ in the horizontal and vertical directions is to produce a resolution ellipsoid in detector space. Adopting normal beam detector geometry of γ in the horizontal plane and ν vertical, the ellipse parameters corresponding to having $(\alpha/\alpha_i)^2 + (\beta/\beta_i)^2 = 1$ are:

$$\alpha = \left(\frac{\sin\nu\sin\gamma(\cos\nu\cos\gamma - 2)}{2(1 - \cos\gamma\cos\nu)^2}\right)d\nu + \left(\frac{\cos\gamma\cos\nu}{2(1 - \cos\gamma\cos\nu)} + \frac{\cos\nu(\cos\gamma - \cos\nu)}{2(1 - \cos\gamma\cos\nu)^2}\right)d\gamma$$

6.1

$$\beta = \left(\frac{(\sin^2\nu - 2)\cos\gamma = 2\cos\nu}{2(1 - \cos\gamma\cos\nu)^2}\right)d\nu - \left(\frac{\sin\nu\sin\gamma\cos\nu}{2(1 - \cos\gamma\cos\nu)^2}\right)d\gamma$$

6.2

For a given γ, ν position and corresponding to particular values of α, β, the $d\lambda$ value is found from the expression:

$$d\lambda = \frac{(\cos\gamma\sin\nu d\nu + \sin\gamma\cos\nu d\gamma)\lambda}{2(1 - \cos\gamma\cos\nu)}$$

6.3

Wilkinson derives expressions for the major and minor axes of the ellipse and its inclination in detector space. Figure 11 shows the results for $\alpha_i, \beta_i = 2/3°$ in the ν, γ range $0 - 30°$ and $0 - 180°$ respectively. He treats the problem of mosaic spread in a similar manner and shows that it corresponds to a second elliptical function in detector space. The combined effects of beam divergence and crystal mosaic is then obtained as the convolution of the two functions, which he evaluates via the product of their Fourier transforms.

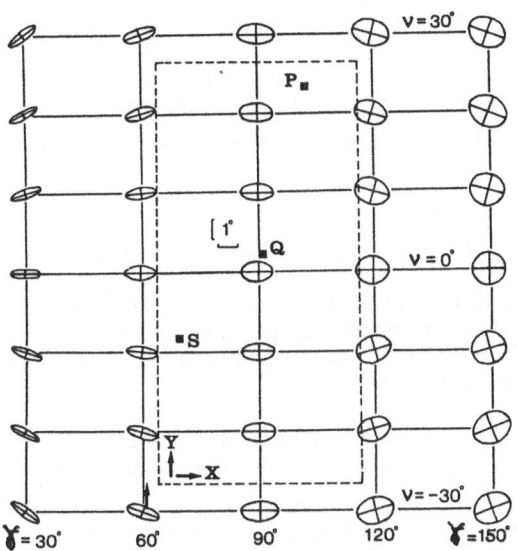

Figure 11. Variation of the shape of the diffracted beam with diffraction angles γ, ν. Isotropic beam divergence of 2/3° FWHH and crystallite mosaic spread of 0.25° FWHH are assumed (Wilkinson 1986).

For negligible sample size, the second moments of the resolution function depend on the time and angular uncertainties in the following way:

$$\frac{\sigma^2(H_x)}{H^2} = \frac{\sigma_t^2}{T^2} + \cot^2\theta \; \frac{\sigma^2(\alpha_i) + \sigma^2(\alpha_f)}{4} \qquad\qquad 6.4$$

$$\frac{\sigma^2(Hy)}{H^2} = \frac{\sigma^2(\alpha_i) + \sigma^2(\alpha_f)}{4} + \eta_s^2 \qquad\qquad 6.5$$

$$\frac{\sigma^2(H_z)}{H^2} = \frac{\sigma^2(\beta_i) + \sigma^2(\beta_f)}{4\sin^2\theta} + \eta_s^2 \qquad\qquad 6.6$$

here, $H = 2\sin\theta/\lambda = d^*$, the reciprocal lattice vector; α and β refer to the angular divergencies in the plane of scattering and perpendicular to it; the subscripts i and f refer to the incident and scattered beam, respectively; η_s is the standard deviation of the mosaic width of the sample; σ_t^2 is the time spread of the neutron pulse and T is the nominal time of flight. The axes are choosen so that x lies along the scattered beam and y is in the plane of scattering.

A feel for the instrumental resolution required for a given size of unit cell can be obtained quite simply. If we postulate that the diffracted intensity is contained within $\pm 2.5\sigma$ of the profile, then the reflections from a cubic lattice of side $\underline{a}$ can only be separated if the following conditions are satisfied:

$$\sigma(H_i) \leq 1/5a \ , \ (i = x,y,z) \qquad\qquad 6.7$$

Reflections are most likely to overlap at the largest value of H, where the dominant contribution to $\sigma(H_x)$ is the time uncertainty. Putting $H_{max}\sigma_t/T \leq 1/5a$, one obtains the condition for the maximum allowable pulse length:

$$\Delta t_{pulse}(FWHM) \leq 238.1 \ L \ (d^2_{min}/a)sin\theta \qquad\qquad 6.8$$

where d_{min} is the minimum interplanar spacing in Å and L is the flight path in metres; the pulse width is in µs. The conditions for the angular separation of reflections impose limits on the allowable beam divergencies and sample mosaicity.

Performance comparisons between a pulsed source and a reactor for the study of a protein to d_{min} = 1.2 Å and a simple substance to d_{min} = 0.3 Å are given by Jauch and Dachs (1984) and the use of the IBR-2 pulsed reactor for the study of biological crystal structures has been discussed by Bally et al (1975). Since the mean value of $|F_{hkl}|^2$ is roughly proportional to V_o, the average intensity of the reflections is inversely proportional to V_o and, for a complex biological crystal, is down by a factor of 10^3 in comparison to that for a simple cell. Incoherent scattering from hydrogen is the principal source of background in biological structure work; the expected incoherent background counts under a Bragg reflection are:

$$I_{inc} = i_0(\lambda)\Delta\lambda(V/V_0^2) \ \Sigma \ \sigma_{inc}(\Delta\Omega_c/4\pi)A(\lambda)t \qquad\qquad 6.9$$

where $\Delta\lambda$ is the wavelength interval contributing to a reflection, $\Delta\Omega_c$ is the angular dimension of a reflection at the detector, $A(\lambda)$ is the transmission factor and t is the counting time.

Repeating the Jauch and Dachs analysis, which they originally applied to the performance of the SNQ design, for the pulse characteristics of ISIS shows that it would be significantly worse than the ILL reactor for the study of protein structures. In particular, in order to achieve the required resolution, the useable wavelength band could not extend below 2 Å so, consequently, the large flux of shorter wavelength neutrons would not be used.

The study of an 8 Å unit cell to a value of d_{min} = 0.3 Å could, however, be carried out on the SXD in a time comparable to a similar study on the D9 instrument at the ILL, supposing both were fitted with the same area PSD and the D9 sudy was carried out at 0.4 Å.

It is clear that further work is required on TOF diffraction to investigate the remaining sources of error in the data it produces.

Typically, R-factors are produced in the region of 4 - 5%, compared to the best reactor studies at a single wavelength with R-factors of 1 - 2%. Some of the difference must undoubtedly lie in the use of PSDs, which have yet to reach the degree of uniformity and stability of the simpler detectors.

7. SOFTWARE

A considerable ammount of effort has already been expended in providing the necessary software to enable TOF diffractometers to operate efficiently and to reduce their data to structure amplitudes and their standard deviations.

Programs are required in a number of areas:

(a) Detector adjustment and calibration
(b) Instrument control, specimen orienting, specimen environment control and data collection
(c) Peak search and auto-indexing
(d) Refinement of the crystal orientation matrix and diffractometer offsets
(e) Display of observed data in planes and along lines in reciprocal space
(f) Peak integration and reduction to observed intensities and structure amplitudes.

Of these, programs in the areas (a) and (b) tend to be Institute-specific, whereas the remainder have an increasing degree of portability. Schultz and Leung (1986) have described the software for the SCD at ANL in the areas (c) - (f).

Much work continues to be done on the problems of treating PSD data and there is a common interest with diffractionists using synchrotron radiation. An International Workshop on Evaluation of Single Crystal Diffraction Data from 2-D Position-Sensitive Detectors was held in Grenoble in 1986 and the proceedings are available (see Schultz and Leung 1986). Of particular interest is the work of Wilkinson (1986), who has shown that his earlier treatment of monochromatic beam, single crystal diffraction data from 2-D PSDs can be adapted to TOF. The technique is related to the $\sigma(I)/I$ method, described by Lehmann and Larsen (1974), and the number of PSD elements summed to estimate the intensity of weak peaks is reduced compared to a the number used for a strong peak in the vicinity. The reduction in weak peak intensity is then corrected using the observed shape of the strong peak. Suppose that the strong reflection is characterized by measuring the intensity $I(p)$ contained in p peak points which lie within a series of shapes set up to approximate to the intensity contours in the peak. Let $I_o(p_o)$ be the "total" intensity of the peak contained within p_o points and let $x(p)$ be the ratio $I(p)/I(p_o)$. Suppose that the weak peak has a total intensity $W_o(p_o)$ and sits on the same background B. Wilkinson and Kharmis (1983) showed that the quantity $\sigma(W)/W$ and also the quantity $\sigma(W/x)$ $[\equiv \sigma(W_o)]$ is minimised when the equation

$$I_o(dp/dI) - 2p/x = W_o/cB \qquad\qquad 7.1$$

is satisfied and therefore corresponds to the minimum error in measurement when the peak is divided up in this way. (The value of the constant c depends on the ratio of the number of background points to the number of peak points at which $\sigma(W)/W$ is a minimum. It is unity when the number of background points greatly exceeds the number of peak points and 2 when these numbers are constrained to be equal.)

Equation 7.1 can be solved for the statistically optimum number of points for integration of the weak peak when the quantities $I_o(dp/dI)$ and $2p/x$ have been obtained from a nearby strong peak. Figure 12 shows their variation with p for a typical case.

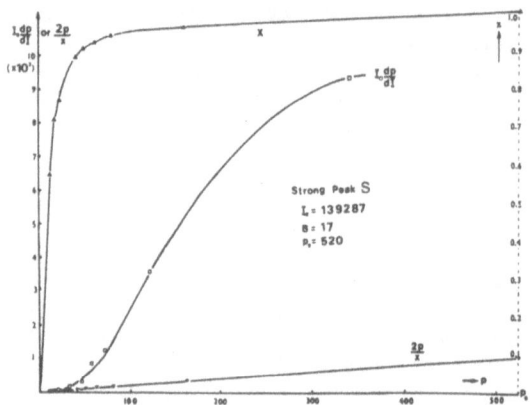

Figure 12. Variation of x, $I_o(dp/dI)$ and $2p/x$ with the number of peak points, p, included within integration contour ellipsoids for a strong diffraction peak. The lowest contour ellipsoid contained 520 points.

Tests on the reliability of the method were made with a series of artifically generated peaks with intensities 0.1, 0.01 and 0.001 of the strong peak, but with the same shape; the signal to noise ratios were 1.6, 0.16 and 0.016 respectively. $\sigma(W)$ was reduced by a factor of 0.7, 0.4 and 0.3 for an average of 100 repeated generations of the peaks with random noise applied when the integration was over 80 points compared to the full 540 points occupied by the strong peak. The improvements agreed well with those predicted by equation 7.1. The method is particularly suitable for TOF data, since on-line processing to establish the appropriate peak integration volumes can proceed as the data statistics build up, and a list of integrated intensities can be established soon after the measurement is finished. Fortran code for the method is available at RAL and has also been installed at ANL.

REFERENCES

Alkire R W, Larson A C, Vergamini P J, Schirber J E and Morosin B (1985) J Appl Cryst 18 145.

Anger H O (1958) Rev Sci Instrum 29 27.

Balagurov A M, Borca E, Dlouha M, Gheorghiu Z, Mironova G M and Zlokazov V B (1979) Acta Cryst A35 131.

Bally D, Chirtoc V, Gheorghiu Z, Popovici M, Stoica A D, Tarina E and Balagurov A M ((1975) Report FN-48-1975 Institute of Atomic Physics, Bucharest Romania.

Becker P J and Coppens P (1974) Acta Cryst A30 129.

Borkowski C J and Kopp M K (1975) Rev Sci Instrum 46 951.

Buras B (1969) Report Nr 1108/II/PS, Inst of Nuclear Research. Warsaw Poland.

Buras B and Holas A (1968) Nukleonika 13 591.

Convert P and Forsyth J B (1981) Eds "The Position-Sensitive detection of Thermal Neutrons" Academic Press London and New York.

Cooper M J and Nathans R (1968) Acta Cryst A24 481.

Darwin C G (1922) Phil Mag 43 800.

Day D H and Sinclair R N (1970) Acta Cryst B26 2079.

Ewald P P (1911) Z Krist 56 129.

Forsyth J B (1987) to be published in J Phys E.

Frank I M and Pacher P (1983) Physica 120B 37.

Jauch W and Dachs H (1984) in "Proceedings of the Workshop on Neutron Scattering Instrumentation for SNQ, Maria Laach, 3-5 Sept.1984" Ed Scherm R and Stiller H. Júlich - Nr. 1954 page 31.

Lehmann M S and Larsen F K (1974) Acta Cryst A30 580.

Lowde R D (1956) Acta Cryst 9 151.

Niimura N, Tomiyoshi S, Watanabe N and Kimura M (1973) in "Proceedings of the joint meeting on pulsed neutrons and their utilization" Ispra, 1971. Commision of the European communities. Luxemburg. EUR 4954 e, page 139.

Schultz A J and Leung P C W (1986) J de Physique, FASC 8, C5-137.

Schultz A J, Srinivasan K, Teller R G, Williams J M and Lukehart C M (1986) J Amer Chem Soc 106 999.

Strauss M G, Brenner R, Lynch F J and Morgan C B (1981) IEEE Trans Nucl Sci NS-28.

Stoica, A D (1975) Acta Cryst 31 193.

Turberfield K C (1970) in "Thermal Neutron Diffraction" ed Willis B T M, Oxford University Press page 34.

Wilkinson C (1986) in "Workshop on Neutron Scattering Data Analysis 1986", ed Johnson M W. Institute of Physics Conference Series Number 81. IOP, Bristol and Boston, page 47 and J de Physique, FASC 8, C5-35.

Wilkinson C (1987) to be submitted to Acta Cryst B.

Wilkinson C and Kharmis H W (1983) in "Position Sensitive Detection of Thermal Neutrons" ed Convert P and Forsyth J B, Academic Press New York, page 358.

DISCUSSION

RESOLUTION AND INCOMMENSURATE STRUCTURES

Whereas synchrotron sources give good resolution over reciprocal space, further improvements are needed in neutron instruments if they are to deal with problems such as incommensurate structures which give satellite peaks close to the Bragg peaks.

Time resolution could be improved by increasing the distance from source to crystal, and that should be combined with the use of a neutron guide to counteract inverse square law attenuation. A 20 m machine would give a 75 Å unit cell capability. Dr Willis's phonon experiments, which used a single crystal in a powder diffractometer, show the high resolution which can be obtained from a long flight path, even with relatively coarse angular resolution. Back scattering would improve the geometric term but only for the longer wavelengths, and consideration should also be given to beam divergence.

SIGNAL TO NOISE RATIO

Several factors contributed to the low signal to noise ratio in one of the examples shown. The sample contained much hydrogen which gave a strong incoherent background. The scintillation detectors used in time of flight instruments are sensitive to γ-rays, the signals from which are very difficult to separate from those due to neutrons. γ-rays were present, as they are a component of neutron beams and are produced when neutrons interact with many materials. This experiment was also done in a pressure vessel, the material of which was close to the sample.

CORRECTIONS FOR EXTINCTION

Unless the crystal is ideally imperfect, it has some degree of perfection which reduces the reflected intensities; this extinction is more marked at higher than at lower reflectivities.

In monochromatic work, corrections are made using the Becker-Coppens formalism, in which a mosaic width and a domain radius are adjusted to give the best fit over the whole range of intensities. In time of flight work, the corrections must also deal with extinction being greater at longer than at shorter wavelengths. In practice, the wavelength dependence implicit in the Becker-Coppens formalism, but not tested in monochromatic work, gives a consistent account of extinction as a function of both intensity and wavelength in time of flight experiments. Thus the more demanding role of these corrections in time of flight than in monochromatic work is compensated by the confidence given by their more rigorous testing.

A similar test could be applied in conventional work by obtaining a second data set on another instrument working at a different wavelength. The question of whether extinction corrections should be isotropic or anisotropic applies equally to both types of work.

PULSED NEUTRON POWDER DIFFRACTION

A. K. Cheetham
University of Oxford
Chemical Crystallography Laboratory
9, Parks Rd., Oxford OX1 3PD, U.K.

Abstract

The nature and scope of powder diffraction with a white, pulsed beam of neutrons is discussed. Analysis of the data by the Rietveld profile technique is described in brief, and a range of applications in solid state physics and chemistry is reviewed, including the use of ultra-high resolution methods to solve crystal structures from powder data. Diffraction with pulsed neutron sources is compared with constant wavelength methods and also with X-ray synchrotron powder techniques.

Introduction

The possibility of carrying out powder neutron diffraction experiments with a pulsed, polychromatic beam, rather than a constant, fixed-wavelength beam, has been recognised for over twenty-five years, and the technique was demonstrated on both a reactor and a pulsed source in the early sixties (Buras and Leciejewicz, 1963; Nitts et al., 1964). But only during the 1980's has the method been utilised to any significant extent, largely due to the construction of several pulsed neutron facilities that are competitive in flux with the modern reactor sources. To date, the bulk of the powder diffraction work has been carried out at the Intense Pulsed Neutron Source (IPNS) at the Argonne National Laboratory, where two diffractometers (GPPD and SEPD) have been in routine operation since 1981. New instruments have recently been built at KENS in Japan, ISIS in the U.K. and at WNR, Los Alamos, and it can be expected that an increasing proportion of powder neutron measurements will be carried out in this mode.

In terms of physics and chemistry, it should be recognised at the outset that the scope of pulsed neutron diffraction with powders is rather similar to that of the constant wavelength method. The main crystallographic advantages are those offered by powders rather than single crystals, and neutrons rather than X-rays. For most structural studies, the use of single crystals would normally be preferred, but sometimes these are not available of sufficient size or quality, and in other cases, the crystals may disintegrate before reaching the

137

M. A. Carrondo and G. A. Jeffrey (eds.), Chemical Crystallography with Pulsed Neutrons and Synchrotron X-Rays, 137–158.
© 1988 by D. Reidel Publishing Company.

temperature or composition of interest. The virtues of neutrons are well documented (eg. Bacon, 1975). One is able to probe the location of light elements, especially hydrogen, in the presence of heavy ones; to differentiate between elements that are close in the periodic table and are therefore difficult to distinguish with X-rays; and to probe the magnetic structures of materials by observing the additional reflections that are found with neutrons. In addition, most elements have extremely low absorption cross-sections for neutrons, thus facilitating the study of samples in special environments such as cryostats, furnaces and pressure cells. Pulsed neutron diffraction has substantial advantages over the constant wavelength method for measurements in special environments, a topic that will be covered in detail in an accompanying paper. In the opinion of many neutron scatterers, it should also offer superior performance in high resolution experiments.

This paper gives a brief description of the basis of powder diffraction with pulsed beams and discusses the experimental procedure. Some of the recent achievements of the technique will be described. We shall try to assess the relative strengths of the constant wavelength and pulsed methods, and to examine their complementarity to powder diffraction at synchrotron X-ray sources. Finally, the future prospects of powder diffraction at pulsed sources will be examined.

Time-of-Flight Diffraction with Pulsed Beams

In a constant wavelength experiment, the different reflections, h k l, are monitored by sweeping a counter, or bank of counters, through a range of scattering angle, 2θ. In terms of the Bragg equation, $\lambda = 2d_{hkl}\sin\theta$, λ is fixed and θ is varied. In the simplest time-of-flight (TOF) experiment, we use a single counter at a *fixed* 2θ value and a pulsed, white neutron beam. For a given reflection, the Bragg condition is then satisfied at a particular wavelength, λ. In order to measure the wavelength, the beam must be pulsed, and λ can then be determined from the time taken for the neutrons to reach the detector, fast neutrons (short λ) arriving earlier than slow ones (long λ). Although in the earliest examples of TOF diffraction, the pulsed beam was frequently obtained from a reactor source by means of a chopper, more recently the emphasis has been on the use of accelerator-based pulsed sources. In this case, white pulses of neutrons are produced when pulses of high energy electrons (from a LINAC) or protons (from a proton synchrotron) strike a metal target. A schematic layout of the ISIS facility is shown in Figure 1.

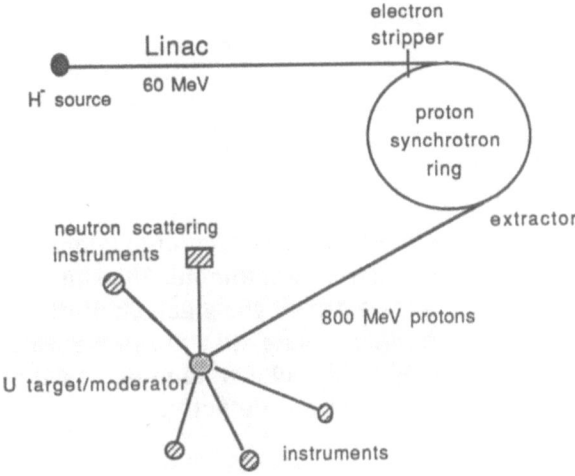

Figure 1: A schematic diagram of the spallation neutron source at ISIS.

It was shown in the late 60's (Meier-Leibnitz and Springer, 1966) that the instrumental resolution was determined by the time of flight (t) and the uncertainty in its measurement (Δt), the flight path (L) and its uncertainty (ΔL), and the scattering angle θ and the width of the detection element ($\Delta \theta$):

$$R = \frac{\Delta Q}{Q} = \left[\left(\frac{\Delta t}{t}\right)^2 + \left(\frac{\Delta L}{L}\right)^2 + (\text{Cot}\, \theta \, \Delta\theta)^2 \right]^{1/2}$$

........................(1).

It can easily be seen, therefore, that the best resolutions will be obtained with a long flight path and with the detector in the back scattering position. A simple experimental set-up, of the type demonstrated by Steichele and Arnold (1973), is shown in Figure 2. Of course, the use of a single detector element is very inefficient, and the data acquisition can be dramatically improved by using the principle of time-focusing (Carpenter, 1967); a bank of counters is used, but reflections that are observed in several detector elements (with different λ's) are merged in the software to appear as if they were obtained from a single counter. The back-scattering counter-bank on the high resolution powder diffractometer (HRPD) at the spallation neutron source ISIS can be seen in Figure 3. Note that the resolution has been enhanced by using an exceptionally long flight path (96 metres!).

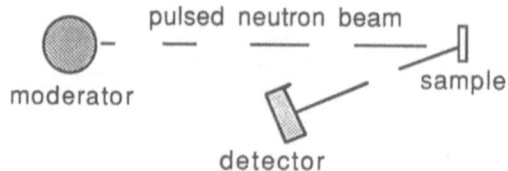

Figure 2: **A schematic diagram of a simple**
powder diffractometer, showing
a detector in the back-scattering
position. Note that the pulses are
longer and closer together as they
approach the detector.

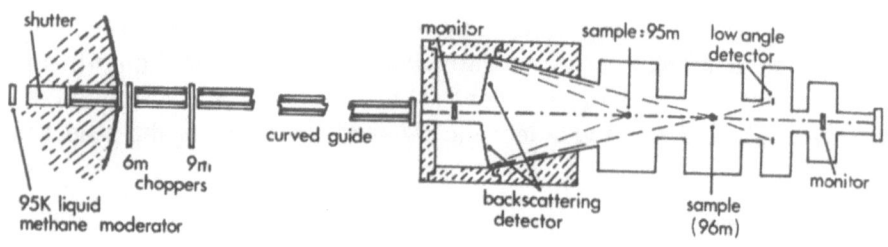

Figure 3; The High Resolution Powder Diffractometer (HRPD).

The experimental procedure is very straightforward. Typically, the sample is loaded into a vanadium can, which may be a cylinder or a flat plate depending upon the geometry of the counter bank, and with the sample placed in the appropriate position, data collection may commence. Unlike the constant wavelength method, there are no moving parts. Only the range of times-of-flight need be specified. Up to about 10g. of material is normally required, although it is perfectly feasible to study very much smaller samples if long counting times are used. Each pulse of neutrons yields a complete difffraction pattern, albeit with poor statistics, and successive pulses are accumulated and summed until the quality of the data is sufficiently good for analysis. It is worth noting, however, that on instruments with very long flight paths (eg. HRPD at

ISIS), overlap may occur between slow neutrons from one pulse and fast neutrons from the next; under these circumstances, it may be necessary to use choppers in order to remove, say, alternate pulses from the incident beam if a full set of data is to be collected without frame overlap. Data collection will take several hours on most of the presently available facilities.

The powder neutron diffraction pattern of $FePO_4$, obtained on the General Purpose Powder Diffractometer (GPPD) at IPNS (Jorgensen and Rotella, 1982) by Battle, Cheetham and Harrison (unpublished results), is shown in Figure 4. The dramatic variation in the background, with high intensity in the short wavelength (epithermal) region, is typical of a pattern from a pulsed source because the primary neutron beam is usually only partially moderated in order to maintain a short pulse length. On the HRPD instrument at ISIS, this epithermal peak is removed by the use of a neutron guide (Meier-Leibnitz and Springer, 1963) between the moderator and the sample (see the pattern of $FeAsO_4$ in Figure 5). On both instruments, the overall shape of the background mimics the spectral distribution of the incident neutron beam.

Analysis of Time-of-Flight Powder Diffraction Data

A dramatic advance in the development of powder neutron diffraction methods took place in 1969 when Rietveld published a paper describing the refinement of low symmetry stuctures from powder data by means of a curve-fitting procedure. This had an immediate impact on crystallography and, even by 1977, Cheetham and Taylor were able to review the application of Rietveld profile analysis, principally at constant wavelength, to over 200 inorganic and organic materials. The extension of the Rietveld method to time-of-flight data collected at a pulsed neutron source was first described by Windsor and Sinclair in 1976 in a study of nickel metal, and a comprehensive package of programmes was made available by Von Dreele et al. (1982). Previous papers in this volume have dealt with the complications that arise because different Bragg reflections are measured with neutrons of different wavelengths, difficulties which include the need to make wavelength-dependent corrections for absorption and extinction (Sabine, 1985). Other problems concerned with the complexity of the peak shape and background have also impeded the rapid development of these computational procedures, but most of the difficulties have now been overcome and the method has become almost as routine as the constant wavelength procedure. A profile fitting of a relatively simple example, the structure of the

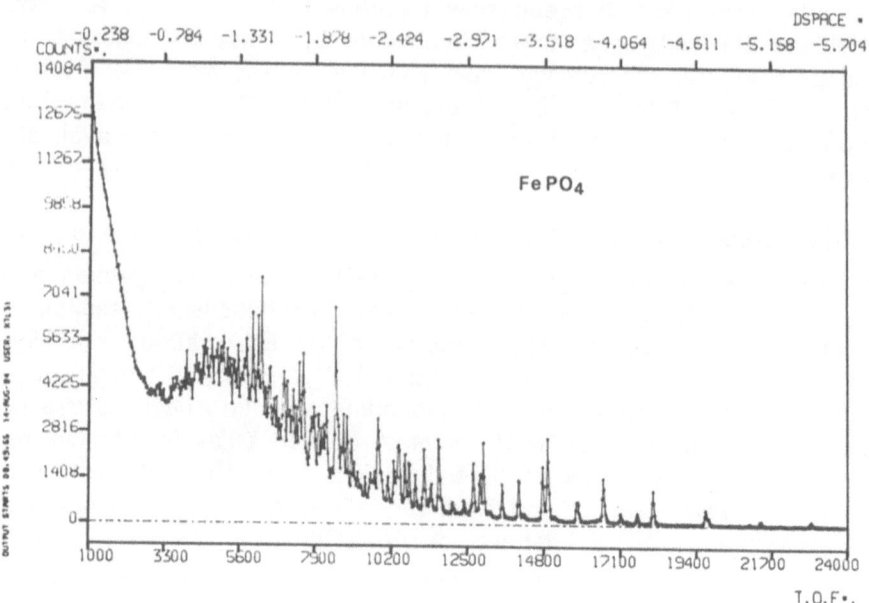

Figure 4: Time-of-Flight Pattern of FePO$_4$ (GPPD at IPNS)

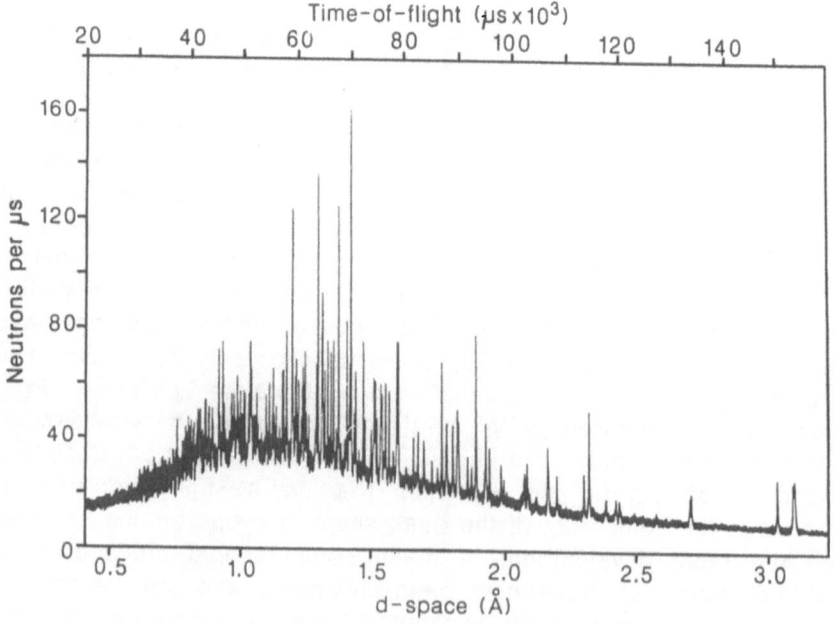

Figure 5: Time-of-Flight Pattern of FeAsO$_4$ (HRPD, ISIS).

perovskite-related phase Sr_2YRuO_6 (P2$_1$/n; a=5.7690, b=5.7777, c=8.1592Å, β=90.23°; 6 atoms in the asymmetric unit) is shown in Figure 6 (Battle and Macklin, 1984). Note the manner in which the density of reflections builds up at the short d-spacing end of the pattern.

Applications - Some Selected Examples.

The range of problems to which the pulsed diffraction method has now been applied is extremely wide, spanning layered phosphate materials such as $ZrKH(PO_4)_2$ (Rudolph and Clearfield, 1985), hydrides of the lower rare earth halides, eg Zr_2Br_2H (Wijeyesekera and Corbett, 1986), zeolites (Parise *et al.*, 1985, Newsam *et al.*, 1986), mixed metal oxide catalysts (Teller *et al.*, 1984), and intercalation compounds of TTF in FeOCl (Kauzlarich *et al.*, 1986). We shall not attempt a comprehensive survey of these and other applications, but rather illustrate the range of science that can be done by means some selected examples.

Location of Light atoms. One of the best known features of neutron diffraction is its sensitivity to very light elements; these may be difficult to locate with X-rays, especially if they are in close proximity to heavy ones. Hydrogen is, of course, the most striking example, and conventional neutron diffraction experiments have made many outstanding contributions to our knowledge of metal hydrides, organometallic hydrides, and hydrogen-bonding in organic, inorganic and biological molecules. Clearly, pulsed diffraction methods can be expected to make an impact in this area, and several interesting examples have already been reported in the literature. Rotella *et al.* (1982) used difference Fourier methods to locate the deuterium atoms in $DTaWO_6$, the heavy atom positions having been previously established with X-rays (deuterium is normally preferred to hydrogen for diffraction studies because of its superior neutron scattering properties). The results were in good agreement with those that had been reported for the closely-related hydrogen tungsten bronzes. In a more recent example, Moini *et al.* (1986) were able to demonstrate that the compound $NaNi_2(OH)(H_2O)(MoO_4)_2$ contains alternate OH and H_2O units, rather than are random distribution of hydrogen. Again, difference Fourier methods were adopted, but the hydrogen, rather than deuterium, compound was studied.

Atomic Distributions. Although the neutron scattering factors of the elements (or scattering lengths, as they are known) show only a small variation with atomic number, thus accounting for the sensitivity to light atoms in the presence of heavy ones, in certain parts of the periodic table,

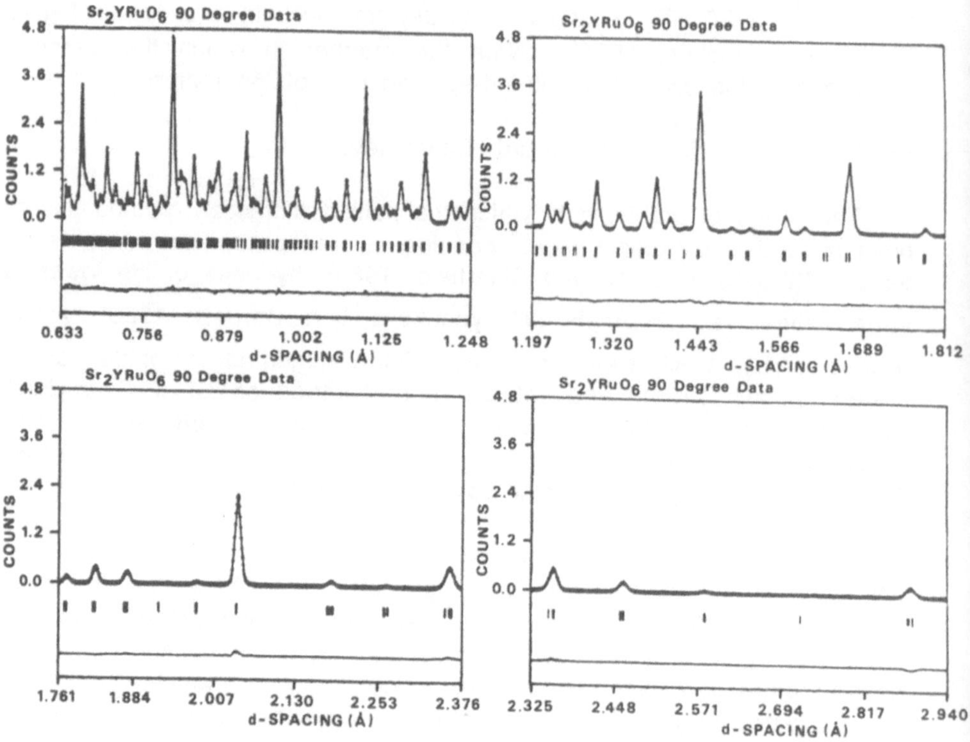

Figure 6: Observed (+), calculated (-) and difference profiles of Sr$_2$YRuO$_6$ at 295K; reflection positions are marked.

Figure 7: Idealised structure of W$_3$Nb$_{14}$O$_{44}$; filled circles-tetrahdral sites; octahedral sites numbered 1-4.

most notably the first transition series, there are some striking variations between adjacent elements. For example, the scattering lengths, b, of Cr, Mn, Fe, Co and Ni are +0.35, -0.36, +0.96, +0.25 and +1.03 (10^{-14}m), respectively (Bacon, 1975). Variations of this magnitude enable neutrons to differentiate between elements that are indistinguishable with x-rays because of their similar scattering factors. For this reason, neutrons have played a central role in advancing our understanding of, for example, site preferences in spinels, order-disorder phenomena in alloys, and the coordinate chemistry of cyanides (b_C=+0.66, b_N=+0.94; 10^{-14}m).

Pulsed diffraction methods have yet to make a real impact in this area, but a recent experiment by Cheetham and Allen (1983) gives an indication of what can be achieved. The tetragonal structure of $W_3Nb_{14}O_{44}$ comprises columns or blocks of corner-sharing MO_6 octahedra, 4x4 octahedra in cross-section and infinite along the unique axis, linked to similar columns by edge-sharing of the octahedra. The arrangement of the blocks is such that they surround tunnels containing tetrahedrally coordinated cations (Figure 7). In the X-ray structure determination of Roth and Wadsley (1965), it was assumed that one tungsten per formula unit occupied the tetrahedral site with the balance randomly distributed over the four crystallographically inequivalent, octahedral sites, but the neutron study performed on GPPD at IPNS revealed a different story. The distribution of tungsten and niobium over the metal sites was determined by refining the scattering length of each site (b_W =0.477 x 10^{-14}m; b_{Nb}=0.705 x10^{-14}m) and the results are presented in Table 1. Also shown are the average potentials at the metal sites, calculated using the program of Van Gool and Picken (1969). The results, which are at variance with the assumptions of Roth and Wadsley, show that the cation distribution is largely controlled by electrostatic interactions, as was found in the

TABLE 1

Cation distribution and calculated site potentials ($Å^{-1}$) for $W_3Nb_{14}O_{44}$ (e.s.d's in parentheses).

Site	% Tungsten	Site Potential
Tet	27(8)	-3.68
Oct(1)	39(4)	-4.17
Oct(2)	23(7)	-3.64
Oct(3)	0(6)	-3.58
Oct(4)	0(5)	-3.53

titanium-niobium oxides (Von Dreele and Cheetham, 1974). The sequence of site potentials reproduces the observed sequence of tungsten occupancies, although there is no obvious, intuitive reason why tungsten should prefer site oct(2) to oct(3). The highest concentration of tungsten is found at site oct(1), which is located within the blocks in an octahedron that shares only corners, as in WO_3 itself. The tungsten occupancy of the tetrahedral site is rather low, reflecting its low co-ordination number.

Magnetic Structure. The neutron diffraction patterns of magnetically ordered materials contain extra Bragg reflections which arise because there is an interaction between the magnetic moments of the neutrons and those of the atoms or ions within the sample. The magnetic peaks are most intense in the large d-spacing region of the spectrum (which corresponds to the low angle region of a constant wavelength pattern) because the magnetic scattering shows a strong form-factor dependence; unpaired electrons are normally in the valence shell, and the scattering from them diminishes with $\sin\theta/\lambda$ in much the same way as with X-ray scattering. Pulsed sources are not really optimised for magnetic studies because they have a high flux of short wavelength neutrons but a low flux in the long wavelength region. In a time-of-flight pattern, the magnetic reflections would be observed in this weak, long wavelength region of the pattern, so for this type of experiment the constant wavelength method is to be preferred. There is, however, some compensation because the integrated intensities, I_{hkl}, are enhanced at longer wavelengths by a factor of λ^4 in the relationship between intensity and structure factor:

$$I_{hkl} = C\lambda^4 \, S(\lambda) \, F_{hkl}^2 \qquad\qquad \ldots\ldots\ldots\ldots(2),$$

where C is a constant and $S(\lambda)$ is the incident flux of neutrons with wavelength λ. Magnetic reflections can, therefore, be observed in some favourable cases, as demonstrated by the work of Lawson et al. (1985) on UGa_2 and UGa_3. Data was collected with a low angle counter-bank ($2\theta=40^{\circ}$) rather than in the backscattering mode. A reflection with a d-spacing of 5Å will be observed with 10Å neutrons in backscattering and with 3.42Å neutrons at $2\theta=40^{\circ}$.

Ultra-High Resolution Diffraction with Pulsed Sources.

Whereas the constant wavelength diffraction method appears to be the best approach for the study of magnetic materials, the pulsed technique is ideal for powder measurements at very high resolution. The combination of a very long flight path, which gives excellent time resolution within each pulse, and the geometrical advantages of the back-scattering mode (see Eqn. 1) produces a performance which is difficult to match at fixed wavelength. In particular, a larger number of short d-spacing reflections are normally accessible by the pulsed technique, thus improving the precision of the subsequent structure refinement. The HRPD instrument at ISIS (Figure 3), which was commissioned during 1986, represents the most advanced, high resolution instrument in the world, and the first experiments have amply demonstrated its power. Two recent examples will be given to underline this view.

Study of hydrogenous compounds. It has already been noted that for studies of hydrogen-containing compounds, the deuterium analogue is normally preferred because of the more favourable scattering properties of the latter nucleus. The disadvantage of hydrogen stems from its large incoherent scattering cross-section, which gives rise to very high background levels. In principle, one of the consequences of increasing the instrumental resolution should be an improvement in the peak-to-background ratio of the data, suggesting that it may be possible to obtain reasonable neutron powder data from hydrogenous materials. Prout and co-workers (1987) have collected high resolution data on two such materials, hydrazine sulphate and diketopiperazine, and their results show a dramatic enhancement in the quality of the data compared with previous attempts on similar materials.

Ab Initio structure determination. One of the other areas where high resolution is expected to play an important role is in the *ab initio* determination of crystal structures from powder data. The main loss of information in a powder pattern compared with a single crystal data-set arises from the overlapping of adjacent Bragg reflections, which leads to ambiguities in their individual intensities. This information is essential if the classical methods of structure solving, the Patterson synthesis and direct methods, are to be employed. The improvement in resolution on instruments such as HRPD leads to a dramatic reduction in the incidence of peak overlap, in spite of the fact that the instrumental resolution is now so good that most samples show line broadening due to particle size and strain effects, so that the *solving* of structures from powder data may now be possible on such an instrument (Cheetham, 1986).

This possibility has been realised in a recent study of $FeAsO_4$ on HRPD using the data shown in Figure 5 (Cheetham *et al*, 1986). The compound had previously been studied by D'Yvoire(1972), who indexed the X-ray powder pattern according to a monoclinic cell and assigned the space group $P2_1/n$.

On the basis of infra red evidence and the facile transformation of the monoclinic modification to one with the $CuSO_4$ structure, it was suggested that the iron atom might be octahedrally coordinated. The correctness of the D'Yvoire cell was first confirmed by auto-indexing of the pattern using the programme FZON (Visser, 1969), and the lattice parameters were then refined and the space group confirmed by examining the systematic absences. Integrated intensities were then obtained manually for 139 reflections, including approximately 60 weak ones, and the structure factor amplitudes, calculated according to Eqn. 2, were used as the input for a direct methods analysis using the programme MITHRIL (Gilmore, 1984). The peak list from the calculation with the highest figure of merit is shown in Table 2. Analysis of the inter-peak distances and

TABLE 2

(a) Peak listing from Direct Methods Solution.

Peak No.	X	Y	Z	Height
1	0.178	0.449	0.758	2691
2	0.075	0.179	0.243	1960
3	0.210	0.080	0.413	1230
4	0.015	0.373	0.369	1199
5	0.898	0.046	0.178	1193
6	0.131	0.256	0.913	1104
7	0.562	0.587	0.726	591
8	0.385	0.617	0.887	590

No. of reflections = 139

Triplets generated from top 66 reflections

Negative quartets generated from top 67 reflections

Figure of merit = 4.0

(b) Final atomic coordinates, with e.s.d's in parentheses

Atom	X	Y	Z
Fe	0.173(2)	0.462(2)	0.763(2)
As	0.073(3)	0.202(2)	0.223(4)
O(1)	0.257(3)	0.101(3)	0.420(4)
O(2)	0.027(3)	0.377(2)	0.384(4)
O(3)	0.889(4)	0.077(3)	0.152(4)
O(4)	0.125(4)	0.267(3)	0.939(5)

Temperature Factor = -3.8(4) A^2

angles confirmed that a chemically sensible solution had been found, with peaks 1 and 2 corresponding to Fe and As, respectively, and peaks 3-6 to oxygen. Note that the peaks appear in order of their neutron scattering lengths and that there is a substantial gap between peaks 6 and 7; the latter indicates the noise level of the map. The coordinates obtained from the direct methods analysis were then refined by integrated intensity methods $(R_I = 6.5\%)$ to yield the structure shown in Figure 8. One of the interesting features of $FeAsO_4$ is the presence of a five coordinated iron site. Attempts to obtain good refinements of the structure by profile analysis have so far been hindered by subtle variations in peak shape which appear to stem from anisotropic strain or particle size effects. This is a problem that requires urgent attention in order for the full scope of high resolution methods to be realised.

Comparison of pulsed and constant wavelength methods.

Two facets of this comparison have already been noted: the advantages of the former for high resolution and the latter for magnetic studies. In an attempt to make a quantitative comparison, Fischer et al. (1986) made a careful study of the structure of terephthalic acid on diffractometers at IPNS (SEPD) and the high flux beam reactor, ILL Grenoble (D1a). The results of their refinements did not confirm the expected superiority of the pulsed method, largely, it seems, because the instrumental resolution was eroded by broadening from the sample, itself. On the other hand, in a similar comparison of the same instruments, Harrison (1986) studied an excellent, polycrystalline sample of $Al_2(MoO_4)_3$. The structure contains six independent MoO_4 tetrahedra in the asymmetric unit, thus affording a good opportunity to assess the relative precisions of the two methods by comparing the range of Mo-O distances. The results show that the spread obtained from the analysis of the pulsed data is almost three times less than that found from the constant wavelength experiment (Table 3). These results probably exaggerate the difference between the two techniques because the aluminium molybdate problem is at the limit of what can be done on these instruments, but they underline the need to take extra care in the synthesis of materials that are to be studied at high resolution. It is the practice of this laboratory to examine our compounds by electron microscopy before neutron data are collected.

Another area in which difficulties have arisen with pulsed diffraction data is in the treatment of the background. It is customary for this to be fitted as part of the profile in the time-of-flight method, largely because it is otherwise difficult to estimate the background in the short d-spacing region of the pattern where the density of reflections is very high. For

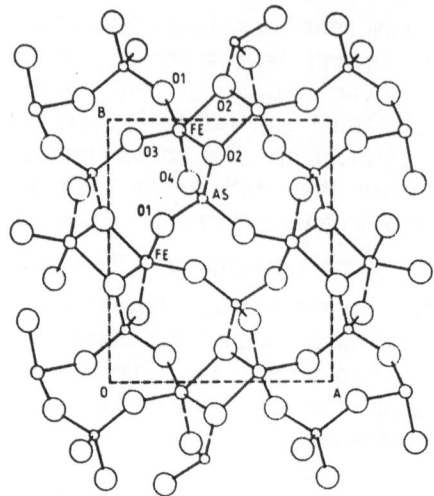

Figure 8: Crystal structure of FeAsO$_4$ projected on *ab* plane.

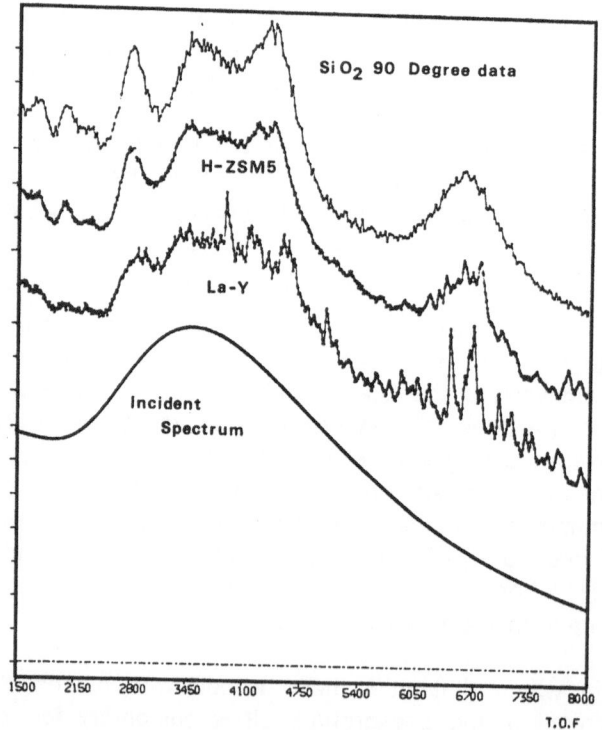

Figure 9: TOF patterns of glassy SiO$_2$, and zeolites H-ZSM5 and La-Y; the incident neutron spectrum is also shown (Eddy, 1986).

most samples this can be done very satisfactorily, but for some, e.g. those containing high concentrations of defects with short range order, the background may fluctuate dramatically and fitting can be difficult, even if a complex polynomial is used. A particular area in which this problem has inhibited progress is in the study of zeolites, where for many samples the background at short d-spacings is almost the same as that of glassy

TABLE 3

Molybdenum - oxygen distances in $Al_2(MoO_4)_3$; a comparison between TOF and constant wavelength studies (Harrison, 1986).

Bond		Distance/Å	
		TOF	Constant λ
Mo(1) -	O(7)	1.79	1.76
	O(9)	1.76	1.86
	O(12)	1.74	1.82
	O(20)	1.74	1.64
Mo(2) -	O(2)	1.77	1.73
	O(3)	1.78	1.68
	O(4)	1.76	1.65
	O(13)	1.76	1.78
Mo(3) -	O(5)	1.80	1.79
	O(6)	1.72	1.85
	O(8)	1.75	1.82
	O(15)	1.74	1.61
Mo(4) -	O(14)	1.80	1.77
	O(16)	1.74	1.70
	O(18)	1.73	1.73
	O(21)	1.70	1.69
Mo(5) -	O(17)	1.74	1.71
	O(19)	1.80	1.92
	O(23)	1.81	1.79
	O(24)	1.77	1.85
Mo(6) -	O(1)	1.79	1.74
	O(10)	1.80	1.89
	O(11)	1.72	1.71
	O(22)	1.76	1.64
Average:		1.761	1.755
Standard deviation:		0.03	0.08

quartz (Figure 9). Recently, however, Richardson et al. (1986) appear to have solved the problem by means of a Fourier filtering technique that facilitates the extraction of a radial distribution function for the non-crystalline component, in addition to the structure of the crystalline phase. In the constant wavelength method, the background changes relatively smoothly and slowly within the accessible d-spacing range.

Notwithstanding these caveats, pulsed neutron diffraction is probably the optimum way of collecting powder data for most materials, a fact that was indeed recognised by Meier-Leibnitz and Springer in their 1966 review, but the full potential will not be realised until some of the outstanding computational problems have been solved, and, even then, only with high quality samples.

Comparison between Neutron and X-Ray Synchrotron Diffraction with Powders

Although this paper is concerned primarily with pulsed neutron diffraction, it is interesting, especially in relation to the emphasis of the meeting as a whole, to compare the neutron method with X-ray powder diffraction using synchrotron radiation. Attfield et al (1987) have studied the structure of α-$CrPO_4$ by both X-ray and neutron powder diffraction, having first determined the structure by the synchrotron method (Attfield et al., 1986). The results are shown in Table 4, where the refined atomic coordinates are compared with those reported in a very recent single crystal x-ray study by Glaum et al. (1987). The single crystal study is, of course, superior on all counts, but the precision of the oxygen atoms is almost as good in the neutron powder study, where the sensitivity of neutrons to light elements is most apparent. The precision of the x-ray powder study is inferior to that of the neutron refinement for all atoms except chromium, where the two are comparable. This lead the authors to conclude that neutron powder techniques are to be preferred for the *refinement* of structures, because of their comparable sensitivity to a wide range of elements, but the x-ray powder method is probably more powerful for *solving* structures because a satisfactory phasing model can often be obtained after locating a small number of dominant scatterers. With neutrons, a high proportion of the atoms must be found before the phasing of the reflections is approximately correct.

Limitations and Future Developments

Pulsed neutron diffraction with powders has already made a significant impact on the physics and chemistry of solids, but an examination of the references to this article reveals that most of the applications have been published in the last five years. Only now is the technique becoming relatively routine, and if it develops with the same vigour as the constant wavelenth method, the future indeed looks bright. The emphasis is likely to be on the refinement of complex structures and the study of materials in special environments. At present, the most complex structure to be refined is that of $Cr_2(MoO_4)_3$ (Battle et al., 1985), with 34 atoms in the asymmetric unit and 102 variable atom coordinates;

TABLE 4

Structural Parameters for α-$CrPO_4$ refined using synchrotron X-ray (marked X) and neutron (N) data in Imma (No. 74) with e.s.d.'s in parenthesis. Values from the reported single crystal study[a] (marked S) are given for comparison.

Atom	Symmetry position	x	y	z	B_{ISO}[b]
Cr(1)	4b	1/2	1/2	0	0.3(2)N 0.283(6)S
Cr(2)	8g	1/4	0.3660(3)X 0.3650(4)N 0.36611(3)S	1/4	0.0(1)N 0.316(4)S
P(1)	4e	1/2	1/4	0.0819(12)X 0.0790(8)N 0.0825(2)S	0.0(1)N 0.30(1)S
P(2)	8g	1/4	0.5738(4)X 0.5739(2)N 0.57358(5)S	1/4	0.47(8)N 0.245(7)S
O(1)	8i	0.3790(10)X 0.3766(3)N 0.3773(2)S	1/4	0.2269(17)X 0.2280(5)N 0.2268(3)S	0.53(8)N 0.42(2)S
O(2)	16j	0.3603(6)X 0.3610(2)N 0.3611(1)S	0.4914(5)X 0.4907(1)N 0.4902(1)S	0.2145(11)X 0.2142(3)N 0.2146(2)S	0.62(6)N 0.42(1)S
O(3)	16j	0.2263(6)X 0.2240(1)N 0.2238(1)S	0.6352(5)X 0.6368(2)N 0.6363(1)S	0.0576(10)X 0.0546(3)N 0.0552(2)S	0.68(5)N 0.56(1)S
O(4)	8h	1/2	0.3509(8)X 0.3486(2)N 0.3496(2)S	-0.0457(15)X -0.0422(4)N -0.0432(3)S	0.31(7)N 0.50(2)S

(a) Glaum et al., 1987.
(b) For the powder X-ray refinement, overall $B_{ISO} = 0.24(7)Å^2$

$Al_2(MoO_4)_3$ is isomorphous (see Table 3). On very high resolution instruments such as HRPD, it may be feasible to refine structures with as many as 400 parameters once the problems of line shape have been overcome, and the inclusion of slack constraints (e.g. Baerlocher, 1982) should increase this figure even further. Even macromolecular systems may come within reach, especially if simultaneous refinements of X-ray and neutron powder data are carried out. For the reasons given in the previous section, however, the determination of structures from powder data is likely to be more feasible with synchrotron X-ray data, notwithstanding the recent success with $FeAsO_4$.

Another important area will be the use of high flux instruments to study reactions and other processes in real time. Pannetier (1986) has already demonstrated the power of the constant wavelenth method when combined with a position-sensitive detector. At present, the data acquisition time is about 10 minutes, but future instruments at high intensity pulsed sources should be capable of reducing this to as little as one second, thus bringing an exciting range of additional science within reach.

References

J.P. Attfield, A.W. Sleight and A.K. Cheetham , Nature 322, 620 (1986).

J. P. Attfield, A. K. Cheetham, D. E. Cox and A. W. Sleight, to be published (1987).

G. E. Bacon, Neutron Diffraction, 3rd edition (Oxford University Press, Oxford, 1975).

Ch. Baerlocher, The X-ray Rietveld System, XRS-82 (ETH Zurich, 1982).

P.D. Battle and W.J. Macklin, J.Solid State Chem. 52, 138 (1984).

P.D. Battle, A.K. Cheetham, W.T.A. Harrison, N.J. Pollard and J. Faber, J.Solid State Chem., 58, 221 (1985).

B. Buras and J. Leciejewicz, Nukleonika 8, 75 (1963); 8, 259 (1963)

J. M. Carpenter, Nucl. Inst. & Methods, 47, 179 (1967).

A. K. Cheetham, Materials Science Forum, 9, 103 (1986).

A.K. Cheetham and N.C. Allen, J.Chem.Soc., Chem. Comm., 1370 (1983).

A.K. Cheetham, W.I.F. David, M.M. Eddy, R.J.B. Jakeman, M.W. Johnson and C. Torardi, Nature 320. 46 (1986).

A.K. Cheetham and J.C. Taylor, J. Solid State Chem., 21, 253 (1977).

M. M. Eddy, D. Phil. Thesis, University of Oxford, Oxford (1986).

F. C. D'Yvoire, C. R. hebd. Seanc. Acad. Sci. , Paris, 275C. 949 (1972).

P. Fischer, P. Zolliker, B. H. Meier, R. R. Ernst, A. W. Hewat, J. D. Jorgensen and F. J. Rotella, J. Solid State Chem. 61, 109 (1986).

C. J. Gilmore, J. Appl. Crystallog., 17, 42 (1984).

R. Glaum, R. Gruehn and M. Moller, Z. Anorg. und allg. Chem., to be published.

W. T. A. Harrison, D. Phil Thesis, Oxford (1986).

J.D. Jorgensen and F.J. Rotella, J.Appl. Crystallog., 15, 27 (1982).

S. M. Kauzlarich, J. L. Stanton, J. Faber, Jr. and B. A. Averill, J. Amer. Chem. Soc., 108, 7946 (1986).

A. C. Lawson, A. Williams, J. L. Smith, P. A. Seeger, J. A. Goldstone, J. A. O'Rourke and Z. Fisk, J. Mag. & Mag. Materials, 50, 83 (1985).

H. Meier-Leibnitz and T. Springer, Reactor Sci. Technology, 17, 217 (1963).

H. Meier-Leibnitz and T. Springer, Ann. Rev. Nucl. Sci., 16, 207 (1966).

A. Moini, P. R. Rudolph, A. Clearfield and J. D. Jorgensen, Acta Cryst., C42, 1667 (1986).

J. M. Newsam, A. J. Jacobson and D. E. W. Vaughan, J. Phys. Chem. 90, 6858 (1986).

V. V. Nitts, Z. G. Papulova, I. Sosnovskaya and J. Sosnovskii, Sov. Phys.-Solid State, 6, 1070 (1964).

J. Pannetier, Chemica Scripta 26A, 131 (1986).

J. B. Parise, L. Abrams, T. E. Gier, D. R. Corbin, J. D. Jorgensen and E. Prince,

J Phys. Chem., 88, 2303 (1984).

C.K. Prout, W.I.F. David and W. T. A. Harrison, unpublished results.

J. W. Richardson, J. Faber, Jr. and R. L. Hitterman, Experimental Report in IPNS Progress Report (1986).

H.M. Rietveld, J. Appl. Crystallog., 2, 65 (1969).

F.J. Rotella, J.D. Jorgensen, R.M. Biefeld and B. Morosin, Acta Cryst. B38, 1697 (1982).

R. S. Roth and A. D. Wadsley, Acta Crystallog., 19, 38 (1965).

P.R. Rudolph and A. Clearfield, Inorg.Chem. 24, 3714 (1985).

T. M. Sabine, Aust. J. Phys., 38, 507 (1985).

E. Steichele and P. Arnold, Phys. Lett., A44, 165 (1973).

R.G. Teller, J.F. Brazdil, R.K. Grasselli and J.D. Jorgensen, Acta Cryst. C40, 2001 (1984).

W. Van Gool and A. G. Picken, J. Mater. Sci., 4, 94 (1969).

J. W. Visser, J. Appl. Crystallog., 2, 89 (1969).

R. B. Von Dreele and A. K. Cheetham, Proc. Roy. Soc. London, Ser. A, 338, 331 (1974).

R.B. Von Dreele, J.D. Jorgensen and C.G. Windsor, J.Appl.Crystallogr., 15, 581(1982).

S.D. Wijeyesekera and J.D. Corbett, Inorg.Chem., 25, 4709 (1986).

C. G. Windsor and R. N. Sinclair, Acta crystallog. A32, 395 (1976).

DISCUSSION

LONGER LATTICE SPACINGS

A new, low angle counter bank will soon be installed on the HRPD. It should give a spatial resolution ($\Delta d/d$) of about 0.004, and so permit the measurement of d-spacings up to at least 20 Å. The neutron flux will be comparable with that now used at backscattering for shorter spacings, and it should also be appropriate for the solution of magnetic structures.

PROTON SITES

The extra, acidic proton in La zeolite Y, produced on heating by the hydrolysis of hydrated lanthanum ions, was located by difference Fourier methods. It is probably attached to the only one of the 4 available oxygen atoms which is not co-ordinated to any of the cations and is therefore the most basic.

THERMAL PARAMETERS

The reliability of thermal parameters depends crucially, in X-ray work on the absorption correction, and in time of flight work on making a proper correction for the background and having a very accurate description of the incident neutron spectrum for normalisation.

In time of flight work with simple backgrounds, the data analysis programs usually fit the total profile of diffraction data and background. Tschebyscheff polynomials are used, and if a sufficient number of terms is included, some fluctuations in the background can be overcome. More sharply varying backgrounds, as for example from a sample with both crystalline and glassy components, can be dealt with by Richardson's method. The data are Fourier transformed, features with periodicities associated with the glassy component are removed by digital filtering, and the separate crystalline and glassy diffraction data are recovered as the reverse Fourier transforms.

In neutron powder diffraction, visual estimation of the background does not give accurate thermal parameters, but polynomial fitting for MgO has given values which agree well with those obtained from lattice dynamical calculations and the best single crystal measurements, even in the presence of heavy extinction.

In the results for chromium phosphate listed in table 4 above, the discrepancy between the isotropic temperature factors, $B_{iso}/\overset{\circ}{A}^2$, of Cr(2) and P(1) obtained by synchrotron X-rays (0.3) and neutrons (0.0) is more apparent than real, because it is within the neutron 3σ (3×0.1), perhaps the result of visual estimation of the background.

In deformation density studies on single crystals, Prof Fuess has found that X-ray and neutron anisotropic temperature factors agree to within 1% for hard materials, but only to within 5% for soft materials. The greater contribution of thermal diffuse scattering, different for X-rays and neutrons, is thought to account for the wider spread for soft materials. With powders it appears that special efforts are needed to extract reliable thermal parameters, even for simple structures.

ISOTOPES WITH DIFFERENT SCATTERING LENGTHS

If an element has isotopes with different scattering lengths, it is normally assumed that the isotopic distribution is uniform and random over all positions for the same element.

DIRECT METHODS FOR STRUCTURE DETERMINATION

Although the atomic scattering amplitude f for X-rays is positive for all atoms, for some atoms the neutron scattering length b is negative, corresponding to a phase change caused by a nuclear resonance. The question therefore arises as to how this difference affects the use of direct methods to estimate the phases needed for structure determination by neutron diffraction.

Some structures containing atoms with negative b values, e.g. Mn, have been solved by direct methods from neutron data. Dr J Karle stated that if less than about 20% of the total scattering is of negative amplitude, normal direct methods are applicable. Otherwise, another method can be used, in which the original data are transformed into the hypothetical data which would be seen if all the atoms acted with amplitudes equal to the squares of their real scattering lengths. Then all scattering appears to be positive and the normal methods are applicable. At Oak Ridge many years ago, a structure which gave a Patterson function from which the hydrogen peaks were missing was successfully solved in this way.

Direct methods were originally developed to solve structures, e.g. organic ones, in which the atoms had comparable scattering powers (equal atom structures). Dr J Karle emphasised that this did not imply that direct methods were applicable only to equal atom structures. In fact, the probabilities that enter into direct methods are strongest where disproportionately heavy atoms are present, because the effective number of atoms is then much smaller and the problem becomes a simpler one.

PULSED NEUTRON DIFFRACTION IN SPECIAL SAMPLE ENVIRONMENTS*

J. D. Jorgensen
Materials Science Division
Argonne National Laboratory
Argonne, IL 60439

ABSTRACT. Neutron diffraction is a powerful tool for structural studies of samples in special sample environments because of the high penetrating power of neutrons compared to X-rays. The neutrons readily penetrate special sample containers, heat shields, pressure vessels, etc., making it unnecessary in most cases to compromise the effectiveness of the sample environment system by providing windows for the incident and scattered neutrons. Pulsed neutrons obtained from an accelerator-based pulsed neutron source offer the additional advantage that many diffraction experiments can be done at a single, fixed scattering angle by the time-of-flight technique. Thus, if windows are needed (e.g., in an especially thick-walled sample vessel such as a pressure cell), they need to cover only a limited angular range. More importantly, in the fixed-angle scattering geometry, shielding and collimation can be optimized in order to access the largest possible sample volume with neutrons while completely avoiding scattering from the surrounding sample vessel. Thus, the data are free from unwanted background scattering which could hamper data interpretation. In this paper, the basic principles of neutron diffraction in special sample environments are discussed and examples of apparatus used for neutron diffraction measurements at low temperature, high temperature, and high pressure are presented. The concepts are illustrated with several unique scientific results taken from the published literature.

1. INTRODUCTION

At the present time the majority of neutron scattering experiments are done under special sample environment conditions. Refrigerators, cryostats, furnaces, and pressure cells are used routinely at all of the major neutron scattering centers for both elastic and inelastic neutron

---
*Work supported by the U. S. Department of Energy, BES-Materials Sciences, under contract W-31-109-ENG-38.

159

scattering experiments.[1] Using proper experimental techniques the
quality of data for a sample in a cryostat, furnace, or pressure cell
can be equivalent to what would be obtained at room temperature and
atmospheric pressure.

The ability to perform neutron scattering experiments in special
environments is a result of the high penetrating power of the neutron.
Because the neutron is a neutral particle and because the total cross
section (scattering plus absorption) is rather low for most materials,
scattering experiments can easily be done in geometries where the sample
is contained in heat shields, pressure cells, reaction vessels, etc.,
through which the incident and scattered neutrons must pass. For
typical thin aluminum heat shields the neutron attenuation is negligible. For thicker sample containers such as aluminum or aluminum oxide
pressure cells, the overall attenuation may be on the order of 50% for a
vessel with 1 cm-thick walls.

1.1. Advantages of Pulsed Neutrons for Diffraction in Special Sample Environments

Pulsed neutron sources offer the additional advantage that complete data
can be obtained at a single, fixed scattering angle by the time-of-
flight technique.[2,3] Thus, if it is necessary to provide windows for
the incident and scattered neutrons in a particularly restrictive sample
vessel (e.g., a large pressure cell), the windows need to cover only a
limited angular range. Furthermore, the scattering angle can be chosen
so that shielding and collimation can be optimized to provide the
maximum effective sample volume while eliminating unwanted scattering
from the sample vessel.

1.2. Experimental Geometry for Diffraction in Special Environments

The goal of any special environment apparatus for neutron diffraction is
to provide clean diffraction data (i.e., data free from unwanted
scattering from the sample vessel) while achieving the desired sample
conditions (temperature, pressure, etc.). Virtually all special sample
environments used at pulsed neutron sources are extensions of the two
basic geometries shown in Fig. 1. These geometries ensure that the
sample is the only volume which is seen by both the incident beam and
the detector. In Fig. 1(a), the incident and scattered beams are both
collimated by a cadmium mask, with windows at selected angles, which
surrounds the sample and heat shield. Thus, even though the incident
beam passes through and scatters from the aluminum heat shield before
and after striking the sample, the detector is prevented from seeing the
unwanted scattering. Of course, a small part of the scattered beam is
also scattered from the heat shield between the sample and the detector,
but this is a second order effect and, moreover, contributes to the
smooth background rather than to sharp features in the data. The
geometry shown in Fig. 1(a) is most useful when the scattering angle can
be chosen to be at or near $2\theta = 90°$, so that the collimation can define
a localized sample volume. For larger or smaller angles it becomes

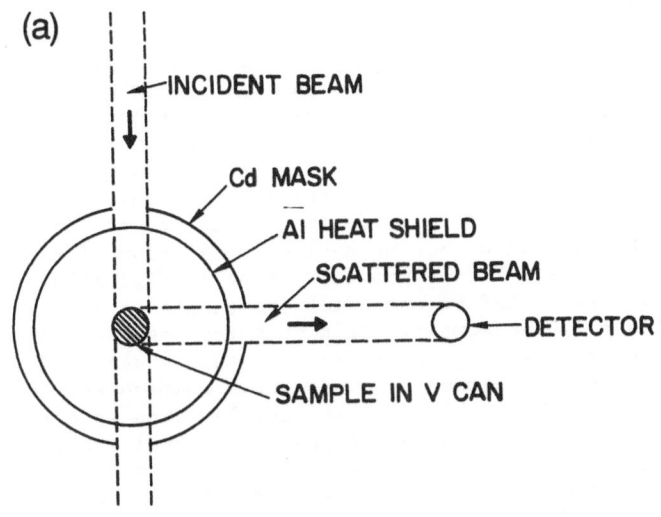

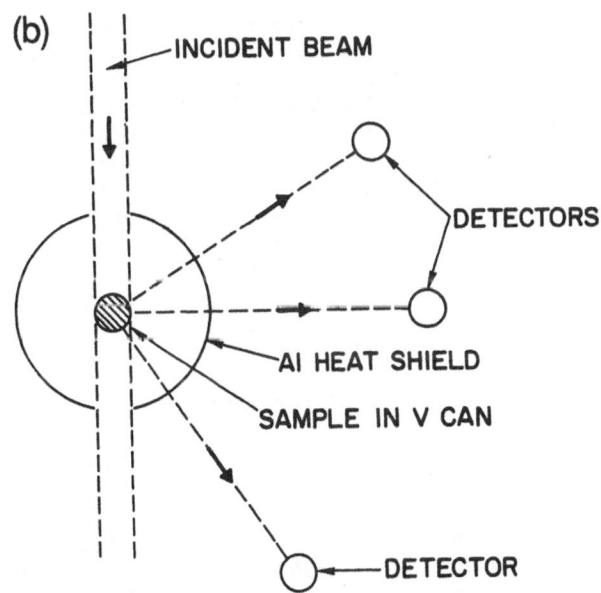

Figure 1. Generalized special environment geometries for time-of-flight diffraction experiments: (a) Collimation of both the incident and scattered beams at a single, fixed scattering angle. (b) The use of windows for the direct beam so that multiple scattering angles can be used.

increasingly more difficult to eliminate scattering from the region surrounding the sample because the probed volume is elongated.

The geometry shown in Fig. 1(b) can be used where small or large scattering angles are needed or where multiple scattering angles are used. Even though a wide range of d spacings can be accessed at a single scattering angle by the time-of-flight technique (e.g., enough data can typically be obtained in back scattering or at $2\theta = 90°$ to perform a Rietfeld structural refinement), some experiments demand that the data range be extended further. For example, small scattering angles must be used to measure magnetic reflections at large d spacing or to obtain the largest d spacing nuclear reflections for determination of unknown structures. In the geometry shown in Fig. 1(b), windows are cut in the heat shield before and after the sample so that the incident neutrons strike only the sample. This geometry has been most effectively used at low temperature or in furnaces designed to reach a few hundred degrees C. Since the window does compromise the effectiveness of the heat shield, care must be taken to ensure that the sample (in its container) does not support a thermal gradient and that the sample temperature is measured correctly. In the case of low temperature experiments, the window can often be covered with several layers of aluminized mylar (sometimes called superinsulation) to provide a degree of heat shielding while introducing a negligible amount of scattering material into the incident beam.

Some commonly used materials for sample containment and shielding are listed in Table I. Vanadium is the most commonly used material for sample cans because of its nearly zero coherent cross section. If the walls of the vanadium can are thin (typically less than 0.2 mm) the Bragg peaks from the vanadium will be below the background levels of most experiments. Vanadium has mechanical and thermal conductivity properties similar to stainless steel and can be extruded into tubes and other useful shapes if the starting material is of sufficient purity.

TABLE I. Commonly used materials for neutron diffraction sample containment and shielding

| Material | Cross Sections (barns) | | Use |
	Coherent	Absorption	
V	0.02	5.1	Sample cans
Al	1.50	0.2	Heat shields
Cd	3.3	2,520	Shielding
B	3.54	767	Shielding
Gd	34.5	48,890	Shielding

Reference: Methods in Experimental Physics, Vol. 23-Part A, Neutron Scattering, edited by K. Sköld and D. L. Price (Academic Press, New York 1986) Appendix.

Permanent vacuum-tight joints can be made by electron beam welding of vanadium to stainless steel or titanium, but the seams may not survive thermal cycling since the thermal expansions are not identical. The best vacuum-tight sample cans have beam-welded vanadium end caps. Vanadium has also been used as a heat shield in some furnace designs. However, if the volume of vanadium in the beam becomes excessive, it may become the dominant contribution to the background, due to the nonzero incoherent cross section (5.2 barns). Moreover, since the thermal conductivity of vanadium is rather poor (similar to stainless steel), a thin-walled heat shield can support a sizeable temperature gradient. In at least one furnace design, a thin-walled vanadium tube (surrounding the sample) is used as the heating element.

The most commonly used shielding material is cadmium. The neutron absorption cross section of cadmium is sufficiently large that sheets on the order of 0.5 mm thickness are essentially opaque to thermal neutrons. Cadmium is a soft and malleable metal, similar to lead, and can be easily formed into useful shapes for shielding. Other useful shielding materials (chosen because of their large absorption cross sections) include boron and gadolinium. Boron is typically used in the form of boron nitride (BN), which can be easily machined into useful shapes and will survive high-temperature environments, and boron carbide (B_4C) which is most commonly available as a coarse (and very abrasive) powder but can be mixed with epoxy resin and cast into useful shapes or contained in metal cans for bulk shielding. Gadolinium is often used in the form of fine gadolinium oxide powder mixed with thin epoxy resin and applied as a paint.

Figure 2 shows a possible design for a neutron diffraction sample can which can be sealed. The can is made from a length of thin-walled (approximately 0.1 mm) vanadium tubing with a solid cap on one end and a flange and removable cap on the other. The removable cap can be sealed with either an elastomer or indium O-ring. An indium seal will be effective at low temperature if the cap is designed in such a way that the indium is under compression at all temperatures. In the design shown in Fig. 2, elongation of the small-diameter screws which secure the cap to the flange around its perimeter (combined with a careful choice of thermal expansions) provides the stored mechanical energy which ensures that the indium remains under compression at low temperatures. The design of high-temperature sealed cans can also proceed along similar lines with an appropriate choice of the gasket material.

2. NEUTRON DIFFRACTION AT LOW TEMPERATURE

Perhaps the most common special sample environment used in neutron diffraction is low temperature. Achieving a low sample temperature can be desirable for a number of reasons. In some cases, the purpose is to reduce thermal vibration and thus increase the signal-to-noise ratio of the data by increasing the Bragg intensity while reducing thermal diffuse scattering. Low-temperature data collection can be particularly important in the case of "soft" materials such as organic systems or

164

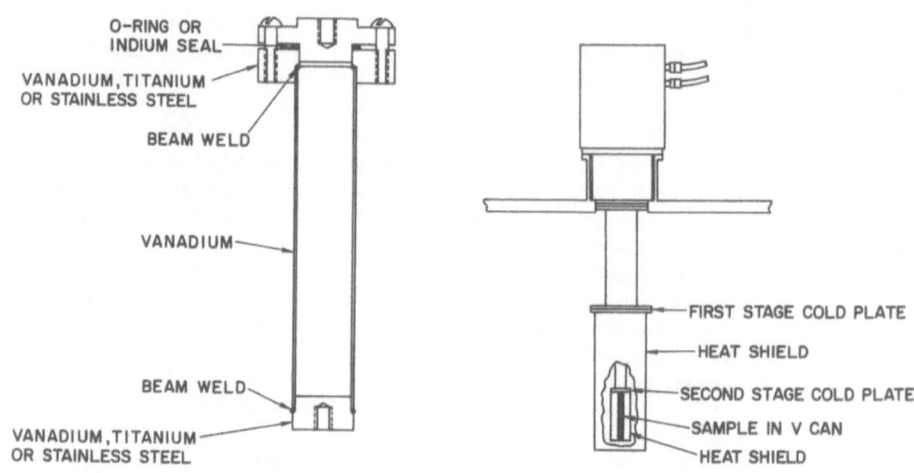

Figure 2. Typical design for a reusable, sealed neutron diffraction sample can.

Figure 3. Closed cycle helium refrigerator (Displex) configured for low-temperature neutron diffraction experiments.

systems containing mobile atoms (e.g., metal hydrides) although in some cases, ordering of the mobile species at low temperature may produce a phase transition to a structure different from the one which is of interest.

Another purpose for low temperature data collection is to enable the study of structural or magnetic phase transitions. In many cases, the relevant order parameter for the transition can be studied as a function of temperature to establish the critical exponent.

Since many other interesting physical phenomena occur at low temperature, another common purpose for low temperature data collection is to establish the structure at the same temperature at which other novel behavior is observed. For example, neutron diffraction at low temperature has been used to determine the structures of systems such as solid helium and hydrogen, as well as to establish important correlations between structure and superconductivity in both organic and inorganic compounds.

2.1. Experimental Methods

Virtually all of the techniques employed in low-temperature physics have been successfully applied to neutron diffraction experiments. The

TABLE II. Experimental methods for neutron diffraction at
           low temperatures.

Method	Min. Temp. (K)
Closed cycle refrigerator (Displex)	10
Helium cryostat	4.2
Pumped helium cryostat: ^{4}He	1.5
^{3}He	0.5
^{4}He-^{3}He dilution refrigerator	0.05
Nuclear demagnetization	0.001

cooling method employed depends on the minimum temperature desired. The
most commonly used techniques, along with the minimum temperatures that
can be achieved, are listed in Table II. The closed-cycle helium
refrigerator (Fig. 3) is a particularly convenient technique for
temperatures down to 10 K because there is no need for periodic transfer
of cryogenic liquids, as is the case for a cryostat. However, a well-
designed cryostat may provide shorter cooldown times in cases where the
cryostat has been designed for insertion of the sample directly into the
cold zone. In both cases, the temperature can be controlled from a
computer-based data acquisition system so that data at a series of
temperatures can be collected according to a predetermined program
without intervention from the experimenter.

If temperatures lower than 4 K are required, pumped helium
cryostats (pumped ^{4}He to 1.5 K and pumped ^{3}He to 0.5 K) can be used
although such systems require more attention while operating. Below
0.5 K, a ^{3}He-^{4}He dilution refrigerator can be employed.[4] The literature
contains a number of important elastic and inelastic neutron scattering
experiments in the range of 0.05-4 K involving the use of dilution
refrigerators. In particular, these techniques have allowed unique
studies of important quantum solid and liquids such as ^{3}He, ^{4}He,
and H.[5-7]

Proceeding below about 0.05 K requires the use of a nuclear
demagnetization stage in combination with a ^{3}He-^{4}He dilution
refrigerator. Neutron scattering experiments have been performed in at
least two laboratories with the use of such techniques. The lowest
temperature neutron diffraction experiment is a recent study of the
nuclear magnetic ordering in solid ^{3}He at 0.0005 K.[8] For this
experiment, careful attention to the choice of materials in the direct
beam was required in order to minimize the neutron and gamma-ray heating
effects. Care was also taken to minimize heating due to vibration of
the experimental apparatus. Figure 4 shows a nuclear demagnetization
system for similar experiments conducted at the Intense Pulsed Neutron
Source at Argonne National Laboratory.[9]

2.2. Examples

A number of straightforward examples from the published literature can
be used to illustrate some of the basic concepts of pulsed-source time-
of-flight neutron diffraction at low temperature.

2.2.1. Mechanism of thermal expansion in α-quartz SiO_2.

On the high
resolution powder diffractometer at the ZING prototype pulsed neutron
source, a simple study of α-quartz SiO_2 versus temperature was performed
in order to demonstrate experimentally that the thermal expansion at low
temperature, which is markedly nonlinear below 100 K, could be explained
in terms of a rigid rotation of corner-linked SiO_4 tetrahedra.[10] The
experimental arrangement used for this experiment (Fig. 5) illustrates
some basic concepts which apply to all neutron scattering experiments at
low temperature. The powder SiO_2 sample was contained in a helium

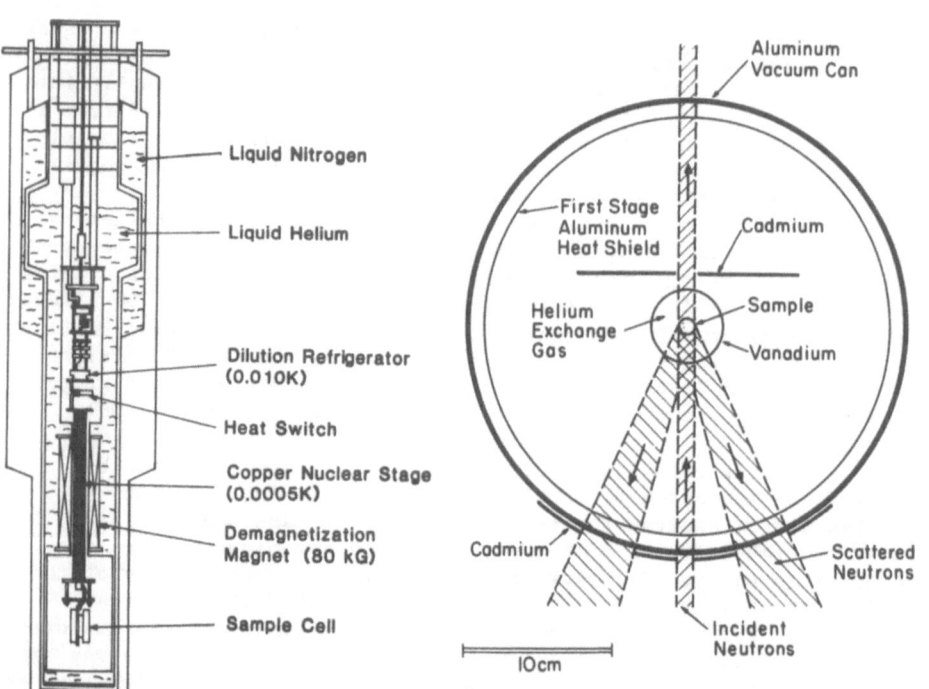

Figure 4. Nuclear demagnetization
system used for ultra-low-
temperature neutron diffraction
measurements at the Intense
Pulsed Neutron Neutron Source.

Figure 5. Schematic drawing of the
experimental arrangement of heat
shields and neutron shielding for
low-temperature neutron diffraction
measurements on a backscattering
(2θ = 160°) instrument.
(Reproduced from Ref. 10.)

exchange gas to ensure rapid and uniform cooling of the powder
particles. Without the use of an exchange gas, it is impossible to
ensure that the powder actually cools at the same rate as the sample can
and the cold plate to which it is mounted. This is especially true for
powders which have poor thermal conductivity at low temperature.
Additionally, without the use of an exchange gas, it may not be valid to
assume that the thermometer, which is typically mounted to the sample
can or cold plate, accurately measures the sample temperature. (It
should be remembered that a thin-walled vanadium sample can is capable
of supporting a significant temperature gradient.) Uniform cooling of a
powder sample can be achieved by containing the sample in a properly
sealed (i.e., with an indium seal) vanadium can along with one
atmosphere (at room temperature) of helium exchange gas.

2.2.2. <u>Low temperature magnetic structures in FeGe</u>. The study of
magnetic structures requires the use of small scattering angles in order
to ensure that the large d spacing Bragg peaks which result from the
magnetic ordering are seen. The low temperature magnetic structures of
FeGe were determined from analysis of data taken by the time-of-flight
technique on the Special Environment Powder Diffractometer (SEPD) at
IPNS at a scattering angle of $2\theta = 57°$.[11] At this scattering angle the
SEPD provides data to about 8 Å d spacing at a resolution of $\Delta d/d \approx 1\%$.
There are two magnetic ordering transitions of the monoclinic ($z = 6$)
structure. At 340 K, some of the Fe atoms order antiferromagnetically
to give a cell doubling along the c axis; at 120 K an additional
incommensurate modulation of spins along the b direction occurs. Raw
time-of-flight neutron data in each of the three phases, with the new
magnetic reflections identified, are shown in Fig. 6. For this
experiment, even though the decrease in flux at long wavelengths
resulted in rather small raw intensities at large d spacing, the 1%
resolution available at $2\theta = 57°$ was especially useful in solving the
incommensurate structure.

2.2.3. <u>Structural instability and superconductivity in the Chevrel
phases (MMo_6S_8)</u>. Recent low-temperature neutron diffraction
measurements on Chevrel phase compounds with composition MMo_6S_8 (M = Sn,
Pb, Ba, Yb) have shown a direct relationship between high superconducting transition temperatures (T_c) and structural instability in
these compounds.[12] Original structural measurements had shown that the
Chevrel phases which exhibited superconductivity (e.g., Sn-, Pb-, and
$YbMo_6S_8$) were in a rhombohedral, $R\bar{3}$, structure. It was later discovered
that similar compounds which showed no superconductivity (e.g., $BaMo_6S_8$
and $EuMo_6S_8$) transformed to a triclinic, $P\bar{1}$, structure at low temperature. The triclinic distortion destroys superconductivity by splitting
a half-filled band and opening a gap at the Fermi energy. The highest
T_c Chevrel phase compounds, $SnMo_6S_8$ and $PbMo_6S_8$, ($T_c \approx 14$ K) had been
repeatedly reported to exhibit evidence for structural instability at
low temperature based on a wide variety of experimental techniques, but
the nature of the low-temperature anomaly was not clear in any of the
experiments. (See the references cited in Ref. No. 12). Neutron powder
diffraction measurements were recently able to identify the subtle low-

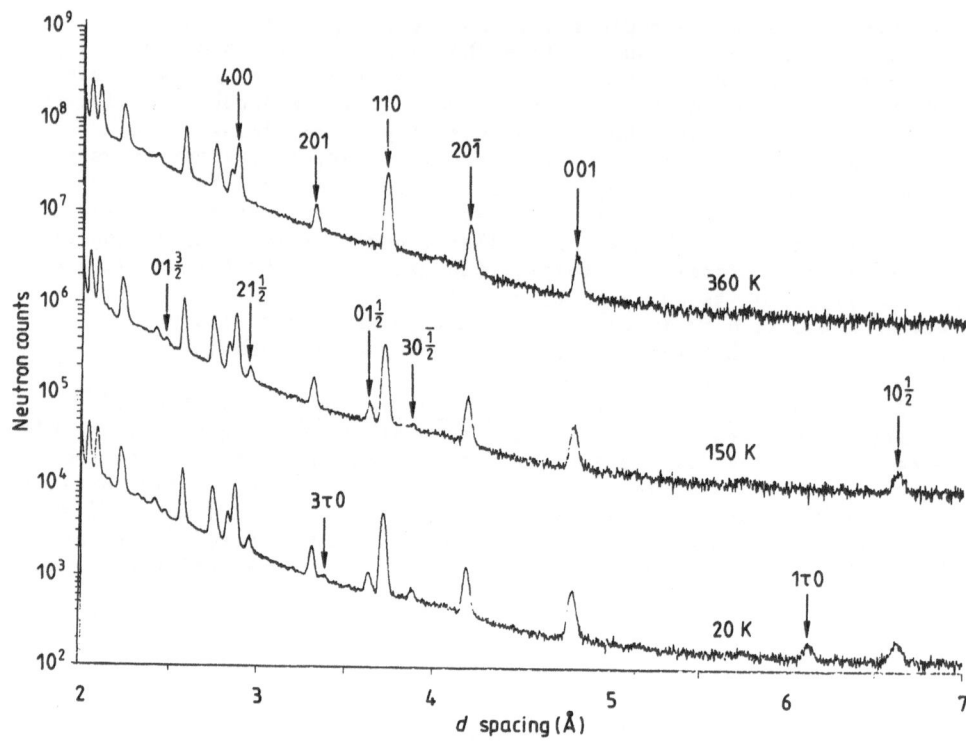

Figure 6. Raw time-of-flight diffraction data for FeGe at 360 K, 150 K, and 20 K taken with the $2\theta = 57°$ detectors of the Special Environment Powder Diffractometer at IPNS. At each progressively lower temperature, only the new (magnetic) reflections are labelled. The data are plotted on a logarithmic scale (x 100 for 150 K and x 10,000 for 350 K) to make the large d spacings more visible since their raw intensities are reduced due to decreasing neutron flux at long wavelengths. (Reproduced from Ref. 11.)

temperature transition in $SnMo_6S_8$ and $PbMo_6S_8$ as the onset of a two-phase ($\overline{R}3 + \overline{P}1$) region (presumably resulting from grain interaction stresses) which is a precursor to the transition to the $P\overline{1}$ phase. Thus, the structural instability which results in the transition to a $P\overline{1}$ phase in $EuMo_6S_8$ and $BaMo_6S_8$ also exists in $PbMo_6S_8$ and $SnMo_6S_8$, although the actual phase transition, which would result in a loss of super-conductivity, is not reached in the latter compounds. The existence of the two-phase region, however, establishes that, at low temperature, $SnMo_6S_8$ and $PbMo_6S_8$ are adjacent to the phase line, leading to the conclusion that high-T_c superconductivity in these systems is related to the instability which gives rise to the symmetry-lowering transition.

Two-phase Rietveld refinements, such as the one shown in Fig. 7 for $PbMo_6S_8$ at 10 K, have established the structural parameters of the coexisting $R\bar{3}$ and $P\bar{1}$ phases and the phase fractions as a function of temperature.

2.2.4. Low-temperature structural ordering and superconductivity in the organic compounds $(TMTSF)_2X$ and $(BEDT-TTF)_2X$. Structural ordering has been shown to be intimately related to superconductivity in organic compounds. Low-temperature single crystal neutron diffraction techniques have played an important role in establishing the ordered structures. The structural ordering usually involves the X anion. In the case of $(TMTSF)_2ClO_4$, which exhibits a T_c of 1 K, the ordering is commensurate and results in a cell doubling along the b axis.[13] However, for $(BEDT-TTF)_2I_3$, the ordering is incommensurate.[14] The

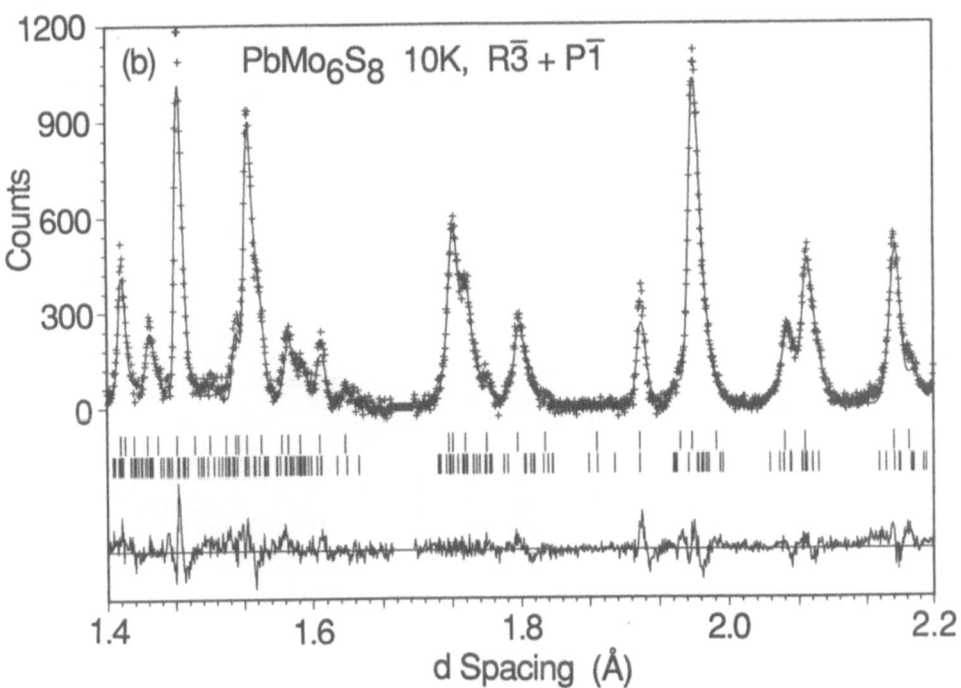

Figure 7. Portion of the Rietveld structural refinement profile based on a two-phase (rhombohedral $R\bar{3}$ plus triclinic $P\bar{1}$) model for $PbMo_6S_8$ at 10 K. Plus signs (+) are the raw data. The continuous line is the calculated profile. At the bottom of the curves, vertical tick marks indicate positions of allowed $R\bar{3}$ (upper tick marks) and $P\bar{1}$ (lower tick marks) reflections. (Reproduced from Ref. 12.)

single crystal time-of-flight Laue technique, which uses a two-dimensional position-sensitive detector, is especially well suited for such studies because new reflections are always seen regardless of their position in reciprocal space.[15] Figure 8 shows a cut of reciprocal space containing the new incommensurate satellite peaks which are associated with the ordering of I_3 anions in $(BEDT-TTF)_2I_3$ at low temperatures.

3. NEUTRON DIFFRACTION AT HIGH TEMPERATURE

High temperature neutron diffraction capabilities extend to temperatures above 2500°C.[16] Diffraction experiments are conducted at high temperature for number of reasons. Structural phase transitions occur in a number of systems. In some cases more than one transition occurs as a function of temperature. If the diffractometer count rate is sufficiently high, a large number of data separated by small increments in temperature can be collected in order to study the behavior in the region of the transition. In some cases, intermediate phases, which exist only over a narrow temperature range have been identified. Several examples of these types of measurements are given elsewhere in this book. (See the paper by J. Pannetier.)

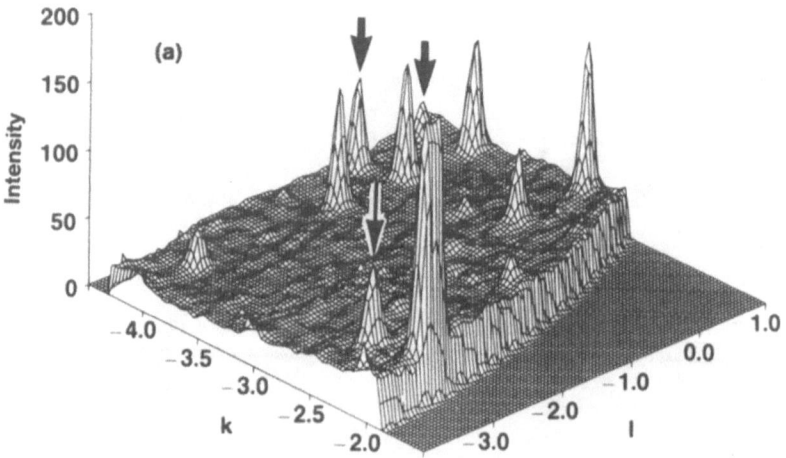

Figure 8. Plot of the time-of-flight neutron diffraction intensity in the 4.92, k, 1 reciprocal lattice plane of $\beta-(BEDT-TTF)_2I_3$ at 20 K. Satellite peaks at $(5, \bar{2}, \bar{3})$-q, $(5, \bar{4}, \bar{1})$-q, and $(5, \bar{4}, 0)$-q (marked with arrows) are clearly visible. (Figure adapted from Ref. 36.)

Neutron diffraction techniques have also been used to study chemical reactions as a function of time or temperature. Such studies have included hydration or dehydration reactions, intercalation, or reactions involving two separate chemical species. Again, for this type of work an instrument providing rapid data collection, with the entire pattern being collected simultaneously is required.

For several systems, diffraction data have been collected at high temperature in order to understand the structural features which are associated with other interesting high-temperature properties. For example, in the case of ionic conductors, it has been possible in some cases to identify structural changes which give rise to enhanced ionic conductivity at high temperature. In this context, it has also been possible to study the temperature-dependent creation of complex defects in systems such as nonstoichiometric metal oxides.

In some cases, high temperature diffraction measurements have been extended beyond the melting point into the liquid phase. Such measurements have provided particularly interesting results on the existence of specific local atomic configurations in liquid metals and melted glasses.

3.1. Experimental Methods

A wide range of furnace designs and heating techniques have been employed for neutron scattering experiments at high temperature at both steady state and pulsed neutron sources. In many cases, details of the furnace designs have been published. Although specific details of furnace geometry are too numerous to cover, all of the designs can be grouped into two general categories, and some general design criteria, particularly those which apply to time-of-flight neutron diffraction will be discussed.

Depending on the requirements of the sample, furnaces for neutron diffraction are either designed to operate with the sample in a vacuum or in a controlled atmosphere. In almost all cases, the heating element is in vacuum. Since the neutrons can easily penetrate the heating element and several concentric heat shields, beam attenuation is usually not a serious problem. However, particular care must be taken to avoid unwanted scattering from the furnace components surrounding the sample. This is usually accomplished through careful collimation or by leaving windows in the heating element and heat shields as shown in Fig. 1. Moving the heater and heat shields away from the sample is often not an acceptable solution since power requirements, and the overall heat load on the surroundings, go up markedly as the dimensions of the hot zone are increased. For this reason, time-of-flight techniques employing a single, fixed scattering angle can offer important advantages for high-temperature diffraction.

Several furnace designs are used routinely on the time-of-flight diffractometers at IPNS. For temperatures up to about 400°C, a simple aluminum can, with one heat shield, surrounding the sample (which is usually enclosed in a vanadium can) with windows for the straight-through beam (as in Fig. 1b) can provide perfectly clean data at all scattering angles with no beam attenuation.

172

For higher temperatures, a furnace without windows must be used in
order to overcome undesirable losses and large thermal gradients in the
hot zone. One successful design used at IPNS (designed and constructed
by A. Howe and N. Wood, University of Leicester, U.K.) consists of a
tubular heater and concentric heat shields made of thin vanadium. This
furnace will reach temperatures of about 900°C and allows data
collection at any scattering angle. Bragg scattering from the furnace
is negligible, but the amount of vanadium in the beam is sufficient to
contribute an amount of incoherent scattering which is larger than the
normal sample background for many experiments.

Figure 9 shows a furnace design which maintains the sample in a
controlled atmosphere. This furnace has been particularly useful for
the study of high temperature oxides where the chemical stoichiometry is
a function of the oxygen partial
pressure of the atmosphere
surrounding the sample. The
sample is supported inside a
closed-end aluminum oxide tube
which contains a steady flow of
the desired atmosphere. The
heater is a spirally wound ribbon
wound directly on the outer
surface of the aluminum oxide
tube in the original design, or
onto a second closely-fitting
concentric aluminum oxide tube in
later designs. Since the aluminum
oxide tube is close to the
sample, scattering is limited to
$2\theta = 90°$ where collimation can be
used to eliminate unwanted
scattering from the furnace. The
collimator is machined from boron
nitride with windows for the
incident beam and the scattered
beam at ±90°. Tantalum heat
shields and zircar insulation are
used outside the heat shield to

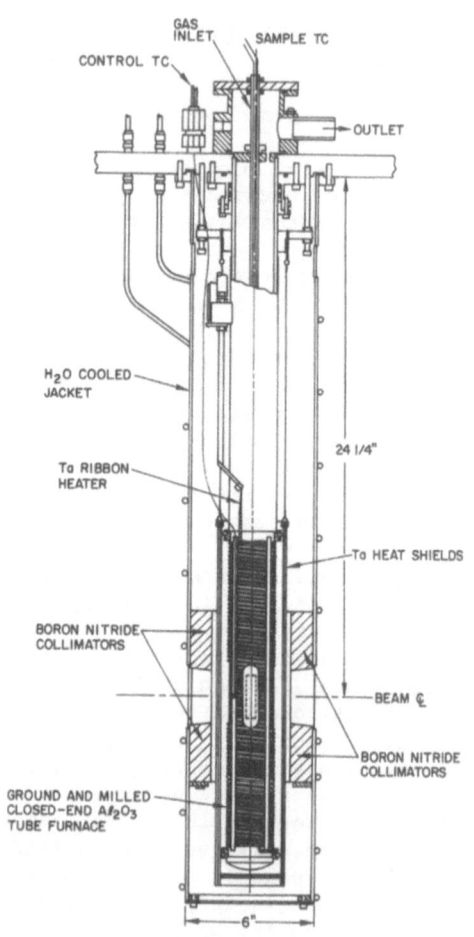

Figure 9. Schematic of the
controlled-atmosphere, restricted
angle ($2\theta = 90°$) furnace used on
the time-of-flight powder
diffractometers at the Intense
Pulsed Neutron Source.

minimize heat losses and to achieve a more uniform hot zone. The entire furnace is mounted inside a water-cooled jacket to control heat losses to surrounding instrument surfaces. This furnace design has been used successfully to temperatures of 1400°C, with approximately 2100 watts of input power.

3.2. Examples

The literature contains a number of high-temperature neutron diffraction experiments performed at both steady state and pulsed neutron sources. The highest temperature furnace designs presently operate at steady state sources where experiments have been performed at temperatures above 2500°C.[17] The multidetector instrument D2B at ILL has also been used for a large number of high data rate experiments involving the study of phase transitions and reaction kinetics at lower temperature. These experiments are described elsewhere in these proceedings. (See the paper by J. Pannetier.) Although such experiments are not currently being done at pulsed sources, the time-of-flight technique also provides the fast, simultaneous data collection required for real-time experiments and will undoubtedly be used for such experiments in the future as appropriate experimental apparatus continues to be developed. In this paper two recent examples of high-temperature diffraction experiments performed at IPNS will be described.

3.2.1. Defects in CeO_2 versus temperature and composition. The controlled atmosphere furnace shown in Fig. 9 has been used to perform a number of experiments on metal oxides as a function of temperature and stoichiometry. For compounds which exhibit nonstoichiometric behavior, in situ experiments are required since quenching techniques are not always successful in freezing in the high-temperature state. In situ studies allow investigation of the equilibrium thermodynamic state in order to determine the structural and bonding properties and the atomic defects responsible for nonstoichiometric behavior.

Recent work on CeO_{2-x} provides an example of the kind of information that can be obtained from high-temperature neutron diffraction in a controlled atmosphere.[18] In the case of CeO_{2-x} the oxygen stoichiometry can be varied over a range which includes the stoichiometric compound ($0 < x < 0.2$) by controlling the oxygen partial pressure surrounding the sample. Previous studies of nonstoichiometric CeO_{2-x} have shown that, when large vacancy defect concentrations are present, the oxygen ions displace from their ideal fluorite-structure sites towards neighboring interstitial cavities in the lattice. These displacements were attributed to the onset of complex defect interactions. A more recent study has concentrated on the stoichiometric compound CeO_2 produced by maintaining a pure oxygen atmosphere on the sample during neutron diffraction measurements to 1200°C. The powder diffraction data were analyzed in a model which includes third and fourth order terms in the Debye-Waller factor. Such an analysis is made possible by the existence of data to very short d spacings in the pulsed source experiment. These higher order terms in the Debye-Waller factors represent anharmonic displacements of the atoms. A plot of the component of the symmetric third rank tensor, C(ijk), that corresponds to cubic displacements of the oxygen atoms from their ideal fluorite positions versus temperature is shown in Fig. 10. The significant increase in this tensor component at high temperature corresponds to oxygen ion displacements toward the neighboring interstitial cavities, i.e., the same displacements that were previously seen as large static displacements in heavily defected samples. This suggests that the static displacements of atoms around oxygen vacancies are mediated by lattice phonons.

3.2.2. Oxygen atom positions in yttria-stabilized zirconia at high temperature and in an applied electric field.

Nonstoichiometric oxides have also been investigated at high temperature by single crystal TOF neutron diffraction techniques. Figure 11 shows a simple furnace which was used to heat a single crystal of yttria-stabilized cubic zirconia.[19] Data were collected at room temperatures and at 1040 K, with and without an electric field, on $Zr(y)O_{1.862}$ in an attempt to study the paths for

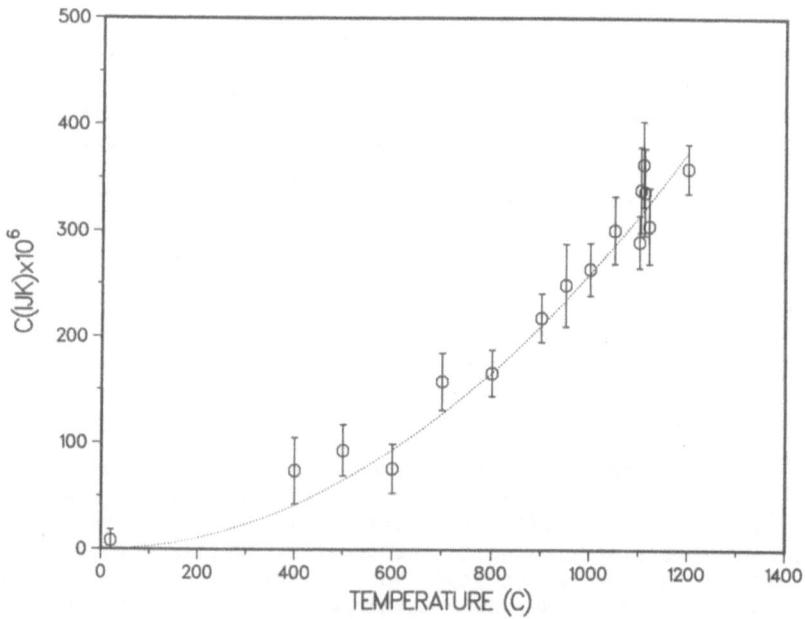

Figure 10. Component of the symmetric third rank tensor, C(ijk), that corresponds to cubic displacements of the oxygen atom from their ideal fluorite positions in CeO_2 plotted as a function of temperature.

oxygen ionic motion leading to high ionic conductivity in this compound. Data for the sample in an electric field showed clear evidence for scattering density displaced from the oxygen sites in the <100> direction, thus suggesting that ionic current occurs with mobile O^{2-} ions moving along <100> directions through vacant oxygen sites.

4. NEUTRON DIFFRACTION AT HIGH PRESSURE

Neutron diffraction, and in particular the time-of-flight technique, has played a unique role in the area of high pressure.[20,21] Since pressure cells necessarily have rather thick walls, a diffraction probe with high penetrating powder must be used. Additionally, in a high pressure

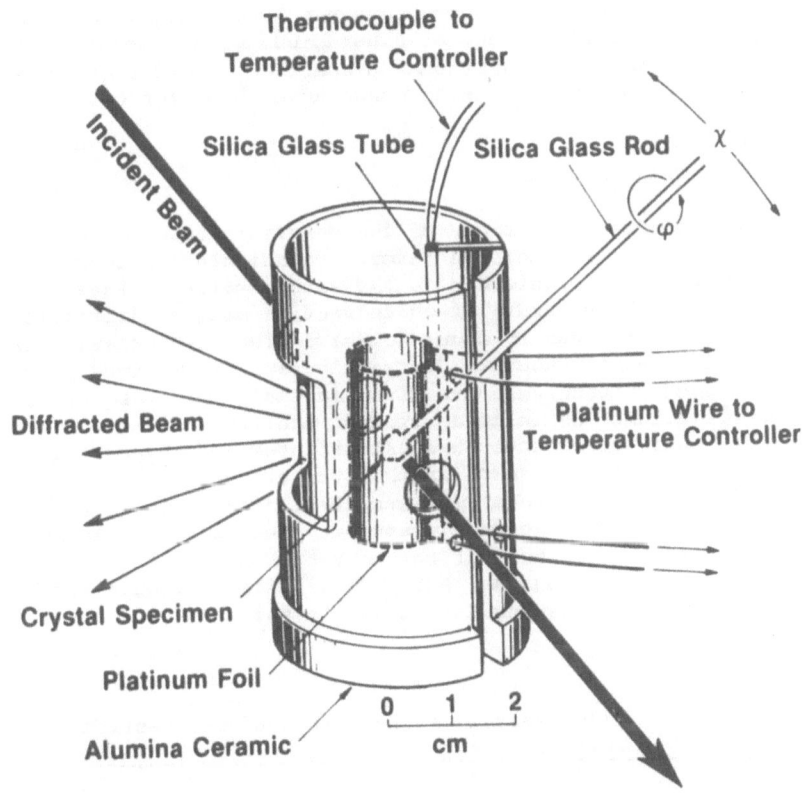

Figure 11. Furnace used for a time-of-flight single crystal neutron diffraction experiment on yttria-stabilized zirconia at high temperature and in an applied electric field. (Reproduced from Ref. 19.)

environment, the sample is supported by the walls of the pressure vessel. Thus, the time-of-flight, fixed-angle diffraction technique offers clear advantages for arranging collimation which will avoid scattering from the pressure vessel while maximizing scattering from the sample.

A number of interesting structural phenomena occur at high pressure. For most compounds the structural changes which occur at modest pressures are several times larger than those which can be produced by varying the temperature over the entire stability range of the compound. Thus, when the structural phase diagram for a compound is extended into the high pressure region, the number of different phases that are accessible is usually many times larger than the number which occur by varying temperature alone. Many of the high pressure studies in the literature report the determination of high pressure structures and the characterization of phase transitions as a function of pressure. Additionally, high pressure has become more common as a research tool for a wide range of other kinds of measurements, e.g., transport, elastic, and vibrational properties; and structural studies have been important in explaining other novel behavior observed at high pressure.

4.1. Experimental Methods

Because of strength limitations of the materials used to construct pressure cells, there exists an inverse relationship between the maximum pressure which can be achieved and the sample volume. These criteria presently limit the pressure at which routine neutron diffraction measurements can be done to about 40 kbar. The high pressure techniques most commonly used for neutron diffraction and the maximum pressure that can be achieved in each case are given in Table III. Gas pressure cells offer sample volumes as large as typical neutron diffraction samples (5 cm^3) and achieve perfectly hydrostatic pressure, even at low temperature, if helium is used as the pressure fluid.[22] However, for safety reasons and because of the practical difficulty in achieving reliable helium seals at higher pressures, the use of helium gas cells for neutron diffraction has not been extended much beyond 8 kbar. Similar pressure cell designs, but with a suitable liquid for the pressure fluid, have been used to about 15 kbar.

TABLE III. High pressure techniques used for neutron scattering experiments.

Method	Max. Pressure
Gas cells	8 kbar
Liquid cells	15 kbar
Supported aluminum oxide cells	40 kbar

Beyond that pressure, the supported aluminum oxide cell is the design most often used for neutron diffraction measurements.[20,23] Such a pressure cell, which has been used for many years to perform neutron diffraction measurements by the time-of-flight technique at both steady state and pulsed neutron sources, is shown in Fig. 12. This cell was originally developed in the late 1960's for high pressure structural measurements at the MTR Reactor in Idaho, where it was used with a chopper-based time-of-flight diffractometer. Even though the time-of-flight technique using a chopper on a reactor source is generally inferior to steady state techniques, the results achieved at high pressure were unique, owing to the ability to obtain data free from unwanted scattering from the pressure cell. The development of pulsed neutron sources has further enhanced this capability, allowing high pressure experiments with higher resolution and count rates.

The supported aluminum oxide design takes advantage of the high compressive strength of sintered aluminum oxide (approaching 500,000 psi for the best materials). A cylindrical sample is contained along with a

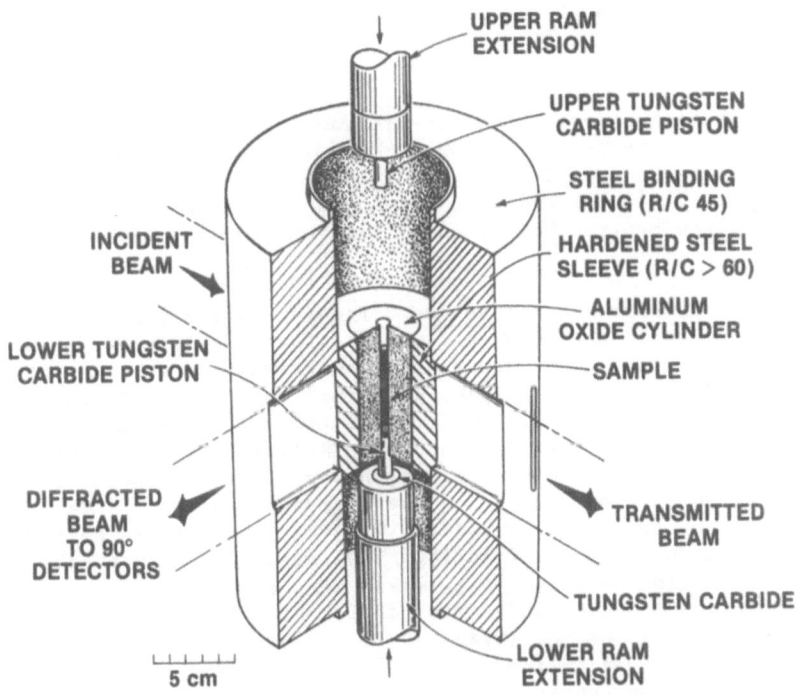

Figure 12. Supported aluminum oxide pressure cell used for time-of-flight neutron diffraction measurements at pressures to 40 kbar.

suitable hydrostatic fluid in a sealed capsule in the bore of a hollow aluminum oxide cylinder. The aluminum oxide cylinder is supported radially by one or more steel binding rings. Windows are cut in the steel, but not in the aluminum oxide, for the incident and scattered neutrons. In the case of the time-of-flight cell shown in Fig. 12, these windows also provide the desired collimation to mask scattering from the aluminum oxide. The neutrons must, of course, pass through the aluminum oxide. Typical attenuation factors are on the order of 50% for a total path of 2 cm of aluminum oxide. The cell is assembled in such a way that, with the sample at zero pressure, the aluminum oxide is at its compressive strength limit. In the design shown in Fig. 12, this is achieved by first loading the aluminum oxide cylinder into the hardened steel sleeve and then driving the sleeve, which is tapered, into the tapered bore of the steel binding ring. After assembling the cell in this way, as the sample pressure is increased, due to force applied to the ends of the sample capsule through tungsten carbide pistons, the compression in the aluminum oxide decreases and eventually passes through zero. The aluminum oxide then fails, by fracturing, when its tensile limit is reached. For typical designs this corresponds to a sample pressure of 40-50 kbar.

4.2. Examples

Even though high pressure neutron diffraction experiments are not as common as those performed at low or high temperature, the literature contains numerous examples and, additionally, several review articles on the subject.[20,21] The examples cited here are from recent work at the Intense Pulsed Neutron Source and illustrate the current state of the art.

4.2.1. High pressure phases of ice.

Ice is presently known to have at least eleven solid phases as a function of temperature and pressure. The structures of many of these phases were studied several years ago by neutron diffraction from samples which had been quenched from high pressure.[24-26] Neutron diffraction, of course, allows the determination of the hydrogen atom positions so that the details of bonding can be studied. Not all of the ice phases, however, can be successfully quenched. During the last three years, there have been three papers reporting the structures of D_2O ice VI, VII, and VIII as determined by in situ neutron powder diffraction. The structure of ice VIII was refined from time-of-flight data taken at a reactor using the pressure cell shown in Fig. 12[27] while that of ice VII was determined using the same pressure cell at the Intense Pulsed Neutron Source.[28] Additionally, the structures of ices VI, VII, and VIII were studied in situ at the Institute Laue Langevin using fixed wavelength techniques.[29] These three recent papers provide a clear example of the relative merits of the three techniques. The two time-of-flight experiments both exhibit data free from pressure cell scattering, with the ice VII data having much higher resolution owing to the superior characteristics of the pulsed neutron source diffractometer. The data taken on the fixed wavelength instrument, by contrast, contained a large amount of Bragg scattering from the aluminum oxide (Al_2O_3) pressure cell. This required an accurate subtraction of the Al_2O_3 pattern from the data in order to proceed with the analysis of the ice structures. A comparison of the data for D_2O ice VII taken by the two methods is shown in Fig. 13. The resulting answers from the data analysis were essentially identical, attesting to the skill of the ILL scientists in achieving an accurate pressure cell subtraction. However, the advantages of the time-of-flight technique are clearly illustrated and are essential for the refinement of more complex structures or for the solution of unknown structures.

4.2.2. Structure of high pressure KNO_3-IV. One unknown high pressure structure recently solved from neutron powder diffraction data is that of KNO_3-IV.[30] This phase of KNO_3 exists above about 3 kbar at room temperature. Data were collected at the Intense Pulsed Neutron Source with the sample in a helium gas pressure cell at 3.6 kbar. The unit cell was determined by autoindexing from a list of the d spacings of the observed Bragg peaks. The systematic absences required orthorhombic space group Pnma or its noncentrosymmetric equivalent. Atom locations were determined by attempting refinements of similar structures having the same space group chosen from the literature, until the correct model was found, and convergence of the Rietveld refinement was achieved. The high pressure structure is related to the atmospheric pressure structure in the novel way illustrated in Fig. 14. Both structures have the same orthorhombic Pnma symmetry. In the high pressure phase, the NO_3 molecules are displaced along the direction of the c-axis by an amount which results in a 26% reduction of the c-axis. This explains the large volume reduction associated with the KNO_3-II to KNO_3-IV transition.

4.2.3. "Compressibility collapse" transition in ReO_3. Recent work on ReO_3 provides an instructive example of the use of both elastic and inelastic neutron scattering techniques to understand the pressure-induced phase transitions in ReO_3. For this material, a continuous transition involving an order-of-magnitude increase in the compressibility was observed at 5 kbar.[31,32] At atmospheric pressure ReO_3 has an ideal ABO_3-type cubic perovskite structure with the A site being empty. It is well known that this structure can exhibit a number of structural instabilities leading to lower-symmetry or larger-cell structures. An obvious, but unproven, explanation for the large increase in compressibility was that the structure had undergone a transition to a symmetry in which coordinated rotations of the corner-linked ReO_6 octahedra could contribute to the cell compression.

This concept was first confirmed by a single crystal time-of-flight neutron diffraction experiment in which the ReO_3 2a x 2a x 2a supercell with ReO_6 octahedra in rotated positions was observed.[33] Subsequently, inelastic neutron diffraction, performed at a reactor source, was used to observe the pressure-induced softening of the M-point phonon associated with the rotation of the ReO_6 octahedra.[34] In the final experiment, ReO_3 was studied by neutron powder diffraction over a range of pressures up to 27.4 kbar in order to investigate the order parameter and critical exponent of the transition.[35] The powder data revealed

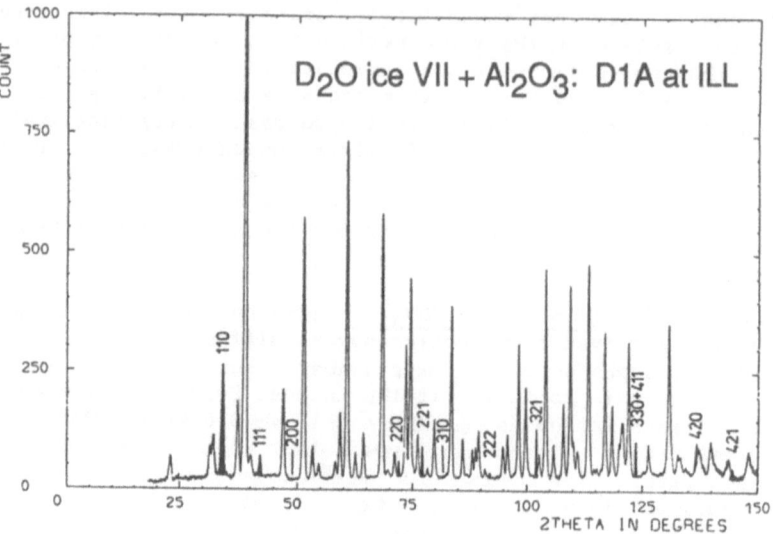

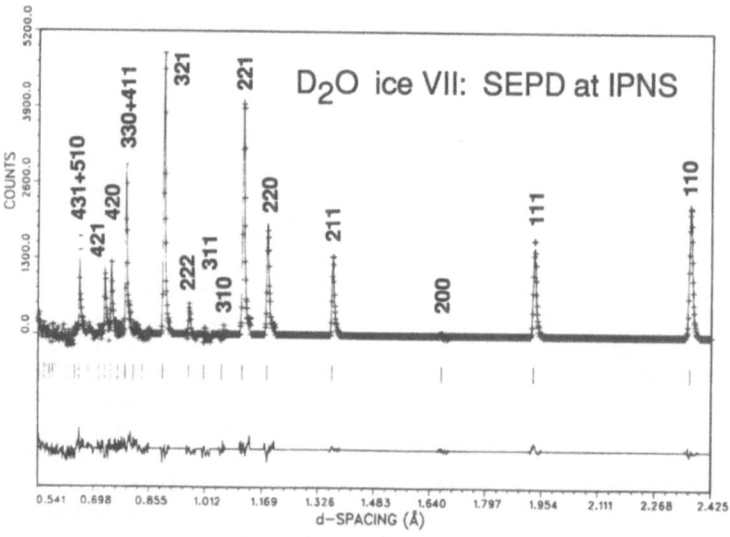

Figure 13. In situ neutron powder diffraction data for D_2O ice VII at approximately 25 kbar collected on the D1A fixed-wavelength diffractometer at ILL (top) and the SEPD time-of-flight diffractometer at IPNS (bottom). The D1A data show a large contribution of Al_2O_3 Bragg peaks from the pressure cell (the shaded peaks which are labelled with Bragg indices are the D_2O peaks in the D1A data). (Reproduced from Ref. 28 and 29.)

KNO₃-II (0 kbar, 295 K)

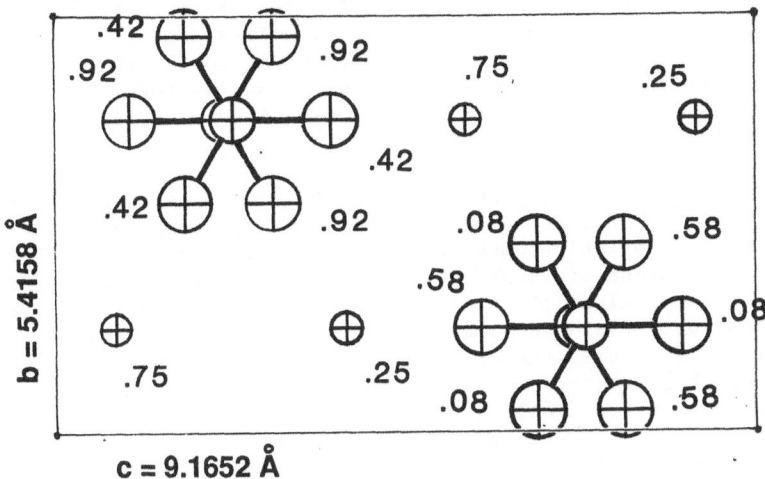

KNO₃-IV (3.6 kbar, 295 K)

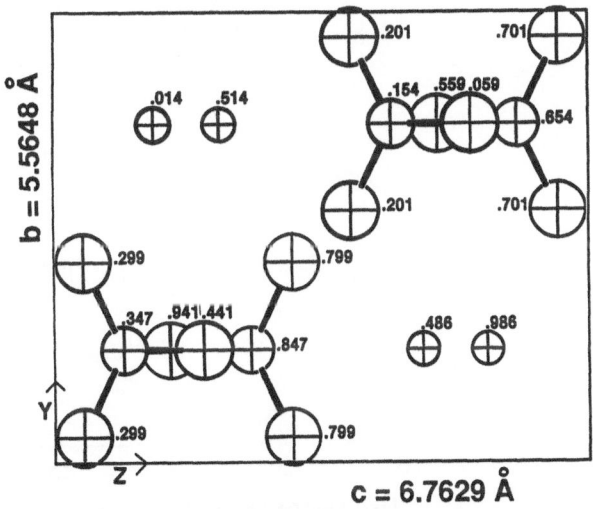

Figure 14. Comparison of the structures of KNO₃-II at atmospheric pressure and KNO₃-IV at 3.6 kbar. The largest symbols are O atoms and the smallest symbols are K atoms. Numbers near the atoms are fractional positions along the a axis. Both structures have orthorhombic Pnma space group symmetry (different origins are used in the two figures). Note how the NO₃⁻ ions slide past one another to substantially shorten the c axis in the high pressure KNO₃-IV structure.

182

that the system actually undergoes a series of transitions resulting from progressive softening of M-point phonons. The ReO_6 octahedra initially rotate round [001] axes, but as the pressure is further increased the axis of rotation shifts (in a first order transition that preserves cell volume) to [111]. The rotation angle, which is an order parameter for the transition, plotted as a function of pressure and fit to a power law yields the unusual critical exponent 0.322 ± 0.005, as shown in Fig. 15.

4.2.4. Structural ordering and pressure-induced superconductivity in (BEDT-TTF)$_2$I$_3$. The importance of structural ordering for super-conductivity in organic compounds was discussed previously in section 2.2.4. In the case of (BEDT-TTF)$_2$I$_3$, which exhibits a superconducting transition temperature of about 1.5 K, the ordering is incommensurate. More recent work has shown that the T_c of this compound suddenly jumps to about 8 K at a pressure of 0.5 kbar. The cause of this change in superconducting properties has been investigated by single crystal time-of-flight neutron diffraction at the Intense Pulsed Neutron Source.[36] At high pressure, the incommensurate satellite reflections shown in Fig. 8 are absent, indicating an even higher degree of ordering associated with the 8 K superconducting state.

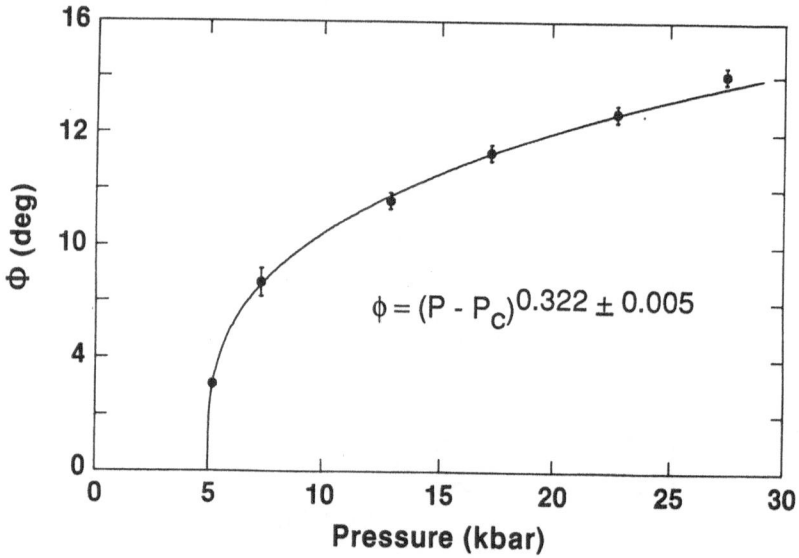

Figure 15. The rotation angle, ϕ, of ReO_6 octahedra in ReO_3 as measured by neutron powder diffraction plotted as a function of pressure. A power law fit to the data yields a critical exponent of 0.322 ± 0.005. (Figure adapted from Ref. 35.)

5. SUMMARY

In this paper the general principles of neutron diffraction experiments in special sample environments have been discussed and illustrated with examples from the recent published literature. The discussion has focussed on low-temperature, high-temperature, and high-pressure techniques since these are routinely used at the major neutron scattering facilities. Space has not permitted inclusion of a discussion of more exotic techniques which could include sample environments such as steady-state or pulsed magnetic or electric fields, uniaxial compression or tensile stress, or in situ reaction vessels, but all of these techniques have been successfully used for neutron scattering experiments. Undoubtedly, the future will bring further developments, coupled with increased neutron flux and resolution at the advanced pulsed neutron sources that are now being proposed.

REFERENCES

1. The report of a recent meeting about special environment techniques for neutron scattering can be found in 'Proceedings of the Workshop on Sample Environments in Neutron and X-ray Experiments, Institut Laue-Langevin, Grenoble, France, Feb. 13-15, 1984,' Revue Phys. Appl. **19**, No. 9 (1984).
2. J. M. Carpenter, G. H. Lander, and C. G. Windsor, Rev. Sci. Instrum. **55**, 1019 (1984).
3. J. Faber, Jr., Revue Phys. Appl. **19**, 643 (1984).
4. P. A. Hilton and N. W. Kerley, Revue Phys. Appl. **19**, 775 (1984).
5. P. E. Sokol, R. O. Simmons, J. D. Jorgensen, and J. E. Jørgensen, Phys. Rev. B **31**, 620 (1985).
6. K. Sköld, C. A. Pelizzari, R. Kleb, and G. E. Ostrowski, Phys. Rev. Lett. **37**, 842 (1976).
7. W. G. Stirling, R. Scherm, P. A. Hilton, and R. A. Cowley, J. Phys. C **9**, 1643 (1976). P. A. Hilton, R. A. Cowley, R. Scherm, and W. G. Stirling, J. Phy. C **13**, L295 (1980).
8. A. Benoit, J. Bossy, J. Flouquet, and J. Schweizer, J. Phys. Lett. (Orsay Fr.) **46**, 923 (1985).
9. P. Roach, et al. (unpublished).
10. G. A. Lager, J. D. Jorgensen, and R. J. Rotella, J. Appl. Phys. **53**, 6751 (1982).
11. G. P. Felcher, J. D. Jorgensen, and R. Wapping, J. Phys. C: Solid State Phys. **16**, 6281 (1983).
12. J. D. Jorgensen, D. G. Hinks, and G. P. Felcher, Phys. Rev. B **35**, 5365 (1987).
13. P. C. W. Leung, A. J. Schultz, H. H. Wang, T. J. Emge, G. A. Ball, D. D. Cox, and J. M. Williams, Phys. Rev. B **30**, 1615 (1984).
14. T. J. Emge, P. C. W. Leung, M. A. Beno, A. J. Schultz, H. H. Wang, L. M. Sowa, and J. M. Williams, Phys. Rev. B **30**, 6780 (1984).
15. A. J. Schultz and P. C. W. Leung, J. Physique **47**, C5-137 (1986).
16. P. Aldebert, Revue Phys. Appl. **19**, 649 (1984).

17. K. Clausen, W. Hayes, M. T. Hutchings, J. E. Macdonald, and R. Osborn, Revue Phys. Appl. **19**, 719 (1984).

18. 'High temperature studies of stoichiometric cerium dioxide,' J. Faber, Jr. and R. L. Hitterman, Proceedings of the Materials Research Society Meeting, Boston, Dec. 1-6, 1986 (in press) and other unpublished work.

19. H. Horiuchi, A. J. Schultz, P. C. W. Leung, and J. M. Williams, Acta Cryst. B **40**, 367 (1984).

20. D. B. McWhan, Revue Phys. Appl. **19**, 715 (1984).

21. C. J. Carlile and D. C. Salter, High Temp.-High Pressures **10**, 1 (1978).

22. J. Paureau and C. Vettier, Rev. Sci. Instrum. **46**, 1484 (1975).

23. J. D. Jorgensen, J. Appl. Phys. **49**, 5473 (1978).

24. E. Walley, in Physics of Ice, edited by N. Riehl, B. Bullemer, and H. Engelhardt (Plenum Press, New York, 1969) p. 19.

25. B. Kamb, Acta Cryst. **17**, 1437 (1964).

26. S. J. LaPlaca and W. C. Hamilton, J. Chem. Phys. **58**, 567 (1973).

27. J. D. Jorgensen, R. A. Beyerlein, N. Watanabe, and T. G. Worlton, J. Chem. Phys. **81**, 3211 (1984).

28. J. D. Jorgensen and T. G. Worlton, J. Chem. Phys. **83**, 329 (1985).

29. W. F. Kuhs, J. L. Finney, C. Vettier, and D. V. Bliss, J. Chem. Phys. **81**, 3612 (1984).

30. T. G. Worlton, D. L. Decker, J. D. Jorgensen, and R. Kleb, Physica B **136**, 503 (1986).

31. B. Batlogg, R. G. Maines, and M. Greenblatt, in Physics of Solids Under High Pressure, edited by J. S. Schilling and R. N. Shelton (North Holland, 1981) p. 215.

32. B. Batlogg, R. G. Maines, M. Greenblatt, and S. DeGregorio, Phys. Rev. B **29**, 3762 (1984).

33. J. E. Schirber, B. Morosin, R. W. Alkire, A. C. Larson, and P. Vergamin, Phys. Rev. B **29**, 4150 (1984).

34. J. D. Axe, Y. Fujii, B. Batlogg, M. Greenblatt, and S. DeGregorio, Phys. Rev. B **31**, 663 (1985).

35. J.-E. Jorgensen, J. D. Jorgensen, B. Batlogg, J. P. Remeika, and J. D. Axe, Phys. Rev. B **33**, 4793 (1986).

36. A. J. Schultz, M. A. Beno, H. H. Wang, and J. M. Williams, Phys. Rev. B **33**, 7823 (1986).

DISCUSSION

HIGH PRESSURE WORK WITH SYNCHROTRON RADIATION

There is much interest in such work with diamond cells, but the latter cannot achieve a powder average and so are unsuitable for high quality powder experiments. Pressures can approach 1 Mbar. diffraction lines and hence phase changes can be seen, but intensities are unlikely to be reliable. The cubic anvil presses developed in Japan have a larger sample volume and have been used satisfactorily for high pressure work with synchrotron radiation. There will always be a place, however, for neutrons in high pressure studies.

ACCURACY OF NEUTRON INTENSITIES IN HIGH PRESSURE WORK

Measurement of the individual reflections in the study of potassium nitrate at 4 kbar shown above would require deconvolution of the overlapping ones, but an accuracy of about 3% should be possible for the best 20% of the reflections. The gas pressure cell contained a large volume sample and Rietveld refinement could be done to 1 - 2% expected statistics.

MEASUREMENT OF HIGH PRESSURES

Three techniques predominate. Gas pressure cells communicate with the pumping station by capillary, and the perfect fluidity of helium gas allows the pressure to be measured out of the pumping station by gauges with good accuracy. In the rhenium trioxide work the relation between volume and pressure was measured accurately by dilatometry in a separate experiment. For silica the pressure was determined by mixing a CsCl calibrant with the sample.

With a helium gas cell pressure and temperature changes can be reversed freely, but piston cylinder cells with Bridgeman seals are often not reversible.

USE OF PULSED NATURE OF NEUTRON BEAM

Very little use has so far been made of this aspect. Dr Jorgensen thought that pulsed neutron sources offered few advantages over reactors for time-resolved work, especially when the time-structure was shorter than the repetition rate. The pulse structure exists only at the source: at the sample the beam is instantaneously monochromatic and its wavelength varies with time.

APPLICATIONS OF SYNCHROTRON X-RAYS TO CHEMICAL CRYSTALLOGRAPHY*

Åke Kvick
Chemistry Department
Brookhaven National Laboratory
Upton, NY 11973 USA

ABSTRACT. Synchrotron radiation generated by relativistic charged particles gives copious x-ray fluxes in wavelength regions suitable to crystallographic research. The radiation is tunable, highly collimated and 3-4 orders of magnitude more intense than existing conventional sources of x-ray radiation. The properties of the synchrotron radiation particularly useful in chemical crystallography are discussed and illustrated by examples from recent research on microcrystals, dynamical scattering, electric field experiments and anomalous dispersion.

1. BASIC PRINCIPLES

Synchrotron radiation is generated when charged particles, i.e. electrons or positrons moving at relativistic speeds follow a curved trajectory. At non-relativistic energies the charges radiate in a dipole pattern; this pattern becomes sharply peaked in the forward direction at synchrotrons as shown in Fig. 1. For general

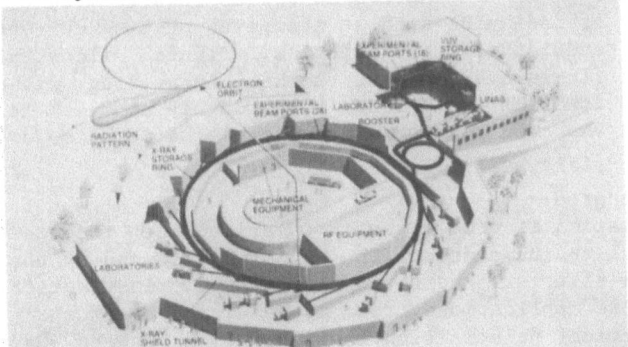

Figure 1. The National Synchrotron Light Source, Brookhaven National Laboratory, Upton, NY.

M. A. Carrondo and G. A. Jeffrey (eds.), Chemical Crystallography with Pulsed Neutrons and Synchrotron X-Rays, 187–203.

references the reader is referred to Synchrotron Radiation Research (1980).

The angular distribution of the radiation emitted by the charged particles moving through the trajectory introduced by a bending magnet is dependent on the energy of the particles and the radiation occurs in a narrow cone of angular width of ~ 1/γ where

$$\gamma = \frac{E}{mc^2}$$ E = energy of particle in GeV.

$$\theta_v \approx \frac{mc^2}{E}$$

The vertical natural collimation thus becomes increasingly narrow with increasing energy of the charged particles and reaches values of ~ 1 mrad at 1 GeV energies. The radiation is continuous in wavelength with half of the radiated power above and half below a critical wavelength λ_c defined by the equation

$$\lambda_c = \frac{4\pi\rho}{3\gamma^3} \text{Å}$$

where ρ is the radius of the circular motion in meters. This can conveniently be expressed in terms of ring energy, (E, GeV) and magnetic field (B, kilogauss).

$$\lambda_c = 186.4 \mid (BE^2)$$

giving a value of λ_c = 2.48 Å for the NSLS X-ray ring working at 2.5 GeV and with a bending magnet field of 12 kGauss. The flux available to the experimenter is usually calculated as photons/sec/mrad/1% $\Delta\lambda/\lambda$/A where mrad corresponds to mrad of horizontal fan of radiation used at A amperes of ring current and using 1% $\Delta\lambda/\lambda$ bandwidth of the radiation. Figure 2 illustrates typical photon fluxes available at the NSLS X-ray ring. It can be noted that the spectrum has a broad peak roughly at $3\lambda_c$. It drops off roughly as the cube root of λ at wavelengths much longer than λ_c but drops rapidly as $(\lambda_c/\lambda)^{1/2} \cdot e^{-\lambda_c/\lambda}$ for much shorter wavelengths.

The radiation is predominantly polarized with the electrical vector parallel to the acceleration vector introduced by the magnetic field (i.e., in the horizontal plane). In practical crystallographic applications from modern synchrotrons the horizontal/vertical polarization usually varies between 90-95%. but should be measured experimentally for each individual installation.

The time structure of the synchrotrons vary with the design parameters of the particular facility; in Table I examples of typical variations for the U.S. X-ray sources in operation or under construction are given.

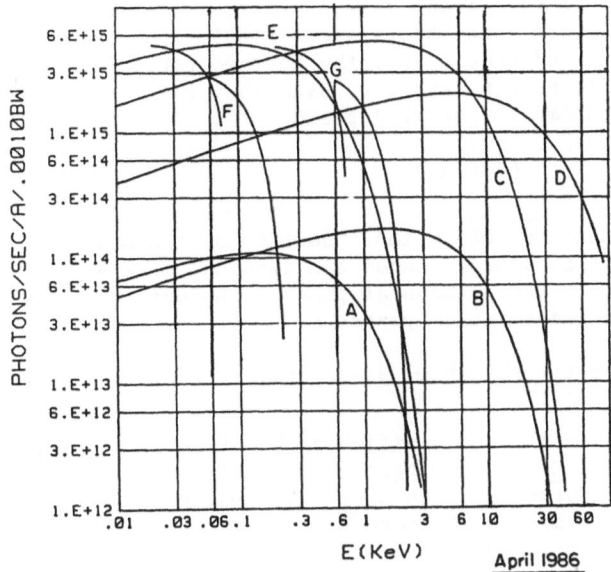

	A	VUV Bend	65 horizontal milliradians
	B	X-ray Bend	3 horizontal milliradians
	C	X25	3 horizontal milliradians (= 1.2 K/γ)
	D	X17	5 horizontal milliradians
	E	U13TOK	65 horizontal milliradians (= 1.2 K/γ)
	F	U5U	Undulator 1st, 3rd harmonics K = 0.5 - 2.5
	G	X1	Undulator 1st, 3rd harmonics K = 0.3 - 2.3

Figure 2. Photon fluxes at the National Synchrotron Light Source

TABLE I[†]

	SSRL	CHESS	NSLS	6-GeV
λ_c	2.64	1.43	2.48	0.78
Orbit period (nsec)	760	2560	568	2670
Number of bunches	1(4)	1(7)	30	1(10)
Bunch duration (psec)	300	160	1700	100
Interpulse period* (nsec)	760(190)	2560(366)	18.9	2670(267)

*Numbers in ( ) refers to alternative operations.
†(Shenoy and Viccaro, 1985).

In the future generation of synchrotron radiation so called insertion devices are going to be increasingly important. In these devices (wigglers and undulators) the charged particles travel through a periodic magnet structure with many magnetic poles generating a sinusoidally varying field.

$$B = B_o \cos(2\pi z/\lambda_m)$$

where z is the distance along the insertion device and the λ_m the magnet period. This field induces a sinusoidal motion of the charged particle in the horizontal plane. The motion can be characterized by a deflection parameter K

$$K = e \cdot B_o \lambda_m/2mc = 0.934 \, \lambda_m(cm) B_o(Tesla)$$

The insertion devices are classified as WIGGLERS if $K \gg 1$ (> 10) and as UNDULATORS if $K \leq 1$. In wigglers the radiation from different parts of the trajectory adds incoherently and produces flux increases proportional to 2N where N is the number of poles in the device. The wigglers may also be used as a wavelength shifter (λ_c) by changing the magnetic field of the device. Curve B in Fig. 2 illustrates the flux from a 1.2 Tesla bending magnet at the NSLS (λ_c = 2.48 Å) as compared to the flux of a super-conducting wiggler (curve D) with 6 poles and a 5 Tesla field giving a λ_c = 0.5 Å (magnet period 17.4 cm) (General ref. Krinsky, Thomlinson, van Steenbergen, 1982).

In undulators the radiation from different periods interferes coherently resulting in sharp peaks at a fundamental wavelength as well as at harmonics (n) of the fundamental. The bandwidth of these peaks decreases with increasing number of poles (N) in the device.

$$\frac{\Delta\lambda}{\lambda} \approx \frac{1}{nN} \qquad (n = 1,2,3,\ldots\ldots)$$

The integrated flux (vertical and horizontal) can dramatically increase at these peaks when compared to a portion of the bending magnet continuum. A comparison of the first harmonic of an undulator with the flux from a 1 mrad bending magnet source at the critical wavelength can be calculated.

$$\frac{I_u(\lambda_1)}{I_B(\lambda_c)} \approx 17500 \cdot \frac{NK^2}{\gamma[1 + (K^2/2)]}$$

for the same bandwidth $\Delta\lambda/\lambda$. At the NSLS X-ray ring γ = 4900 and assuming N = 100, K = 1 the flux radiation from 0.6 mrad horizontal

191

undulator is 238 times that radiated into 1 mrad by the bending magnet.

The undulator radiation is highly collimated in the horizontal plane and they are very efficient in producing a high ratio of useful photons to the total power of the beam, since the power generated by an insertion device is

$$P(kw) = 3.9 \ B^2 \cdot LI$$

where B is the magnetic field, L the length of the device and I the current in the storage ring. Undulators are usually low field, small magnetic period devices as compared to the high magnetic field wiggler devices.

II. AVAILABILITY OF SOURCES

Table 2 lists the synchrotron and storage rings in operation, which produce X-rays in the region useful for crystallographic applications. There are at least four future installations in the

TABLE 2. Storage rings with synchrotron radiation in an energy range suitable for crystallography.

Location	Ring	E(GeV)	E_c(KeV)[†]	Comment
China				
Beijing	BEPC(IHER)	2.2–2.8	~3.8 keV	C,Parasitic
England				
Daresbury	SRS	2.0	3.2	O,Dedicated
France				
Orsay	DCI	1.8	3.4	O,Partly dedicated
Germany				
Hamburg	DORIS	5.0	23	O,Partly dedicated
Italy				
Frascati	ADONE	1.5	1.5	O,Partly dedicated
Japan				
Tsukuba	Photon factory	2.5	4.1	O,Dedicated
USA				
Ithaca	CESR	5.5	11.5	O,Dedicated
Stanford	SPEAR	3–4	~4.7	O,Partly dedicated
Brookhaven	NSLS II	2.5	5.0	O,Dedicated
USSR				
Novosibirsk	VEPP-3	2.2	4.3	O,Partly dedicated
	VEPP-4	7	4.6	O,Parasitic

C = Under Construction; O = Operational; † = $\lambda_c(Å) = 12.4/E_c(keV)$

planning or construction phase in Europe (ESRF), USA (6-GEV synchrotron source; ALS) and Japan. Three of these are high-energy machines strongly geared towards the use of high-brightness insertion devices. These sources will however not be in operation for the next five years. Most of the sources listed in Table 2 are open to outside users on a proposal basis and access can be obtained by contacting the facilities.

III. SYNCHROTRON RADIATION AND CRYSTALLOGRAPHY

The main properties of synchrotron radiation of use to a crystallographer can be summarized as:

- High intensity
- Excellent collimation
- Wavelength tunable over a wide range
- Highly polarized
- Time-structure in the μsec-ns region

Most of these properties have already been utilized in early experiments but the crystallographic applications other than protein crystallographic data-collection using film as a detector medium are still in early stages and rapid development is expected as dedicated crystallography beam lines with stable beam conditions become available. Some details and examples are given below.

III,1. INTENSITY AND ENERGY RESOLUTION

The intensity available for a crystallographic experiment at a synchrotron depends critically on the following factors:

(1) Current in the storage ring
(2) Bandwidth ($\Delta\lambda/\lambda$)
(3) Horizontal and vertical focussing of the radiation
(4) Reflectivity of optical elements, i.e. mirrors and monochromators
(5) Choice of wavelength and critical wavelength of storage ring

Figure 3 illustrates the set-up of the Crystallography Station at the NSLS designed for single-crystal work in the wavelength region 0.6-3.1 Å with an energy resolution of a few eV to allow the use of the rapidly varying anomalous dispersion factors (f' and f") in the vicinity of an absorption edge.

Here the high intensity is sacrificed somewhat by choosing perfect Si crystals ($w_0 \approx 7$ sec of arc for Si_{111} at 1.5 Å) as

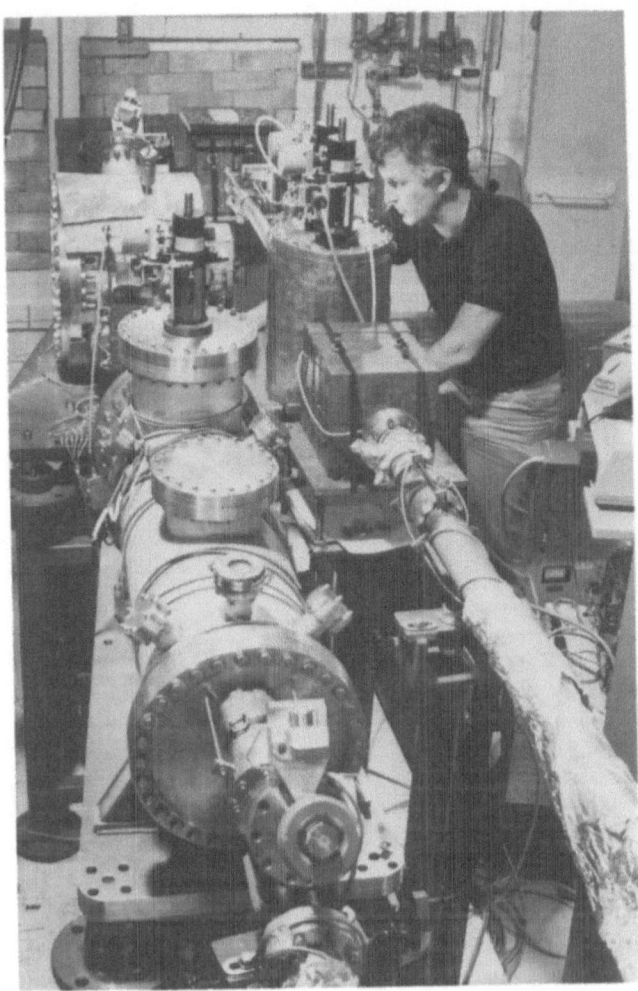

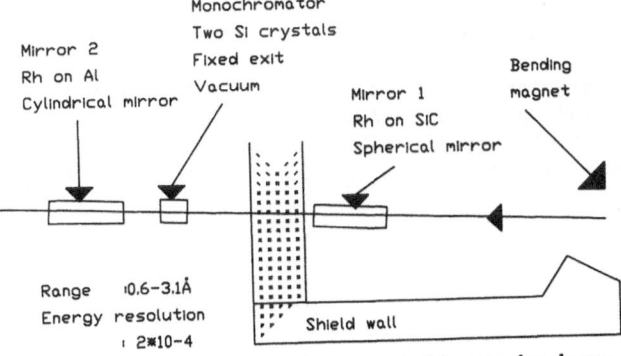

Figure 3a. Optical design of the crystallography beam line at the NSLS.

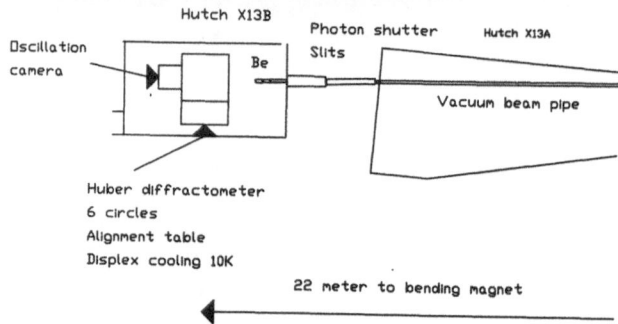

Figure 3b. Experimental facilities at the crystallography beam line at the NSLS.

monochromators to obtain an energy resolution of $\Delta E/E \sim 2 \cdot 10^{-4}$. Other installations commonly use Ge as a monochromator thus increasing intensity at the expense of the energy resolution.

The available intensity for the set-up illustrated in Fig. 3a corresponds to $\sim 10^{11}$ photons/sec/mm^2 at 100 mA of current at $\lambda \sim 1$-1.5 Å. Typical intensity fluxes range from 10^{12}-10^{10} photons/sec/mm^2 with bandwidths in the range 10^{-3}-10^{-4}.

The fluxes available from rotating anodes and sealed tubes are in the range 10^9–10^7 photons/sec/mm^2 giving the synchrotron sources intensity gains of 3 to 4 orders of magnitude. In addition the synchrotron radiation has a high natural collimation in the vertical plane (40' of arc for the x-13B setup) which enhances the peak/background ratio. The collimation makes it possible to resolve fine features in the diffraction process. Figure 4 illustrates the multiple diffraction superimposed on the weak Si_{222} reflection encountered during rotation around the 222 diffraction vector (Kvick, Chen, Post, 1986).

Multiple refl Si 222 1.325Å

A =   910568.

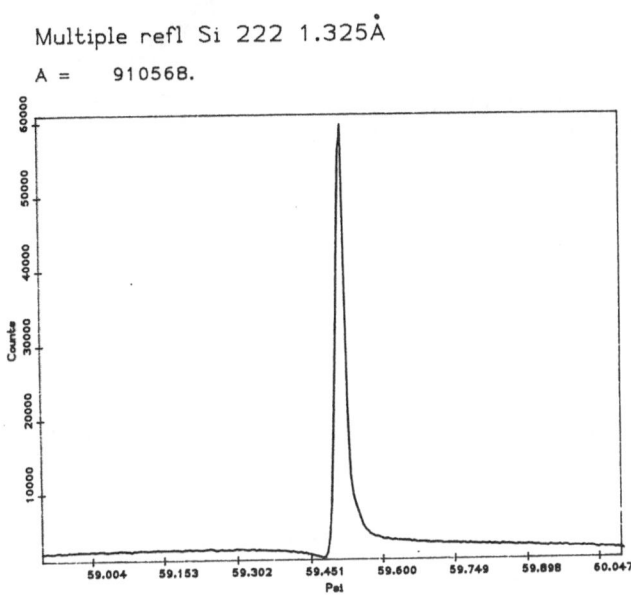

Figure 4. Multiple diffraction peak in Si.

Further analysis shows details of the coherent dynamical interaction in the X-ray multiple diffraction. The asymmetry of the peak associated with the phases of the structure factors involved (Post, 1979) are now clearly experimentally accessible (see Figure 5).

The collimation and high beam intensity will also make it possible to study extremely small crystals, resolve minute changes in cell dimensions or gather statistically significant data in short time intervals.

Multiple refl Si 222 1.325Å

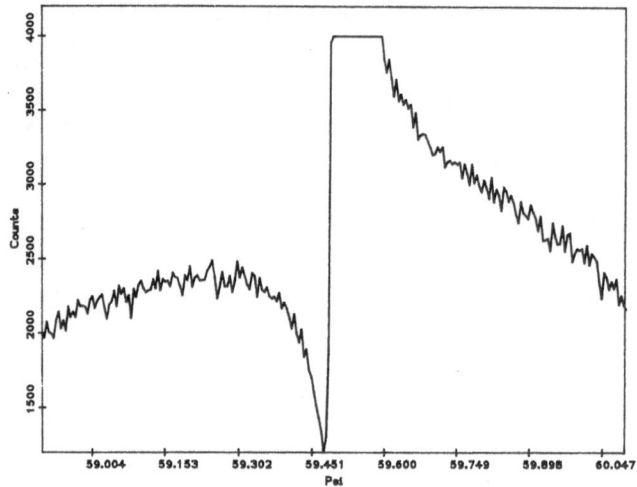

Figure 5.   Details of the dynamical three beam interaction in Si.

The integrated intensity from a single crystal is given by the formula:

$$I = \left[\frac{e^2}{mc^2}\right]^2 \cdot \frac{1}{\sin 2\theta} \cdot \left|\frac{F_{hkl}}{V_{cell}}\right|^2 \cdot \lambda^3 \cdot V_{cryst} \cdot I_o \cdot \frac{1}{w}$$

where e^2/mc^2 is the classical electron radius, $1/\sin 2\theta$ is the Lorentz factor and V_{cell} and V_{cryst} are the cell volume and crystal volume respectively. I_o is the intensity of the primary beam and w the speed of rotation through the peak.

   The fluxes from the synchrotron sources thus makes it possible to study extremely small crystals. A single crystal of CaF_2 with a crystal size of ~200 µm^3 has been studied at Hasylab, Hamburg (Bachmann, Kohler, Schulz and Weber, 1985). Recently several complex zeolite structures from microcrystals (< 50 µm on edge) have been studied at the NSLS. Typical count rates and spatial resolution is illustrated in Figure 6 (Kvick, 1986). This example shows the 10,0,0 reflection from a monoclinic ZSM-5 zeolite crystal (V_{cell} = 5400 Å^3, λ = 1.51 Å, beam current = 80 mA, E = 2.5 GeV) of approximate size 50 x 10 x 5 µm. The observed FWHM in the Bragg

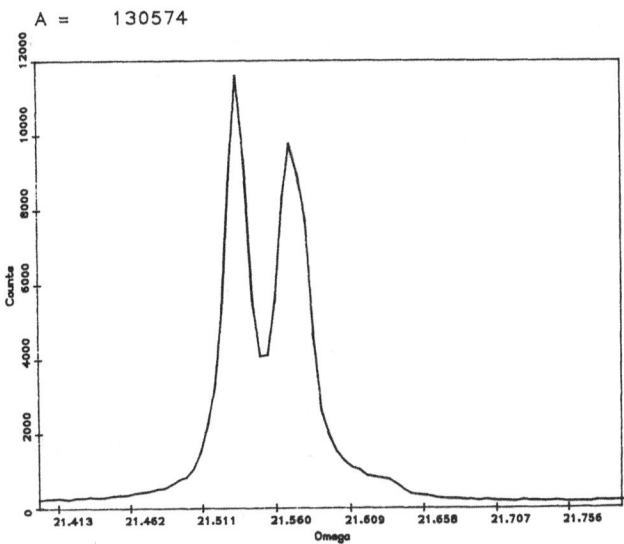

ZSM-5 10 0 0 Lambda=1.51Å 50x10x5 μm

A = 130574

Figure 6. Diffraction profile of the 10,0,0 reflection from a
twinned microcrystal of the zeolite ZSM-5.

peaks of the two crystallites are ~0.015° with a spatial
separation of ~0.03°. The count rates are in excess 10,000 cps
with a peak/background ratio of > 130/1.

From the present studies it is clear that statistically
significant data may be obtained from single crystals of sizes down
to 1–10 μm on edge.

Further examples of the use of intensity and resolution is
given in the electric field studies of $LiNbO_3$ (Kvick, Stahl,
Abrahams, 1986). The experimental setup is illustrated in Figure 7
and the observed results are illustrated in Figure 8. Here we
observe Bragg peak FWHM values of ~0.007° allowing changes in cell
dimensions as a function of applied electric field to be followed
well into the $\Delta d/d \sim 10^{-6}$ region. The excellent counting
statistics furthermore makes it possible to precisely measure
extremely small intensity differences as a result of the field.

198

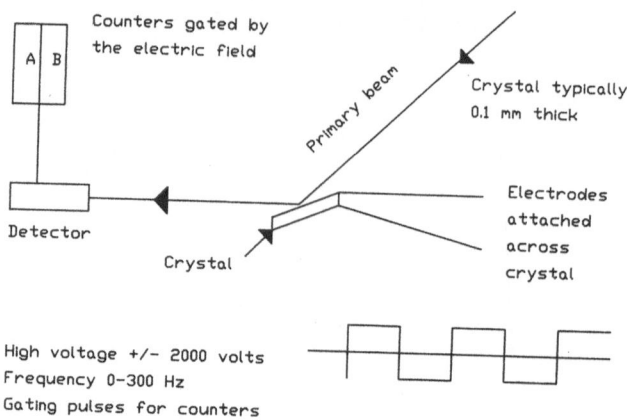

Figure 7. Experimental set-up for studies of single crystals in electric fields.

III,2. WAVELENGTH TUNABILITY

Existing bending magnet and wiggler sources produce a continuous spectrum with high intensities from ~ $\lambda_c/4$ up to wave lengths without practical importance for a crystallographer. The Cornell High Intensity Synchrotron Source A2 beam line on a 6-pole wiggler ($\lambda_c = 0.55$ Å) thus produces useful fluxes well below 0.2 Å whereas the NSLS bending magnet sources ($\lambda_c = 2.48$ Å) are usable down to 0.6 Å. The upper limits in the range is generally 3.5-4 Å because of the adsorption in the Be walls commonly used to separate the machine vacuum from the outside environment. It is therefore possible to tune to K absorption edges for elements with Z $\simeq$ 20-70 and L edges between Z = 46-100.

The tunability gives the crystallographer the possibility to choose a convenient wavelength. The scattering factor,

$$f = f_o + f' + if''$$

where the anomalous scattering factors f' and f" are wavelength dependent, may thus be changed selectively. This may be used to solve the phase problem (see for instance Karle, 1985 and references therein), to study X-ray dichroism and birefringence (Templeton and Templeton, 1980, 1985, 1986) or to separate elements with similar Z values.

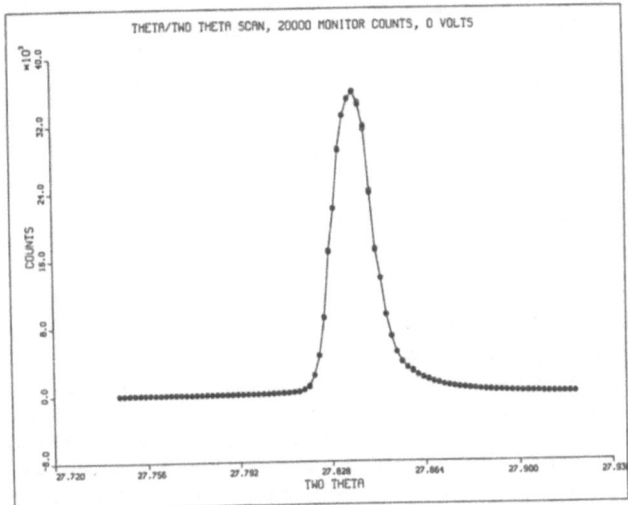

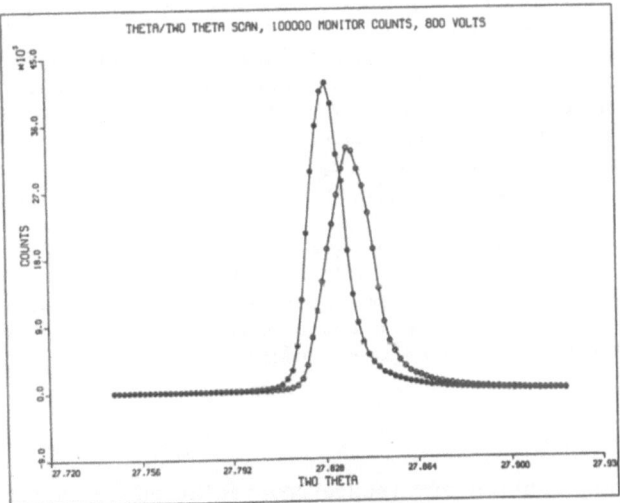

Figure 8. Changes in $\Delta d/d$ for the 006 reflection in LiNbO$_3$ as a function of applied ± V square wave electric field (50 Hz) along the C direction (crystal thickness 0.2 mm).

The anomalous scattering terms are related to the absorption cross section σ by the relations (James, 1962; Wagenfeld, 1975)

$$f''(w) = mcw\sigma(w)/4\pi e^2$$

which can be measured and related by a Kramer-Kronig relationship to

$$f'(w) = (2/\pi) \int_0^\infty [w'f''(w')/(w^2-w'^2)]dw'$$

The values can be calculated quite accurately away from the absorption edges (Cromer, Liberman, 1981). Close to the edges, however, deviations may occur and for precise work close to the edges the values should be determined experimentally (Templeton and Templeton, 1985; Begum, Hart, Lea, Siddons, 1986). The f" values show a sharp jump at the absorption edges whereas the f' values have a sharp dip at the edge. It should be noted that the precise location of the edge may shift somewhat depending on chemical environment.

The possibility of using an intense source at short wavelengths has significant advantages if high-resolution data free from systematic errors such as absorption and extinction is essential for the experiment.

Experiments of this type notably include precise charge density determinations. A feasibility study was recently performed at CHESS, the Cornell High Intensity Synchrotron Source, to study the electronic structure of hexaammine chromium(III) hexacyanochromate(III) (Nielsen, Lee, Coppens, 1986). An example of the resulting X-X deformation maps in the $Cr((CN)_4)$ plane is given in Figure 9.

This experiment used $\lambda = 0.302$ Å with a total of 7224 intensities out to $\sin\theta/\lambda = 1.15$ Å^{-1} collected. 1968 independent reflections and 44 variable parameters gave a R = 0.029 showing that the synchrotron radiation sources now are sufficiently stable to allow precise diffraction data to be collected.

IV. CONCLUDING REMARKS

The pioneering application of synchrotron radiation to chemical crystallography are very encouraging. The experimental set-ups have proven to be sufficiently stable to allow extensive data collections under optimized conditions. The high intensity has been used to study extremely small samples, to improve counting statistics and to probe the structures during short time intervals. The excellent resolution has been used in new experiments of crystals perturbed by electric fields as well in preliminary studies of new methods for

structure determinations. The tunability has been used for
anomalous scattering applications to study the phase problem as well
as to selectively study single elements. Further improvements in
synchrotron sources and particularly in the development of new
detectors promises to further expand the applications into new areas
of time-resolved diffraction.

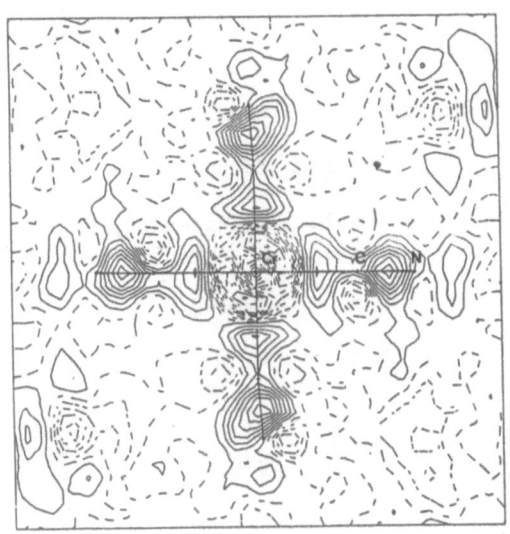

Figure 9. X-X Deformation map in the $Cr(CN)_4$ plane of $Cr(NH_3)_6$ $Cr(CN)_6$.

ACKNOWLEDGMENT
      Work performed at Brookhaven National Laboratory under
contract DE-AC02-76CH00016 with the U.S. Dept. of Energy and
supported by its Division of Chemical Sciences, Office of Basic
Energy Sciences.

REFERENCES

Synchrotron Radiation Research, (H. Winich and S. Doniach, Ed.)
Plenum Press, New York and London (1980).

Shenoy, G. K. and Viccaro, P. J. An Overview of the Characteristics
of the 6 GeV Synchrotron Radiation: A Preliminary Guide for Users',
Argonne National Lab. Special Publ. ANL-85-69 (1985).

Krinsky, S.; Thomlinson, W. and van Steenbergen, A. 'An Overview of
Undulators and Wigglers for the NSLS', Brookhaven Natl. Lab. Special
Publ. BNL 31989 (1982).

Kvick, Å.; Chen, E. M. and Post, B. Private communication (1986).

Post, B. Acta Cryst. A35, 17-122 (1979).

Bachmann, R.; Kohler, H.; Schulz, H. and Weber, H. P. Acta Cryst.
A41, 35-40 (1985).

Kvick, Å. Private Communication (1986).

Kvick, Å.; Ståhl, K. and Abrahams, S. C. Private Communication
(1986).

James, R. W. The Optical Principles of the Diffraction of X-rays,
Cornell University Press, Ithaca, NY (1962).

Karle, J. Acta Cryst. A41, 387-394 (1985).

Templeton, D. H. and Templeton, L. K. Acta Cryst. A36, 237-241
(1980).

Templeton, D. H. and Templeton, L. K. Acta Cryst. A41, 133-142
(1985).

Templeton, D. H. and Templeton, L. K. Acta Cryst. A42, 478-481
(1986).

Wagenfeld, H. Anomalous Scattering, (S. Ramaseshan and
S. C. Abrahams, Eds.), Munksgaard, Copenhagen (1975).

Cromer, D. T. and Liberman, D. H. Acta Cryst. A37, 267-268 (1981).

Begum, R.; Hart, M.; Lea, K. R. and Siddons, D. P. Acta
Cryst. A42, 456-464 (1986).

Nielsen, F. S.; Lee, P. and Coppens, P. Acta Cryst. A42, 359-364
(1986).

DISCUSSION

DETECTOR DEVELOPMENT

Optimum detectors for high intensity synchrotron radiation are
not yet available.

The best conventional detectors now measure 10^6 photons per
second, and their ultimate potential of about 10^7 photons per second
is insufficient. Dr Kvick is developing a linear counter with
separate electronics for each wire: it will count up to 5×10^7
photons per second per wire, and with 128 wires a total of $10^9 - 10^{10}$
photons per second. The construction of large area detectors will
be very expensive, and progress is slow because manpower is limited.

A recent innovation is a Japanese detector consisting of a barium
fluorobromide film containing europium. After exposure to X-rays,
it is scanned by laser to emit light in proportion to the X-ray dose.
The film has small pixels, a dynamic range of 10^5, and its
background is lower than that of photographic film by a factor of
about 100. The changes produced by the X-rays decay, so the film
must be scanned within about 8 - 24 hours. The device is attractive
for protein work, but some film changing mechanism will be needed
for time-resolved studies. The film, now commercially available,
is reuseable, so the main cost is that of the associated measuring
equipment. (For further details, see section 3.3 of the paper by
W Saenger and C Betzel - Editor)

SYNCHROTRON X-RAY AND NEUTRON RADIATION IN PROTEIN CRYSTALLOGRAPHY - PRESENT AND FUTURE

Wolfram Saenger and Christian Betzel
Institut für Kristallographie, Freie Universität Berlin
Takustr. 6, D-1000 Berlin 33, FRG, and EMBL Outstation,
DESY Hamburg, Notkestr. 85, 2000 Hamburg 52, FRG

ABSTRACT. Synchrotron radiation has opened a new field in protein crystallography owing to the to the high brilliance X-ray radiation, the tuneability of wavelengths and the excellent collimation. New instrumentation like area counters and powerful software speed up protein crystallography tremendeously. Concerning neutron radiation, it is useful in the study of proton positions which are important in mechanistic investigations and when flexibility of proteins has to be determined. These advances and recent development in genetic enginee- ring which allows the production in large amounts of proteins with natural and modified sequences, have made the future of protein crystallography bright and encouraging.

1. INTRODUCTION

Since the first protein crystal structure was determined by diffraction methods in the early 1960s, great advance has been made in crystallo- graphy, in biochemistry, and in molecular biology. The most dramatic change occurred when genetic engineering was introduced which now enables us to produce large quantities of proteins which might be only available in very small quantities from their natural sources. Also, genetic engineering allows us to modify proteins as we wish, and a new field of research called protein design has emerged. In order to modi- fy a protein intelligently, one has to know its three dimensional structure to be able to pinpoint certain amino acids or segments in the protein which one desires to change. This has on the other hand a tremendous influence on protein crystallography which has developed rapidly in the past ten years due to fast and less expensive compu- ters, to the construction of area sensitive counters, and to new methods for the solution of the phase problem. In addition, radiation sources like the synchrotron have changed our ways of data collection, and neutrons allow us to investigate H-atom positions which yield insight into mechanistic problems and give a measure of the flexibili- ty of proteins. In the following we shall briefly discuss the present

M. A. Carrondo and G. A. Jeffrey (eds.), Chemical Crystallography with Pulsed Neutrons and Synchrotron X-Rays, 205–216.
© *1988 by D. Reidel Publishing Company.*

use of synchrotron and neutron radiation and then we shall discuss the future aspects of these sources in protein crystallography (1).

2. ADVANTAGES OF SYNCHROTRON RADIATION

Below, a brief survey is given describing the advantages of synchrotron X-ray radiation, with emphasis on the use in protein crystallography.

Properties of synchrotron radiation	Advantages in protein crystallography
High intensity	Rapid data collection possible Small crystals can be used Weak, high resolution data can be measured
Tuneable wavelength	Data can be collected near absorption edges of heavy atoms High resolution data can be collected on film of manageable size
White radiation	Laue method Very rapid data collection kinetic measurements possible
High collimation	Well defined reflections Higher resolution data obtainable
Pulsed time structure	Use in kinetic measurements
Polarization of radiation	Not yet used in protein crystallography

In recent work in protein crystallography, the high flux, good collimation and variable wavelength have been used to great advantage. Let us briefly illustrate our own experience with the synchrotron radiation on two proteins: proteinase K, the structure of which we have determined at 1.5Å resolution and ribulose-1,5-bisphosphate carboxylase/oxygenase (Rubisco), the structure of which has been described recently at 5Å resolution.

1.1. The Use of Synchrotron Radiation in the Collection of High Resolution Data

Proteinase K is a proteolytic enzyme of molecular weight 27 000 and consists of 278 amino acids. It crystallizes readily from a 10 % solution with the addition of 1 M $NaNO_3$ and 0.01 M Tris, pH 6.5. The crystals have the tetragonal space group $P4_32_12$, with cell constants a = b = 68.3Å, c = 108.3Å. There is one molecule in the asymmetric unit. The structure was solved by conventional methods using three heavy

Table I.

Parameters for a typical high resolution data collection of a protein crystal (proteinase K) using synchrotron radiation (DESY Hamburg, storage ring DORIS)

Experimental setup	Arndt-Wonacott rotation camera at line X11, EMBL
Power DORIS	5.1 GeV, 30-40 mA
Film	KODAK DEF-2
Film cassettes	flat, max. radius 60 mm
Wavelength used	0.8655Å
Collimator	0.35 mm
Crystal-to-film distance	90 mm
Rotation angle per exposure	0.5°
Max. resolution	1.5Å
Number of films	282 (94 film packs with 3 films each)
Exposure time per film pack	~8 min
Total exposure time	12 hrs.
Total rotation angle	47° (space group tetragonal, mounted along c-axis
Space group	$P4_32_12$
Cell constants	a = b = 68.3Å, c = 108.3Å

Table II

Results of the data collection on proteinase K

Total number of reflections		116,040
Independent reflections		36,039
R-factor (Intensities)		8.8 %
Reflections with negative intensity		1,058
Number of reflections	> 1σ	36,014
	> 2σ	35,349
	> 3σ	32,096

Refinement with restrained least squares:

Total number of cycles	113
Number of refined parameters	8,773
Number of reflections used	30,812
Resolution range	5.0 - 1.5Å
Number of atoms	2,194
solvent molecules	174
Mean temperature factor	11,1Å²
Final R-factor (based on F)	16.7 %

atom derivatives ($HgCl_2$, $PbCl_2$, $SmCl_3$), and the structure was determined at 3.2Å resolution (2).

Even with conventional methods data collection, we noticed that the diffraction of the crystals was superior and the crystals lasted for about one weak in the X-ray beam without too much damage. Using a four circle diffractometer, we collected data to 2.5Å resolution but found that this was much too slow to go to higher resolution. Measurements at the DESY in Hamburg using an Arndt-Wonacott-oscillation camera on the line X11 made clear that the data collection could be speeded up tremendeously and at the same time, by choosing a suitable wavelength, we were able to collect high resolution data. The initial idea to collect data at 1.1Å resolution to which fresh, new crystals actually diffract, was found too amibitious because the diffraction at that high resolution decayed after only about one hour and it was therefore decided to collect the data at 1.5Å resolution. Some specifications for the data collection are given in Table I. Since the radiation from the synchrotron is not constant with time but decreases steadily after injection, the incoming radiation in the primary beam is measured, and after a certain number of quanta have been counted, the shutter is closed and a new exposure is taken.

In the high resolution data collection of proteinase K, we have made use of the high flux, the good collimation, and the variable wavelength. The high flux allowed rapid data collection so that the whole data set could be measured with only two crystals. The high collimation produced very well defined reflections in all resolution ranges which is of course especially valuable with the high angle data. The variable wavelength finally was important to make optimum use of the film method. With the conventional wavelength of 1.542Å, the highest resolution attainable is usually about 2.0Å with flat film cassettes. With the shorter wavelength of 0.8655Å, we were able to place the film at a distance from the crystal where the problems with parallax are not yet severe. The evaluation of the film data was done with an Optronics P-1000 film scanner, the digitalized data were processed by the program system MOSCO, and the refinement was done using a computer linked graphics display Evans & Sutherland PS 300 in combination with the restrained least squares refinement program from Hendrickson & Konnert. The refinement process is illustrated in Table II and led to a final R-factor of 16.7 % for the 30,812 data greater than 3σ in the range 5.0Å to 1.5Å.

1.2. The Low Resolution Structure of Ribulose-1,5-bisphosphate carboxylase/oxygenase (Rubisco)

The situation with rubisco is very different compared to proteinase K. Rubisco is the CO_2-fixing protein that occurs in all plants and consists of 8 large (L) and 8 small (S) subunits of molecular weight 51 000 and 15 000 each so that the total molecular weight of the L_8S_8 protein is 528 000. The rubisco we are using, from the bacterium Alcaligenes eutrophus, crystallizes from ammoniumsulfate solution in

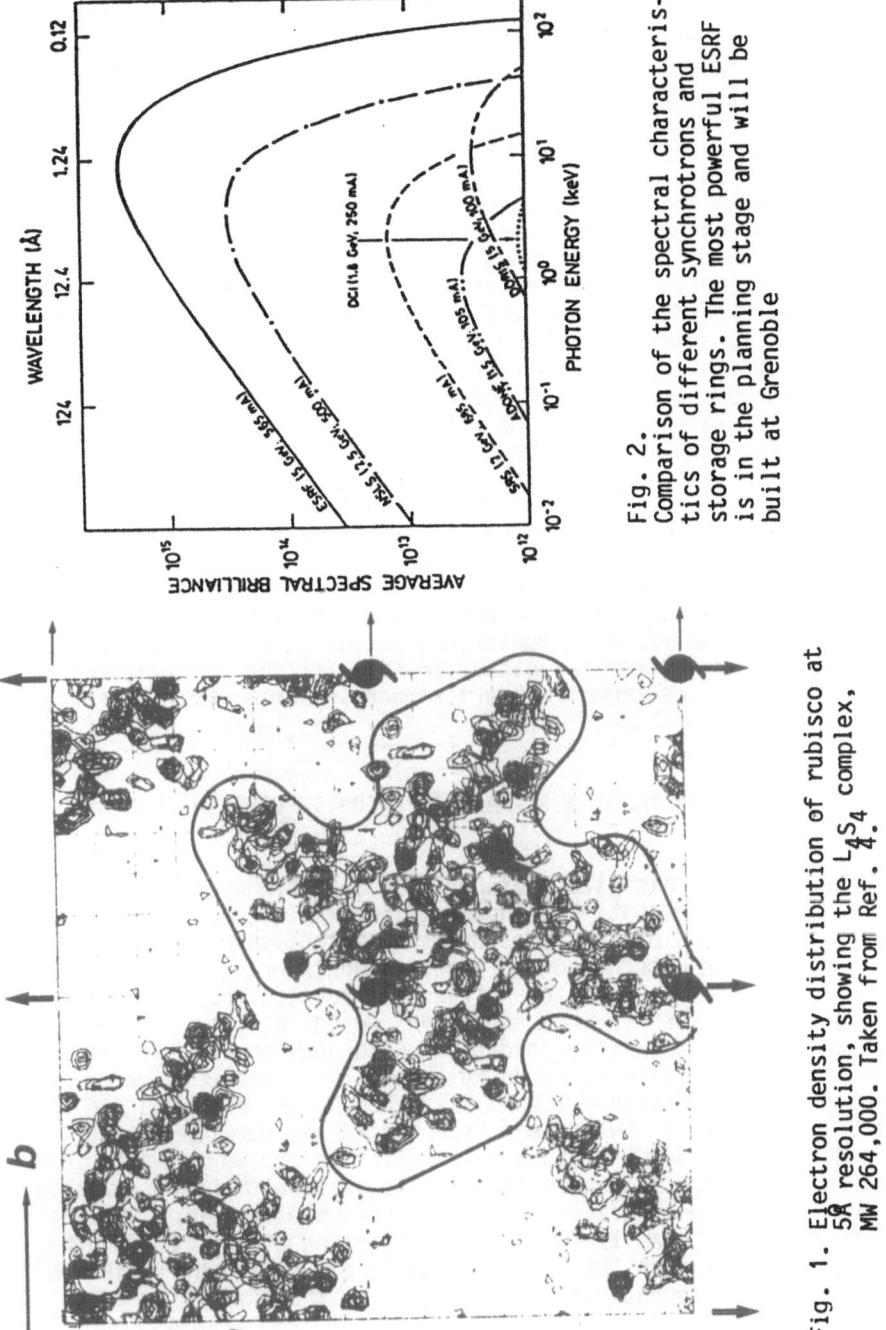

Fig. 2.
Comparison of the spectral characteristics of different synchrotrons and storage rings. The most powerful ESRF is in the planning stage and will be built at Grenoble

Fig. 1. Electron density distribution of rubisco at 5Å resolution, showing the L_8S_8 complex, MW 264,000. Taken from Ref. 4.

the orthorhombic space group C222$_1$ with cell constants a = 159Å, b = 161Å, c = 201Å. There is half a molecule (L$_4$S$_4$) in the asymmetric unit; the results reported here were obtained by Drs. R. Tokuoka, K. Harata, T. Ishida and G.P. Pal, see Ref. 4.

Due to the size of the asymmetric unit (264 000 Da), the diffraction from the crystals is rather weak but can still be measured to 3.5Å resolution using conventional X-ray generators with rotating anode. Since there was problems with heavy atom derivatives which displayed changes in the c-axis, it was tried to solve the crystal structure with only one heavy atom derivative, HgCl$_2$, of which the data were collected on a rotating anode X-ray generator. Making use of the anomalous contribution of the Hg atom, which at 1.542Å is only 8 electrons, an electron density map was obtained which clearly had some kind of fourfold symmetry looking in the c-axis direction, indicating the presence of the L$_4$S$_4$ unit. This was so encouraging that solvent flattening methods were applied and the powerful method of noncrystallographic symmetry which gave the electron density map presented in Fig. 1 (4). The next step was to collect new X-ray data of the heavy atom derivative at the synchrotron at Hamburg, where a wavelength of 1.0Å was used, very close to the absorption edge of Hg. The contribution from the anomalous scattering is now optimised, and the data are superior with respect to the data measured with the conventional X-ray generator, also because the number of crystals used to collect the data set is reduced. It is clear that in the case of rubicso, once the structure is solved at a medium resolution of about 3Å, the collection of more data from the native protein will be continued at the synchrotron where, as known from preliminary experiments, the resolution is as good as 2.2Å.

3. FUTURE USES OF SYNCHROTRON X-RAY RADIATION IN PROTEIN CRYSTALLO-GRAPHY

The future of protein crystallography in combination with synchrotron X-ray radiation is bright. This general statement is based on the experience which our and other laboratories had with synchrotron radiation, and on the new techniques of genetic engineering which are able to produce proteins in large quantities and require that we know many more protein structures in the future than we do at present. The uses and advantages of synchrotron radiation which have been mentioned in the two cases of proteinase K and rubisco will continue to be be employed. At the new designed line X11 at EMBL on the positron-side of the storage ring DORIS (where the brightest X-ray beam in Europe is produced), routine X-ray protein data collection is now a matter of one hour. This short measuring time is also a result of the automatic beam and intensity optimization which reduces the measuring time by up to 50 %. At the beam line X31 on the electron side of DORIS, the instrumentation is such that the wavelength can easily be varied and can be used advantageously to collect data at the absorption edges of heavy atom derivatives. Because the anomalous signal is improved considerably compared with conventional CuKα radiation, this will be sufficient in combination with solvent flattening and, if applicable, with noncrystallographic symmetry approaches, to determine the crystal structures from only one heavy atom derivative, if no more derivatives are available (1).

3.1. The Laue-Method

The white radiation from synchrotron sources is smooth and does not exhibit the characteristic spikes which X-ray radiation emerging from anodes has. Therefore, it can be used in the Laue method and since no monochromatisation is needed, the flux is increased considerably, reducing the exposure time to the sec or even msec range. The diffraction patterns are recorded from stationary crystal (good for cooling devices) and stationary detector, and the white radiation automatically integrates over wavelength. A disadvantage of the Laue method is that for a given reflecting plane, several orders of reflections can overlap, and that the detector should, ideally, allow both spatial and energy resolution (which precludes multiwire proportional counters).

The number of data that can be recorded in a single exposure is of the order of 10.000, depending on the unit cell constants. For crystals with high symmetry space groups, only one exposure might be sufficient to collect the whole data set, otherwise a few settings at, say, 30° intervals have to be used.

At present, there are still some problems with data evaluation from Laue photographs, but first progress with the solution of a small molecule crystal structure has been reported (M. Harding, S. Maginn, P. Machin, J. Campbell and I. Clifton). If these problems are overcome, the Laue method will prove especially valuable in cases where kinetic investigations are done by crystallographic techniques, which are already in progress at the Daresbury Synchrotron (1,5).

3.2. Use of the Pulsed Time Structure

The pulsed time structure which is inherent in the synchrotron X-ray radiation can be used to do kinetic experiments. The principles have been investigated by Dr. H.D. Bartunik and his group and he will outline the results in his lecture. Therefore, I do not want to go into this matter any further.

3.3. Future Instrumentation in Protein Crystallography

There are three main classes of instruments which the protein crystallographer has to rely on. One is a graphics display to do model building, the other is fast computers to do the calculations which become especially time consuming if refinement of proteins at high resolution is to be done, and the third is data collection. Here, we shall only be concerned with the aspects of data collection because this is where the synchrotron radiation is concerned.

The last 5 years have seen the development of three area counters which are now commercially available: One of them is based on a video camera system whereas the other two are based on multiwire chambers and are therefore proportional counters. Moreover, an area detector based on the multiwire system has been constructed at EMBL (6). These systems are producing good data which have been used to solve crystal structures of proteins. Here is not the place to discuss the advantages and disadvantages of these counters and to compare them, but it appears to us that in combination with a synchrotron which has a variable wavelength, the counter based on video screen is of advantage because its sensitivity depends to a lesser extent on the wavelength of the incoming radiation than that of the proportional counters. Therefore, it will be more profitable in such cases where the wavelength is tuned so that one can measure close to the absorption edge of a heavy atom, and it will be of advantage if the Laue method is employed. Moreover, this instrument has a much shorter dead time than the proportional counters and has a wider dynamic range, both of advantage if synchrotron radiation is used (7,8).

A new development has been made in Japan. It is based on a film which is presently used in medical applications. This film is about tenfold more sensitive than normal films and has an enormous linearity with respect to the X-ray intensity. It consists of a 200 x 250 mm "imaging plate", i.e. of a flexible plastic sheet coated with a fine photostimulable phosphor, $BaFBr$ doted with Eu^{2+}. By UV or X-ray light excitation, Eu^{2+} is ionized to Eu^{3+}, the electrons are liberated and trapped at F^{+} centers. The image is thus temporarily stored (with a half lifetime of about 10 hrs) and can be read out by a He-Ne laser beam ($\sim$ 390nm) of 150 μ diameter which scans over the film and liberates the trapped electrons from the F-centers. They return to Eu^{3+} ions and reduce them to Eu^{2+} with emission of luminescence which can be detected by a photomultiplier. If exposed to a large dose of visible light, the latent X-ray picture can be erased and the film be used again (9).

The advantages of this system are obvious. The screen has near-ideal quantum efficiency in the wavelength range used by crystallographers, a spatial resolution of better than 0.2 mm (full width at half maximum) in horizontal and vertical directions, practically no local distortion of the image, a dynamic range of 1 : 10^{5} and practically no counting rate limitations. These properties make the system especially useful for applications with the synchrotron, and ideal in combination with the Laue method. An automatic system has been constructed where two films are used in tandem, one being exposed while the other is automatically screened and the data are processed by a computer.

3.4. New Synchrotron Sources

In the future. synchrotron X-ray radiation will continue to be of importance for protein crystallography because new sources are being constructed and present sources are being upgraded with wigglers. This means that their output will be in the wavelength range needed by crystallographers, so that measurements at the absorption edge of almost all heavy atoms can be carried out. One of the instruments that are being upgraded is DORIS as dedicated synchrotron radiation source in the early 1990s at Hamburg. A new instrument with excellent characteristics as shown in Fig. 2, will be built, the European synchrotron radiation facility (ESRF) at Grenoble. These developments in Europe show already that synchrotrons are no longer exotic instruments which can only be used by high energy physicists, but their side product, the X-ray radiation, becomes more and more important and interesting for other scientific applications.

4. NEUTRON RADIATION IN PROTEIN CRYSTALLOGRAPHY

Until now, only a few proteins have been investigated by neutron radiation (10-12). The reason for this sparse application of neutron radiation in protein crystallography is mainly the low flux and the long counting times which have to be employed. The construction of area counters at Brookhaven, Washington and Grenoble has brought us into a position where we can now speed up data collection considerably so that more proteins can be investigated. What can we learn from protein neutron crystallography?

4.1. Evaluation of Proton Positions and of Proton Exchange in Proteins

One of the main problems in protein neutron crystallography is the large amount of protons in the proteins. Since protein crystals consist to about 50 % of mother liquor, it is advantageous to exchange H_2O by D_2O. which can easily be done by soaking the crystals in heavy water. On the other hand, if proteins are produced by bacteria, one can grow the bacteria in D_2O and so obtain proteins which are deuterated. Such kind of experiments are done successfully at Berlin and a new laboratory has been installed at Grenoble who are involved in the same procedures. Here again genetic engineering becomes of great importance.

If a protein has been crystallized from H_2O and is then transferred into D_2O, the H/D exchange will not necessarily be complete. This is because there are regions in the protein structure in which the secondary structure elements like α-helix and ß-pleated sheet which are stabilised by hydrogen bonding between peptide groups are either very stable or the peptide N-H groups, which otherwise would easily exchange to form N-D, are not accessible by the solvent so that the H/D exchange does not occur.

This was illustrated in all the proteins which were investigated thus far. In trypsin, the H/D exchange occurred mainly in the loop regions and in an α-helix, but the pleated sheet regions are so stable that they did not allow the exchange to occur. This suggests that there are very stable regions in the protein, and others that are more flexible which, on the other hand, are suggestive of amino acid exchanges which one could make in order to increase the stability of a protein. Such kind of studies will become more and more important, especially in view of commercial applications of (stable) proteins.

Also, knowledge of the location of individual protons in a protein can be of importance if, for instance, the chemical mechanism of an enzyme is not clear. This occurred in trypsin, where the protonation stage of the histidine in the active site, which was elucidated by neutron diffraction, indicated that the charge-relay-mechanism postulated long time ago from the first X-ray structure analysis, had to be modified. Also, in X-ray crystallography, it is often impossible to distinguish, in asparagine and glutamine side chains, where the amide carbonyl and where the amino groups are, and so hydrogen bonding interactions to and from these groups might be obscured.

4.2. Future of Protein Neutron Crystallography

The future of protein neutron crystallography is, with conventional reactors, not as bright as with the synchrotron X-ray radiation. The reason is that the flux is still several orders of magnitude lower than with X-rays and the long counting times and the few reactors available are discouraging. Nevertheless, it is necessary to study more proteins by neutron diffraction so that we learn about the fluctuations and the mobilities within the folding of the polypeptide chain.

The development of pulsed neutron sources offers a new field for protein crystallography because the flux is increased considerably compared with conventional reactors. Will we witness a development that parallels that of synchrotron X-ray sources? The relative fluxes of the pulsed X-ray and neutron sources are still in favour of X-rays, but there will be special applications where only neutron diffraction studies can give an answer.

5. SUMMARY

In this short review, we tried to highlight the present and future uses of synchrotron X-ray and of neutron radiation in protein crystallography. The development of new crystallographic techniques with, on the hardware side, fast computers, graphics displays and area counters and, on the software side, of solvent flattening techniques, maximum entropy approaches and the use of non-crystallographic symmetry, have tremendous impact especially on protein crystallography. The main problem that remains to be solved is to make the alchimistic art of crystal growing and of producing heavy atom derivatives less hazardous and to design techniques which allow us to more efficiently grow crystals suitable for diffraction studies. Concerning the availability of proteins in suitable quantities for crystallographic (and other) studies, genetic engineering has opened a new field and protein crystallography will be one of the main tools to widen our knowledge in this fascinating branch of science.

Acknowledgements

These studies were supported by Sonderforschungsbereich 9 (Teilprojekt A 7), by Bundesministerium für Forschung und Technologie (FKZ 05 313 IAB, FKZ 03-B72C05-0), and by Fonds der Chemischen Industrie.

References

1. J.R. Helliwell, J. Mol. Struct. 130, 63-91 (1985).
2. A. Pähler, A. Banerjee, J.K. Dattagupta, T. Fujiwara, K. Lindner, G.P. Pal, D. Suck, G. Weber, W. Saenger, EMBO J. 3, 1311-1314 (1984)
3. Ch. Betzel, Ph. D. Thesis, Freie Universität Berlin, 1986. To be published.
4. A. Holzenburg, F. Mayer, G. Harauz, M. van Heel, R. Tokuoka, K. Harata, T. Ishida, G.P. Pal, W. Saenger, Nature in press (1987)
5. J.R. Helliwell, The Rigaku Journal 3, 3-12 (1986)
6. J. Hendrix, H. Lentfer, Nuclear Instr. and Meth. in Phys. Res. A252, 246-250 (1986)
7. U.W. Arndt, D.J. Gilmore, J. Appl. Cryst. 12, 1-9 (1979)
8. J. Miyahara, K. Takahashi, Y. Amemiya, N. Kamiya, Y. Satow, Nucl. Instr. Meth. Phys. Res. A246, 572-578 (1986)
9. B. Schoenborn, Ed. Neutrons in Biology, Plenum Press, New York (1984)
10. A.A. Kossiakoff, Nature 296, 713-718 (1982)
11. A. Wlodawer, L. Sjölin, Proc. Natl. Acad. Sci. USA, 79, 1418-1422 (1982)

DISCUSSION

BACKGROUND INTENSITY

The wide variations of background intensity seen in the study of
proteins at various synchrotron sources are probably mainly affected
by differences in the protein crystals and the content of the mother
liquor surrounding the crystals, provided that the beam line,
monochromator, mirrors and slits are in the best possible arrangement.

PROTEIN STRUCTURE

Proteins are thought to have essentially the same tertiary structure
in the crystal and in solution. When the same protein has been
studied in the crystal by X-rays and in solution by nmr no major
differences have been found, normally only a few side chains being
differently oriented at the periphery.

There is a limit to the size of a protein molecule for which two-
dimensional nmr can be used to study the hydrogen atoms: a relative
molecular mass of 14,000, for example, is on the high side.

Vibrational motion of atoms in proteins has been investigated by
the refinement of temperature factors in high resolution work. The
α-helices and pleated sheets are very rigid, but there is more
flexibility at the outside of the molecule, especially in loops
and side chains.

SYNCHROTRON RADIATION FOR ELECTRON DENSITY STUDIES

Pierre J. BECKER
Laboratoire de Cristallographie, associé à l'USTMG
CNRS, 166 X, 38042 GRENOBLE Cedex, France

The fundamental interest of one-particle electron density functions is now well established. It has also been recognised that the experimental determination of such properties is a unique test of detailed models for electronic structure of crystallised materials. The interplay with theory is very demanding since it refers to a delocalised function in real or momentum space.

The principles of these experiments are rather simple, but their success depends on both technological progress and detailed understanding of scattering processes. Is synchrotron radiation expected to have a strong impact on this field of research. Experience is rather limited but it is possible to discuss the progress in already well established techniques and also the development of new methods. So far pulsed neutrons did not lead to any significant result and we shall not discuss this case.

After a short summary of the important properties of charge, spin and momentum densities and their accessibility by X Ray or neutron scattering, we shall discuss a few relevant results obtained by the use of synchrotron radiation. We shall in particular discuss experiments in momentum space, which are the most spectacular in this respect.

There have been several conferences and schools on electron densities in the last decade, some of them being published[1-5].

FUNDAMENTALS OF ELECTRON DENSITIES

We will be dealing with a N-electron system. The charge density $\rho(r)$ is a local measure of the electron content per unit volume.

$$\rho(r) = <\sum_{j}^{N} \delta(r - r_j)> \qquad (1)$$

where j refers to a given electron.

The momentum density $n(p)$ measures the number of electrons with a given momentum p, per unit volume in momentum space.

217

M. A. Carrondo and G. A. Jeffrey (eds.), Chemical Crystallography with Pulsed Neutrons and Synchrotron X-Rays, 217–245.
© 1988 by D. Reidel Publishing Company.

$$n(p) = < \sum_{j}^{N} \delta(p - p_j) > \qquad (2)$$

The current density $j(r)$ is thus given by :

$$j(r) = \frac{1}{2} < \sum_{j}^{N} [p_j \delta(r - r_j) + \delta(r - r_j) p_j] > \qquad (3)$$

Now consider the density of electrons with a given spin state, ρ_+, ρ_- .

$$\rho = \rho_+ + \rho_- \qquad (4)$$

Consider the function :

$$s(r) = 2 < \sum_{j}^{N} s_{jz} \delta(r - r_j) > = \rho_+(r) - \rho_-(r) \qquad (5)$$

It is the spin density, proportional to the z component of the spin magnetisation density $M_s(r)$, where

$$M_s(r) = -2\mu_B < \sum_{j}^{N} s_j \delta(r - r_j) > \qquad (6)$$

The orbital component of magnetisation $M_l(r)$ satisfies :
$$\text{rot}(M_l(r)) = 4\pi j(r) \qquad (7)$$

$s(r)$ is called the spin density. The same spin analysis can be performed in momentum space, where one gets :

$$n(p) = n_+(p) + n_-(p)$$

$$\sigma(p) = n_+(p) - n_-(p) \qquad (8)$$

It occurs that these density functions which have just been defined are measurable through X-ray or neutron scattering. Given such density functions, any one-electron property can be calculated. For instance, the kinetic energy T is simply :

$$T = \int n(p) \frac{p^2}{2m} dp \qquad (9)$$

and through the virial theorem, the energy is :

$$E = -T \qquad (10)$$

From $\rho(r)$, one can obtain all Coulombic contributions in both electron- nuclear and electron- electron interactions. The combination of $n(p)$ and $\rho(r)$ should in principle lead to a measure of exchange-correlation contributions.

Spin density is of course the leading function in magnetism, but cannot lead by itself to exchange interactions : combination with other functions is needed.

All two-particle operators, such as electron-electron interaction, involve pair correlation functions. However, the part that is really specific of electron-electron correlation is hardly measurable and involves coherent inelastic X-Ray scattering, which is not practically operational. It should be noticed that high energy electron scattering in the gas phase can lead to such information [6]. Suppose $P(r_1,r_2)$ is the probability density for two electrons to be at r_1 and r_2, $\rho(r)$ is simply given by :

$$\rho(r_1) = N \int P(r_1,r_2)\, dr_2 \qquad (11)$$

It may seem paradoxical to insist on one particle density functions which only represent a contraction from a N-electron wavefunction onto a one elctron space. Remember, if Ψ is the wavefunction in real space (ω_j being a spin variable) :

$$\rho(r_1) = N \int |\Psi(r_1\omega_1,r_2\omega_2,.....,r_N\omega_N)|^2\, d\omega_1 d\omega_2..d\omega_N dr_2..dr_N \qquad (12)$$

and is therefore a projection over (N-1) electron space.

However, there are two very strong theorems that prove the one particle densities to be the leading information relative to any electron system.

Hohenberg-Kohn theorem [7] states that given a charge density $\rho(r)$, there can only exist one N-electron wave-function Ψ from which it derives. As a consequence, any property of the ground state is totally characterized by the charge density ρ, which thus carries all the useful information.

Much effort has been devoted to the practical determination of properties as a functional of ρ : in particular the energy. By assuming approximate functional form for exchange and correlation, it has become possible to determine ρ and electronic properties without any reference to the N-electron wave-function. This density functional method[8,9] is very popular and powerful, and allows one to do quantum chemistry on systems of large size or systems in interaction with their environment[10,11]. The theorem generalises for open shell systems, where ρ_+ and ρ_- become the leading density functions.

It is a rather unusual property, which can be visualised by the following indirect and picturial argument. Given the charge density, the points of its cusps represent the nuclear positions. The radial behaviour of ρ at those points provide us with the nuclear charges. From there, one can construct the conventional Hamiltonian, and from its solution, one obtains uniquely the electronic properties.

A second and very famous theorem is the Hellman- Feynman theorem[12] valid within the Born- Oppenheimer scheme. It is related to the electronic response to nuclear motion. It states that the electronic contribution to the force acting on a given nucleus is given by the Coulombic classical expression using as a charge density the electron density $\rho(r,R)$, where the symbol R refers to a given fixed geometry. This theorem is also called the electrostatic theorem, since for calculations of forces on atoms, the electronic system acts just by its electron density at this given geometry. Though very well known, this theorem does not have as many applications as it could. One has to realise that the "dynamic" charge density $\rho(r,R)$ carries all the information about nuclear motion and thus chemical reactivity (which takes place along the easiest paths for nulear motion). This density function connects the electronic structure with the atomic skeleton (that is really the central interest in physics and chemistry). It should be noticed that on practical grounds, the knowledge of vibrational properties could be used to derive or test models concerning the dynamic density $\rho(r,R)$.

From the preceding discussion, we see that charge density and more generally one electron densities are essential information to the static and dynamic behaviour of any material. Since these functions are basically measurable, it is now important to understand how they are related one to the other, and how they are expandable in atomic components. This is the purpose of the next paragraph, when we assume a fixed geometry .

Let $\Psi(1,2,...,N)$ be the ground state. We define the one particle density matrix as :

$$\mathcal{G}(1,1') = N \int \Psi^*(1',2,3,...,N)\Psi(1,2,3,...,N) \; d2d3...dN \qquad (13)$$

If $\mathcal{Q}$ is a one electron property, its expected value is :

$$<\mathcal{Q}> = \text{tr}(\mathcal{Q}\;\mathcal{G}) \qquad (14)$$

In real space representation, since Ψ is an eigenstate of S_z, the density matrix can be expanded as :

$$\mathcal{G}(1,1') = \mathcal{G}_+(r,r')\,\alpha(\omega)\,\alpha^*(\omega') + \mathcal{G}_-(r,r')\,\beta(\omega)\,\beta^*(\omega') \qquad (15)$$

where α and β are the two spin states.

In momentum space representation, we use the Dirac-Fourier transform and get :

$$\mathcal{H}_{\pm}(p,p') = (2\pi h)^{-3} \iint e^{-i(p.r - p'.r')/h} \, \mathcal{G}_{\pm}(r,r') \, dr \, dr' \qquad (16)$$

$$\mathcal{H}(1,1') = \mathcal{H}_{+}(p,p') \, \alpha(\omega) \, \alpha^{*}(\omega') + \mathcal{H}_{-}(p,p') \, \beta(\omega) \, \beta^{*}(\omega') \qquad (17)$$

The spinless operators are :

$$\mathcal{g}(r,r') = \mathcal{G}_{+}(r,r') + \mathcal{G}_{-}(r,r')$$

$$\qquad (18)$$

$$\mathcal{R}(p,p') = \mathcal{H}_{+}(p,p') + \mathcal{H}_{-}(p,p')$$

The densities ρ and n are the diagonals of the matrices $\mathcal{g}$ and $\mathcal{R}$.

$$\rho(r) = \mathcal{g}(r,r) \qquad n(p) = \mathcal{R}(p,p) \qquad (19)$$

In particular, the momentum density $n(p)$ is given by :

$$n(p) = (2\pi h)^{-3} \int \mathcal{g}(r,r') \, e^{-ip.(r - r')/h} \, dr \, dr' \qquad (20)$$

Let $B(t)$ be the auto-correlation function of the one particle density matrix :

$$B(t) = \int \mathcal{g}(r,r+t) \, dr$$

$$n(p) = (2\pi h)^{-3} \int B(t) \, e^{ip.t/h} \, dt \qquad (21)$$

In a similar way, we can write :

$$F(q) = \int \mathcal{R}(p,p+q) \, dq \qquad (22)$$

$$\rho(r) = (2\pi h)^{-3} \int F(q) \, e^{-iq.r/h} \, dq \qquad (23)$$

These relationships show that real and momentum space electron densities are connected through autocorrelation functions and their Fourier transforms.

The same properties apply to spin dependent components ± of the density matrices.

$$s(r) = \mathcal{G}_+(r,r) - \mathcal{G}_-(r,r)$$

$$\sigma(p) = \mathcal{H}_+(p,p) - \mathcal{H}_-(p,p)$$

$$\rho_\pm(r) = \mathcal{G}_\pm(r,r) \qquad\qquad n_\pm(p) = \mathcal{H}_\pm(p,p) \qquad (24)$$

After this rather abstract summary, we have to describe how density functions are expandable in terms of atomic-centred orbitals. $\mathcal{G}(r,r')$ is an hermitian matrix, whose eigenvectors are the natural orbitals $\psi_{j\pm}(r)$ and eigenvalues are $n_{j\pm}$, with the following relations :

$$\mathcal{G}_\pm(r,r') = \sum_j n_{j\pm} \, \psi_{j\pm}^*(r') \, \psi_{j\pm}(r)$$

$$\sum_j n_{j\pm} = N_\pm \qquad (25)$$

$n_{j\pm}$ is the occupation number of the natural orbital $\psi_{j\pm}$. One can show that $1 \geq n_{j\pm} \geq 0$. When the system is non magnetic (closed shell),

$$\mathcal{G}_+ = \mathcal{G}_- \qquad (26)$$

and the natural orbitals for either spin are equivalent (double occupancy for non degenerate orbitals). For magnetic systems, this is no more true and owing to the disymmetry of exchange, spacial orbitals of + or - spin are spacially slightly decoupled : this is the spin polarisation effect. Notice also that any unitary transformation among the $N_\pm$ orbitals of a given spin has no effect on the one-particle matrices.

A very important approximation corresponds to the case where $\mathcal{G}_\pm^2 = \mathcal{G}_\pm$. In that case, the occupation numbers are 0 or 1, and there are exactly $N_\pm$ orbitals occupied. In this approximation,

$$\mathcal{G}_\pm(r,r') = \sum_j^{N_\pm} \psi_{j\pm}^*(r') \, \psi_{j\pm}(r) \qquad (27)$$

This constitutes the Hartree-Fock approximation, or independent elctron model. The N-electron wavefunction is expressible as a Slater determinant constructed on the N spinorbitals and all properties are exactly calculable from the one particle density matrix. The orbitals $\psi_{j\pm}$ are the eigenfunctions of a mean field one particle hamiltonian, the Fock operator.

Correlation effects result in the breakdown of (27); excited orbitals must be included, at the expense of the occupied states of the Hartree-Fock model.

In practice, one works in a subspace of the full Hilbert space, and this subspace is defined by atom-like orbitals $\varphi_m(r - R_m)$, where m refers to a given nucleus.

$$\psi_{j\pm}(r) = \sum_m^M c_{jm\,\pm} \varphi_m(r - R_m)$$

$$\qquad (28)$$

$$\varphi_m(r) = \zeta_m^{3/2} R_m(\zeta_m r) \, \mathcal{Y}_m(r/r)$$

ζ_m is a screening constant which is considered as adjustable, depending on the charge transfer with adjacent atoms ($\zeta_m = 1$ for free atoms). R_m is a radial function (generally combination of Slater type functions or Gaussian functions) and $\mathcal{Y}_m$ is a spherical harmonic describing the symmetry of the given atomic orbital. M is the size of the basis set.

In momentum space, we get similarly :

$$\mathcal{f}_{j\pm}(p) = \sum_m^M c_{jm\pm} \, e^{ip.R_m/h} \, \mathcal{R}_m(p) \qquad (29)$$

$\mathcal{R}_m(p)$ being the Dirac-Fourier transform of $\varphi_m(r)$.

$$\mathcal{G}_\pm(r,r') = \sum_{m,n}^M \mathcal{P}_{mn\pm} \, \varphi_m^*(r'-R_m) \, \varphi_n(r-R_n)$$

$$\mathcal{H}_\pm(p,p') = \sum_{m,n}^M \mathcal{P}_{mn\pm} \, e^{i(p.R_n - p'.R_m)/h} \, \mathcal{R}_m^*(p') \, \mathcal{R}_n(p) \qquad (30)$$

224

$$\mathcal{P}_{mn\pm} = \sum_j^{N_\pm} n_{j\pm} c_{jm\pm}^* c_{jn\pm}$$

A detailed and picturial discussion of various aspects of density matrices can be found in [14].

The smallest basis set that can be used (minimal basis set) is made from those atomic orbitals which are occupied in the free atom. But it is a general result that the inclusion of excited state atomic orbitals is very important in order to obtain an adequate description of the density. This is rather evident since the potential created by neighbouring atoms polarises a given atom, which can only be accounted for by including orbitals of angular momentum higher than in the free state.

We developed the basic expressions that are the frame of any research on one particle-densities. These expressions are essential if one wishes to interconnect different experiments on a given system.

We finally have to worry about the fact that a given system is always surrounded by other systems and at a definite temperature. This implies the excitation of vibrational modes and the definition of thermodynamic average of density functions. The observable charge density will for instance be :

$$\rho(r) = \int \rho(r,R) \, P(R) \, dR$$

with :

$$P(R) = \sum_n P_n \, |\chi_n(R)|^2$$

$$P_n = e^{-\beta \varepsilon_n} / Z \qquad\qquad (31)$$

$$Z = \sum_n e^{-\beta \varepsilon_n} \qquad\qquad \beta = \frac{1}{kT}$$

CHARGE DENSITY STUDIES

X Ray elastic scattering by a single crystal leads to Bragg diffraction. If **H** is the Bragg vector, the structure factor in the single scattering Born approximation is:

$$F(H) = \int \rho(r) \, e^{2\pi H.r} \, dr \cdot \qquad (32)$$

where $\rho(r)$ is the thermally averaged density, as defined in (31). The Fourier inversion leads to :

$$\rho(r) = \frac{1}{V} \sum_H F(H) \, e^{-2\pi i H.r} \qquad (33)$$

V is the unit cell volume. Within the same approximation, the kinematic integrated intensity is :

$$I_k(H) = s \, Q \, v \qquad (34)$$

s is a scale factor, v the crystal volume and :

$$Q = \left| \frac{\lambda F(H) r_0}{V} \right|^2 \frac{\lambda}{\sin 2\theta} \left\{ \frac{p_H \cos^2 2\theta + p_V \cos^2 2\theta_M}{p_H + p_V \cos^2 2\theta_M} \right\} \qquad (35)$$

p_H and p_V are the fractions of photons with a polarisation horizontal or vertical with respect to the plane of diffraction. r_0 is the classical radius of the electron.

In reality, the observed intensity is :

$$I(H) = I_k(H) \, A \, y \, (1 + \alpha) \qquad (36)$$

where A is the absorption factor :

$$A = < e^{-\mu l} > \qquad (37)$$

μ is the linear absorption coefficient and l the local optical path length.

α is the contribution from thermal diffuse scattering and y is the most troublesome factor, the extinction factor, that accounts for the breakdown of the simple Born approximation. It takes into account the effect of multiple scattering, and thus the interaction between the incident and diffracted beams. It is a difficult term, which is a function of F and of the imperfections present in the crystal. Several theories exist to correct for y [15,16]. But these theories are approximate and assume a simple statistical distribution of defects. Arguments have been developed in γ -Ray diffraction [17], to correct for extinction from measured reflection profiles, provided one can access them. There are basically two effects leading to extinction. The local distortions lower the overall scattered amplitude, like a static Debye-Waller factor. But these distortions also create optical incoherence between the waves at different points in the sample and the lost coherent power appears as an incoherent component. Let $\Lambda = (d / Q)^{1/2}$ be the so called extinction length. If the size of an undistorted domain $t > \Lambda$, the scattering is essentially coherent and it is the range of validity of dynamical theory. The second crucial parameter is the coherence length τ , the distance over which there appears a complete lack of phase coherence in the waves. When $\tau < \Lambda$, the theory developed in [15] applies correctly : it is the domain dominated by secondary extinction (the mosaic spread is simply d / τ). If $\tau > \Lambda$, the mosaic spread would be smaller than the dynamical width d / Λ , which is of course absurd. No correct theory exists for this case.

Now notice that $\Lambda \approx V / (\lambda F r_0)$. There is thus a great advantage in working at a small wavelength : this reduces primary extinction and gives more security in the available models. One should however be cautious about pathological cases [18].

Secondary extinction becomes more severe when the crystal size increases, though primary extinction depends only on the internal degree of perfection inside the sample. Thus, those reflections that have contributions from valence electrons, i.e. occuring at low scattering angle, are the most affected by extinction. It is especially important to reduce the temperature, in order to get measurable intensities at higher Bragg angles : this leads to an improved resolution, but reduces Λ and thus increases the importance of extinction. Therefore, it is interesting to reduce the wavelength in order to reduce extinction. Moreover, in any case, it is important, at least for strong reflections, to be able to do the experiment at different wavelengths : after processing them, one should obtain the same estimate for the structure factor, and this is a unique test of validity of a given scattering theory [19] : synchrotron radiation in that sense is a very favourable source, owing to its wavelength tunability.

Let us now draw some practical conclusions about this extinction problem. Secondary extinction is a very general effect in neutron scattering (low flux and thus large crystals). With synchrotrons, it is possible to work with small samples. There has been an experiment on CaF_2 using a 6 μm single crystal [20], and extinction was shown to be essentially absent. The same authors have shown [21] that owing to the small divergence of the beam, it is possible to take advantage of the width of a reflection profile to determine experimentally its mosaic spread : this allows for a better check of the validity of implied scattering model. The use of synchrotron radiation is therefore very favorable for the treatment, both experimental and theoretical, of extinction [22].

Another advantage of synchrotron radiation is the reduction of thermal diffuse scattering correction. It is due to the reduction of the profile width of the reflection, compared with standard sources. As a consequence, TDS can be mainly considered as a background [20,21,23].

There are however some problems to be considered very seriously when using synchrotron for diffraction. First, the scattered intensity is highly dependent on the state of polarisation of the beam (35). This is not always an easy parameter to refine, especially when using a short wavelength where most reflections occur at small Bragg angles [23]. It may be experimentally estimated but is variable. An incorrect estimate of polarisation leads to biased Debye-Waller parameters [20,21].

The source being pulsed, it is necessary to estimate correctly the deadtime correction for the counter [23,24]

The radiation being a white spectrum, $\lambda / 2$ contamination has to be precisely corrected for [23].

The peaks being very narrow, it is important to use unequal step scans, rather than the usual isometric grid [23].

Is μ always known accurately enough for a precise absorption correction ? The same question applies of course to anomalous scattering contributions.

There is only one full three dimensional study of charge density, made by Coppens et al. [23], at room temperature, on $Cr(NH_3)_6Cr(CN)_6$. The wavelength was .302 A . Using high angle data, it was possible to determine positional and thermal parameters, since only core (non polarisable) electrons scatter at high $\sin\theta / \lambda$. This allows for a calculation of " independent atom model " structure factors $F_a(H)$, and for a calculation of the deformation density :

$$\Delta\rho\,(r)\ =\ \frac{1}{V}\ \sum_{H}^{H_0}\ [\,F\,(H\,) - F_a(H\,)\,]\ e^{-2\pi iH.r} \qquad (38)$$

where H_0 is the higher limit of measurement in reciprocal space. This well known function measures the polarisation of the electron density due to interatomic forces. The structure being centrosymmetric, one assumes the signs of structure factors to be unchanged from F_a to F . For non centric cases, accurate phasing is highly dependent on the implied modeling for the deformation density $\Delta\rho$. Sections of the deformation density by planes containing Cr atoms are shown on Fig.1. The resolution of this map is as good as in most low temperature experiments with standard sources. One should in particular realize that it is a (X-X) technique, whereas most standard studies involve a neutron scattering determination of positional and thermal parameters. It is a great advantage of using wiggler radiation to be able to probe reciprocal space far enough to be able to obtain accurate structural parameters with X Ray high order data. There is always an advantage in using the same crystal, under the same experimental conditions, for a complete study. Problems concerning the adjustment of neutron Debye-Waller factors to the X Ray experiment are famous and very confusing.

It does not seem obvious that the use of synchrotron is of absolute necessity for measuring all the structure factors corresponding to a given electron density. It seems that it is essential when there is severe extinction for the conventional experiment. It also looks very important in order to measure very low reflections, in particular "forbidden" reflections. Those forbidden reflections have long been a challenge to crystallography : in particular the $h+k+l = 4n+2$ reflections of diamond like structures. Their occurence can only be explained by charge polarisation and anharmonicity. The complementarity with neutron diffraction has been considered important to separate the two contributions properly. Recently, those reflections have been carefully measured at CHESS [25,26]. On Fig.2, we show a plot of the (442) intensity of silicon as a function of T, taken from

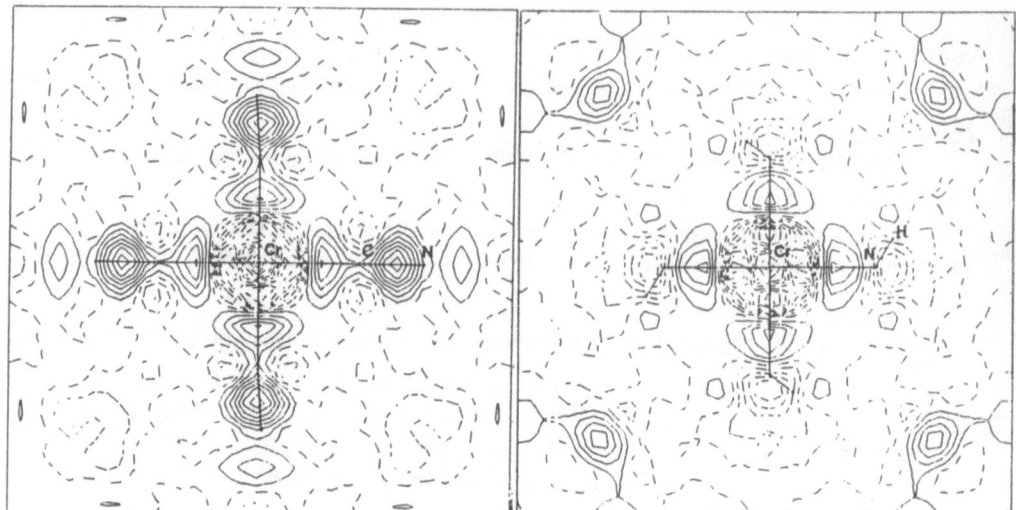

Fig.1.   Deformation density maps for $Cr(CN)_6Cr(NH_3)_6$, (from ref. 23) ,after averaging among chemically equivalent directions.
a. $Cr(CN)_4$ plane

b. Plane containing Cr and four N atoms of the ammino groups
    Contours at .05 A^{-1} , negative contours broken.

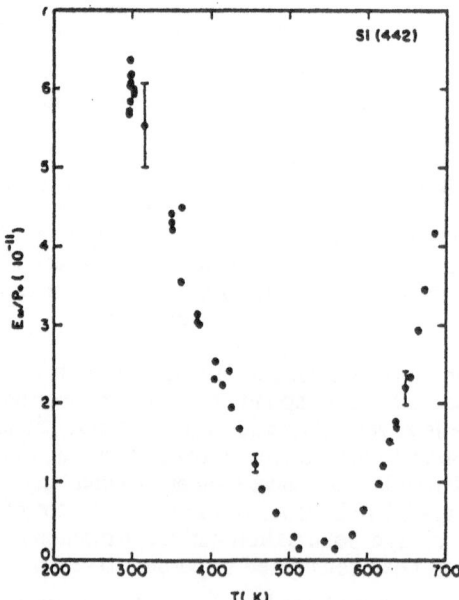

Fig.2.   Measured integrated intensity of Si (442) reflection versus T (from ref. 25)

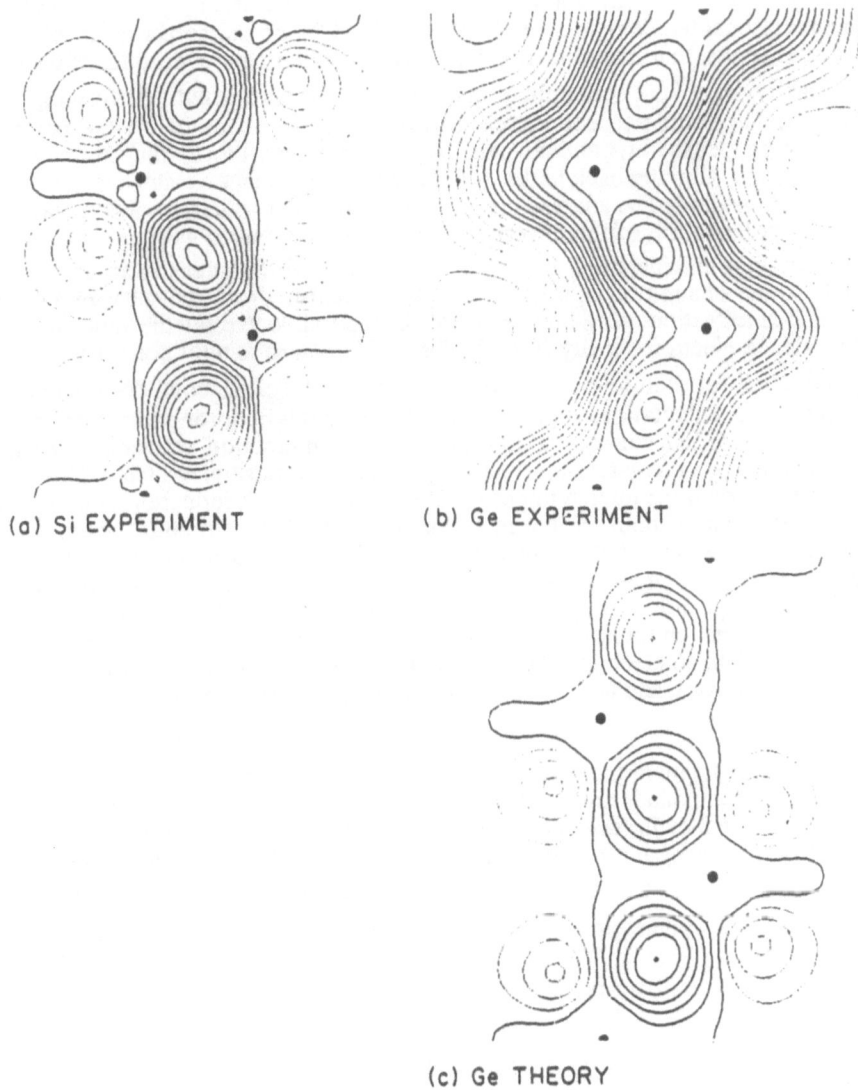

(a) Si EXPERIMENT

(b) Ge EXPERIMENT

(c) Ge THEORY

Fig 3.   Contour plots of the antisymmetric part of the valence charge distribution for Si and Ge (from ref. 25)

[25]. The authors of this experiment insist on the importance of multiple reflection effect, when three reciprocal lattice points H,H',H" (H+H'=H") lie on the Ewald sphere. For those forbidden peaks, multiple reflection can lead to much bigger contributions than the genuine intensity. The λ/2 contamination has also to be eliminated. Performing a φ scan, it has been possible to calculate the reflectivity in terms of all the possibly interfering Bragg peaks. Dynamical calculations are sensitive to the phase of the forbidden reflection. Coming back to the (442) case, it occurs that its sign changes when T increases. The results can be analysed in terms of a model in which one incorporates an antisymmetric polarisation of the charge density which competes with an anharmonic Debye-Waller factor that increases as T^2. The minimum of the intensity around 520°K corresponds to the change of sign of the structure factor. From the measurement of several forbidden reflections of Si and Ge, it has been possible to draw the antisymmetric part of the valence electron density (Fig.3). The result is surprisingly good, based on X Ray only measures for such low structure factors. In [25] Collela et al. discuss the fact that such dynamical calculations with N beams are only possible for exceptional cases. They propose a modified method, taking into account interaction between very small peaks, and thus using kinematic theory, and applicable to mosaic crystals. They show that, taking into account the interaction of (442) with other weak reflections, one gets an unambiguous estimate of both the phase and magnitude of the structure factor. Those studies are examplar about what can be achieved with present techniques. Even though such measurements are long and difficult, they lead to a direct observation of bonding.

Finally, let us mention the possibility of measuring changes of structure factors under applied perturbation (pressure, electric field...). It should also be possible to detect effects associated with an electronic excitation. Using short wavelength allows to do the analysis without neutrons, and to probe the density closer to the nuclei.

In conclusion, the main interest in charge density work will be to probe electronic structure in simple compounds on the one hand, and to look for electron reorganisation in complex systems like metallo-organic compounds : in this last case, there may exist two states very close in energy but of different symmetry, and diffraction can discriminate between those two[28]. It is also of great interest to combine charge and spin density measurements on those interesting compounds [49,50].

COMPTON SCATTERING

Inelastic interaction between the incident photon and the electrons can be separated into collective excitations (like plasmons in solids) and individual excitations (Compton effect). The energy transferred to the electron being in general higher for Compton scattering, this effect has been the only detectable inelastic process with X or γ Rays, the situation being different with electrons [6]. A very good review of Compton scattering has recently been published [30].

First consider the electrons as quasi free particles, with momentum p . The interaction with a photon (k_0, ω_0) leads to a photon (k_f, ω_f) . The recoil energy transmitted to the electron is $\Delta E = h\omega = h(\omega_0 - \omega_f)$. The scattering vector is $K = k_f - k_0$. The basic equations for conservation of momentum and energy are :

$$p_f = p - hK$$

$$h\omega_o = h\omega_f + \frac{h^2 K^2}{2m} - h\frac{K.p}{m} \qquad (39)$$

The direction of K will be denoted by z . Let φ be the scattering angle, we have :
$$K = (4\pi / \lambda) \sin \varphi/2$$

$$\frac{\omega}{\omega_o} = 2\frac{h\omega_o}{mc^2}\sin^2\varphi/2 - 2\sin\varphi/2\frac{p_z}{mc} = \frac{\Delta E_o}{E_o} + \frac{\Delta E'}{E_o} \qquad (40)$$

The first term is the well known Compton shift. The second is the Doppler broadening effect, leading to a finite width of the Compton line. It is first obvious that large φ are favorable $(\varphi \approx 145°)$. In the case of X Rays, $\Delta E_o/E_o \approx 10^{-2}$ to 5.10^{-2} , this factor being much larger with γ Rays. ΔE_o is typically 500 eV, much larger than electronic transitions involving valence electrons. This observation justifies the neglect of binding energy of outer electrons, the so called impulse approximation. Nevertheless, the width of the Compton line, given by $(\Delta E'/E_0)$ is a significant fraction of the basic shift.

It is easy to prove that, in the impulse approximation, the cross section is proportional to :

$$J(p_z) = \int n(p)\, dp_x dp_y \qquad (41)$$

which is the projection of the momentum density onto the scattering vector direction. $J(p_z)$ is the directional Compton profile. Its Dirac-Fourier transform is the function $B(0,0,z)$ with $t = (r.K/K)K/K$. By changing the direction between K and the frame of the sample, one can thus in principle reconstruct n(p) or B(t) [29].

In order to get an ejected electron, the transferred energy must be larger than the binding energy E_B. If this is the case with valence electrons, whose profile is concentrated in the low p_z region, it is not obvious for core electrons, especially with

232

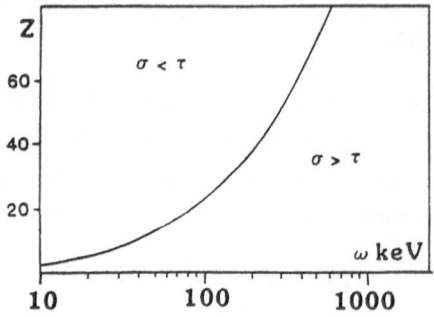

Fig.4.   *Cross sections for photoelectric effect* τ *and Compton scattering* σ.Z *is the atomic number and* ω *the incident photon energy (from ref. 34)*

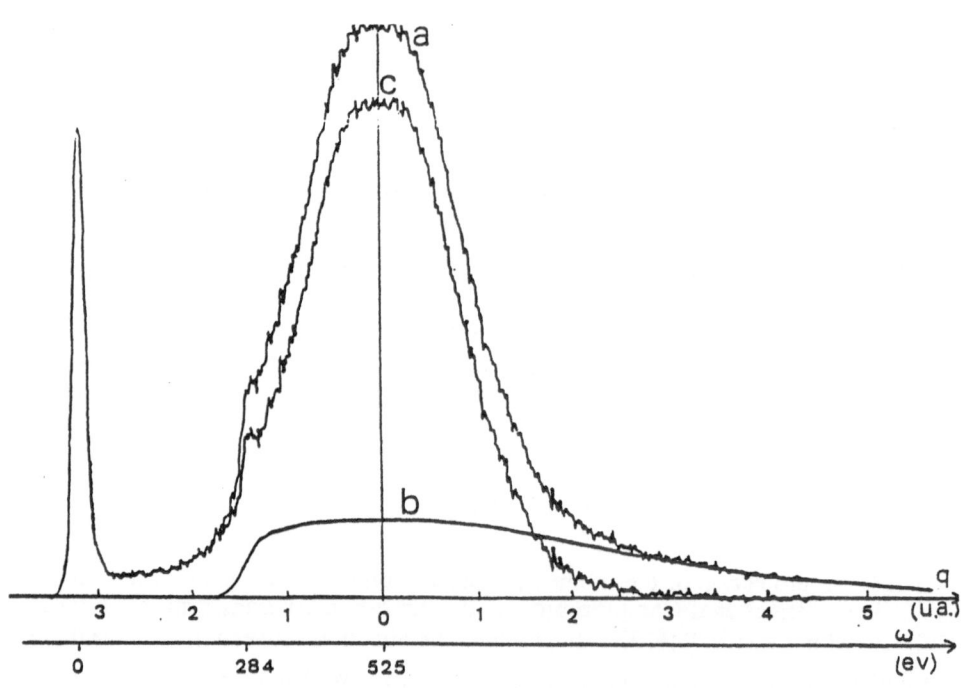

Fig.5.   *Compton profile of Be, along* c . *Scales in momentum and energy loss units (from ref. 31).*
a. Total profile
b. Core profile
c. Valence profile

heavy elements. As a consequence, the Compton profile is not symmetrical, and is limited in the short energy loss direction.

With adjustable wavelengths as provided by synchrotron sources, one can adjust at any convenience the position of the binding energy limit (Raman departure line) with respect to the centre of the Compton line, and get rid of a part of core electron Compton scattering. The situation is depicted on Fig.4, on the example of Be [31].

The major question is : how well can one measure Compton profiles ? Remember that this incoherent process can compete with photoelectric absorption. An interesting plot (Fig.5) compares the two cross sections, σ for Compton, τ for absorption, as a function of atomic number and incident photon energy. The interest of high energy (γ Rays or wiggler radiation) is evident in order to probe elements with Z higher than 15, the present limit with conventional X Rays. Up to now, transition elements could only be studied by γ Ray Compton scattering.

The main limitation is the resolution of the experiment. If $p_z = h q$, typical q values for valence electrons is of the order of 2 A^{-1} (1 au = 1.89 A^{-1}). With γ Rays, an intense source, the resolution is very poor, $\approx$ .45 au. The great advantage of synchrotron radiation is resolution. The low divergence of the beam (10^{-4}rd) is similar to the rocking curve width of strong Si perfect crystal reflections. This leads to both high reflectivity of the monochromator (channel cut Si crystal) and a very good resolution. Δq is presently of about .1au, and could be decreased to .05au with harder photons (wiggler radiation). This effect is essential since, even though the resolution function can be estimated, it damps high q tails of the profile, and leads to ridiculous statistics for large momentum transfer. With a resolution of the order of one half the width of valence electron profile, an enormous amount of information becomes hazardeous.

The apparatus presently used at LURE DCI is depicted on Fig.6, where a curved focusing analysing crystal is used, together with a position sensitive detetctor. Present statistics are poorer with X Rays than with γ Rays, but this is not of major importance.

If profiles are compared for different orientations of the sample with respect to K , one compares experiments done under the same conditions. Differences between two such profiles, revealing anisotropies of the momentum density, are indeed free from most systematic errors, compared to indivudual profiles, and the characteristics of such difference profiles are believed to be a very good signature of the electronic structure in momentum space. Core electron contributions subtract out. Such anisotropies can be checked towards theoretical calculations and are very sensitive to the level of approximations made in the theory, such as correlation.

Before reviewing a few examples, we should say that Compton profiles are much more sensitive to valence electrons than structure factors. However, it is an incoherent effect, so a low scattering process (problems with statistics). Therefore, the analysis of a Compton experiment for a molecular solid is highly questionnable, though some progress has been made [32,14]. Compton scattering is the favorite tool for very delocalized electrons, and diffraction can very easily distinguish between different sites ans deal with more localized electrons. The complementarity of the two techniques should be exploited, when possible.

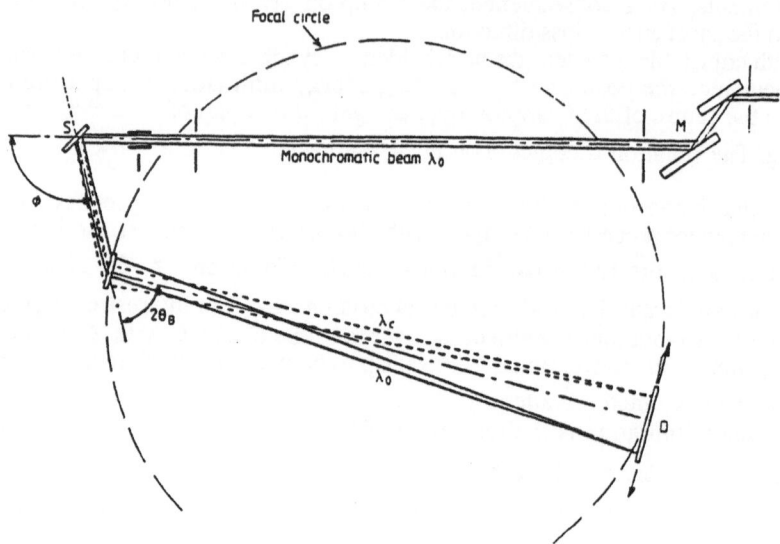

Fig.6. Compton scattering apparatus at LURE DCI (from ref. 30)

Synchrotron radiation has been essentially used by Loupias et al. at LURE DCI and Schulke at DESY. In the case of Be metal, many experiments have been done with conventional X Rays, γ Rays and synchrotron [30,33]. The general features of the various anisotropy profiles agree rather well. It is not possible on this example to quote a spectacular improvement using synchrotron, except for resolution. Comparisons with theory give the preference to density functional pseudopotential calculation : the difference between theory and experiment can be further lowered by 50% if a correction for correlation is incorporated. In this case as in other examples, $(J_{th} - J_{exp})$ is positive for small q, and becomes negative beyond the Fermi momentum. This might be due to the fact that theories do not treat correctly the core electrons : core-valence interaction is certainly not negligible (just think of core-valence orthogonalisation in metals).

Another example is that of LiH crystal. In this case the synchrotron experiment [34] led to anisotropies different from other experiments and to a much better agreement with a theory based on interacting closed shells for Li^+ and H^-. The experiment also favours 2p type contributions on Li atom. The experiment is highly complementary from an accurate diffraction experiment [35] and it is a case where one should be able to use a common model to analyse both experiments.

Another very interesting situation is that of graphite intercalation compounds, and particularly LiC_6 [36]. The crystallinity is poor, preventing from any diffraction, whereas Compton scattering is possible. A famous model for interpreting conducting properties of these compounds was the rigid band model where the valence electron of Li is donated to the rigid band of graphite. In such an hypothesis, the difference between Compton profiles for LiC_6 and graphite, normalised to one electron, should be of $p\pi$ character. Another extreme possibility would be a conduction due to the 2s electrons of sheets of lithium (the difference profile thus being of s character). The experimental result is shown on Fig.7, where the two extreme models are drawn for reference. The experiment thus appears decisive concerning the electronic structure of this compound. The rigid band model is wrong. This experiment led to an interaction with theorists [37] and it was found that the bands of graphite are significantly transformed, and that both s and p types are present in the conduction band structure. One can also infer the interest of Compton scattering to study hydrogen in metals, as suggested by Loupias.

In conclusion, Compton scattering is a difficult but rich type of measurement. Combination with diffraction may enlarge its domain of application to more molecular species. Synchrotron radiation leads to a considerably increased resolution and the tunability of the wavelength can be used profitably. It is also possible to think about experiments revealing more collective effects, like plasmons.

236

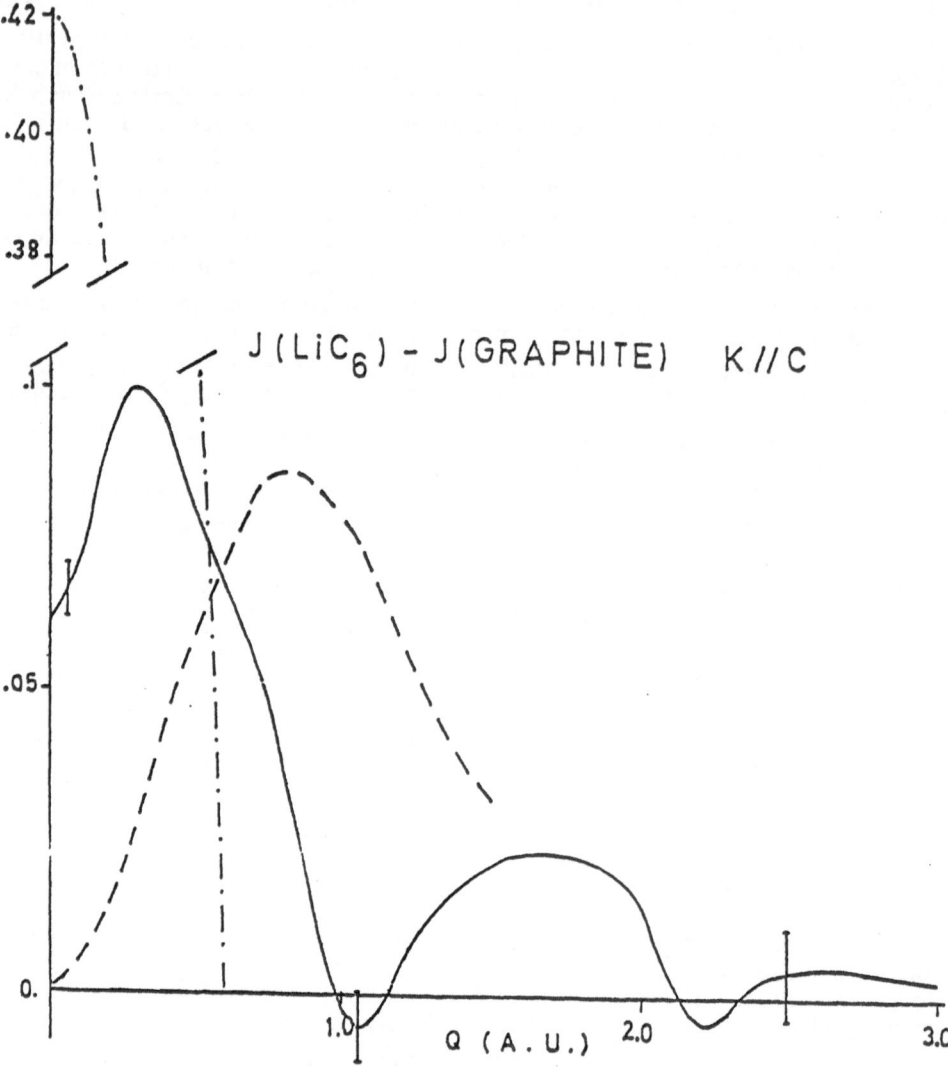

Fig.7.   Difference profile between LiC_6 and graphite, along c .
(———) . p type hypothetical profile
(_._._.). s type hypothetical profile

NON DIAGONAL MOMENTUM DENSITY

Schulke [38,39,40] recently introduced a very fascinating extension of the conventional Compton scattering, which is presently only accessible to quasi-perfect crystals : the measurement of non diagonal parts of the momentum space density matrix.

It is well known that in a perfect crystal at Bragg condition, there is a coherent coupling between the incident and diffracted waves, with a definite phase shift (which depends essentially on the departure from exact Bragg condition). Let k_0 and $k_H = k_0 + H$ be the wavevectors of the two coherent waves. Since the wavefield is a coherent superposition of the two waves, the target can absorb either a photon of k_0 or k_H, and through the interaction reemit an other photon of wavevector k_f. If k_f is such that :

$$| k_f - k_0^{\cdot} | = | k_f - k_H |$$

then for the same energy transfer, one can get either the normal Compton effect , or a non trivial term that mixes the momentum p with $p + H$ for the electron. If one takes a one electron orbital $\hbar(p)$, it can be shown that the cross section involves the product $\hbar^*(p+H)\hbar(p)$. Generalisation leads to an additional profile :

$$J_H(p_z) = \int \hbar(p, p + H) \, dp_x dp_y \qquad (42)$$

This gives access to the non diagonal part of the density matrix, and this contains information about the phase of various natural orbitals : such an information is lost in diagonal terms, like ρ or n. The example of $H = (1,1,1)$ for silicon is shown on Fig.8 [40]. We recall from (22,23) that :

$$\int J_H(p_z) \, dp_z = F(H) \qquad (43)$$

This type of experiment is really intermediate between the coherent and incoherent scattering processes. It might give information about the phase of a given structure factor. In fact, for (1,1,1) , Schulke finds by (43) a value in fair agreement with a direct structure factor measurement. The advantages of synchrotron radiation (low divergence, high brightness, tunability) are even more pronounced than for normal Compton scattering.

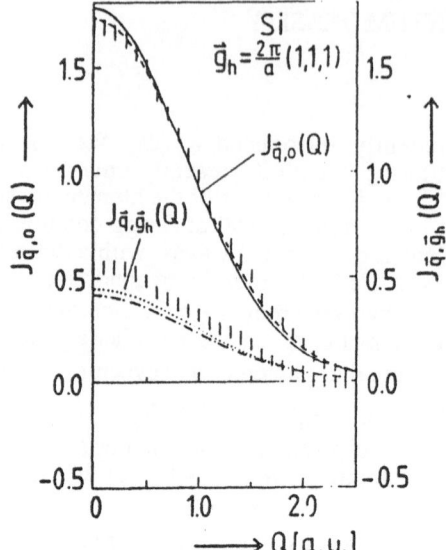

Fig.8.   Upper curve : Compton profile for Si (20,-18,1) direction. (___) corresponds to free atom $3s^2 3p^2$, (----) to a free atom $3s^1 3p^3$.

Lower curve. J_H for (111) reflection. (......) for $3s^2 3p^2$, (-----) for $3s^1 3p^3$ (from ref. 40)

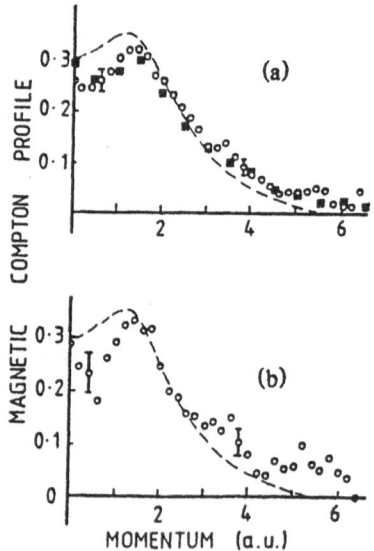

Fig.9.   Magnetic Compton scattering for Fe (from ref. 45)
a. Low energy : 46.4 kev
b. High energy : 61.9 kev
The dotted line corresponds to an APW calculation convoluted by the resolution function

MAGNETIC X-RAY SCATTERING

We will conclude this paper by a short discussion of a new process, the magnetic X-Ray scattering. This technique has been pioneered by de Bergevin and Brunel [41,42] in the case of diffraction, Sakai and Ono [46] and Cooper [45] in Compton scattering. Such studies were heroic with standard sources, but become more routinely accessible with synchrotron sources. They compete (or complement) with neutron magnetic scattering, the celebrated technique for measuring spin density.

The simplest theory can be found in a recent paper by Blume [43] and Platzman and Tsoar [44]. There is first a standard dipolar interaction between the electron spins and the magnetic field of the incident wave. Moreover there is a spin orbit coupling, proportional to $[\, s.E \times \{\, p - e/c \, A\}\,]$, where the magnetic component of E is $[\, -1/c \, \partial A/\partial t\,]$. Finally the term $(p. A)$ in the modified kinetic operator gives a third contribution. These terms lead to a cross-section that is dependent on both spin and orbital magnetic moments of the electron. The final result is :

$$\frac{\partial^2 \sigma}{\partial \Omega \partial \omega} = r_0^2 \left[\sum_{n,m} < n \mid \sum_j M_j e^{iQ.r_j} \mid m >^2 \delta(E_n - E_m - \hbar\omega) \right]$$

$$M_j = A + \frac{C.p_j}{mc} + i \, B.s_j \frac{\hbar\omega_0}{mc^2}$$

$$A = \varepsilon_0.\varepsilon_f \qquad \text{(Thomson scattering)}$$

$$\tag{44}$$

$$B = -\frac{\hbar\omega_0}{mc^2} \left[\varepsilon_0.\varepsilon_f (k_0 \times k_f) - \frac{1}{2}(q.q)(\varepsilon_0 \times \varepsilon_f) - q \times (q \times \varepsilon_0 \times \varepsilon_f) \right]$$

$$C = (\varepsilon_0.k_f)\,\varepsilon_f + (\varepsilon_f.k_0)\,\varepsilon_0$$

$$q = \frac{k_f}{k_f} - \frac{k_0}{k_0}$$

B and C are vectors which only depend on the polarisation states and the scattering geometry, in a complicated form. The important remark is that magnetic terms are both of order 10^{-2} with respect to A, the standard component. If no special care is taken concerning polarisation, no coupling term will exist between magnetic and Thomson amplitudes, and the intensity will be the sum of the Thomson and the magnetic intensities.Since the ratio of magnetic to total number of electrons is of order 10^{-1}, the magnetic intensity will be of the order of 10^{-6} of the Thomson intensity. The magnetic effect can only be detected when magnetic diffraction occurs at different places from nuclear diffraction.

However, Blume has shown that with a synchrotron source, the ratio between X Ray and neutron magnetic intensities is of order 1, proving the relevance of this effect.

The complexity of polarisation dependence can be seen as follows. A photon with circular polarisation has a definite angular momentum ±1. Its coupling with electron spin or orbital momentum will result in different terms. Moreover the coupling with + or - spins will be different. Linear polarisation is a mixture of two opposite circular polarisations and does not lead to spin state differentiation.

We also observe that the spin dependent component is in quadrature with Thomson scattering. As a consequence, if there is some anomalous scattering, the magnetic amplitude will couple directly with anomalous part of the Thomson amplitude, leading to a magnetic contribution of 10^{-3} of the total intensity. It may become possible to measure spin densities by X Rays. Since the polarisation dependence is different for spin and orbit, it is in principle possible to separate experimentally spin and orbital parts of magnetisation density : this is totally impossible with neutrons.

Finally, it is clear that by using circularly polarized photons, in an Compton scattering experiment, one can obtain a spin dependent momentum density. In order to get circularly polarized beam, it is necessary to use radiation out of the plane of the orbit in the storage ring [45]. By reversing the sign of magnetisation and taking the difference Compton profile, one gets the "magnetic Compton profile" :

$$J_m(p_z) = \int \sigma(\mathbf{p}) \, dp_x dp_y \qquad (45)$$

This experiment was performed by Cooper et al. at Daresbury on Fe, and one result is shown on Fig.9. It agrees rather nicely with an APW calculation, though the resolution is rather poor. The dip at $p_z = 0$ corresponds to an sp negative polarisation, also observed by neutron magnetic diffraction in the spin density. This experiment is essential and opens a new field.

Synchrotron radiation is absolutely necessary to obtain circular polarisation and to adjust the wavelength in order for example to get an adequate amount of anomalous scattering (to measure magnetisation density). There is a nice potential application in the domain of itinerant magnetism, where these experiments may play a crucial role.

In conclusion, we see that synchrotron radiation opens many important new fields for experiments concerning electron densities in condensed matter. Wiggler radiation and definite polarisation are very important properties. Owing to the importance of the concept of electron density in the description of any property, a strong theoretical effort should be made to analyse materials in terms of observable densities rather than artificial orbitals. A great effort was made by Bader [48] in that respect : he analyses molecules in terms of uniquely defined fragments of density, which may be transferable from one situation to another one. It is the opinion of the author that the level of understanding of various possible features of electron densities is still at a very naive stage.

REFERENCES

1. P.Becker, Editor (1980). Electron distributions in molecules and crystals. Plenum, N.Y.

2. K.Kurki-Suonio, Editor (1977). Physica Scripta, **15**, 67–162

3. F.L.Hirshfeld, Editor (1977). Israel J. of Chem.,**16**, 87-230

4. P.Coppens, M.B.Hall, Editors (1982). Electron Distributions and the Chemical Bond, Plenum, N.Y.

5. I.Olovsson, Editor (1986). Sagamore VIII. Chimica Scripta,**26**.

6. R.A.Bonham, M.Fink (1974). High Energy Electron Scattering, Van Nostrand

7. P.Hohenberg, W.Kohn (1964). Phys. Rev.,**B136**, 864

8. W.Kohn,L.J.Sham (1965). Phys. Rev.,**A140**, 1133

9. S.Lundqvist, N.H.March (1983). Theory of the Inhomogeneous Electron Gas, Plenum, N.Y.

10. P.Becker, C.Cohen-Addad, B.Delley, F.L.Hirshfeld, M.S.Lehmann (1986). Applied Quantum Chemistry, 361, Reidel.

11. B.Delley, P.Becker,B.Gillon (1984). J. Chem. Phys., **80**, 4286

12. R.P.Feynman (1939). Phys. Rev.,**56**, 340

13. P.O.Lowdin (1955). Phys. Rev.,**97**, 1474

14. R.Erdhal, V.H.Smith Jr (1987). Density Matrices and Density Functionals, Reidel.

15. P.Becker, P.Coppens (1974). Acta Cryst., **A30**, 129
                        (1975). Acta Cryst.,**A31**, 417

16. N.Kato (1980). Acta Cryst., **A36**, 763

17. J.Schneider (1974). J.Applied Cryst.,7, 541

18. P.Bastie, P.Becker (1984). J.of Phys.,C17, 193

19. P.Becker (1977). Acta Cryst.,A33, 243

20. R.Bachmann,H.Kohler, H.Schulz, H.P.Weber (1985). Acta Cryst.,A41, 35

21. H.Hoche, H.Schulz, H.P.Weber, A.Belzner, A.Wolf, R.Wulf (1986) Acta Cryst., A42, 106

22. P.Suortti, J.D.Jennings (1977). Acta Cryst.,A33, 1012

23. F.S.Nielsen, P.Lee, P.Coppens (1986). Acta Cryst.,B42, 359

24. U.W.Ardnt (1978). J.of Phys.,E11, 671

25. J.Z.Tischler, B.W.Batterman (1984). Phys. Rev.,B30, 7060

26. J.Z.Tischler, Q.Shen, R.Collela (1985). Acta Cryst.,A41, 451

27. J.Dumond (1933). Rev. Mod. Phys.,5, 1

28. K.Tanaka, E.Elkaim, L.Li, Z.Jue, P.Coppens, J.Landraum (1986) J. Chem. Phys.,84, 6969

29. P.E.Mijnaerends (1977). in Compton Scattering, B;Williams, Ed., McGraw Hill.

30. M.J.Cooper (1985). Rep. on Progr. in Phys.,48, 415

31. G.Loupias (1985). Séminaire Dautreppe sur le rayonnement synchrotron, Grenoble.

32. W.Weyrich, P.Pattison, B.G.Williams (1979). Chem. Phys.,41, 271

33. M.Y.Chou, P.K.Lam, M.L.Cohen, G.Loupias, J.Chomilier, J.Petiau (1982). Phys. Rev. Lett.,49, 1452

34. G.Loupias, J.Chomilier (1986). Z. Phys.,D2, 297

35. J.P.Vidal, G.Vidal-Valat (1986). Acta Cryst.,B42, 131

36. G.Loupias, J.Chomilier (1984). J. Phys. Lettres.,45, L301

37. J.Chomilier, G.Loupias, J.Felsteiner (1985). Nucl. Inst.& Meth.,A235, 603

38. W.Schülke (1981). Phys. Lett.,A83a, 451

39. W.Schülke, U.Bonse, S.Mourikis (1986). Phys. Rev. lett.,47, 1209

40. W.Schülke, S.Mourikis (1986). Acta Cryst.,A42, 86

41. F.de Bergevin, M.Brunel (1972). Phys. Lett.,**A39**, 141

42. F.de Bergevin, M.Brunel (1981). Acta Cryst.,**A37**, 314

43. M.Blume (1985). J. Appl. Phys.,**57**, 3615

44. P.M.Platzman, N.Tzoar (1985). J. Appl. Phys.,**57**, 3623

45. M.J.Cooper, D.Laundry, D.A.Cardwell, D.N.Timms, R.S.Holt (1986). Phys. Rev.,**B34**, 5984

46. N.Sakai, K.Ono (1977). J. Phys. Soc. Japn.,**42**, 770

47. D.Gibbs, D.E.Moncton, K.L.D'Amico (1985). J. Appl. Phys.,**57**, 3619

48. R.F.W.Bader (1985). Acc.for Chem. Res.,**18**, 9

49. P.Becker, P.Coppens (1985). Acta Cryst.,**A41**, 177

50. P.Coppens, T.Koritsanzky, P.Becker (1986). Chim. Scripta,**26**, 463

DISCUSSION

EXTINCTION

Reduction of sample size, as is possible for intense synchrotron radiation, diminishes secondary extinction, which is very important for organic crystals, but not primary extinction. Wavelength is a more important variable as decrease of wavelength diminishes extinction. In neutron diffraction by powders with very small crystal sizes, extinction is still visible, especially at the longer wavelengths.

There are samples for which the assumptions of the Becker-Coppens corrections are not valid, and others for which no adequate theory of extinction exists. Interpretation of data for neutron diffraction by silicon powder by a more sophisticated approach, which included a different angular dependence for primary and secondary extinction, gave Debye-Waller factors very different from those obtained from conventional corrections.

THE HALF-WAVELENGTH PROBLEM

This problem can be avoided. For Si(111) it is a $\lambda/3$ problem, because (222) is forbidden, and a focussing mirror at a grazing angle of a few milliradians will remove the high energy part of the beam.

DEFORMATION DENSITIES

An advantage of X-N maps is that when determined at liquid nitrogen temperature the same Debye-Waller factors are often obtained from X-rays and neutrons, suggesting that the more important systematic errors may have been eliminated. It would be even better to use 15 K, but X-ray diffractometers will not support the Displex refrigerator needed for that temperature.

In the X-X experiment, the assumption that the systematic errors are the same for the lower and the higher order reflections may not always be valid.

DECONVOLUTION OF THERMAL MOTION

There is no practically useful or theoretically rigorous method of deconvoluting the effects of thermal motion from the diffracted intensity of X-rays, because it is extremely difficult to predict how electronic wavefunctions vary with changes in nuclear position. Any crude but realistic model which predicted the variation of charge density with nuclear geometry would open the door to the understanding of chemical reactivity.

SCHWINGER SCATTERING

Brown and Forsyth have recently measured the aspherical part of the 2 electron density difference between Ga and As in GaAs by means of the Schwinger scattering of neutrons. In this process the neutrons interact with the electric field in the crystal, and so with both the nuclear charge and all the electrons (the X-ray form factor). Intensity is down by a factor of 10^{-3}, so the effect is normally seen only as an interference term with a major scattering process, such as nuclear scattering and in this case the (222) reflection of GaAs.

ELECTRON DIFFRACTION

High energy electron scattering can give the electron pair correlation function, but, on account of vibrational problems, only for homonuclear diatomic molecules.

It might be possible to do magnetic experiments by electron diffraction, by coupling the weak magnetic interaction with a strong conventional reflection in a double scattering process.

STRUCTURAL APPLICATIONS OF X-RAY ABSORPTION SPECTROSCOPY (EXAFS AND XANES) IN COORDINATION CHEMISTRY.

J. Goulon[1,2], M. Loos[1], P. Friant[1] and M. Ruiz-Lopez[1]

1) Laboratoire de Chimie Théorique, U.A. CNRS n° 510, Université de Nancy I , B.P. 239 – F54506 – VANDOEUVRE-LES-NANCY Cedex – France.
2) LURE – L.P. CNRS/CEA/MEN , Université de Paris-Sud, Bât. 209d – F91405 – ORSAY Cedex – France.

ABSTRACT. The aim of these lecture notes is to make the reader more familiar with the conditions under which accurate structural information can be extracted from both EXAFS and XANES spectra. Following a recent and elegant theoretical contribution due to Natoli et al. [14,25] a unifying interpretation scheme of the whole X-ray absorption spectrum is given using a single electron excitation theory with multiple scattering of the photoelectron by a cluster of atoms. XANES spectra are thus shown to be governed by scattering resonances (cage resonances) of the cluster and by a so-called "full multiple scattering" regime. Conversely, EXAFS spectra are dominated by single scattering processes but higher correlations can also be detected and used for practical purposes under conditions which we discuss. The successive steps of standard analyses of the EXAFS spectra are then detailed. It is shown how "extended continuum" MSW-Xα calculations can be helpful for extracting specific information from edge or XANES spectra. Cases of ill-conditioned EXAFS problems are discussed and it is proven, using several illustrative examples taken from porphyrin chemistry, that difference FT analyses taking specific structural perturbations into account open new paths to unravel the hidden structural information from EXAFS spectra. Selected applications concerning biomimetic complexes of cytochrome P-450 or polymeric chains of organic conducting materials are presented. These lecture notes close with considerations regarding experimental techniques. Exciting time resolved experiments are now becoming possible with energy dispersive spectrometers but emphasis will be laid on the specific advantages of energy scanned spectrometers, i.e. the possibility of recording X-ray excitation spectra using radiative (non radiative) secondary emissions: X-ray fluorescence for low dilution and X-ray excited luminescence for site selectivity.

1. INTRODUCTION

Synchrotron radiation is produced by all existing e^+/e^- storage rings

247

M. A. Carrondo and G. A. Jeffrey (eds.), Chemical Crystallography with Pulsed Neutrons and Synchrotron X-Rays, 247–293.
© 1988 by D. Reidel Publishing Company.

operating at beam energies ranging from a few hundred MeV up to several GeV. Over the past decade, access to such powerful sources of radiation has considerably renewed the interest in X-ray absorption spectroscopy and initiated its mutation into a reliable tool for determining the local structure around any selectively excited absorbing atom [1-8]. As single crystals are not required, there is a wide field of application immediately coming to mind in materials science, industrial catalysis, inorganic chemistry and biochemistry. It is certainly not the aim of these short lecture notes to give an exhaustive review of all the recent applications of X-ray absorption spectroscopy. We simply wish to make the reader more aware of some real strengths but also of some specific limitations of this technique as a practical complement to chemical X-ray crystallography. For the convenience of our presentation, illustrations produced here have all been selected in the very rich domains of metalloporphyrin chemistry and homogeneous catalysis, these particular fields being of direct interest to our group. Frustrated readers or those who are looking for more details on any specific application are kindly referred to recent books or general reviews on the subject [1-8].

X-ray absorption spectroscopy is concerned with the excitation of electronic transitions originating from atomic like deep core states (K for 1s electrons, L for 2s, 2p electrons, ...). For a given sample, the X-ray absorption spectrum thus exhibits a series of well separated K, L... absorption edge singularities at energies which are characteristic of each atomic species present in the sample : each one of these elements can thus be selectively excited... at least in principle. It is now essential to introduce a common distinction between different features of the spectra :

(i) The structures detected in the close vicinity of a given edge singularity are classically referred to as X-ray absorption near edge structures (XANES) [3]. Of particular interest to the chemist are the pre-edge structures which are associated with electronic transitions towards unoccupied molecular orbitals. According to well known electric multipole selection rules, the presence (absence) in the XANES spectra of such characteristic signatures is often a useful indication of the symmetry of the absorbing site [9]. Furthermore, provided that the final state can be unambiguously defined, the relative energy shift of these pre-edge structures can be correlated with variations of the charge (valence) of the absorbing atom [10].

(ii) On the high energy side of the absorption edge singularity, the absorption spectra exhibit oscillatory structures now commonly referred to as EXAFS (Extended X-ray Absorption Fine Structure). These oscillations can be detected in practical cases of interest here up to 1200 eV (sometimes even more) above the edge with a magnitude of ca. 10% of the purely atomic contribution. EXAFS oscillations have a structural content related mostly to the radial distribution of the different atoms surrounding the absorbing center [1]. EXAFS oscillations are indeed completely absent from spectra taken from monoatomic gases.

Indeed the EXAFS oscillations have been known for a very long time as they were discovered by R. de L. Kronig [11] early in this century. However, their potentiality as a structural tool was recognized only

recently by E. Stern and collaborators [12,13] who also developed the
basic theoretical background required for a quantitative analysis of
the phenomenon. It was realized that the final state of the excited
photoelectron has to be represented by the interference of an outgoing
(photoelectron) spherical wave with the incoming contributions due to
the complicated backscattering of the outgoing photoelectron by the
whole cluster of atoms surrounding the absorbing center. As this interference process is basically energy dependent, it gives rise to the
observed modulation of the absorption coefficient on varying the energy
of the incident photons. Clearly EXAFS is a kind of electron scattering
experiment where the involved electrons are generated "in situ" by the
X-ray absorption process. However as explicited in the next section,
only closed scattering paths beginning from and ending at the photo-
absorbing site O are concerned here because one is measuring the X-ray
total cross-sections with the initial state well localized at site O :
it is this peculiarity that entails the site specificity of X-ray absorption spectroscopy [14].

The outline of these lecture notes is then as follows. The next
section is dedicated to the difficult problem of extracting structural
information from both EXAFS and XANES spectra. Following a recent and
elegant theoretical contribution due to Natoli et al. [14,25], we introduce first a unifying interpretation scheme of the whole X-ray absorption spectra. This scheme is based on a single electron excitation
theory but in the general framework of multiple scattering of the photoelectron by a cluster of atoms [14]. Then we shall detail the successive steps of standard analyses of EXAFS spectra. We wish also to show
what kind of structural information can be extracted from edge and
XANES spectra using "extended continuum" MSW-Xα theoretical calculations. We next consider cases of ill-conditioned problems where standard analyses of EXAFS spectra usually fail to give reliable results.
Several illustrative examples taken essentially from porphyrin chemistry will be produced in order to show that difference FT analyses taking specific perturbations into account open new paths to unravel the
hidden structural information of EXAFS spectra.

The last section should be concerned with experimental techniques
using synchrotron radiation sources. Exciting time resolved experiments
are now possible with energy dispersive spectrometers but special emphasis will be laid also on a specific advantage of energy scanned
spectrometers, i.e. the very open possibility of recording X-ray excitation spectra using radiative (non radiative) secondary emissions.

2. THE INTERPRETATION OF X-RAY ABSORPTION SPECTRA

We do not intend to derive in full detail here the formulae currently
used for analyzing EXAFS spectra. Readers in search of more information
on the theoretical foundations of EXAFS and XANES spectroscopies are
strongly recommended to refer to specialized papers [1,14-27]. We wish
simply to link the practical problem of extracting structural information from real experimental spectra with what the usual single-elec-
tron excitation theories can tell us. In this discussion we shall rely

only on the standard knowledge of quantum mechanics.

2.1. Single electron excitation theory based on multiple scattering

Starting from Fermi's golden rule, one can write the X-ray absorption cross section of a representative cluster of atoms as :

$$\sigma = 4\pi^2 \alpha \hbar \omega \sum_f |<\psi_f|\vec{\epsilon}.\vec{r}|\psi_{in}>| \rho(E_i + \hbar\omega) \tag{1}$$

$$\alpha = \frac{e^2}{4\pi\epsilon_o c\hbar}$$

where $\hbar\omega$ is the photon energy, $\vec{\epsilon}$ its linear polarization. This expression is consistent with the usual electric dipole approximation. Recent calculations [22] have confirmed that the neglect of higher order terms (e.g. the electric quadrupole term) remains justified in the X-ray range for absorbers of low atomic number. Now, there are three different approaches for calculating σ :
(i) the scattering wave approach where ψ_f is the time reversed scattering wave $\psi_{\vec{E}}^-$ of a photoelectron with the energy $E = \hbar\omega - I_o$, I_o being the relevant ionisation energy ,
(ii) the Green's function approach whereby the problem reduces to building the solution of the operator equation $(E-H)G^- = I$ with appropriate boundary conditions ,
(iii) the band structure approach for periodic systems where the scattering states are replaced by Bloch states.
According to the generalized forms of the optical theorem established by Lloyd et al. [28] and more recently by Natoli et al. [14,25], all three approaches are equivalent in the framework of the multiple scattering formalism originally introduced by Johnson [29] and others [30]. As illustrated by figure 1 , there is a stringent parallelism between the elegant formalism of Natoli et al. and the well-known theory for electron scattering by an atomic potential of spherical symmetry. The X-ray absorption cross section is thus rewritten as [25,28] :

$$\sigma = -4\pi\alpha k\hbar\omega \sum_{L,L'} M_L M_{L'} \; Im\left\{\left[(\underline{I} - k\underline{T}_a\underline{H})^{-1} \, k\underline{T}_a\right]_{L\,L'}^{o\,o}\right\} \tag{2a}$$

where : $k = [2m(E - V_I)]^{1/2}/\hbar$ and $M_L = <\underline{R}_L Y_L|\vec{\epsilon}.\vec{r}|\psi_{in}>$ (2b)

In equation (2a), $\underline{T}_a$ is an atomic transition matrix (according to the general definition) and contains all the phase-shift informations of each scatterer whereas $\underline{H}$ is a generalized Hankel propagator which contains all the structural, geometrical informations. Within the general Green's function approach, the operator $\underline{\tau} = [\underline{I} - k\underline{T}_a\underline{H}]^{-1} \, \underline{T}_a = [\underline{T}_a^{-1} - k\underline{H}]^{-1}$ is a "scattering path" operator [27]. Of course , equation (2) becomes much simpler for K edge X-ray absorption spectra because $M_L = M_{L'}$ and

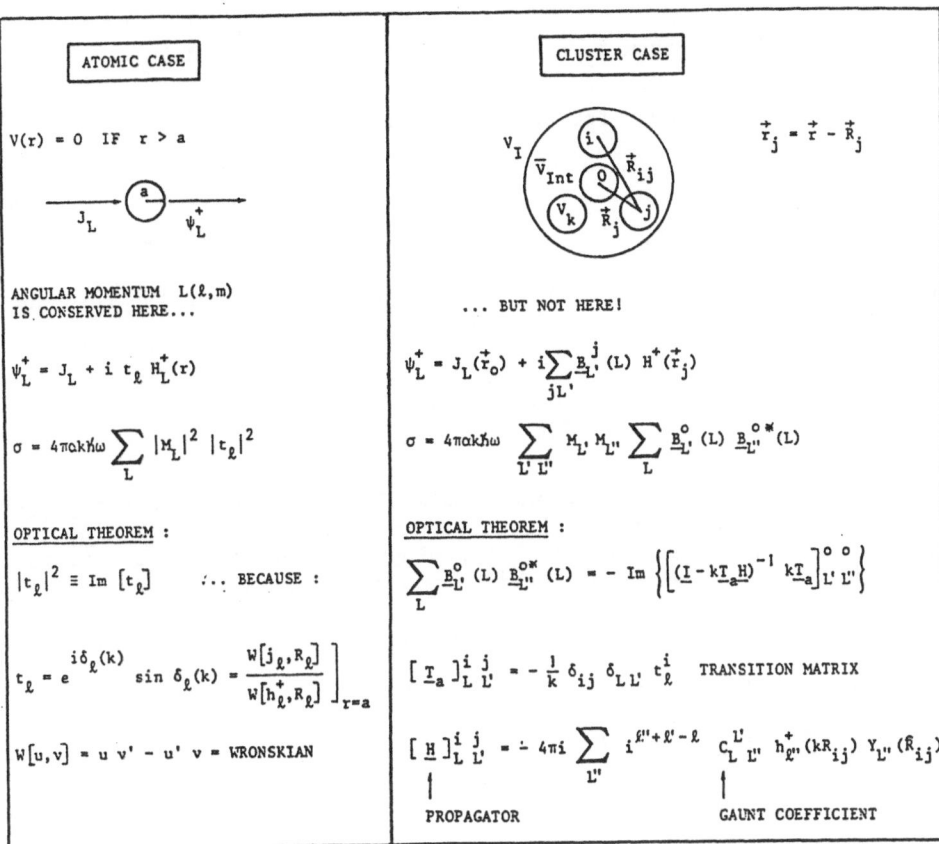

Figure 1 : Single scattering by an atomic spherical potential and multiple scattering by a cluster of atoms.

one single index L is retained. Now following Natoli et al. [14,19], again three important regimes can be considered :

2.1.1. "Shape" or "Cage" resonances [19]. These resonances are nothing more than the cluster analogues of the scattering resonances which are well known in the atomic scattering case and are associated with the presence of some effective repulsive potential (V > 0) : this potential creates a sort of cage that traps the final state electron in a quasi-bound state decaying away with a lifetime $\tau = \hbar \, \Gamma_2^{-1}$ connected with the tunneling probability through the barrier. A well known example is the centrifugal barrier whose height $\sim \ell(\ell+1)/R^2$ gives rise to strong atomic resonances for $\ell_f \geqslant 3$. In the usual atomic formalism, these resonances are thus associated with a singularity of the reactance matrix K defined as :

$$K_\ell^{-1} = k \cotan \delta_\ell = k(t_\ell^{-1}+i) \quad \text{or} \quad K_\ell = \frac{1}{k} \tan \delta_\ell \qquad (3)$$

Singularities thus occur for $\delta_\ell = \pi/2$. Assuming that :

$$\tan \delta_\ell \simeq \frac{\Gamma_r}{2(E_r-E)} \qquad (4a) \quad \text{then :} \quad t_\ell = \frac{\Gamma_r}{2(E_r-E)-i\Gamma_r} \qquad (4b)$$

and one obtains the often called Breit-Wigner formula for the resonant X-ray cross section :

$$\sigma_L \simeq 4\pi\alpha\hbar\omega \, k \, |M_L|^2 \, \frac{\Gamma_r^2}{4(E_r-E)^2 + \Gamma_r^2} \qquad (4c)$$

In the cluster case, resonances are also to be expected for singularities of the scattering path operator rewritten as $\underline{\tau} = [\underline{M}-i\underline{\Delta}]^{-1}$ where $\underline{M}$ and $\underline{\Delta}$ are real matrices. By analogy with equation (4b), it is clear that the resonances are defined now by the general condition [19]:

$$\text{Det } \| \underline{M} \| = 0 = \text{Det } \| \underline{K}_a^{-1} - k \, \underline{N} \| \qquad (5)$$

where we used the classical definitions :

$$\underline{K}_a^{-1} = \underline{T}_a^{-1} - i \, k \qquad (6a)$$

$$\underline{H}_{LL'}^{ij} = N_{LL'}^{ij} - i \, J_{LL'}^{ij} \qquad (6b)$$

taking advantage of the important property that the atomic reactance matrix is hermitian but also __real__ symmetric when using a basis of real spherical harmonics. $N_{LL'}^{ij}$ denotes the associated Neumann propagator.

The implicit relation $F(\cotan \delta_\ell^i, k_r R_i, \hat{R}_{ij}) = 0$ is not always easy to handle in practical cases. However, if one is comparing two different systems corresponding to the same basic geometrical arrangement ($\hat{R}_{ij}$ invariant) and with __transferable__ phase-shifts (δ_ℓ also invariant), then one is led to Natoli's rule [19]:

$$\left[k_{r_1} \cdot R_i\right]^{(1)} = \left[k_{r_2} \cdot R_i\right]^{(2)} = \text{constant} \qquad (7)$$

where k_r is related to the resonance energy by $k_r = \left[2m(E_r-\overline{V}_I)\right]^{1/2}\hbar^{-1}$.

This rule is of great practical interest for the soft X-ray domain where the EXAFS oscillations do not extend over large energy range.

2.1.2. __Full multiple scattering regime__ [14]. In this regime, some "pathological" phase shifts, although not resonating, still cross the $\pi/2$

value before decaying to zero according to the Levinson's theorem (e.g. as for antiresonances). The Natoli-Born series defined below does not converge : only direct but tedious calculations of $\underset{\sim}{\tau}$ can be done in this energy range which is fortunately restricted, in practice, to a few eV above the edge.

2.1.3. <u>Single scattering and higher correlations</u> [14]. As mentioned above, it is convenient to consider also the Natoli-Born series expansion of the scattering path operator :

$$\underset{\sim}{\tau} = (I - kT_a H)^{-1} T_a = \sum_{n=0}^{\infty} (kT_a H)^n T_a \tag{8}$$

Its convergence is ascertained as soon as the convergence radius satisfies the condition $\rho(kT_a H) < 1$ where $\rho(\underset{\sim}{A})$ denotes the maximum eigenvalue of $\underset{\sim}{A}$. Now, it is easy to check that the first term (n=0) of the series leads to the purely atomic cross section $\sigma^{(0)}$ and that the next one is always zero since $H_{1m,1m}^{o\ o} \equiv 0$.

At this stage, we may introduce the EXAFS modulations $\chi^\ell(k)$ of the X-ray absorption spectrum. According to the electric dipole selection rules, the polarization averaged absorption coefficient μ is given by :

$$\mu = n_{abs}.\cdot\sigma = (\ell_0+1)\ \mu_o^{(\ell_0+1)}\ \left[\chi^{(\ell_0+1)} + 1\right] +$$

$$+ \ell_0\mu_o\left[\chi^{(\ell_0-1)} + 1\right] \tag{9}$$

where ℓ_0 indicates the angular momentum of the core initial state (e.g. $\ell_0 = 0$ for K-edge excitation). We have for $\ell_f = \ell$:

$$\chi^\ell = \frac{1}{(2\ell+1)}\ \operatorname{Im}\left\{e^{2i\delta_0^\ell} \cdot \sum_m \left[(I-kT_a H)^{-1}\ kT_a\right]_{1m,1m}^{o\ o}\right\} \tag{10}$$

and as a consequence of the Natoli-Born series expansion :

$$\chi^\ell(k) = \sum_{n=2} \chi_{(n)}^\ell(k) \tag{11}$$

Clearly, χ_n^ℓ represents the partial contribution coming from all processes where the photoelectron emanating from the absorption site 0 is scattered (n-1) times by the surrounding atoms and returns to site 0 before escaping to free space. The first term χ_2^ℓ is associated with the standard single scattering picture of EXAFS and therefore gives information only about the radial distribution of the surrounding scatterers whereas more complete information about the stereogeometry of the system is available, at least in principle, from the higher order correlation terms $\chi_3^\ell(k)$, $\chi_4^\ell(k)$ .. An important question is to know whether these higher order correlations can be detected experimentally :

- Far above the edge, i.e. for very large k values, the scattering amplitude becomes very weak and multiple scattering processes of high order should have little or no contribution. Near the threshold, however, i.e. where the scattering is quite strong, this argument fails.

- There is a need to average the energy for any single particle calculation in order to account for the finite experimental energy resolution ($\simeq$ 1 eV) but also for the lifetime of the core hole ($\Gamma_h \simeq 0.5$ eV) and for the lifetime of the photoelectron itself. The latter contribution $\Gamma_e(k)$ can be estimated from the imaginary part of the complex energy dependent potentials (e.g. Hedin-Lundquist potentials [31]). Both long multiple scattering paths and also distant single scattering events should produce no structure that survives a severe energy broadening because each χ_n^ℓ is to be multiplied by a damping function $D_n^\ell(\Gamma)$ which under the simplest approximations is to be written as [32] :

$$D_n^\ell(\Gamma) = \exp\left\{ -\frac{\Gamma_t}{k} \frac{\hbar^2}{m} \left[R_{np} + \Delta_n^\ell(k) \right] \right\}$$ (12a)

where : $\Gamma_t = \Gamma_{exp} + \Gamma_h + \Gamma_e(k)$ $R_{np} = \sum_{i=0}^{n-1} R_{i(i+1)}$ (12b)

whereas Δ_n^ℓ is the first derivative of the total phase shift :

$$\Delta_n^\ell = \frac{\partial}{\partial k} \left[2\delta_o^\ell(k) + \phi_n^\ell(k) \right] \qquad (\Delta < 0)$$ (12c)

- The angular dependence of the various paths R_{np} involved in the definition of χ_3^ℓ or χ_4^ℓ ...is very determinant. For instance, one would check easily that χ_3^1 is proportional to $P_{\ell=1}(\vec{R}_i \cdot \vec{R}_j)$ and therefore $\chi_3^1 = 0$ if $\vec{R}_i \cdot \vec{R}_j = 0$ (e.g. as in an undistorted octahedral symmetry). Conversely, it has been known for a long time that any colinear or slightly bent arrangement of the central absorber with two (or even more) scatterers results in a so-called "shadowing effect" [1,15], i.e. in a dramatic enhancement of the contribution of the most distant scatterers (Figure 2). Several popular ligands used in coordination chemistry : CO , CS , NO , CN$^-$, SCN$^-$, R-C$\equiv$N , etc...give rise to such spectacular effects which can be explained by the fact that (i) forward scattering is always most intense, (ii) the χ_2^ℓ , χ_3^ℓ , χ_4^ℓ

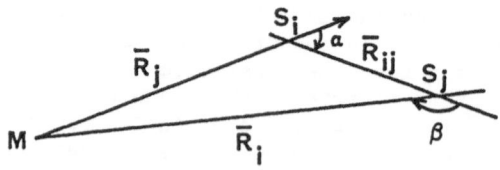

<u>Figure 2</u> : Multiple scattering path. "Shadowing" effect is expected to occur for $\alpha \simeq 0°$ and $\beta \simeq 180°$.

contributions are overlapping as $R_{2p} \simeq R_{3p} \simeq R_{4p}$. In that situation, corrections for χ_3^{ℓ} and χ_4^{ℓ} are absolutely essential but very complicated due to the interference of χ_3^{ℓ} and χ_4^{ℓ} which cannot be resolved.

- What is actually measured is never χ^{ℓ} but the configurational average $\langle\chi^{\ell}\rangle$ as a consequence of the thermal disorder affecting the whole cluster configuration. As illustrated below, the functional expression of χ_n^{ℓ} can always be written as:

$$\chi_n^{\ell} = D_n^{\ell}(\Gamma) \; |A_n^{\ell}(\hat{R}_{ij}, k, R_{np})| \; \sin\left[kR_{np} + 2\delta_0^{\ell} + \phi_n^{\ell}(k, R_{np}, \hat{R}_{ij})\right] \qquad (13)$$

and it is true that the first order correction due to thermal disorder is concerned with the calculation of $\langle\exp\{-ikR_{np}\}\rangle$. Boland and Baldeschwieler have addressed this problem and derived through a normal mode analysis a generalized Debye-Waller factor for a 3-atom cluster with C_{2v} symmetry ($A_1 + B_1$ stretching modes + A_1 bending mode) or $D_{\infty h}$ ($\Sigma_u^+ + \Sigma_g^+$ stretching modes + Π_u bending mode) [33]. They concluded that the damping due to thermal disorder is always less intense for χ_3^1 or χ_4^1 than for χ_2^1 (second shell) and is highly dependent upon the bridging angle and temperature. One has however to realize that the complex scattering amplitudes $f_{n>2}^{\ell}$ are also affected by the bending modes as already suggested by Alberding and Crozier [34]. In other words, a configuration average should be taken on the successive products of the propagators H_{LL}^{ij} expressed as function of the normal coordinates Q_p. It is our opinion that this problem should be analyzed in detail for quantitative uses of $\langle\chi_3^{\ell}\rangle$ and $\langle\chi_4^{\ell}\rangle$.

For applications to practical analyses, we need to explicit χ_2^{ℓ} ... in a more tractable form and discuss successive levels of approximations. Neglecting the damping term D_2^{ℓ}, one obtains according to Schaich [23] or Gurman et al. [24]:

$$\chi_2^{\ell} = (-1)^{\ell} \; \mathrm{Im}\left\{ e^{2i\delta_0^{\ell}} \sum_{j\neq 0, \ell'} (2\ell'+1) \; t_{\ell'}^{j} \; (-1)^{\ell'} \times \right.$$

$$\left. \times \sum_{\ell'} (2\ell''+1) \left[\begin{pmatrix} \ell & \ell & \ell'' \\ 0 & 0 & 0 \end{pmatrix} \underline{h}_{\ell''}^{+}(kR_j) \right]^2 \right\} \qquad (14)$$

where the classical Wigner's 3j-symbol has been used. "Reduced" Hankel functions can also be defined:

$$\underline{h}_{\ell}^{+}(\rho) = i^{(\ell+1)} \; h_{\ell}^{+}(\rho) = \frac{e^{i\rho}}{\rho} \; \underline{\underline{h}}_{\ell}^{+}(\rho) \qquad (\rho = kR) \qquad (15)$$

and the complex scattering amplitude f_2^{ℓ} is to be rewritten as:

$$f_2^{\ell}(\pi, k, \rho) = \frac{1}{k} \sum_{\ell'} (2\ell'+1) \; t_{\ell'}^{j} \; \underline{\underline{H}}^{+}(\ell, \ell') \qquad (16)$$

with for K and L shell excitations :

$$\underline{\underline{H}}(1,\ell'\S) = \frac{\ell'+1}{2\ell'+1} \left[\underline{\underline{h}}^+_{(\ell'+1)}\right]^2 + \frac{\ell'}{2\ell'+1} \left[\underline{\underline{h}}^+_{(\ell'-1)}\right]^2 \tag{17a}$$

$$\underline{\underline{H}}(0,\ell') = \left[\underline{\underline{h}}^+_{\ell'}\right]^2 \tag{17b}$$

$$\underline{\underline{H}}(2,\ell') = \frac{3}{2}\frac{(\ell'+2)(\ell'+1)}{(2\ell'+3)(2\ell'+1)}\left[\underline{\underline{h}}^+_{(\ell'+2)}\right]^2 +$$

$$+ \frac{\ell'(\ell'+1)}{(2\ell'+3)(2\ell'-1)}\left[\underline{\underline{h}}^+_{\ell'}\right]^2 + \frac{3}{2}\frac{\ell'(\ell'-1)}{(2\ell'+1)(2\ell'-1)}\left[\underline{\underline{h}}^+_{\ell'-2}\right]^2$$

$$\tag{17c}$$

and finally :

$$\langle\chi^\ell_2\rangle = \sum_j \frac{N_j}{k R_j^2} |f^\ell_2(\pi,k,\rho)| \; D^\ell_2(R_j,\Gamma,\Delta^\ell) \; e^{-2\sigma_j^2 k^2} \times$$

$$\times \sin\left[2kR_j + 2\delta^\ell_o(k) + \phi^\ell_j(k,\rho)\right] \tag{18}$$

Of course, one may also write formulae equivalent to equations (14) → (18) for the higher terms $\langle\chi^\ell_3\rangle$ or $\langle\chi^\ell_4\rangle$ but the relevant expressions are far more cumbersome and should involve 6j- and 9j-symbols. Now in order to speed up the calculations one may think of using two levels of approximation :

(i) $\underline{\underline{h}}^+_\ell(\rho) \equiv 1$ is the plane wave approximation (pwa).

(ii) $\underline{\underline{h}}^+_\ell(\rho) \xrightarrow[\rho\to\infty]{} e^{i\ell(\ell+1)/2\rho}\left[1 + \frac{\ell(\ell+1)}{2\rho^2}\right]^2$ is the fast spherical

wave approximation (fswa) which has been quite recently used by Rehr et al. [26]. A major difference is that in the former case (pwa), both the modulus and phase of $f^\ell_2(\pi,k)$ become independent of $\rho = kR_j$ and thus can be tabulated as long as the atomic t^ℓ_j are not really dependent of the cluster geometry via the potential definition. Clearly, the pwa formulae have been most extensively used over recent years with unequivocal success which can be ascribed to several reasons :

- the approximation becomes more and more justified for large ρ values which are most often heavily weighted in standard numerical analyses (e.g. FT analyses),
- the phase shift $\phi^\ell_j(k,\rho)$ is usually smaller than $2\delta^\ell_o(k)$ so that the dependence of the total phase shift upon ρ is weak and smooth : the errors induced on the phases by the pwa can thus be partially reduced or masked by an artificial variation of the unfixed ionization energy.

However, detailed numerical calculations have clearly established that the pwa introduces at low/medium k values quite severe distortions of $\chi^\ell_2(k)$ [14] while the real signal is well reproduced both in phase and amplitude with the fswa. As expected the situation gets worse for the higher order term χ_3^ℓ or χ_4^ℓ : obviously pwa definitely has to be abandoned and even the fswa fails to reproduce the amplitudes correctly. These calculations reported by Natoli et al. [14] could help to explain why the former attempts to reproduce quantitatively the shadowing effects have not yet met convincing success.

We draw attention onto the fact that the Debye-Waller factor introduced in equation (18) is normally consistent only with the plane wave approximation. Under this restriction σ_j^2 can be expressed as a function of the relevant normal modes Q_p [33] :

$$\sigma_j^2 = <[\vec{R}_j \cdot (\vec{u}_j - \vec{u}_o)]^2> = \sum_p [\vec{R}_j \cdot (\vec{q}_j^p - \vec{q}_o^p) <Q_p^2>] \tag{19}$$

where the $\vec{q}_j^p$ are the mass weighted normal coordinates of the unit displacement vector $\vec{u}_j |\vec{u}_j|^{-1}$ in the pth normal mode and :

$$<Q^2> = \frac{\hbar}{2\omega_p} \coth\left[\frac{\hbar\omega_p}{2k_B T}\right] \tag{20}$$

Thus σ_j^2 can in principle be calculated from the normal mode frequencies ω_p^j at least for simple clusters. Note also that σ_j^2 is different from the Debye-Waller defined in X-ray crystallography because the cross term $<\vec{u}_j \cdot \vec{u}_o>$ usually does not vanish [35].

In the fast spherical wave approximation, the Debye-Waller factor has to be recalculated directly from the quantities $<\frac{e^{2i\rho}}{\rho^2} \underset{=}{H}^+ (\ell, \ell')>$.

Using the same asymptotic expression of $\underset{=}{h}_\ell^+(\rho)$ we obtained again :

$$< \frac{e^{2i\rho}}{\rho^2} \underset{=}{h}_\ell^{+2}(\rho)> \simeq \frac{e^{2i\bar{\rho}}}{\bar{\rho}^2} \cdot \underset{=}{h}_\ell^{+2}(\bar{\rho}) \cdot \exp\left\{-2\sigma_j^2 \left[k^2 + \frac{\sigma_j^2}{\bar{R}_j^2} L^2 (\frac{L^2}{4\bar{\rho}^2} - 1)\right]\right\} \tag{21}$$

where $\bar{\rho} = k\bar{R}_j$ and $L^2 = \ell(\ell+1)$. This result clearly indicates that the scattering amplitude $|f_\ell^2|$ becomes an explicit function of $(k, \bar{\rho}, \sigma_j^2/\bar{R}_j^2)$. Numerical calculations have confirmed that this simple correction should not be neglected for large L^2 values [36].

However, as pointed out by Eisenberger and Brown [37], disorder effects result always in a phase perturbation $\Sigma_j(k, \bar{R}_j, \sigma_j^2, \sigma_j^3 \ldots)$ because the quantity $<\frac{e^{2i\rho}}{\rho^2} \underset{=}{h}_\ell^{+2}(\rho)>$ is usually complex as a consequence of the fact that the functions :

$$y(\delta R_j) = g_j(\bar{R}_j + \delta R_j) \cdot \underset{=}{h}_\ell^{+2}(\delta R_j) \frac{1}{k^2 \bar{R}_j^2 (1 + \delta R_j / \bar{R}_j)^2} \tag{22}$$

where $g_j(R)$ denotes the pair (o,j) distribution function, are not even functions of δh. As shown by these authors, the phase correction $\Sigma_j(k,\bar{R}_j,\sigma_j^2)$ remains practically negligible for a gaussian distribution within the pwa. This conclusion is not true for very disordered systems with anharmonic potentials anymore. A generalization of the results given by Eisenberger and Brown for the fswa falls beyond the scope of the present discussion [36].

2.2. Standard analyses of EXAFS spectra.

Given the basic EXAFS equations (18) , there is clearly a large number of approaches for extracting information either on interatomic distances or about phase-shifts, scattering amplitudes and Debye-Waller factors. Our own strategy is summarized in figure 3. It relies on a fine comparison of the EXAFS spectra of the unknown system and of relevant model compounds. Of course, these parallel analyses have to be carried out in exactly the same way and with consistent theoretical calculations of the scattering amplitudes and phase-shifts. The question relative to the need for more accurate phase-shifts will be discussed below.

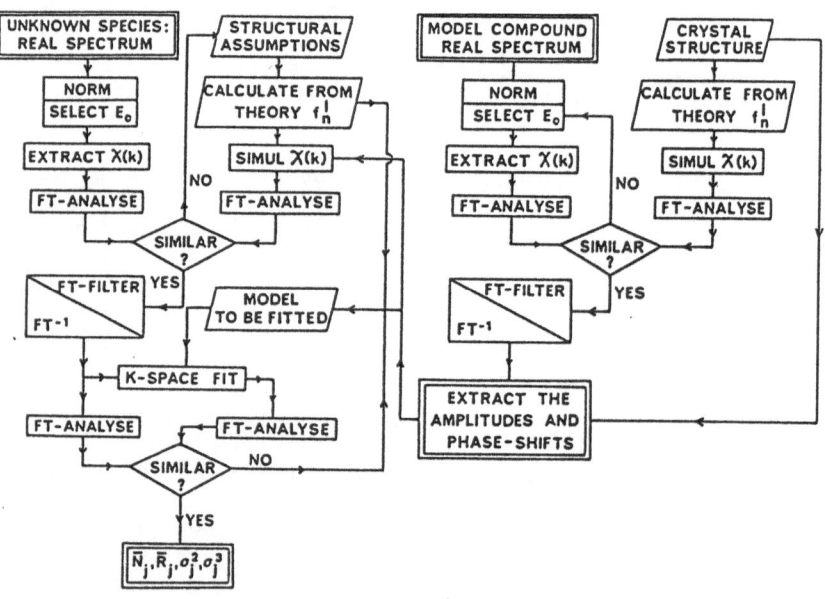

Figure 3. Block diagram of standard EXAFS analyses.

2.2.1. Preparations of the data. As illustrated by figure 3, anyone of our analyses starts with "preparations" which have to be done with extreme care : (i) removal of the pre-edge absorption, (ii) normalization of each spectrum with respect to some corrected edge jump, (iii) selection of the ionization energy $E_o = I_o$ defining the photo-electron wavevector : $k = \left[(2m_e/\hbar^2)(E-E_0)\right]^{1/2}$, (iv) extraction of the renormalized EXAFS modulations $\chi^\ell(k)$ which have to be properly corrected for the k-dependence of the atomic background $\mu_0(k)$. Details of the numerical procedures we use are to be found elsewhere [38].

2.2.2. Fourier transform step. The next step in the analysis is a mathematical transformation. The FT spectra $\tilde{\chi}_j(R)$ discussed in these lecture notes are all corrected for the phase shifts and scattering amplitudes of the dominant shell j according to the prescription of Lee and Beni [17] :

$$\tilde{\chi}_j(R) = \int_0^\infty dk \; W(k) \; \chi^\ell(k) \; \frac{k \; \bar{R}_j^2}{|f_2^\ell(\pi,k,\bar{R}_j)| \; D_2^\ell(R_j,\Gamma) \; S^2(k)} \times$$

$$\times \exp\left[2 \; \sigma_j^2 k^2 - 2ikR - i\psi_j(k) \right] \qquad (23)$$

where : $\quad \psi_j(k) = 2\delta_0^\ell(k) + \phi_{2j}^\ell(k,\bar{R}_j)$

       W(k) is a Kaiser-Bessel window function aimed at minimizing the side lobes of the observed signals,

       $S^2(k)$ is a small but semi-empirical correction accounting for the k-dependent losses due to the inelastic emission of photoelectrons [38].

Indeed the quantity $\tilde{\chi}_j(R)$ is complex and it is common practice to display either $|\tilde{\chi}_j|$ or $\mathrm{Im}[\tilde{\chi}_j]$. We found several advantages in using $\mathrm{Im}[\tilde{\chi}_j]$: (i) the resolution of the latter spectra is obviously much better. The price to be paid for this advantage is that the EXAFS signatures are convolved with $\tilde{\omega}(R)$ and exhibit two negative side lobes. It is our experience, however, that under specific circumstances we can use these two negative lobes to advantage since where they are symmetric the main peak must be well resolved and E_0 adequately selected. (ii) This representation preserves an important phase information which can be helpful for a discrimination between scatterers of different types, e.g. {C,N,O} v.s. {P,Cl,S}. (iii) Multiple scattering calculations are implicitly dependent upon some "constant" interstitial potential $\bar{V}_{int.}$ which is a poor representation of reality neglicting chemical bonding. It is now a rather well established practice to adjust E_0 in order to let the theoretical calculations better match the experimental spectra. According to Lee and Beni [17], a convenient criterion for adjusting E_0 for an isolated shell j is to achieve the coincidence between the maxima of $|\tilde{\chi}_j|$ and $\mathrm{Im}[\tilde{\chi}_j]$.

    For the particular shell j, the FT spectrum $\mathrm{Im}[\tilde{\chi}_j]$ should thus peak at the right distance R_j if the calculated or transferred phase shifts are accurate enough. In principle, this is not true for the other shells anymore. In practice however, for systems like metalloporphyrins where the successive scatterers are low Z atoms (4N, 8C$_A$, 4C$_{meso}$,...), neither

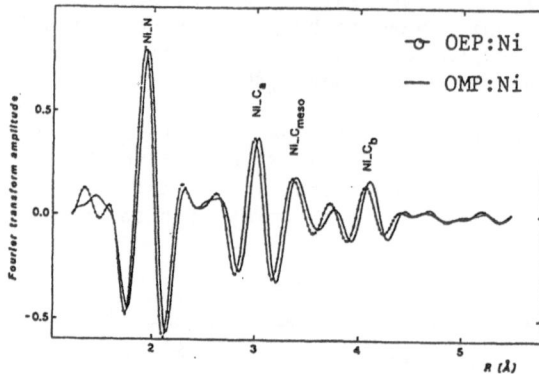

Figure 4. Comparative plot of the FT spectra of two nickel porphyrins OMP:Ni and OEP:Ni. For the former compound, the macrocyclic core is perfectly planar while it is typically ruffled for the other one (see section 2.4.)

the phase distortions nor the differences between the observed peak positions and the true interatomic distances are radical...as long as multiple scattering paths do not contribute (figure 4). This result is not too surprising as in equation (18) the data recorded at high k-values are indeed more heavily weighted : we already know that under these circumstances the phase distortions associated with the pwa are considerably reduced. Nevertheless, for the sake of greater precision, we have developed a new algorithm which makes it possible, although more time consuming, to calculate corrected $\hat{\chi}_j(R)$ spectra fully consistent with the spherical wave approximation. In other words, special corrections for the $\bar{R}_j$ dependence of $f_{\frac{\pi}{2}}(\pi,k,\rho)$ can be made [36]. We hope that these corrected FT spectra should make more apparent the distortions now due only to multiple scattering paths.

2.2.3. Structural refinements using k-space data fitting techniques. The next step can be to filter out some selected portion of the spectrum $\hat{\chi}_j(R)$ and to transform partial contributions back into the k-space. Making use of the basic EXAFS equation (18) several fitting procedures have been developed in order to refine the determination of the interatomic distances $\bar{R}_q$ and to quantify the information about the number N_q of equivalent atoms pertaining to the shell q and about the moments σ_q^2, σ_q^3 ... which characterize the thermal or structural disorder. A safe practice is, however, to keep the number of adjustable parameters in the fit much lower than the maximum number of degrees of freedom which is determined, according to Brillouin [39], by the length of the data set $\Delta k = k_{max} - k_{min}$ and the width $\Delta R(filter)$ of the filter window :

$$N_{free} = \frac{2}{\pi} \Delta R(filter).\Delta k \approx 7 \quad for \begin{cases} \Delta R \approx 1 \ \mathring{A} \\ \Delta k \approx 12 \ \mathring{A}^{-1} \end{cases} \quad (24)$$

This result clearly points out the futility of trying to fit filtered data with overcomplicated models. It is therefore essential to keep a number of parameters fixed making reference to model compounds.

Often people seem to believe that it is possible to improve the R-space resolution by fitting in the k-space such filtered/backtransformed data. However, as shown many years ago by Martens et al. [40], there is practically no chance of resolving in k-space the contributions of two EXAFS signals if one cannot measure accurately the beating half period defined by the condition :

$$2 k \Delta R + \Delta \psi (k) = n\pi \tag{25}$$

Thus, if the two scatterers are identical $\Delta \psi \equiv 0$, the condition for observing the first beating minimum is $k_{max} > \pi/2\Delta R$. Conversely, in principle, the lowest resolution is given by $\Delta R_{min} = \pi/2k_{max} \approx 0.1$ Å for $k_{max} \approx 16$ Å^{-1} . In practice however, the resolution clearly depends upon the accuracy with which the beating minimum can be localized and therefore a practical limit of 0.15 Å has to be retained. This is also the typical position of the negative feet in $Im[\tilde{\chi}_1(R)]$. For different scatterers $\Delta \psi \neq 0$, the situation is even worse. As far as we are concerned with structural studies of metalloporphyrins, this resolution remains too large for an accurate characterization of the axial ligands and this is the reason why we had to develop differential EXAFS methods (see section 4).

Below this limit of 0.15 Å, another way to proceed is to use a single shell fit but with an additional structural disorder characterized by the successive moments :

$$\bar{N} \, \sigma_s^{(n)} = \sum_j N_j \, (R_j - \bar{R})^n \tag{26a}$$

$$\chi_2^\ell(k) = \frac{\bar{N}}{k\bar{R}^2} \, |f_2^\ell(\pi,k,\bar{R})| \, e^{-\sigma_{th}^2 k^2} . Im\{e^{2ik\bar{R}+\psi(k,\bar{R})}\} . [A-iB] \tag{26b}$$

where :

$$A \approx 1 - 2k^2\sigma_s^{(2)} + 3 \frac{\sigma_s^{(2)}}{\bar{R}^2} + 4 \frac{\sigma_s^{(3)}}{\bar{R}^3} + 4 \frac{\sigma_s^{(3)}}{\bar{R}} k^2 +... \tag{26c}$$

$$B \approx 4k \frac{\sigma_s^{(2)}}{\bar{R}} \left[1 - \frac{3\sigma_s^{(3)}}{2\sigma_s^{(2)}\bar{R}} \right] + 4 k^3 \frac{\sigma_s^{(3)}}{3} + ... \tag{26d}$$

Notice that the phase correction due to the spread in distances is essentially of the order $\sigma^{(3)}k^3$ and so, for data with a limited range in k, a small σ_s produces only a very small effect on the phase.

Fitting procedures have proven to be very useful in the analyses but one should never forget that fits can also be delusive : there is often the temptation to generate fit with a larger set of parameters even though one knows that the price to be paid will be a loss of structural information due to the higher level of parameter correlations. A safe and helpful practice might be to produce the correlation matrix together with the results of a fit.

Recently, a different approach has been proposed by Livesey [41] : it uses the concept of maximum entropy reconstruction of the radial distribution function around the absorber. The basic idea is to select in a feasible set the radial distribution function which maximizes the regularizing Shannon-Jaynes entropy subject to several constraints on the data or from prior knowledge of the system. If the results of simulations are very impressive, analyses carried out on real data are still far from the required level of reliability. More work is clearly necessary before one can judge the real potential of this exciting approach when multiple scattering path are not anymore neglected.

2.2.4. <u>Phase-shifts and backscattering amplitudes</u>. It has hitherto been assumed that the complex scattering amplitude $f_n^y(k, \rho_j, \sigma_j^2)$ can be calculated "ab initio" i.e. from the first principles of physics. The reliability of these calculations has been discussed elsewhere [42] and is continuously improving. Convenience rather than fundamental physical reasons can explain why tabulated phases and scattering amplitudes of Teo and Lee [43] are most often used with pwa analyses. The present development of fswa calculations is however stimulating systematic reevaluations of these functions for each particular system to be investigated.

On the other hand, $\bar{R}$ dependent phase-shifts and scattering amplitudes can also be extracted from the experimental spectra of model compounds. The question of the transferability of these quantities from one system to another is now well documented [1]. It is our experience that up to now, slightly better fits can be generated with these tabulated data as compared to the results obtained with Teo and Lee's amplitude and phase functions in the context of the pwa (figures 5a/5b).

We would like to draw attention onto a "novel" scheme for structure determination in EXAFS without resorting to model compounds nor theoretical calculations [44]. The method rests on the point that the composite phase-shift function $\psi(k)$ should decay smoothly to zero for large

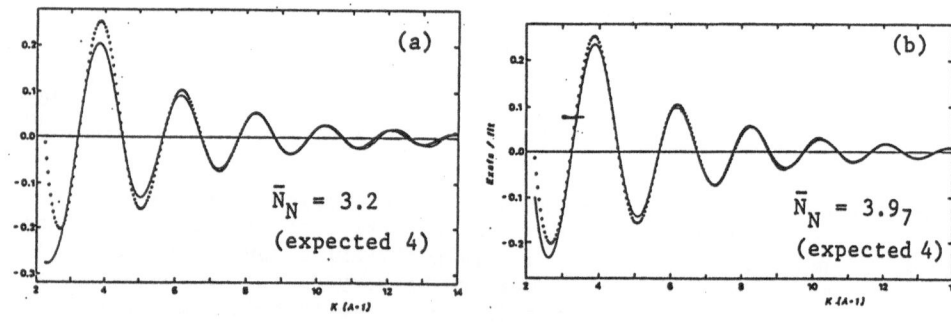

Figure 5. k-space fits of the Ni^{+}...N shell of OMP:Ni ,
(a) using the pwa and scattering amplitudes / phase-shift
    functions calculated by Teo and Lee [43],
(b) using amplitude and phase-shift functions numerically
    extracted from the EXAFS data of OEP:Ni and then
    re-injected into the analysis of OMP:Ni.

k values : this is a consequence of Levinson's theorem which states :
$\delta_\ell(0) - \delta_\ell(\infty) = n_\ell \pi$. Accordingly $\psi(k)$ cannot contain a linear term in k
and therefore the only linear component in the total phase is due to the
path difference $2k\bar{R}_j$. This restriction on the functional form of $\psi(k)$
allows a differential equation to be set up in terms of the unknown
phase function $\psi(k)$ only . We are investigating the potentiality of this
interesting approach but using numerical integration techniques.

2.3. Towards quantitative analyses of edge and XANES spectra.

Until recently, only qualitative interpretations of the edge spectra
were proposed : most often it was attempted to find out semi-empiric
correlations between the valence state of the absorbing metal and the
relative shifts of the prepeaks or of the edge itself [10]. Systematic
investigations of series of compounds have revealed that the presence
(absence) of prepeaks but also of other structures of the XANES spectra
can be indicative of a specific site symmetry [9,10], spin state [45]
or structural distortions [46,47]. However, theoretical calculations
can now be of considerable help in clarifying the origin of the principal spectral features and their possible dependence on a specific
structural parameter to be determined : it should also be possible to
detect the presence of multielectron excitations and thus to learn more
about the response of the multi-electron system. There are basically
two different approaches :
(i) "Ab initio" Hartree-Fock calculations with configuration interaction [48] are time consuming and very expensive. Electron-electron
interactions are, however, treated with far more generality and the relative energy location of the bound states and multi-electron excitations should be most reliable. Conversely the method does not model
properly the continuum states of the photoelectron.
(ii) Multiple scattering calculations with SCF-Xα potentials or with
the energy dependent Hedin-Lundquist or Dirac-Hara potentials are much
less expensive and can account for the gross features of the continuum
XANES spectrum. Effects of screening due to the very slow outgoing
electron just above threshold are probably underestimated even with the
Hedin-Lundquist prescription : further amplitude distortions of the
calculated spectral features might have to be taken into account. However, the relative energy location of these features should not be
affected and is already quite informative : correlations with simulated
changes of structural parameters can be quantitatively exploited.
As an example, we have reproduced in figure 6 the vanadium K-edge
XANES spectra of oxo- and thio-vanadyl porphyrins. In both compounds
the metal is in the formal oxidation state (IV) and is penta-coordinated with site symmetry C_{4v} . For both systems one observes a well resolved pre-peak which is to be assigned to transitions from the metal
1s core level to 3d-like molecular orbitals (M.O.) of allowed symmetry
a$_1$ and e . Unrestricted Hartree-Fock INDO/S calculations (without configuration interaction) have confirmed that for these systems the a$_1$
and e levels are very close lying and cannot be resolved experimentally. Furthermore these levels shift together by c.a. 1.2 and 0.8 eV
between the oxo- and thio-derivatives : this prediction is indeed fully
consistent with a measured shift of $\sim$ 1.0 eV [49] (figure 7). The

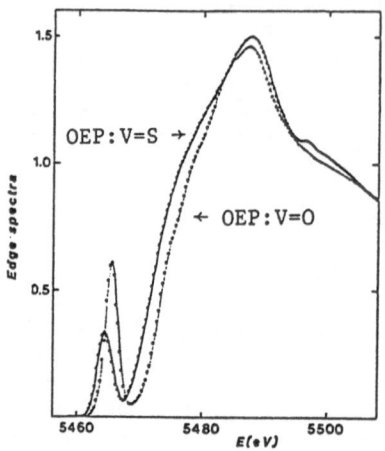

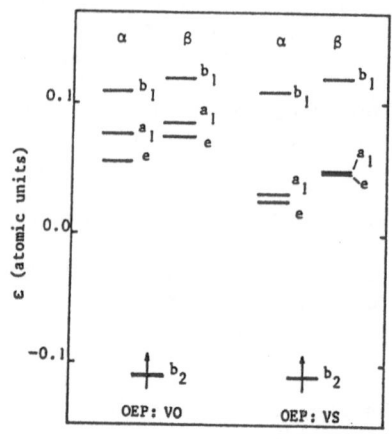

Figure 6. Experimental XANES spectra of OEP:V=O and OEP:V=S.

Figure 7. Unrestricted Hartree-Fock INDO/S molecular orbitals of OEP:V=O and OEP:V=S in the electronic ground state (without any core hole).

creation of the core hole induces orbital relaxation processes which, however, do not seem to affect considerably this difference of energy as indicated by further SCF-Xα calculations [50]. Theoretical simulations of the XANES spectra were also carried out in the "extended continuum" multiple scattering Xα framework [51]: final states with ener-

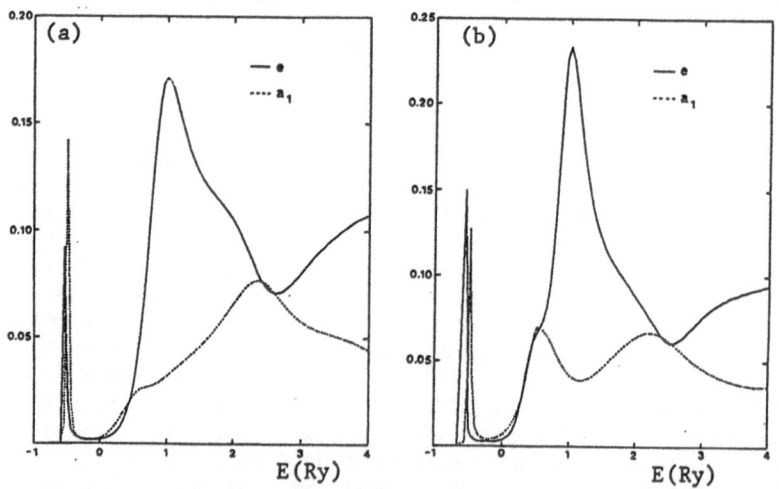

Figure 8. "Extended continuum" MSW-Xα calculations of the XANES spectra for $\vec{e}_{\parallel}$ (a_1) and $\vec{e}_{\perp}$ (e) polarizations. Note the relative contributions of $1s \rightarrow a_1$ and $1s \rightarrow e$ transitions to the observed pre-peak. (a) OEP:V=O . (b) OEP:V=S .

gies greater than the interstitial potential $\bar{V}_I$ are treated artificially as continuum resonances even though the $\bar{V}_I < E < 0$ levels correspond to bound states in SCF calculations. Calculated XANES spectra with the polarization vector either perpendicular or parallel to the porphyrin plane are shown in figure 8 [50] : they give a direct illustration of the relative contributions of the transitions $1s \rightarrow a_1$ ($e\perp$) and $1s \rightarrow e$ ($e/\!/$) to the observed pre-peak. One can now easily understand why a complete extinction of the pre-peak can never be observed when polarized XANES spectra are recorded [52,53] on single crystals of these (or similar) compounds : the contribution of the $1s \rightarrow e$ transition always remains sizeable even though it declines when the metal is axially expelled from the macrocyclic cavity. On the other hand, it has been proposed rather recently to combine both XANES and ESR experiments for extracting some information about the spin-orbit coupling [49] : obviously polarized XANES experiments with $\vec{e}/\!/$ are needed because the link between the two techniques involves accurate determinations of the shifts of this specific e level (ESR experiments being insensitive to the location of the a_1 level).

2.4. Difference EXAFS spectroscopy in porphyrin chemistry.

2.4.1. Ill-conditioned EXAFS problems. In practical EXAFS applications, the user is quite often facing complex structural problems where the aforecited standard analyses simply fail to give reliable results. Major sources of difficulties are : (i) destructive interferences of EXAFS modulations with a consequent loss of structural information, (ii) inadequate radial resolution of the FT spectra resulting in a dramatic alteration of the structural information, (iii) lack of site selectivity wherever a given absorber is present in a sample with variable proportions of unequivalent environments.

It is our intention to try to convince the reader that some of these difficulties can be circumvented by using appropriate difference methods. To begin with, we wish first to give rather illustrative examples of situations (i) and (ii). Thus we have reproduced in figure 9a the FT spectra $Im[\tilde{\chi}(R)]$ of the aforecited oxo- and thio-vanadyle porphyrins : OEP:V=X where OEP = 2,3,7,8,12,13,17,18-octaethylporphyrinato and X = O , S. Similarly one may compare in figure 9b the FT spectra of the oxo- and seleno- derivatives. The typical signatures of the carbons C_A , C_{meso} and C_B of the porphyrin macrocycle are easily identified and exhibit identical intensities for the whole series OEP:V=X . Quite contrarily, there appears to be a strong distortion of the signal of the four nitrogens in the spectra of OEP:V=S and OEP:V=Se. For the former of these two compounds, this effect can perfectly be explained by the fact that the V...N and V...S shells are nearly equidistant while the scattering phase-shifts of nitrogen and sulfur are known to differ by c.a. π [54,55] : under these circumstances a destructive interference can induce a partial cancellation of the V...N and V...S EXAFS oscillations. The same interpretation, however, does not hold true for OEP:V= Se because the difference between the scattering phase-shifts of selenium and nitrogen is close to 2π whereas the V...Se bond length is now slightly longer than the V...N distance : we are simply facing a typical

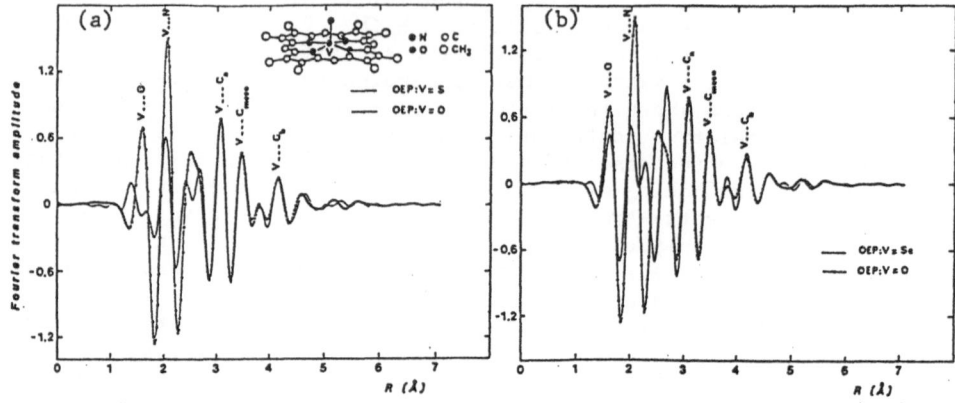

Figure 9. Comparative plots of the FT spectra Im[$\tilde{\chi}$(R)] of
(a) OEP:V=S (full line) v.s. OEP:V=0 (dotted line)
(b) OEP:V=Se (full line) v.s. OEP:V=0 (dotted line).

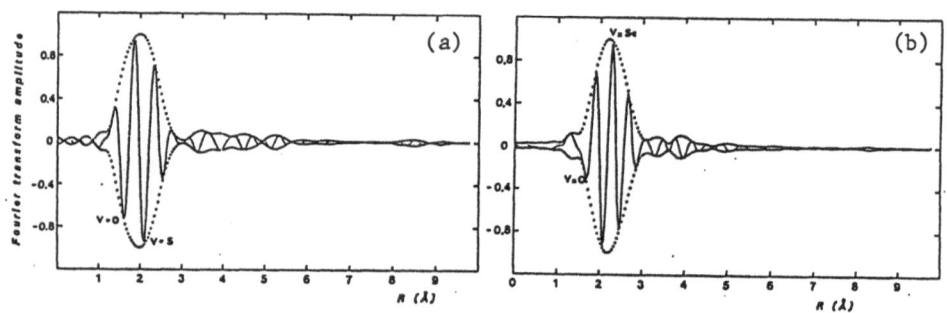

Figure 10. Difference Fourier spectra : dotted line = |$\tilde{\chi}$(R)|
and full line = Im[$\tilde{\chi}$(R)].
(a) OEP:V=S minus OEP:V=0 . (b) OEP:V=Se minus OEP:V=0 .

example of situation (ii) which can be summarized by the overlap of the
positive V...Se peak and of a negative side lobe of the V...N signature.
However, insofar one is interested only in accurate determinations of
the axial V...X bond length, the problem can still be solved by Fourier
transforming the difference spectra : $\chi(k)$ [(OEP:V=S)-(OEP:V=0)] or
$\chi(k)$ [(OEP:V=Se)-(OEP:V=0)]. The results are illustrated by figure 10a/b:
a reasonable cancellation of all signatures of the porphyrin macrocycle
(e.g. carbons C_A, C_{meso}, C_B) is observed and one is left with the difference spectra of the axial ligands : (V...X)-(V...0). As the V...O bond
length (1.62 Å) is much shorter than the distance V...S or V...Se, the
desired structural information is then easily recovered: R(V...S) = 2.06
± 0.02 Å and R(V...Se) = 2.19 ± 0.02 Å . The self-consistency of the
latter result can easily be evaluated by recording the EXAFS spectrum
of OEP:V=Se at the selenium K-edge : the Se...V signal is unambiguously
well resolved and leads to the independent result R(Se...V) = 2.18$_5$ ± 0.02

Å. One should also notice that the apparent phases of the V...S and
V...Se signatures are consistent with the aforequoted differences between the scattering phase shifts of nitrogen, sulfur and selenium. We
shall see in the next section that more can be gained from the FT difference spectra.

2.4.2. <u>Difference FT analyses with structural perturbations</u>. Direct
difference Fourier analyses of EXAFS spectra were first introduced by
Cramer et al. [56]. These authors wanted to localize the contribution
of the unbound oxygen O' in the oxy-picket fence model of oxyhemoglobin
from a direct difference FT analysis of the spectra of the oxy- v.s.
deoxy- complexes. The results were not really convincing and the method
was then abandoned to the benefit of fitting procedures. Elam et al.
met more success in analyzing the direct difference spectrum of deoxy-
v.s. oxy-hemerythrin but their investigation was restricted to the
coordination shell only [57]. As regards the stereochemistry of porphyrins more specifically, a serious objection raised by Cramer himself
against direct difference FT analyses was that variations of the axial
displacement of the metal were neglected. Indeed when such variations
do occur, it becomes impossible to have a perfect cancellation of the
successive signatures of the porphyrin macrocycle and misleading residual signals are obtained [58].

We have shown [58], however, that these difficulties can easily be
overcome by applying some appropriate structural perturbation to the FT
spectrum $\tilde{\chi}(R)$ of the reference compound before carrying out any difference. An illustrative block-diagram of the "perturbed difference" analyses is shown in figure 11. The basic requirement is to select some

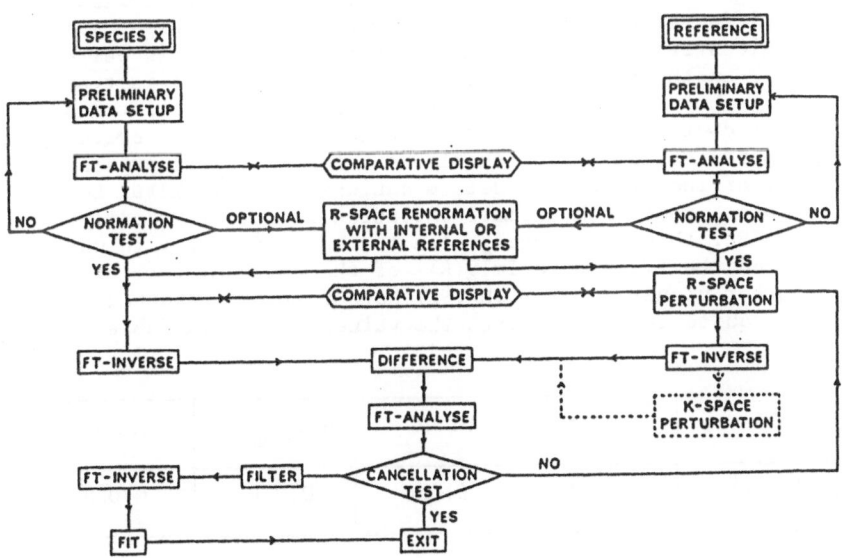

Figure 11. Block diagram of the perturbed difference analyses

analytical transformation $R' = f(R)$ taking into account the relevant structural changes of the reference. All what we then need is an accurate interpolation procedure in order to linearize the sampling of the function $\overset{\sim}{\chi}(R')$.

From a compilation of the enormous amount of crystal structure data on metalloporphyrin stereochemistry, we have been led to retain three major types of perturbation :

(i) <u>an axial shift δh of the metal</u> with respect to the average plane of the porphyrin macrocycle :

$$R'^2 = R_{CT}^2 + (H+\delta h)^2 = R^2 + 2H \, \delta h + \delta h^2 \qquad (27)$$

(ii) <u>A variation of the macrocyclic cavity radius</u> :

$$R' = R + \Delta_{cav} \qquad (28)$$

is the indirect consequence of the "ruffled" conformations of the pyrrole rings. As pointed out by Hoard [59], small amplitude rotations ϕ of the pyrrole subunits around the Metal...N_{eq} axes are possible in accordance with S_4 symmetry provided that the bond parameters are satisfying the exact relationship (figure 12a) :

$$a_N + b \cos \frac{\alpha}{2} - c \sin \psi = (b \sin \frac{\alpha}{2} + c \cos\psi) \cos \phi \qquad (29)$$

More often, the adjustments in bond parameters are largely taken into account by the Metal...N_{eq} bonds so that an apparent contraction of the cavity radius is observed. In the first approximation, we usually assume Δ_{cav} = constant. This is quite reasonable for the N_{eq}, C_A, C_B shells but less satisfactory for the C_{meso} signatures. The nickel porphyrin OEP:Ni is a good candidate for an evaluation of this approximation because two different crystal structures are known for this compound: in the triclinic structure [60], the porphyrin macrocycle is perfectly planar while in the tetragonal form [61] the macrocycle is typically "ruffled". We have summarized in Table I the average variations ΔR_j of the relevant distances R_j between these two structures. Obviously, the proposed approximation is reasonable as referred to the present limits of accuracy of the structural determinations made by EXAFS. Let us, however, emphasize that such a small variation of Δ_{cav} is readily detectable by EXAFS : this was clearly illustrated by figure 4 in which we have compared the FT spectra $\text{Im}[\overset{\sim}{\chi}(R)]$ of the trigonal form of OEP:Ni and of another planar complex OMP:Ni. Note that the measured core contraction is quite consistent with the values quoted in Table I.

shells j	Ni...N	Ni...C_A	Ni...C_{meso}	Ni...C_B
ΔR_j (Å)	0.026	0.024	0.010_5	0.028

Table I. Average contraction of the R_j distances between a ruffled porphyrin core and a planar macrocycle.

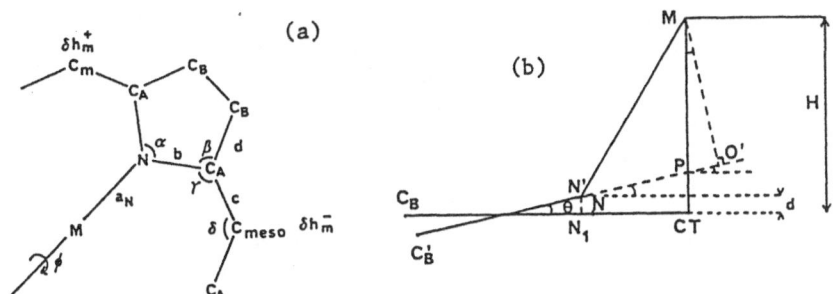

Figure 12. Basic deformations of a porphyrin core :
(a) "Ruffling" conformation . (b) "Doming" conformation.

(iii) "Doming" effects can be easily generated by slightly tilting the
pyrrole rings around axes which are now perpendicular to the Metal...N_{eq}
bonds (figure 12b). Such distortions of the porphyrin core make it possible to keep short Metal...N_{eq} bond lengths while allowing a quite large
displacement of the metal above the 4 N_{eq} plane. Useful parameters for
describing this perturbation are the distance d separating the 4 N_{eq}
plane from the mean plane of the porphine skeleton and the less commonly
given tilt angle θ (figure 12b) :

$$R'^2 = R^2 + H'^2 - H^2 + 2R \ (1 - \frac{1}{2} \frac{H^2}{R^2}) (a'_N - a_N) + (a'_N - a_N)^2 \qquad (30)$$

with : $a'_N = a_N \cos\theta + (H-d) \sin\theta$ and $H' = (H-d) \cos\theta - a_N \sin\theta$

"Doming", "antidoming" or "saddle shaped" conformations of the macrocycle simply differ by the sign of θ and d for each pyrrole subunit.
     An interesting question is whether or not difference EXAFS analy-
ses can discriminate between a change δh of the axial displacement H of
the metal and a variation of the cavity radius Δ_{cav}. In principle, the
answer is affirmative because the two perturbations exhibit a quite different R dependence, as illustrated by equation (31) :

$$R' = R + \Delta_{cav} \ (1 - \frac{H^2}{2 R^2}) + H \frac{\delta h}{R} + \frac{1}{2} \frac{\delta h^2}{R} \qquad (31)$$

where second order terms $(\Delta_{cav})^2$, $(H^2/R^2)^2$ are neglected for the sake
of simplicity : this means that it is not possible to achieve a good
cancellation of all signatures of the porphyrin core, i.e. the N_{eq}, C_A,
C_{meso}, C_B, C_ϕ without taking into account both corrections. Accuracies
of the order $\pm$ 0.05 Å for δh and $\pm$ 0.015 Å for Δ_{cav} have been obtained
in favourable cases. Unfortunately in practice "doming" and related ef-
fects often add correlations degrading the quality of the structural
information available. There are however cases where "doming" like corrections are of definitive requirement in order to obtain an acceptable
cancellation of all signatures of the porphyrin core : a typical example
was provided by the comparison of the spectra reproduced in figure 13a

for $1 = TMP:Mn(V)\equiv N$ and $2 = TMP:Mn(III)-Br$. A paradox, not to be accounted for by equation (31) is that the shift between the two spectra increases with R. Figure 13b shows that the perturbed difference spectrum generated with "doming" like corrections ($\delta h = 0.15$ Å , $\Delta_{cav} = 0.025$, $\Delta d = 0.15$ Å) exhibits the positive Mn...N and negative Mn...Br signatures at the expected distances while there is obviously a nearly perfect extinction of all signals of the porphyrin core. As the crystal structure of 2 is not known yet, we have been unable to refine the determinations of $H_1 \simeq 0.45$ Å (?) and $d_1 \simeq 0.1$ Å (?) but the distances Mn...N (1.50 ± 0.01 Å) and Mn...Br (2.60 ± 0.02 Å) are fully consistent with other crystal structure data [62,63] or EXAFS investigations at the Br K-edge.

Perhaps of greater relevance to biostructural problems is the comparison of the spectra $Im[\tilde{\chi}(R)]$ reproduced in figure 14a and which concerns $4 = TPP:Fe(II)(N-MethylImidazole)_2$ and the naked complex $3 = TPP:Fe(II)$. According to its known crystal structure [64], 3 has a ruffled core with rather short $Fe...N_{eq}$ distances (1.97_1 Å) and thus the perturbed difference spectrum $(4-3)$ shown in figure 14b was performed with a slightly expanded cavity radius ($\Delta_{cav} = 0.015$ Å). It is noteworthy that this spectrum reproduces all the expected features associated with the axial 2-methylimidazole ligands : the $Fe...N_{3,3'}$ and $Fe...C_{A,A'}$ signals peaking at $R_1 = 2.01 \pm 0.01$ Å and $R_2 = 3.02 \pm 0.05$ Å respectively, and even more characteristic the enhanced multiple scattering of the $Fe...N_3...N_1$ or $Fe...N_3...C_B$ paths peaking with an inversed phase at distances $R_3 \simeq 3.8$ Å which are c.a. 0.3 Å short of the exact values. Indeed this perturbed difference spectrum exhibits a remarkable analogy with the FT spectra of copper or zinc tetrakis(imidazole) complexes which have been used as reference compounds for multiple scattering calculations over these recent years [65,66]: calculations of that complexity would have been beyond the scope of current endeavour on the original data of 4 while they are becoming quite possible _via_ our perturbed difference spectrum. These results also make us rather confident of the reliability of our analyses.

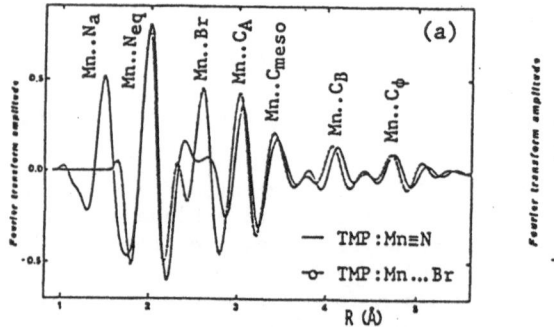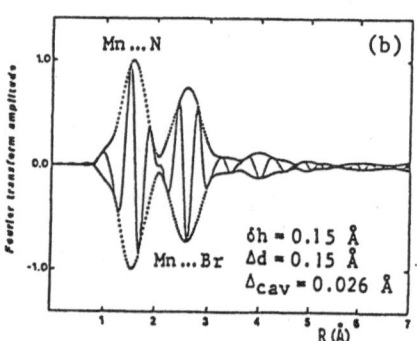

Figure 13. (a) Comparison of the FT spectra $Im[\tilde{\chi}(R)]$ of $1 = TMP:Mn\equiv N$ and $2 = TMP:Mn-Br$. These spectra are implicitly corrected for the scattering amplitude and phase shifts of the dominant Mn...N shell. (b) Perturbed difference FT spectrum of $(1 - 2^*)$.

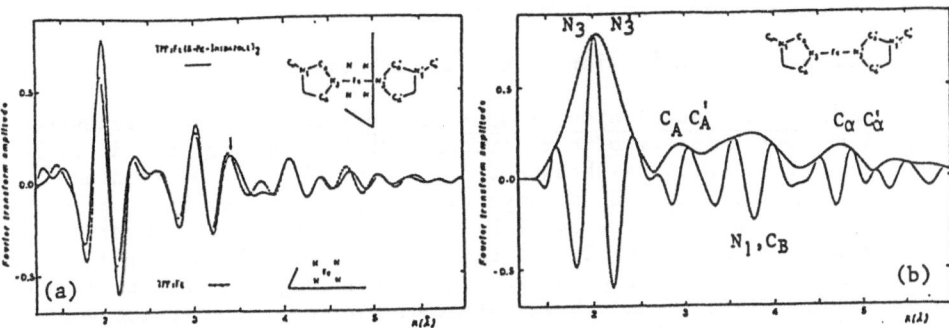

Figure 14. (a) Comparison of the FT spectra of 3 = TPP:Fe and 4 = TPP: $\overline{\text{Fe}}$(N-Methyl-Imidazole)$_2$. The spectra are implicitly corrected for the Fe...N shell. (b) Perturbed difference FT spectrum of $(4 - 3^{*})$.

2.4.3. <u>Strengths and limitations</u>. It may be useful to summarize briefly here the strengths of the differential methods :

. The FT difference spectra are far less complicated to analyze if one obtains a perfect cancellation of the signatures of the porphyrin macrocycle.

. Identification of the axial scatterer type becomes possible by considering the phase of a now well resolved signal.

. Structural information lost by destructive interferences can be recovered.

. The spectral limits of resolution can be transgressed.

. The cancellation tests are very sensitive and very discriminant: minor structural perturbations can be more easily detected.

. Approximations regarding the spherical wave propagation of the photoelectron do not affect in principle the cancellation procedure.

. Some multiple scattering paths do not cancel out and thus become easier to detect.

. As opposed to fitting methods, the difference analyses are much more reliable. Indeed fitting procedures can also be used in order to refine the structural content of a difference spectrum.

Of course, there are also a number of difficulties which limit the applicability of these difference analyses :

. First of all, one needs some "a priori" knowledge of the most probable structure of the system under investigation. One needs to select appropriate reference compounds and relevant structural perturbations.

. The synthesis and chemical stability of the desired reference compounds can be also a source of major troubles.

. The method is most demanding from the quality of the experimental data and also from the reliability of the procedures used for the background extraction and for the normation of the spectra.

3. SELECTED EXAMPLES OF APPLICATIONS

3.1. Structural investigation of biomimetic complexes.

3.1.1. High-valent metal oxo- or oxo- like complexes.
A good illustration of the potentiality of our perturbed difference analyses is to be found in a paper which has been published recently [9] and relates the structural characterization of high valent manganese porphyrin complexes. As full details are available elsewhere, we shall summarize shortly here the major results of this study which was initiated by the isolation of these reactive species at Toulouse [9]. Oxidation by NaOCl of 5 = TMP:Mn(III)Cl gives a high valent complex 6 = TMP:Mn(O)(solv.) which is capable of selective epoxidation of olefins. We produced convincing evidence that in 6 the metal is lying in or very close to the mean porphyrin plane and is hexacoordinated. Using 2 again as a reference, a perturbed difference analysis indicated that of the two axial bonds, only one is well defined (R = 1.84 Å ± 0.02 Å) and is thus to be unambiguously assigned to a Mn...O single bond whereas the second is provisionally assigned to a more loosely bound water molecule at R ≃ 2.3 Å (Figure 15). Thus at least for this specific system, our EXAFS results rule out the presence of a real Mn(V)=O compound.

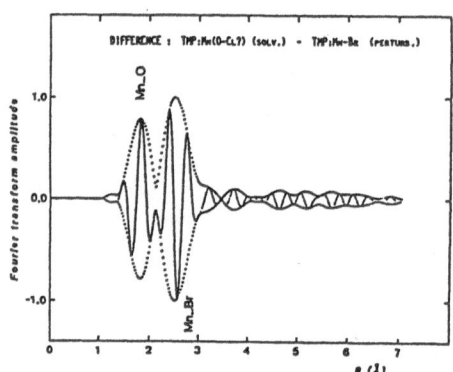

Figure 15. Perturbed difference FT spectra obtained for the difference $\chi(6) - \chi(2)^*$ with the metal shifted back to the mean porphyrin plane by 0.2 Å in 2 .

3.1.2. Carbenes, nitrenes and σ-bonded aryl Fe(II) complexes.
It has been shown by Mansuy and his collaborators [67-69] that the reaction of iron(III) porphyrins with polyhalomethanes in the presence of a large excess of reductant produces a variety of porphyrin-iron bonded carbene species. Up to the present, however, only few structural investigations of iron carbene complexes have been reported and it was especially attractive to check if our perturbed difference analyses can yield accurate determinations of the iron...carbon double bonds.

A typical example of this class of structure is the vinylidene carbene complex 7 = TPP:Fe=C=C(Ar-Cl)$_2$ which was prepared by reaction of the naked complex 3 with the insecticide DDT = Cl$_3$CCH(Ar-Cl)$_2$ and is the most stable of all prepared iron carbene entities [68]. Reference is also made to 7 in order to explain synergistic action of insecticides

[67]. Another example to be also considered here is the dichlorocarbene 8 = TPP:Fe=CCl$_2$ prepared by reaction of **3** with carbon tetrachloride [69]. Perturbed difference FT spectra of **7** and **8** were calculated using **3** as a reference and are reproduced in figures 16 and 17. Both complexes are penta-coordinated and the metal is in both cases slightly shifted out from the mean porphyrin plane with H$_7$ = 0.20 ± 0.05 Å and H$_8$ = 0.30 ± 0.03 Å. The porphyrin core is certainly "ruffled" in the case of **7** (Δ_{cav} = 0.00 Å) but slightly less in the case of **8** (Δ_{cav} = 0.015 Å). The axial Fe...C bond lengths were thus obtained with an excellent accuracy : R$_7$(Fe...C$_1$) = 1.66 ± 0.02 Å , R$_8$(Fe...C$_1$) = 1.70 ± 0.02 Å . The former value is quite comparable to the Fe...C bond length of 1.67$_5$ Å found in the crystal structure of the μ-carbido dimer **9** = TPP:Fe=C=Fe:TPP [70] which is involving a formally dicarbenic carbon atom bridging two metal centers. A preliminary perturbed difference analysis of the EXAFS spectrum of **9** is reproduced in figure 18 and gave exactly the same bond length of 1.66 ± 0.02 Å. As regards **8**, our estimation of the Fe...C bond length is definitively shorter than the value of 1.83 Å found in the crystal structure of the hexa-coordinated species TPP:Fe(CCl$_2$)(H$_2$O) [71] but such a difference is quite consistent with the fact that **8** was prepared in anhydrous non-coordinating solvents.

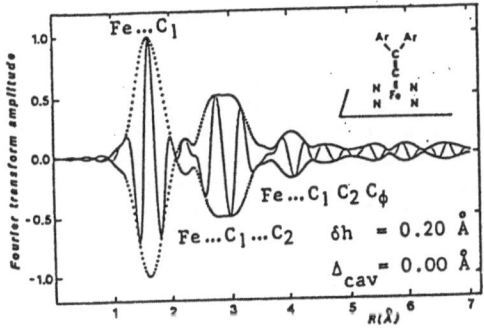

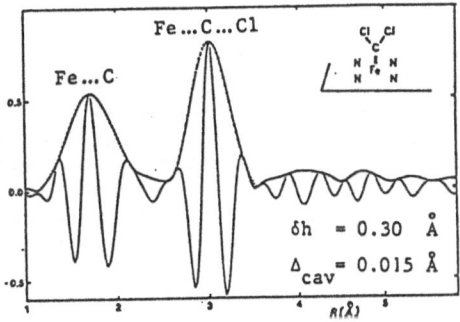

Figure 16. Perturbed difference FT spectrum of (**7** – **3***) where **7** = TPP:Fe=C=C(Ar-Cl)$_2$, δh = 0.20 Å , Δ_{cav} = 0.00 Å .

Figure 17. Perturbed difference FT spectrum of (**8** – **3***) where **8** = TPP:Fe=CCl$_2$, δh = 0.30 Å , Δ_{cav} = 0.015 Å .

Spectacular multiple scattering effects have to be taken into account if one wishes to analyze the signatures observed beyond the first shell : their phase is typically inversed with respect to what we expect in the absence of multiple scattering whereas amplitude enhancements are quite evident. Indeed, these effects are not at all surprising for **7** or **9** because the sequences Fe...C$_1$...C$_2$ (R = 2.39 ± 0.03 Å) or Fe...C... Fe are linear. It is our interpretation that in **8** the Fe...C...Cl multiple scattering contributions might become dominant because the low frequency bending modes of the bent Fe...C...Cl sequence should damp out selectively the direct Fe...Cl scattering contribution [33]. Most convincing is also the case of the μ-nitrido dimer **10** = TPP:Fe-N-Fe:TPP illustrated by

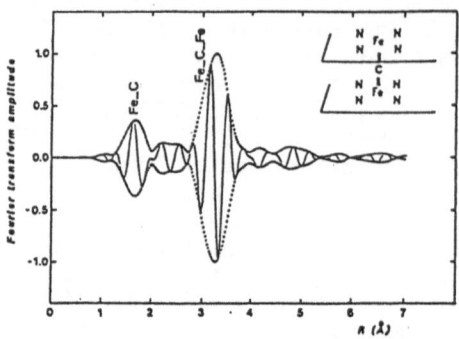

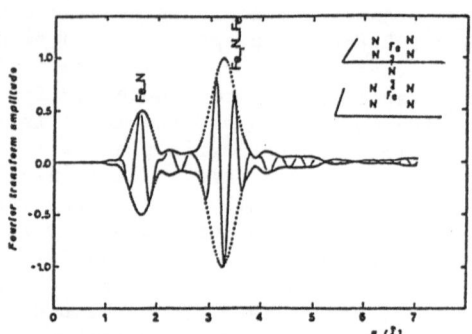

Figure 18. Perturbed difference FT spectrum of $(9-3^*)$ where $9 = \mu-(TPP:Fe=)_2C$, $\delta h = 0.25$ Å $\Delta_{cav} = 0.00$ Å .

Figure 19. Perturbed difference FT spectrum of $(10-3^*)$ where $10 = \mu-(TPP:Fe\underline{=})_2N$, $\delta h = 0.30$ Å $\Delta_{cav} = 0.015$ Å .

figure 19. Again the short Fe...N bond length (1.66 ± 0.01 Å) and the axial shift $\delta h = 0.30 \pm 0.05$ Å found by our method are perfectly consistent with the known crystal structure [72]. The multiple scattering effects relative to the Fe...N...Fe linear sequence are also spectacular.

An important question is whether difference EXAFS analyses can also provide us with accurate determinations of the Fe...C bond lengths in the case of σ-aryl or σ-alkyl bonded iron porphyrins. Thus, we decided to record the EXAFS spectra of one of the most stable representatives of these short living compounds, i.e. $11 = TPP:Fe-C_6H_5$. A puzzling result was that apart from some sort of a different scaling, the two FT spectra of 11 and 3 looked very similar...Fortunately, the crystal structure of 11 was solved recently by Doppelt [73] and it becomes immediately clear from the structural data quoted in Table I that this problem is rather ill-conditioned as regards difference EXAFS analyses: (i) The metal is only slightly displaced from the porphyrin plane ($\delta h \simeq 0.17$ Å), this effect being partially compensated by a ruffled core

SHELL	11 = TPP:Fe-C_6H_5 [73]			3 = TPP:Fe [64]		
j	SCATTERERS	$\bar{N}_j/\bar{N}_1$	$\bar{R}_j$ (Å)	$\bar{R}_j$ (Å)	$\bar{N}_j/\bar{N}_1$	SCATTERERS
1	4 N_{eq} + C_{Ph1}	1	1.96_1	1.97_1	1	4 N_{eq}
2	8 C_A + 2 C_{Ph2}	2	2.99_5	3.01_5	2	8 C_A
3	4 C_{meso}	0.8	3.41_8	3.40_7	1	4 C_{meso}
4	8 C_B + 2 C_{Ph3}	2	4.23_1	4.23_2	2	8 C_B
5	4 C_ϕ + C_{Ph4}	1	4.88	4.90_8	1	4 C_ϕ

Table I. Predicted radial distributions for 11 and 3 according to their crystal structures.

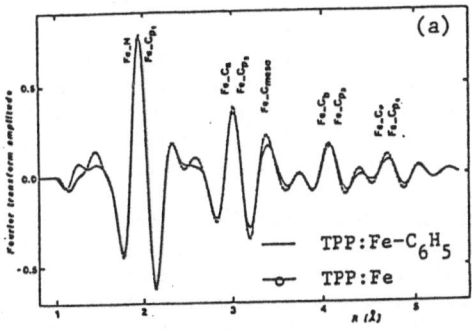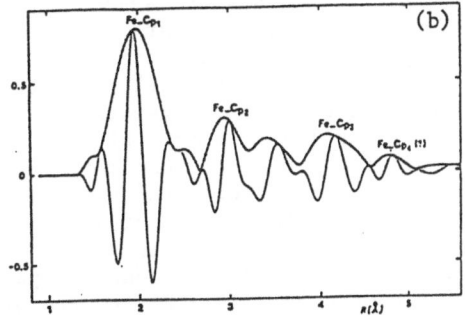

Figure 20.(a) Comparison of the FT spectra of $11 = TPP:Fe-C_6H_5$ and $3 = TPP:Fe$ corrected implicitly for the Fe...N shell. For the sake of consistency with Table I, both spectra have been renormalized to unity with respect to the dominant shell. (b) Perturbed difference FT spectrum of $(11-3^*)$ where $11 = TPP:Fe-C_6H_5$, $\delta h = 0.20$ Å, $\Delta_{cav} = 0.015$ Å.

contraction ($\Delta_{cav} = -0.020$ Å). (ii) Due to the nature and geometry of the axial ligand, the current tests used for evaluating the cancellation of the porphyrin signatures become extremely ambiguous.

As predicted by Table I and nicely illustrated by figure 20a, the only discrimination test is the amplitude of the C_{meso} signal. Indeed, EXAFS results and crystal structure data are fully consistent, the lack of an easy discrimination between the two spectra being in itself indicative of the fact that the iron...carbon bond length (1.95_5 Å) has to be close to the average distance $\bar{R}_1 = 1.96$ Å. Furthermore, the perturbed difference spectrum shown in figure 20b ($\Delta_{cav} = 0.015$ Å, $\delta h = 0.20$ Å) is already quite encouraging and led to the exact Fe...C distance : $R(Fe...C_{p1}) = 1.96 \pm 0.02$ Å. From our analysis, the axial shift of the metal should be of the order of 0.20 ± 0.05 Å. This result is comparable with the crystal structure determination but is associated with a slightly positive value for Δ_{cav} instead of the above mentioned core contraction ($\Delta_{cav} < 0$).

3.2. Stacked or bridge-stacked polymeric chains of metalloporphyrins.

Intensive efforts are directed toward the synthesis of highly conducting molecular materials with metallomacrocyclic complexes as fundamental units. Such "molecular metals" were prepared recently by partial oxidation of nickel or fluorogallium phtalocyanines (Pc) or related porphyrins. A prerequisite for obtaining a high electric conductivity is the formation of a stacked or bridge-stacked polymeric structure. It is our purpose, now, to show that EXAFS can be used in order to check if newly synthesized materials are featuring the required polymeric structure or not.

Most often conventional analyses of EXAFS fail to give reliable answers to this question because the strong signatures of the porphyrin macrocycle dominate the spectra and are masking the much weaker intermolecular signals or, even worse, interfere with them. The difference Fourier analyses appear again as the most suitable method for solving such problems. In fact, originally, we have developed the difference analysis with structural perturbation essentially because we were asked to ascertain the bridge-stacked polymeric structure of a fluorogallium porphyrin [58]. As this study is reported in details elsewhere, we shall focus here on some more recent work on stacking and slipped stacking of nickel phtalocyanines [74].

The EXAFS spectra of a series of oxidized/unoxidized nickel porphyrins and phtalocyanines have been recorded. The corrected FT spectra $Im[\tilde{X}(R)]$ of both Pc:Ni and the partially oxidized system Pc:Ni/I are reproduced in figure 21 and exhibit rather small differences in the range of intermolecular distances. According to its X-ray powder diffraction pattern, the unoxidized Pc:Ni was identified to the β-phase for which short intermolecular distances $Ni...Ni_{meso}$ are expected between

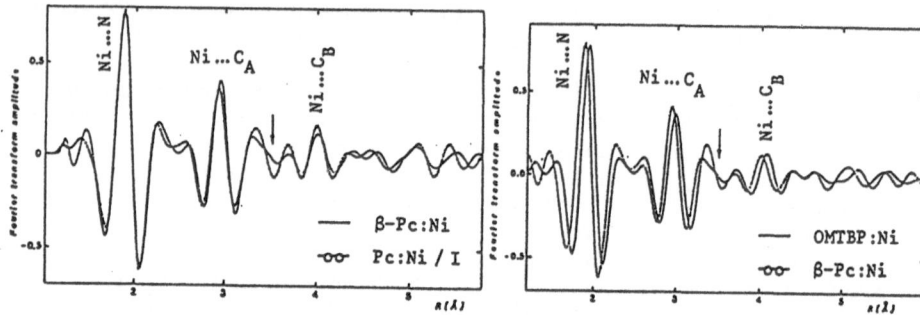

Figure 21. Corrected FT spectra $\overline{Im[\tilde{X}(R)]}$ of β-Pc:Ni and Pc:Ni/I. The arrow indicates the range of first intermolecular distances.

Figure 22. Corrected FT spectra $\overline{Im[\tilde{X}(R)]}$of β-Pc:Ni and OMTBP:Ni. The arrow indicates the range of first intermolecular distances.

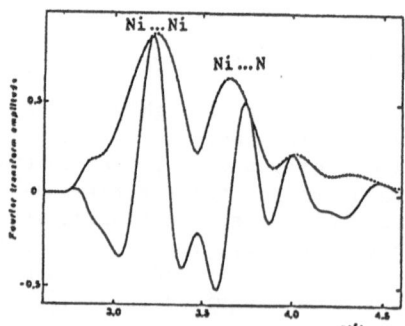

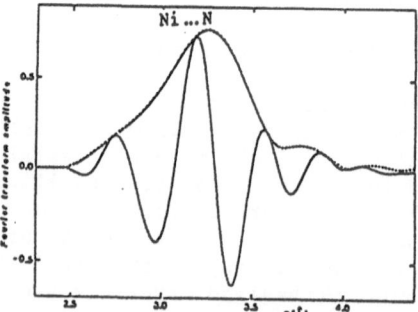

Figure 23. "Extended" difference FT spectrum of (Pc:Ni/I) - (OMTBP:Ni)*

Figure 24. "Extended" difference FT spectrum of (β-Pc:Ni) - (OMTBP:Ni)*

3.2 - 3.5 Å. The FT spectrum of β-Pc-Ni is also compared in figure 22
to the FT spectrum of octamethyltetrabenzporphyrinato nickel(II) =
OMTBP:Ni for which no short intermolecular distances should interfere
in this range of distances.

The two complexes exhibit enough structural similarities [75,76]
to make possible a first order cancellation of all signatures of the
macrocycles when two successive perturbations of the FT spectrum of
OMTBP:Ni are made: (i) a uniform contraction of the macrocycle cavity,
(ii) an additional shift toward larger distances of the signal of the
meso carbons within some appropriate gaussian window.

It becomes then possible to extract by differential FT analysis
the contribution of the weak intermolecular signals in Pc:Ni/I (figure
23) and β-Pc:Ni (figure 24). For Pc:Ni/I, the observed Ni-Ni distance
(3.2_2 Å) agrees quite well with the known crystal structure data (3.2_4
Å) [75]. In the case of β-Pc-Ni, the signal provisionally assigned to
the axial azamethine nitrogens is peaking at ca. 3.20 ± 0.05 Å : such a
distance is of the order of magnitude of the distance found by X-ray
crystallography for other phtalocyanines (M = Mn , Fe , Cu , Co , Zn...) [77].

4. EXPERIMENTAL TECHNIQUES

4.1. Synchrotron radiation sources

The experiments reported in these lecture notes were all carried out
at L.U.R.E. (the French National Synchrotron Radiation facility) using
the synchrotron radiation emitted by a relativistic positron beam
($1.72 \leqslant E_{e+} \leqslant 1.85$ GeV) when it is deflected in the storage ring DCI by
the intense magnetic field ($B_M \simeq 1.56$ T) of so-called "bending magnets".
The synchrotron radiation is thus emitted tangentially along the positron orbit in the machine. It is collected in vacuum pumped beam lines
(e.g. D1, D2, D4) which are isolated from the storage ring itself by
beryllium windows (250 μm thick). The angular spectral density of this
emission is given by a universal relationship of the form :

$$\frac{d^2F}{d\beta \, d\psi} = \frac{6}{\pi} \frac{r_0}{\lambda_{ce}} \frac{I}{e} \gamma^2 \frac{\Delta E}{E} \, F_S(\xi, \gamma\psi) \tag{32}$$

where e, r_0 and λ_{ce} refer to the electron charge, radius and Compton
wavelength respectively, whereas :

$$F_S = F_S^\sigma + F_S^\pi = \frac{\xi^2}{1+\gamma^2\psi^2} \left[K_{2/3}^2(\xi) + \frac{\gamma^2\psi^2}{1+\gamma^2\psi^2} K_{1/3}^2(\xi) \right] \tag{33}$$

with :

$$\xi = \frac{E}{2E_c} \left[1+\gamma^2\psi^2 \right]^{3/2} \tag{34} \qquad \gamma = \frac{E_{e+}}{m_0 c^2} \tag{35}$$

whereas :

$$E_c(\text{critical energy}) = 0.665 B_M E_{e+} \quad (\sim 3.68 \text{ keV for DCI}) \tag{36}$$

In equation (33), $K_{2/3}$ and $K_{1/3}$ represent modified Bessel functions of the second kind of order 2/3 and 1/3 which describe the polarization in components parallel and orthogonal to the orbit plane. Hence it is easy to check that synchrotron radiation is linearly polarized only in the orbit plane defined by $\psi = 0$: this classical result makes it possible to record linearly polarized XANES or EXAFS spectra on oriented crystals. On the other hand, the available flux can be evaluated on integrating relation (32) over the whole vertical plane $\psi \in [-\infty, +\infty]$ and over some fraction of orbit $\Delta\beta$. As an indication, figure 25 compares the calculated fluxes F(E) delivered for 1 mrad of orbit and $\Delta E/E = 0.1$ % by the bending magnets of DCI ($E_{e+} = 1.85$ GeV) and SUPER-ACO ($E_{e+} = 0.8$ GeV), i.e. the two storage rings now operating at L.U.R.E.. For the sake of reference, we also added in figure 25 the continuous background emission of a conventional X-ray tube. It is well known that even more intense X-ray photon fluxes can now be generated using specific insertion devices : multipole wigglers and undulators (on 5-6 GeV machines). Both linearly and circularly polarized light could, in principle, be obtained with these new devices [78], a presentation of which, however, falls beyond the scope of the present notes.

Luckily enough, synchrotron radiation is not uniformly emitted in the vertical plane : at a given photon energy, the angular emission distribution decreases very quickly with $\gamma\psi$. Although this is not strictly a gaussian distribution, one defines the natural spreading of the emission by :

$$\sigma_{nat}^{\psi}(E) = \frac{0.565}{\gamma} \left[\frac{E_c}{E}\right]^{0.425} \simeq \frac{1}{2\gamma} \quad \text{if} \quad E = E_c \tag{37}$$

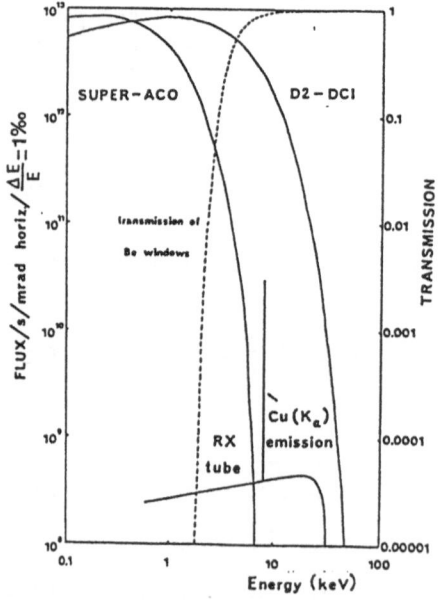

Figure 25. Comparative plots of the X-ray photon fluxes delivered by the bending magnets of the storage rings SUPER-ACO and DCI (ORSAY). Dotted line is the transmission of the Be windows on beam line D2.

and the emission thus has the appearance of a thin cloak centered on the orbital plane. Unfortunately, in a real machine the positrons do not move in ideal orbits, nor do they have constant energy E_e+. Due to the betatron oscillations they are distributed in transverse phase space about a mean trajectory : position (σ_x, σ_y).angular divergence (σ'_x, σ'_y) according to gaussian distributions. The quality of a synchrotron source is thus described by its vertical or horizontal emittances: $\varepsilon_x = \sigma_x \sigma'_x$ or $\varepsilon_y = \sigma_y \sigma'_y$ and the relevant spectral brilliance $B = F/\varepsilon_x \varepsilon_y$. For the D2 beam line on DCI typical values are : $\sigma_x = 2.65$ mm, $\sigma_y = 0.7$ mm, $\sigma'_y = 0.25$ mrad. It is essential to realize that these parameters are most often the limiting factors as regards the optical properties of the EXAFS/XANES spectrometers : monochromator resolution and size of the refocusing spots.

Figure 25 can also help to resolve an apparent contradiction: if the maximum available flux from a machine like DCI is emitted at $E \simeq E_c = 3.6_8$ keV, why should it then be so difficult to record spectra at thresholds of lower energy $(E < 5$ keV) ? The explanation follows immediately: the absorption of the beryllium windows situated behind the vacuum chamber and also in front of the monochromators cut off the low energy photons beyond reprieve.

4.2. Scanning and dispersive spectrometers.

The monochromator is actually the key component of X-ray spectrometers. In the X-ray range, monochromatisation is achieved by one or more Bragg reflections on silicon or germanium monocrystals. We must in fact consider two distinct situations : (i) The Bragg angle remains constant at any point of the reflecting crystals. Forgetting the higher order reflections, the output photons all have the same energy which can be changed on rotating the reflecting crystals. This is a classical energy scanning monochromator. (ii) If the Bragg angle varies continuously over the surface of the crystal, then this crystal selects a finite, continuous range of photon energies and reflects these photons in specific directions. This monochromator is termed an energy dispersive monochromator.

One should emphasize that for constant illuminated surface, the two types of monochromator should provide essentially the same flux of reflected photons. Each one has given rise to a particular type of spectrometer:
(i) As far as synchrotron radiation sources are used, energy scanned spectrometers are equipped with two parallel monocrystals and give a monochromatic exit beam propagating in the same direction as the incident beam. Furthermore, it is possible to keep both the entrance and exit beam positions fixed if the spacing of the two crystals is continuously changed during their rotation [79]. Another advantage of using two independent crystals is that the second reflection can be slightly mistuned from the first : this technique can be used to some extent to diminish the level of harmonics because of the narrower Darwin width of the higher order reflections. The detuning due to the thermal heating of the first crystal can also be compensated. The price to be paid is in the need of a very stable mechanism because there is no exit beam

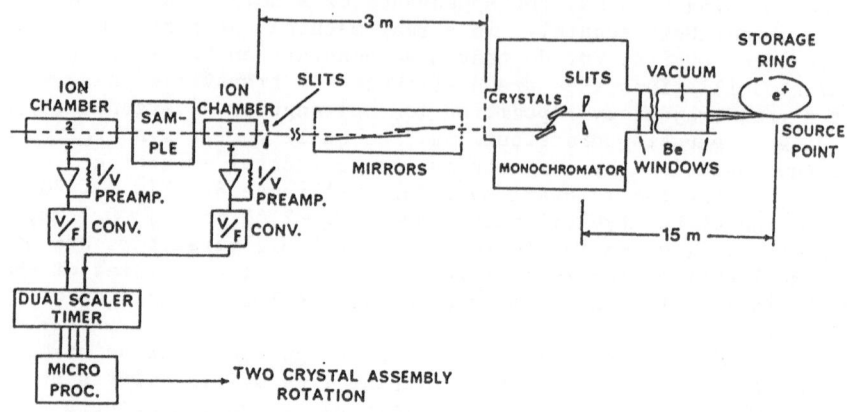

Figure 26. Schematic representation of a conventional energy scanning spectrometer (EXAFS-II station/L.U.R.E.)

unless the two crystals can be set and kept parallel to within less than an arcsecond. Yet another advantage of the two separate crystal setup is that the first crystal can be slightly bent so as to achieve sagittal refocusing of the beam. A schematic presentation of the energy scanned EXAFS-II spectrometer at L.U.R.E. is shown in figure 26.
(ii) The principle of a dispersive setup is illustrated by figure 27. A bent triangular (shaped) crystal is located at ca. 20 meters from the synchrotron radiation source and refocuses the reflected polychromatic beam at an image point of the source where a small sized sample can be

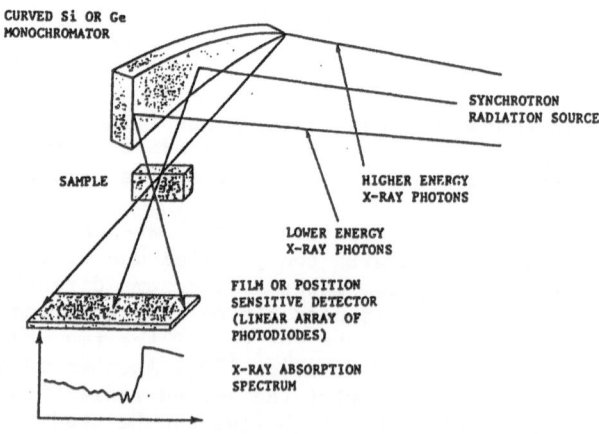

Figure 27. Schematic apparatus for energy dispersive absorption spectroscopy.

placed. Whereas the mechanical problem has been tremendously simplified because one is using a single crystal with fixed orientation, the detection problem has become more involved here because one needs a fast linear position sensitive detector with a high spatial resolution and a fast read-out circuitry and logic. This approach of X-ray absorption spectroscopy has been successfully developed by two groups simultaneously [80-83]. Since there is no mechanically mobile component, XANES and EXAFS spectra can be accumulated here at a rate of one every 8 ms whereas a single spectrum taken over the same energy range requires several minutes on energy scanned spectrometers. Obviously, energy dispersive spectrometers offer spectacular possibilities for kinetic studies even though some difficulties still subsist:

. energy calibration is not straightforward,

. for microheterogeneous samples (e.g. powders, supported catalysts etc...) the photons of different energies do not follow the same optical paths,

. the detection yield and dark current of each individual photodiode may be different and require diode specific corrections.

There are, however, other sources of limitations which indeed concern both types of spectrometers:

1. The <u>energy resolution</u> which in the case of a two crystal energy scanning spectrometer is given by [79]:

$$\frac{\Delta E}{E} = \left[(\omega_D \cot \theta_B)^2 + (\beta_y \cot \theta_B)^2 + (\frac{1}{2}\beta_x^2)^2 \right]^{1/2} \qquad (38)$$

where ω_D is the angular width of dynamical diffraction, β_x and β_y are the horizontal and vertical divergence of the incident beam, θ_B being the Bragg angle. Obviously, the best resolution is to be obtained for Bragg angles close to 90°. For XANES experiments requiring the finest energy resolution, it is of common practice to keep the Bragg angle as large as possible (i.e. $\theta_B > 45°$) by selecting reflecting planes with large enough Miller indices (e.g. 311, 331, 511, ...). Furthermore as ω_D is decreasing as $(h^2+k^2+l^2)^{-1}$, it usually becomes possible to satisfy the condition : $(\omega_D \cot\theta_B)^2 \leqslant (\beta_y \cot\theta_B)^2$ which indicates that the ultimate energy resolution is limited by the emittance of the source as already mentioned in the previous section. New storage rings fully dedicated to synchrotron radiation have much lower emittances and then offer quite significant advantages.

2. The <u>contamination of the exit beam with harmonics</u> has been recognized for many years to be a major source of distortion of the EXAFS/XANES spectra, especially for highly absorbing samples or in the soft X-ray range [84,85]. Harmonics are also known to enhance the presence in the EXAFS spectra of artefactual "glitches" when multiple Bragg reflections are satisfied simultaneously at a given energy [1]. The basic reason for these dramatic effects is that the two detection channels of figure 26 will have a different response in the presence of harmonics. It is our experience that it is of paramount importance to keep the level of these unwanted harmonics as low as possible. A considerable gain of the quality of the spectra, and even of the sensitivity of the spectrometer in the soft X-ray range, resulted from the adjunction behind the

monochromator of an efficient harmonic rejector made of two flat, parallel SiO_2 mirrors [85].

3. Glitches. As mentioned above, artefactual "glitches" sometimes pollute the EXAFS spectra especially when there are differences in the detection channels, e.g. due to detector non linearities, harmonics, incorrect beam or sample positioning...etc...[1]. Even for two identical detection channels, a glitch can still arise because the multiple Bragg reflection changes the energy distribution profile of the transmitted beam : due to the dispersive effect of the sample derivative like glitches can be observed. The only hope of reducing their contribution might be to preserve the best energy resolution.

4. Real (μ_1) and apparent (μ_A) absorption coefficients [84]. Any kind of leakage of the incident radiation, e.g. around a mispositioned sample or through a broken pellet with holes, has to be avoided because it can result in erroneous analyses. Formally, very similar effects are associated with the transmission of harmonics or any other unwanted radiation (e.g. spurious fluorescence excitations). The apparent absorption coefficient is then given by :

$$\mu_A \bar{d} = - \ln \frac{I}{I_o} = \mu_1 \bar{d} - \ln(1+\xi) - \ln A_o \qquad (39)$$

with: $$\xi(\mu_1) = \exp \{\mu_1 \bar{d}\} \sum \eta_p \qquad (40)$$

Here I and I_o refer to the currents measured by the two ion chambers of figure 26, A_o being a scaling term related to the linearity of the two detectors whereas the η_p's are describing the relative level of harmonics or the proportion (%) of holes. Clearly, the correction term ξ can become quite large for highly absorbing materials or in the soft X-ray range. The consequences on the EXAFS data analyses become more apparent if one expands $\mu_A(\chi)$ in Taylor series as [84]:

$$\mu_A(\chi) = \mu_A(0) + \chi \mu_A' + \frac{1}{2} \chi^2 \mu_A'' + ...$$

where the partial derivatives are easily calculated analytically [84]. It can be shown, then, that the measured amplitude of the EXAFS signal can be reduced in spectacular proportions for highly absorbing samples if the measurements are not carried out carefully enough.

5. Reflectivity measurements (REFLEXAFS). According to Snell's law, X-ray are totally reflected by an "optically" planar surface if the glancing angle θ is smaller than the critical value $\theta_c \simeq \sqrt{2\delta}$, where δ is related to the complex refractive index n^* by $n^* = 1 - \delta - i\beta$. For $\theta < \theta_c$ the reflectivity R of the surface is lower than unity and it can be shown that R exhibits a sensitive dependence upon the EXAFS modulation χ if θ is adjusted close to θ_c (e.g. $\theta \simeq 0.95 \ \theta_c$) [84]. Of course, this method is very promising for structural studies of species close to a given interface as the penetration depth can be adjusted with the glancing angle and it can be used either in the energy scanned mode or in the energy dispersive mode [86].

4.3. Excitation spectra using secondary emissions.

There is a wide variety of relaxation processes following the creation
of an unstable deep core hole :(i) Radiative processes include the
well known X-ray fluorescence emissions (figure 28) and allow the hole
to move step by step toward the external shells (e.g. $K^o \rightarrow L^o_{II,III} \rightarrow \dots$).
Cascade processes can thus result in optical luminescence. (ii) Non
radiative processes are also well studied and most often can be explained by multielectron transitions (e.g. Auger and Coster-Kronig
transitions). (iii) There are still several examples of complicated
processes involving both photon emission and bielectron transitions, as
in the case of "radiative" Auger processes.

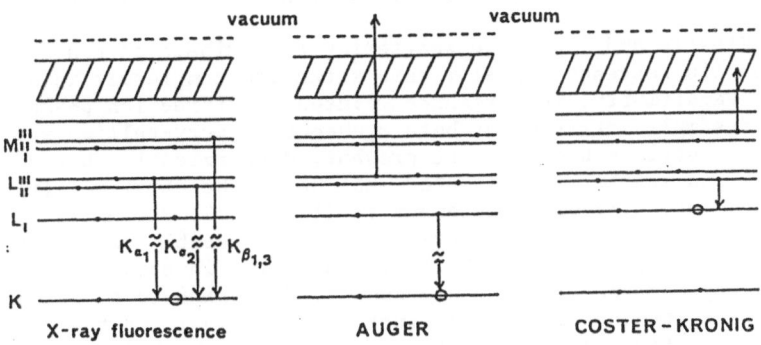

Figure 28. Examples of radiative and non radiative
relaxation processes for the core hole.

Indeed a discussion of these various relaxation mechanisms would
fall beyond the scope of this lecture. However, the point we are concerned with, is the following : excitation spectra can be recorded experimentally by monitoring the intensity of the relevant emissions (X-
ray or optical photons, Auger electrons, etc...) as a function of the
energy of the incident X-ray photon and it is a now well documented
result that these excitation spectra reproduce systematically all the
known features of the XANES and EXAFS spectra when classically recorded
in the transmission mode. As shown below, there are specific advantages
in measuring excitation spectra.

4.3.1. X-ray fluorescence. It is now well appreciated that measuring a
fluorescence excitation spectrum is more efficient than the normal
transmission experiment : (i) at dilute concentrations [1,2,87] or (ii)
when the sample is too absorbing, a situation which often prevails for
soft X-ray measurements (i.e. E < 5keV) [88,89]. The observed contrast
enhancement of the X-ray fluorescence excitation spectra results from
two phenomena :
    . Matrix atoms fluorescence occur most often at a much lower energy

(e.g. below 3 keV) than the fluorescence of the host atom one is interested in. The contribution of the matrix fluorescence can therefore often be filtered out or neglected.

. The fluorescence yield of low Z atoms (e.g. C, N, O) is usually much smaller than that of the host atom.

There are, however, practical difficulties for optimizing the quality of the X-ray fluorescence excitation spectra :

. The first problem stems from the fact that the X-ray fluorescence photons are emitted isotropically in at least 4π. The first consequence is that fluorescence excitation spectra cannot be recorded with the dispersive spectrometers described in the previous section because the fluorescence photons no longer have any memory of their respective excitation energies. Another consequence is that one needs special detectors collecting the fluorescence photons over the largest possible solid angles.

. Elastic and inelastic scattering contaminate or can even dominate the fluorescence signal : this situation can result in a dramatic loss of sensitivity. The presence of intense diffraction peaks for 2D-oriented samples or even for heterogeneous polycrystalline samples can also cause serious problems. At present, discrimination between fluorescence and scattered/diffracted photons can be obtained: (i) by using metallic microfoils of the element (Z-1) or (Z-2) acting as selective filters of the scattered radiation. This is possible only if the absorption edge of the (Z-1)/(Z-2) element falls in-between the excitation energy and the fluorescence energy. The price to be paid is a significant attenuation of the fluorescence signal and the excitation of the X-ray fluorescence of the filter itself which is not easily suppressed [90]. (ii) By using energy dispersive solid state detectors [87]: Unfortunately, the energy resolution of Si or Ge semiconductor devices is known to drop off very rapidly at the fast counting rates corresponding to the rather high level of scattered radiation. The only way to overcome this difficulty is to use multidetectors with a large number of channels. This solution, however, has not yet gained any wide interest on grounds of cost because each channel requires individual charge sensitive preamplifiers, pulse shaping electronics and has to be coupled to expensive multiplexer modules. (iii) As Compton and elastic scattering intensities are basically polarization dependent, the ratio fluorescence / scatter can be optimized for specific detector configurations.

On the other hand, one should always keep in mind that even though the excitation spectra have the same structural content as the absorption spectra, the relative amplitudes of the XANES and EXAFS structures can be significantly different when comparing the spectra recorded with both detection schemes. Indeed, in a geometrical arrangement where the incident and emitted radiations make the angle θ_i and θ_F with the planar surface of the excited sample, the counting rate in the particular direction θ_F is given by :

$$N_F = \sum_j \frac{N_\theta^j \, \sigma_F \, \varepsilon_F \, \Omega(4\pi)}{\mu_j(E) + \alpha_F \, \mu(E_F)} \left\{ 1 - \exp\left[-\frac{\mu_j \bar{d}}{\sin\theta_i} - \frac{\mu_F \bar{d}}{\sin\theta_F} \right] \right\} \quad (42)$$

with : $\alpha_F = \sin \theta_i / \sin \theta_F$ ($\simeq$ 1 in practical cases).

In relation (42), summation is over the fundamental and harmonics of the incident beam, σ_F denotes the absorption coefficient associated with the primary absorption process resulting in the fluorescence emission with quantum yield ε_F. If one is decomposing $\mu_1(E)$ as $\mu_1 = \mu^b + \mu^o(1+\chi)$ where μ^b is some continuous pre-edge background absorption including the matrix absorption whereas μ^o is the absorption edge jump, then the apparent edge jump in the excitation spectrum is given by [84]

$$N_F^o = \frac{\mu^o N^1}{\mu^o + \mu^b + \alpha_F \mu_F} \left\{ \varepsilon_F - \frac{\sigma_{scatt.}}{(1+\alpha_F)\mu^b} \left[1 + \frac{\alpha_F \mu_F}{\mu^o + \mu^b} \right] \right\} \qquad (43)$$

It becomes then clear that the apparent edge jump of X-ray fluorescence spectra decreases rapidly if the matrix is strongly absorbing either at the excitation energy E or fluorescence energy : experiments on dilute samples in the soft X-ray range appear then to be very difficult. Equation (43) also indicates that the presence of large amounts of scattered radiation can also reduce the apparent edge jump [84]. N_F can also be expanded in Taylor series as μ_A in equation (41) :

$$N_F(\chi) = N_F(0) + \chi N_F' + \frac{1}{2} \chi^2 N_F'' + \dots \qquad (44)$$

and the sensitivity to the EXAFS modulation is defined as :

$$S_F = \frac{1}{N_F^o} N_F' \simeq 1 - \frac{\mu^o}{\mu^o + \mu^b + \alpha_F \mu_F} + \frac{\sigma_{scatt.}}{\varepsilon_F(1+\alpha_F)\mu^b} \frac{\alpha_F \mu_F \mu^o}{(\mu^o + \mu^b)^2} + \dots \qquad (44)$$

Correct amplitudes or coordination number can be obtained only if :
(i) $\mu^o \ll \mu^b + \alpha_F \mu_F$ i.e. a typical situation for dilute systems and
(ii) if the scattered radiation is properly filtered out, i.e. $\sigma_{scatt} \simeq 0$.

4.3.2. X-ray excited optical luminescence (XEOL). It has been known for a long time that several systems exhibit a strong optical emission in the UV-visible range when excited by X-ray photons. It has been also recognized that under appropriate experimental conditions, XEOL can be used for recording XANES/EXAFS excitation spectra [91,92]. Clearly, this detection scheme has not gained any wide interest because the quality of the XEOL excitation spectra recorded on dilute samples is still rather poor in comparison with the quality of the X-ray fluorescence excitation spectra recorded from the same samples. XEOL excitation spectra, however, offer a specific advantage : the, so far, unique possibility for site selective EXAFS/XANES experiments when the material under study contains a mixture of non equivalent sites, the contribution of which are simply averaged in all other detection schemes. Conditions for site selectivity to be preserved have been discussed by Goulon et al. [84,92,93].

The first site selective experiment using XEOL excitation spectroscopy was also reported by Goulon et al. [93] for a mixture of powders, only one of which had a XEOL emission. Very recently, Pettifer et al. [94] have demonstrated that site selectivity can still be observed in mixed powders where both materials luminesce. It is our opinion that this technique is likely to be of interest in near-surface studies and for structural investigations of biological materials where the excitation centers are well separated in a low mobility solid or molecule.

4.3.3. <u>Auger electrons, partial or total electron yields.</u> Due to the low mean free path of electrons (typically between 5-100 Å), the electron yield detection is a powerful detection scheme for surface studies (SEXAFS) [1]. Again a detailed discussion of surface related applications of EXAFS or XANES spectra would fall beyond the scope of this lecture but motivated readers are referred to a fairly complete review of SEXAFS developements during the past decade [95]. For a long time, electron detection has been regarded as possible only in ultra-high vacuum until Kordesh and Hoffman have shown [96] that this detection scheme can still be used for a characterization of thick samples in a more realistic environment : energetic Auger electrons emitted from an X-ray excited sample can be converted in a bunch of low energy electrons when crossing an helium atmosphere and all what one has to do is to collect these charges on a biased electrode. This detection scheme exhibits a surprisingly high sensitivity. There is no doubt that the method should find considerable applications in surface studies in the soft X-ray range where the sensitivity of X-ray fluorescence excitation spectra becomes limited by poor quantum yields, strong absorption of the matrix and bad discrimination of fluorescence v.s. scattered radiation.

<u>Abbreviations used</u> :

OMP   : 2,3,7,8,12,13,17,18-octamethylporphyrinatodianion.
OEP   : 2,3,7,8,12,13,17,18-octaethylporphyrinatodianion.
TPP   : meso-5,10,15,20-tetraphenylporphyrinatodianion.
TMP   : meso-5,10,15,20-tetramesitylporphyrinatodianion.
Pc   : phtalocyaninedianion.
OMTBP: octamethyltetrabenzoporphyrinatodianion.

<u>Acknowledgements</u> :

The authors wish to thank Pr. C.R. Natoli for several illuminating discussions during his stay as a visiting Professor at the University of Nancy (1986) and for the permission of using his programs for theoretical multiple scattering calculations.
    Pr. R. Guilard, Dr. J.P. Battioni, Dr. B. Meunier and their collaborators had also a determinant contribution for the synthesis of all the compounds mentioned in this work.

6. REFERENCES

1. P.A. Lee, P.H. Citrin, P. Eisenberger and B.M. Kincaid
   Rev. Mod. Phys. $\underline{33}$ 769 (1981)
2. B.K. Teo and D.C. Joy Editors
   "EXAFS Spectroscopy - Techniques and Applications", Plenum Press
   NEW-YORK (1981)
3. A. Bianconi in
   "EXAFS for inorganic systems", Edited by C.D. Garner and S.S.
   Hasnain - Daresbury Lab. S.E.R.C. DL/SCI/R17, $\underline{13}$ (1981)
4. A. Bianconi, L. Incoccia, S. Stipcich Editors
   "EXAFS and Near Edge Structure II", Springer Series in Chemical
   Physics $\underline{27}$ - Springer Verlag (1983)
5. K. Hodgson, B. Hedman and J.E. Penner-Hahn Editors
   "EXAFS and Near Edge Structure III", Springer Proceedings in Physics
   $\underline{2}$ - Springer Verlag (1984)
6. A. Bianconi and A. Congiu-Castellano Editors
   "Biophysics and Synchrotron Radiation", Springer Series in Bio-
   physics $\underline{1}$ - Springer Verlag (1986)
7. D.C. Koningsberger and R. Prins
   X-ray Absorption : Principles, Applications, Techniques of EXAFS,
   SEXAFS and XANES, John Wiley Sons, Inc. (1987)
8. P. Lagarde, D. Raoux and J. Petiau Editors
   "EXAFS and Near Edge Structure IV", Vol.1 and 2, J. de Physique $\underline{47}$
   Colloque C8 (Dec. 1986) Les Editions de Physique
9. O. Bortolini, M. Ricci, B. Meunier, P. Friant, I. Ascone and
   J. Goulon
   New Journal of Chem. $\underline{10}$ 39 (1986)
10. J. Wong, F.W. Lytle, R.P. Messmer and D.H. Maylotte
    Phys. Rev. $\underline{B30}$ 5596 (1984)
11. R. de L. Kronig
    Z. Phys. $\underline{70}$ 317 (1931)
12. E.A. Stern
    Phys. Rev. $\underline{B10}$ 3027 (1974)
13. E.A. Stern, D.E. Sayers and F.W. Lytle
    Phys. Rev. $\underline{B11}$ 4838 (1975)
14. C.R. Natoli and M. Benfatto
    in reference [8] p. C8-11
15. P.A. Lee and J.B. Pendry
    Phys. Rev. $\underline{B11}$ 2795 (1975)
16. C.A. Ashley and S. Doniach
    Phys. Rev. $\underline{B11}$ 1279 (1975)
17. P.A. Lee and G. Beni
    Phys. Rev. $\underline{B15}$ 2862 (1977)
18. P.J. Durham, J.B. Pendry
    Comp. Physics Comm. $\underline{25}$ 193 (1982)
19. C.R. Natoli
    in reference [4] 43 (1982)
20. B.K. Teo
    J. Am. Chem. Soc. $\underline{103}$ 3990 (1981)

21. J.J. Boland, S.E. Crane and J.D. Baldeschwieler
    J. Chem. Phys. 77 142 (1982)
22. J.E. Muller and J.W. Wilkins
    Phys. Rev. B29 4331 (1984)
23. W.L. Schaich
    Phys. Rev. B29 6513 (1984)
24. S.J. Gurman, N. Binsted and I. Ross
    J. Phys. C : Sol. State Phys. 17 143 (1984)
25. C.R. Natoli, M. Benfatto and S. Doniach
    Phys. Rev. A34 4682 (1986)
26. J.J. Rehr, R.C. Albers, C.R. Natoli and E.A. Stern
    Phys. Rev. B34 4350 (1986)
27. C. Brouder and C.R. Natoli
    unpublished lecture notes on Multiple Scattering (Dec. 1986)
28. P. Lloyd and P.V. Smith
    Adv. Phys. 21 69 (1972)
29. K. Johnson
    Adv. Quantum Chem. 7 143 (1973)
30. J.L. Dehmer and D. Dill
    J. Chem. Phys. 61 692 (1974)
31. L. Hedin and B.I. Lundquist
    J. Phys. C4 2064 (1971)
32. D.B. Tran Thoai and W. Eckardt
    Sol. State. Comm. 40 269 (1981)
33. J.J. Boland and J. Baldeschwieler
    J. Chem. Phys. 80 3005 (1984)
34. N. Alberding and E.D. Crozier
    Phys. Rev. B27 3374 (1983)
35. G. Beni and P.M. Platzman
    Phys. Rev. B14 1514 (1976)
36. M. Loos and J. Goulon
    unpublished results
37. P. Eisenberger and G.S. Brown
    Sol. State Comm. 29 481 (1979)
38. P. Friant
    Ph. D. thesis, University of NANCY I (1986)
39. L. Brillouin
    "Science and Information Theory" 2nd Ed. (1962)
40. G. Martens, P. Rabe, N. Schwentner and A. Werner
    Phys. Rev. Lett. 39 1411 (1977)
41. A.K. Livesey
    Ph. D. thesis, Cambridge University (1984)
42. R.F. Pettifer and A.D. Cox
    in reference [4] p. 66
43. B.K. Teo and P.A. Lee
    J. Am. Chem. Soc. 101 2815 (1979)
44. J.J. Boland, F.G. Halaka and J.D. Baldeschwieler
    Chem. Phys. Lett. 129 1 (1986) and in reference [5] p. 80
45. H. Oyanagi, T. Itzuka, T. Matsushita, S. Saigo, R. Makino,
    Y. Ishimura
    in reference [8] p. C8-1147 (1986)

46. A. Bianconi, A. Congiu-Castellano, M. Dell'Aricia, A. Giovanelli
    P.J. Durham, E. Burattini and M. Barteri
    F.E.B.S. Lett. 178 165 (1984)
47. A. Bianconi, A. Congiu-Castellano, P.J. Durham, S.S. Hasnain and
    S. Philips
    Nature 318 685 (1985)
48. R.A. Bair and W.A. Goddard III
    Phys. Rev. B22 2767 (1980)
49. M.F. Ruiz-Lopez, D. Rinaldi, C. Esselin, J. Goulon, J.L. Poncet and
    R. Guilard
    in reference [8] p. C8-637 (1986)
50. M.F. Ruiz-Lopez, C.R. Natoli, J. Goulon and R. Guilard
    to be submitted to Chem. Phys.
51. S. Doniach, M. Berding, T. Smith and K. Hodgson
    in reference [5] p. 33
52. J.E. Penner-Hahn, M. Benfatto, K.O. Hodgson, B. Hedman, T. Takahashi,
    S. Doniach, J.T. Groves and K.O. Hodgson
    Inorg. Chem. 25 2255 (1986)
53. S. Stizza, M. Benfatto, A. Bianconi, J. Garcia, G. Mancini and
    C.R. Natoli
    in reference [8] C8-691 (1986)
54. J. Goulon, C. Goulon-Ginet, P. Friant, J.L. Poncet, R. Guilard
    J.P. Battioni and D. Mansuy
    Proc. of the IV Int. Conf. on the Organic Chemistry of Selenium
    and Tellurium, p. 379 (1984) - Editors : F.J. Berry and W.R. Mc
    Whinnie
55. J.L. Poncet, R. Guilard, P. Friant, C. Goulon-Ginet and J. Goulon
    New Journal of Chem. 8 583 (1984)
56. S.P. Cramer, T.K. Eccles, F. Kutzler, K. Hodgson and S. Doniach
    J. Am. Chem. Soc. 98 8059 (1976)
57. W.T. Elam, E.A. Stern, J.D. Mc Callum and J. Sanders-Loehr
    J. Am. Chem. Soc. 105 1919 (1983)
58. J. Goulon, P. Friant, C. Goulon-Ginet, A. Coutsolelos and R. Guilard
    Chem. Phys. 83 367 (1983)
59. J.L. Hoard
    Ann. N.Y. Acad. Sci. 206 18 (1973)
60. D.L. Cullen, E.F. Meyer Jr.
    J. Am. Chem. Soc. 96 2095 (1974)
61. E.F. Meyer
    Acta Crystallograph. B28 2162 (1972)
62. C.L. Hill and F.J. Hollander
    J. Am. Chem. Soc. 104 7138 (1984)
63. J.W. Buchler, C. Dreher, K.L. Kay, Y.J.A. Lee and W.R. Scheidt
    Inorg. Chem. 22 888 (1983)
64. J.P. Collman, J.L. Hoard, N. Kim, G. Lang and C.A. Reed
    J. Am. Chem. Soc. 97 2676 (1975)
65. R.F. Pettifer, D.L. Foulis and C. Hermes
    in reference [8] C8-545 (1986)
66. R.W. Strange, S.S. Hasnain, N.J. Blackburn and P.F. Knowles
    in reference [8] C8-593 (1986)

290

67. D. Mansuy, J.P. Battioni, J.C. Chottard and V. Ullrich
    J. Am. Chem. Soc. 101 3871 (1979)
68. D. Mansuy, M. Lange, J.C. Chottard
    J. Am. Chem. Soc. 100 3213 (1978)
69. D. Mansuy, M. Lange, J.C. Chottard, P. Morliere and D. Brault
    J. Chem. Soc. Chem. Comm. 648 (1977)
70. V.L. Goedken, M.R. Deakin and L.A. Bottomley
    J. Chem. Soc. Chem. Comm. 607 (1982)
71. D. Mansuy, M. Lange, J.C. Chottard, J.F. Bartoli, B. Chevrier and
    R. Weiss
    Angew. Chem. Int. Ed. 17 781 (1978)
72. W.R. Scheidt, D.A. Summerville and I.A. Cohen
    J. Am. Chem. Soc. 98 6623 (1976)
73. P. Doppelt
    Inorg. Chem. 23 4009 (1984)
74. M. Loos, P. Friant, I. Ascone, J. Goulon, J.M. Barbe, A. Coutsole-
    los and R. Guilard
    in reference [8] p. C8 (1986)
75. C.J. Schramm, R.P. Scaringe, D.J. Stojakovic, B.M. Hoffman,
    J.A. Ibers and T. Marks
    J. Am. Chem. Soc. 102 6702 (1980)
76. T.E. Phillips, R.P. Scaringe, B.M. Hoffman and J.A. Ibers
    J. Am. Chem. Soc. 102 3435 (1980)
77. J.F. Kirner, W. Dow and W.R. Scheidt
    Inorg. Chem. 15 1685 (1976)
78. J. Goulon, P. Elleaume and D. Raoux
    Nucl. Instrum. and Methods A254 192 (1987)
79. J. Goulon, M. Lemonnier, R. Cortes, A. Retournard and D. Raoux
    Nucl. Instrum. and Methods 208 625 (1983)
80. A.M. Flank, A. Fontaine, A. Jucha, M. Lemonnier, D. Raoux and
    C. Williams
    Nucl. Instrum. and Methods 208 651 (1983)
81. R.P. Phizakerley, Z.V. Rek, G.V. Stephenson, S.D. Conradson,
    K.O. Hodgson, T. Matsushita and H. Oyanagi
    J. Appl. Cryst. 16 220 (1983)
82. E. Dartyge, C. Depautex, J.M. Dubuisson, A. Fontaine, A. Jucha,
    P. Leboucher and G. Tourillon
    Nucl. Instrum. and Methods A246 452 (1986)
83. H. Oyanagi, T. Matsushita, U. Kaminaga and H. Hashimoto
    in reference [8] p. C8-139 (1986)
84. J. Goulon, C. Goulon-Ginet, R. Cortes and J.M. Dubois
    J. Physique 43 539 (1982)
85. J. Goulon, R. Cortes, A. Retournard, A. George, J.P. Battioni,
    R. Frety and B. Moraweck
    in reference [5] p. 449 (1984)
86. E. Dartyge, A. Fontaine, G. Tourillon, R. Cortes and A. Jucha
    Phys. Lett. 113A 384 (1986)
87. J. Jaklevic, J. Kirby, M.P. Klein, A.S. Robertson, G.S. Brown and
    P. Eisenberger
    Sol. State Comm. 23 679 (1977)

88. F.W. Lytle, R.B. Greegor, D.R. Sandstrom, E.C. Marques, J. Wong, C.L. Spiro, G.P. Huffman and F.E. Huggins
Nucl. Instrum. and Methods $\underline{226}$ 542 (1984)

89. A. Retournard, M. Loos, I. Ascone, J. Goulon, M. Lemonnier and R. Cortes
in reference [8] p. C8-143 (1986)

90. E.A. Stern and S. Heald
Rev. Sci. Instrum. $\underline{50}$ 1579 (1979)

91. A. Bianconi, D. Jackson, K. Monahan
Phys. Rev. $\underline{B17}$ 2021 (1978)

92. J. Goulon, P. Tola, J.C. Brochon, M. Lemonnier, J. Dexpert-Ghys and R. Guilard
in reference [5] p. 490 (1984)

93. J. Goulon, P. Tola, M. Lemonnier, J. Dexpert-Ghys
Chem. Phys. $\underline{78}$ 347 (1983)

94. R.F. Pettifer and A.J. Bourdillon
J. Phys. C : Sol. State Phys. $\underline{20}$ 329 (1987)

95. P.H. Citrin
in reference [8] p. C8-437 (1986)

96. M.E. Kordesch and R.W. Hoffman
Phys. Rev. $\underline{B29}$ 491 (1984)

DISCUSSION

SAMPLE THICKNESS

The polymeric metalloporphyrins were examined as powder compressed into pellets. For normal transmission experiments, in which the incident X-ray beam was monitored after it has passed through the sample, the thickness of the sample was adjusted to give about 20% transmission. For experiments in which the X-ray fluorescence of the absorber was monitored, there was no restriction on the sample thickness.

THEORETICAL ASPECTS

Although there were some problems in the calculation of phases from the plane wave approximation used in the late 1970s, correct results can now be obtained from the curved wave approximation using the same potential. Accuracy of phase determination is not critical to EXAFS, because the ionisation energy of the photo-electron is treated as an adjustable parameter. More complicated potentials have been developed, but their effect on the final results obtained from the data analysis has been relatively small.

Current theoretical work is much concerned with the difficult problems of making allowance for many-body effects and inelastic processes. Theoretical XANES spectra with more realistic amplitudes should result, but structure determination is unlikely to be greatly improved because it depends on peak positions, and the factors determining the latter are now thought to be adequately understood.

Dr J Karle drew attention to another approach to phase calculation for EXAFS, recently published in Physical Review Letters. For the phases it uses an essentially heuristic function which must obey some statistical characteristics in relation to the positions of minima and maxima. It gives results in good agreement with those obtained by other experimental methods of structure determination.

BIOLOGICAL APPLICATIONS

Biological systems containing iron at concentrations as low as 100 ppm can now be studied by EXAFS, using X-ray fluorescence. In this way Dr Goulon has found the Fe-N distance in cytochrome-c to be less than 1.98 Å, a value typical of low spin iron compounds; older X-ray work had estimated this bond length to be greater than 2.0 Å.

ZIEGLER-NATTA CATALYSTS

Dr Goulon's EXAFS results for specific industrial hydrogenation catalysts suggest that accepted ideas on their structure mode of activity may be wrong.

EXAFS AS A STRUCTURAL TOOL

In its 10 year life, EXAFS has passed from initial over-optimism, through a period of some unacceptable results, and now, with improvements in theory, it has reached a plateau where reasonable work gives reasonable results. Most information is obtained when EXAFS is used comparatively, but when no model is available, interpretation is more difficult.

"USE OF THE RIETVELD PROFILE ANALYSIS FOR CRYSTAL STRUCTURE DETERMINATION
AND REFINEMENT"

A. Albinati

Istituto di Chimica Farmaceutica
Università di Milano
42, Viale Abruzzi
I-20131 Milano
Italy

ABSTRACT. In the Rietveld method of analysing powder diffraction data,
the crystal structure is refined by fitting the entire profile of the
diffraction pattern to a calculated profile. There is no intermediate
step of extracting structure factors. The method has been used successfully for the treatment of results obtained with neutrons from steady
or pulsed sources and X-rays both from conventional laboratory equip-
ments and synchrotron sources.

1.Introduction

    In the Rietveld method of refining powder diffraction data, the
profile of the entire diffraction pattern is calculated and fitted
to the experimental profile. There is no need to extract integrated
intensities first and so patterns containing many overlapping Bragg
peaks can be analysed.The method was applied originally by Rietveld
(1,2) to the refinement of neutron intensities recorded at a fixed
wevelength. Subsequently, it has been successfully used for analysing
powder data from all types of experimental diffraction techniques, i.e.
with neutrons or X-rays as the primary radiation and with measurement
of the scattered radiation at a fixed wavelength (and variable angle)
or at a fixed angle (and variable wavelength).
    A structural refinement based on powder data is likely to be
inferior to one using good single crystal data, nevertheless, it may
not be possible to obtain single crystals, and with a powder sample,
twinning, absortion and extinction may be neglected.
The theory of the Rietveld method its limitations and applications
have been recently discussed in various papers (3-5). In this article we
shall be concerned with a number of fundamental questions about the
Rietveld method such as: peak-shape, significance of the structural
parameters, limitations in the applicability.

M. A. Carrondo and G. A. Jeffrey (eds.), Chemical Crystallography with Pulsed Neutrons and Synchrotron X-Rays, 295–312.
© 1988 by D. Reidel Publishing Company.

2.Experimental Techniques Basic Intensity Formulae.

(a) Neutron powder diffraction: fixed wavelength

A schematic arrangement for a typical experiment is shown in Fig.1; the neutron wavelength is usually chosen within the range 1.0-2.5 Å, and the detector records the intensity which is scattered by the sample at different Bragg angles 2θ. For a cylindrical sample, fully bathed by the beam, the number of neutron diffracted in a Debye-Scherrer ring can be calculated (8) for a given Bragg angle:

$$N(2\theta) = \frac{\varphi_0 \lambda^3 LV\rho' j_{hkl} N_c^2 F_{hkl}^2}{8\pi R\rho \sin\vartheta_B \sin2\vartheta_B} \delta(\vartheta - \vartheta_B) \tag{1}$$

where φ_0 = incident flux, λ = wavelength, V = specimen volume ρ' and ρ are measured and theoretical densities respectively; j_{hkl}=multiplicity of the hkl family of planes; N_c = no. of unit cell per unit volume; R = specimen to detector distance.
It has been assumed that the non-Bragg intensity has been largely sub-tracted from the elastic part with the background (9). Finite collimation $(\alpha_1 \neq \alpha_2 \neq \alpha_3 \neq 0)$, and the nature of the sample (particle size, strain etc.) give a broadening of the Bragg peak, so that to obtain the diffracted intensity $I(2\theta)$ eq.(1) has to be convoluted with a broadening function $f(2\theta)$ (peak shape function).
The simplest choice, valid for low or medium resolution diffractometers (using neutron or Synchrotron Radiation) is a Gaussian peak shape; thus we can write:

$$f(2\vartheta) = \frac{2}{H}\left(\frac{\ln 2}{\pi}\right)^{1/2} \times \exp\left(-\frac{4\ln 2}{H^2}(2\vartheta - 2\vartheta_B)^2\right) \tag{2}$$

where H is the full-width at half-maximum (FWHM) of the peak and the normalizing constant ensures that the peak has unit area, indipendent of H; from the convolution of eq. (1) and eq. (2) we obtain:

$$I(2\vartheta) = \frac{\varphi_0 \lambda^3 LV\rho' j_{hkl} N_c^2 F_{hkl}^2}{8\pi R\rho \sin\vartheta_B \sin2\vartheta_B} \frac{2}{H}\left(\frac{\ln 2}{\pi}\right)^{1/2}\exp\left\{-[(4\ln2)/H^2](2\vartheta - 2\vartheta_B)^2\right\} \tag{3}$$

Using the notation of Rietveld (2) this equation can be rewritten in the form:

$$y_i = \frac{cj_k L_k F_k^2}{H_k} \exp\left(-4\ln2\left[(2\vartheta_i - 2\vartheta_k)/H_k\right]^2\right) \tag{4}$$

where y_i is the contribution of the Bragg reflection k to the intensity measured at position $2\theta_i$, c is a constant for a given sample and diffraction geometry, L_k is the Lorentz factor, $2\theta_k$ is the calculated position of the Bragg peak, and H_k is the FWHM of the peak.

(b) Neutron powder diffraction: fixed scattering angle.

A steady state reactor can be replaced, as a neutron source, by a pulsed source. Pulsed neutrons may be obtained both by a mechanical chopper or produced by electron Linacs or proton synchrotrons (see Ref.10). A schematic experimental arrangement is given in Fig. 2.
The incident neutron pulse contains a wide range of energies (wavelengths) and the detector can be placed at a fixed scattering angle; the scattered intensity is determined as a function of the wavelength by measuring the time of flight (TOF) of the individual neutrons in a pulse.
TOF, wavelength and d spacings are related through the equation

$$ \text{TOF} = \left(\frac{m\ l}{h}\right)\lambda \quad \text{or} \quad \text{TOF} = \left(\frac{2\ m\ l\ \sin\vartheta}{h}\right)d \quad (5) $$

where m is the neutron mass, h the Planck's constant, l the total flight path.
The flux distribution of the incident neutrons is different for a reactor and a pulsed source. In fact while for the former case a Maxwellian distribution is appropriate

$$ \varphi(\lambda) = \text{constant} \times T^{-3/2}\lambda^{-5}\exp(-h^2/2mk_B T\lambda^2) \quad (6) $$

(where T is the effective neutron temperature and k_B is Boltzmann's constant) for the latter the spectrum is given by the superposition of a Maxwellian and a slowing-down spectrum representing the behaviour of the epithermal flux (see Fig.3 and Fig.4, for the treatment of the spectral distribution of the incident beam see Ref.11).
An equation similar to eq.(4) can be obtained that gives the contribution of the kth reflection to the intensity y_i which is registered in the ith time channel:

$$ y_i = j_k d_k^4 \varphi(\lambda_i)\sin\vartheta_k f(\lambda_i)F_k^2 \quad (7) $$

where $f(\lambda_i)$ is the peak shape function appropriate for a TOF experiment.

(c) X-ray powder diffraction: fixed wavelength.

The application of the Rietveld method in the X-ray case has been limited mainly because the description of the line-shapes, which is crucial for a successfull profile refiniment, is much more difficult than for the neutron case; this is due to instrumental factors and the presence of the overlapping α_1, α_2 doublet.

Malmros and Thomas (13) applied successfully this method to photographic data obtained with a Guinier-Hägg camera, and Khattak and Cox (14) used a powder diffractometer and Cu Kβ radiation, to avoid the α_1, α_2 doublet.

A different approach, giving excellent fit, has been followed by Parrish and coworkers (15) using a deconvolution method to obtain integrated intensities.

Powder diffractometers can be used with synchrotron sources (16). In this case the shape of the reflections can be described by simple analytical functions (17-19): this method shows, indeed, great promises because of resolution and the high intensity of the incident radiation.

(d) X-ray powder diffraction: fixed scattering angle.

Another way of collecting X-ray diffraction data, in a form suitable for analysis by the Rietveld method, is to employ an energy-dispersive technique. The full white spectrum (Bremsstrahlung radiation) from an X-ray tube or a sinchrotron falls on the sample, and a solid-state detector, held at a fixed scattering angle and connected to a pulse height analyser, performs an energy analysis of the diffracted beam (see Fig.5).

There are a number of problems connected with this method: (i) the limited spectral distribution of the ordinary X-ray tubes, (ii) the calculation of the polarisation term, (iii) absorption and extinction are wavelength dependent.
The use of sinchrotron radiation overcames the first two difficulties, and offer the advantage of very high intensities and the possibility of collecting data to high values of sin ϑ/λ . The spectral distribution of the incident beam can be deduced from the diffraction pattern of a standard sample (e.g. Si, SiO$_2$) or calculated from the machine parameters of the storage ring.
Although the resolution of this technique is poor compared to fixed wavelength techniques, and determined by the intrinsic resolution of the detector (at present $\sim$ 150 eV), the high counting rates make this method suitable for the study of time-dependent phenomena.

3. Basic Theory of the methods.

Let I_k the calculated integrated intensity for the kth reflections; the intensity y_i(calc) at any given point is given by

$$y_i(\text{calc}) = \sum_k I_k G_{ik} \qquad (8)$$

where G_{ik} is the peak shape function (cfr.eq.(3) and eq.(4) where G_{ik} is a Gaussian). Thus given G_{ik} the intensity contributed at point i, from a reflection centered at k, is univocally defined. The summation in eq. (8) includes all reflections making a contribution to y_i (calc). In the Gaussian approximation the contribution of the single reflection k to the calculated intensity at $2\vartheta_i$ is given by eq. (4). It should be note that the FWHM of the peaks is a function of the scattering angle. This dependence can be written in simplified form (22) as:

$$H_k^2 = U \tan^2\vartheta_k + V \tan\vartheta_k + W \qquad (9)$$

where U, V, W are half width parameters independent of ϑ_k . This treatment ignores intrinsic sample broadening and these parameters must be treated as adjustable variables in the refinement.
The number of overlapping reflections at any point $i(2\vartheta)$ is readily determined from the $2\vartheta_k$ values and the corresponding widths H_k .
The function minimised in the least-squares refiniment is then:

$$M_p = \sum_i^N w_i \left[(y_i(obs) - b_i) - 1/c \; y_i(calc) \right]^2 \qquad (10)$$

where y_i (obs) is the intensity measured at point i of the diffraction pattern, c the scale factor, w_i is the weight and b_i is the contribution of the background; the summation is over all points N at which the pattern is recorded. Assuming a zero background and that the errors in the measurement are from counting statistics alone we have:

$$w_i = \left(y_i(obs) \right)^{-1} \qquad (11)$$

The least squares parameters are of two kinds. The first group contains ths usual structural parameters: fractional coordinates of each atom in the asymmetric unit and the corresponding thermal factors (isotropic or anisotropic). The second group represents "profile parameters" which are not found in the least-squares refinement of single crystal data: the half-width parameters, the unit-cell constants and in some programs a background function. Further parameters may be added, for example, to allow for asymmetry of the peaks (caused by the finite height of the detectors) (23).
The background may be determined by measuring regions of the pattern which are free from Bragg peaks and making a linear interpolation or fitting a polynomial. However this procedures assume that the background varies smoothly with $\sin \theta/\lambda$, whereas TDS rises to a maximum at peak

positions (9,24). No satisfactory general calculation of TDS in a
powder pattern has yet been carried out. However it may be necessary
to include a background function in the refinement,to avoid a systematic error that affects the statistical validity of the method (see
§4).

The maximum number of parameters that can be refined is largely
determined by the quality of the diffraction pattern;intrinsic line
broadening set a limit,with the existing diffractometers,to c/a 250
parameters (6). Sometimes more than one diffraction pattern occours
in the observed intensity profile;the simultaneous refinement of two or
more phases is a feature that has been successfully incorporated in
a number of programs (4,6,38).

4. Statistical validity

The precision of a refinement is given by the calculated errors
while the presence of systematic errors in the refined model,leads to
incorrect values of the e.s.d.'s.
Cooper in a series of papers (25-27) criticised the Rietveld method on
the ground that residuals $\left[y_i(\text{obs}) - y_i(\text{calc})\right]$ related to the same
Bragg peak are correlated to one another. The existence of correlated
residuals for a resolved Bragg peak is illustrated in Fig.6.The solid
line represents the calculated peak as fitted by the Rietveld method,
assuming a structural model,while the broken line is the best fit of
the given peak shape(e.g. Gaussian) through the experimental points.
By changing a parameter,i.e. $\bar{B}$,points on the calculated peak will all
move together leading to a positive correlation between pairs of resi-
duals.The more closely spaced are the experimental points,the greater
the correlation is likely to be.It was therefore concluded that this
correlation makes the calculated errors statistically unsound. Comparing the Rietveld positional parameters with those from single crystal
and integrated intensities refinements,no systematic errors can be
detected (28-32) but the calculated e.s.d.'s are 2 - 3 times less than
those derived from the integrated intensities refinement of the same
data.
Moreover Prince (33) and Hewat and Sabine (34) showed that the e.s.d.'s
obtained by the Rietveld method are correct if there are no systematic
errors in the refined model;it may be noted that,eventhough the Rietveld positional parameters are unbiased,the error matrix reflects precision rather than accuracy (35).

It has been demonstrated that the weighthing scheme used in the
Rietveld refinement put too much enphasis on statistics:using simulated
data and considering the effect of counting time (36),it can be shown
that the final errors are dominated by the presence of systematic errors.

Nevertheless it should be stressed that the uncertainty about the significance of the e.s.d.'s does not seriously affect the merits and usefulness of the Rietveld method. In those cases in which the exact knowledge of the errors is critical, adjustments to the calculated e.s.d.'s may be made. Pawley (68) has suggested that to obtain an "empirical correct" estimate of the Rietveld e.s.d.'s the actual number of observations N should be divided by the number of observations n in the average FWHM of well resolved peaks. This is equivalent to multiply the e.s.d.'s by a factor tipically in the range 2 to 3. Formula to correct the e.s.d.'s have also been given by Scott (35).

5. The peak-shape function (PSF)

The correct form of the PSF to be used in the Rietveld analysis varies with the nature of the experimental technique and a wide range of PSF's, both analytical and numerical, has been suggested (37). A Gaussian PSF (eq.2) has been widely used, it is easily handled and satisfactorily describes instrumental contributions; however the tails of a Gaussian (see Fig.7) fail off too rapidly to account for particle-size broadening. Therefore the peak shape may be better described by the convolution of a Gaussian and a Lorentzian function: the Voigt function (39-42).
Given the general formula for a Gaussian (G) and a Lorentzian (L):

$$\left[A_1 (k_1) \exp(-x^2/k_1^2)\right] \text{ and } \left[A_2\left(1 + k_2 x^2\right)^{-1}\right] \text{ respectively } (A_i \text{ is the nor-}$$

malizing constant and k_i an adjustable constant), the two parameters Voigt function is defined as:

$$V = A_v(k_1,k_2) \int_{-\infty}^{\infty} L(x')G(x-x')dx \qquad (12)$$

The actual expression to be used in the refinement is complex and is given for example in Ref. 39,41.
In some programs because of its simplicity and functional similarity to the Voigt function a "pseudo-Voigt" function has been used:

$$pV = A(k_1,k_2,\eta)\left[\eta L + (1-\eta)G\right] \qquad (13)$$

which can be satisfactorily parametrized to represent, to very good approximation, a Voigt function (42).
It is noteworthy that most of the intensity in the tails of the reflections is taken into account by the Voigtian profile, in contrast to the Gaussian fit, and this affects,substantially,the magnitudes of the thermal parameters giving more reliable values and allowing a better estimate

of the background; on the other hand it has been found that the use of a Gaussian instead of a Lorentzian shape has little effect on atomic positional parameters (43).

Other PSF's have also been proposed: the Pearson VII, with two variable parameters (38), which can represent peak shapes ranging from pure Gaussian to pure Lorentzian and " general functions" which are the ratio of two polynomia (44).

When X-ray powder diffraction is used the resulting peak shapes are difficult to describe and neither Gaussians nor Lorentzians fit the profile (14).
Malmros and Thomas (13) employed, for the refinement of data from a Guinier-Hägg camera, a modified Lorentzian function:

$$ML(2\theta) = \left[1 + C_K (2\vartheta - 2\vartheta_k)^2 \right]^{-2} \quad (14)$$

where: $C_k = 4(\sqrt{2} - 1)/H_k^2$

The best results in interpreting data from a powder diffractometer (CuKβ radiation) have been obtained using the function (Intermediate Lorentzian):

$$IL(2\vartheta) = \left[1 + C_k (2\vartheta - 2\vartheta_k)^2 \right]^{-1.5} \quad (15)$$

with $C_k = 4(2^{2/3} - 1)/H_k^2$

A good fit of the X-ray peaks, even in the presence of the α_1, α_2 doublet, has been obtained using the sum of several Lorentzians (to account for instrumental and sample contributions) by Parrish and co-workers (44,45,46).
Non analytical PSF's can be exploited; for example Baerlocher (7,4) has used in his program a function of the form

$$G(\Delta 2\vartheta, H, A) = H^{-1} G_s (\Delta 2\vartheta/H) \left[1 - A G_a (\Delta 2\vartheta/H) \right] \quad (16)$$

where H and A are the half-width and asymmetry parameters respectively; G_s and G_a, the normalized symmetric and anti-symmetric parts of the PSF, are non-analytical functions determined, in numerical form, from a single resolved peak in the pattern.

PSF's for instruments using synchrotron radiation are usually simpler and depend on the experimental set up. Gaussian (19) or pseudo-Voigt (17) functions have been successfully used.

A pulsed neutron source gives an asimmetrycal line shape arising

from the fast rise and slow decay of the neutron pulse: to approximate
this shape analytically the convolution of two or more functions (4)
is used.

It can be assumed (11) that the moderator pulse as a function of time
is a rising exponential up to some time, followed by an exponential
decay

$$P(t) = \frac{\alpha \beta}{\alpha + \beta} \exp(\alpha t) \qquad t \leqslant 0 \qquad (17)$$

$$P(t) = \frac{\alpha \beta}{\alpha + \beta} \exp(-\beta t) \qquad t \geqslant 0, \qquad (18)$$

and that the instrumental contribution is a Gaussian. Thus the observed
line shape of a reflection, as a function of the TOF difference from
the Bragg position may be described as:

$$F(\Delta) = \frac{\alpha \beta}{2(\alpha + \beta)} \left[\exp(u) \mathrm{erfc}(y) + \exp(v) \mathrm{erfc}(z) \right] \qquad (19)$$

where $u = \alpha/2(\alpha \sigma^2 + 2\Delta)$, $v = \beta/2(\beta \sigma^2 - 2\Delta)$ and erfc is the complex error
function. α, β and σ are parameters which depend themselves on TOF; the
relevant formulae are discussed in Ref.11.

6. Sample effects

Preferred orientation is a formidable problem which can drastically affect the intensities. A simple, empirical correction of the form:
$\left[1 - A(\Delta 2\vartheta)^2 S \cot \vartheta \right]$ (where A is the asymmetry parameter and S the sign
of the difference $\Delta 2\vartheta$) was proposed by Rietveld (2) for plate-like
morphology. Some progress towards calculating the effect of preferred
orientation for polar-axis contribution has been achieved by Jarvenin
et al. (48) and Pesonen (49).

Peak profiles are also sensitive to strain and crystallite size, and
methods to determine these parameters have been suggested (50-52), but
in some cases it may be difficult (or unphysical) to relate refined
Rietveld profile parameters to crystallite sizes (39).

A treatment of the effect of size and strain effects, based on
Fourier analysis,has been developed by Le Bail and included in a multi-
pattern Rietveld program (72).

7. Performance of the Powder Methods in structure analysis

The Rietveld method has been widely used, for example, in the
study of inorganic compounds, minerals,small organic molecules, catalysts, hydrogen bonding and phase transitions; more than a thousand

papers appeared in various journals in the period 1974–1984 (6); it can be expected that the applicability of the method and the complexity of the problems that can be solved, will continue to grow with the new generation of instrumentation for neutron and synchrotron radiation sources.

Nevertheless it should be stressed that the Rietveld method is a refinement method, applied to powder data once the pattern has been indexed and a starting model is known .

Before the analysis an unknown structure can commence, the unit cell must be found with indexing programs (52–55); this is not a straight forward step requiring the accurate measurement of spacing of many low index lines (2ϑ with an accuracy of $\pm0.02°$) (55).

A satisfactory starting model can be found from isostructural compounds, by trial and error, from packing considerations or by Patterson methods (58), using high resolution X–ray powder data, for structures containing heavy atoms. More recently direct methods have been used to solve unknown crystal structure. Both the Patterson and the direct methods approach require the knowledge of the values of some ($\sim$100 or more) accurate integrated intensities obtained by a deconvolution method, as the one proposed, for example, by Parrish (15) or the method for extracting I(hkl), proposed by Pawley (59); in this method to each (hkl) in the pattern is given an I value which is allowed to vary until the profile is fitted. However the presence of peaks that overlap completely causes instability in the least squares and constraints should be introduced; peak shape description may also affect the deconvolution. A more detailed discussion of the problems encountered in the solution of unknown structures can be found in Refs. 60 and 61. Recent examples of "Ab Initio" determinations from X–ray and (or) neutron data are: $(AlPO_4).(CH_3)_4NOH$ (orthorombic $P2_12_12_1$) (62),

$FeAsO_4$ (monoclinic $P2_1/n$)(63) and $\alpha-CrPO_4$ (orthorombic Imma) (64).

It is worth note that when powder refinement results are compared with those from "standard" single crystal analysis the standard deviations of the lattice parameters are much smaller (4–5 times better) (32); on the other hand the e.s.d.'s on the positional parameters are smaller for single crystal data (31,32,65,67). The value of the thermal parameters is strongly correlated with the choice of the background and, as pointed out earlier, of the PSF; therefore thermal factors may be unreliable and their e.s.d.'s much higher than those in the single crystal case. Nonetheless it has been possible to apply TLX constraints to the refined model, obtaining good agreement with single crystal refinement (66). The study of the temperature dependence of thermal factors can also be carried out, as in UO_2 (67), eventhough in this example the use of cumulants, to account for anharmonicity, did not improve significantly the agreement.

A powerful technique in powder refiniment is the use of geometric (shape) constraints; many example can be found in the literature, in particular for organic molecular crystals and when disorder is present (68,71,73).

We note that the Rietveld method has been mainly applied to "inorganic compounds", while has not been widely used in the refinement of small organic molecules: this may be partly due to problems in the treatment of background and thermal factors which seem more difficult when dealing with molecular crystals.

The use of powder diffraction in structural studies is now a well established technique. The important contribution made by Rietveld was to recognize that structures could be refined from powder data, even when severe overlapping of diffraction peaks is present, using the whole pattern; in spite of limitations or problems discussed above, this is a powerful technique that will play an important role in crystallography.

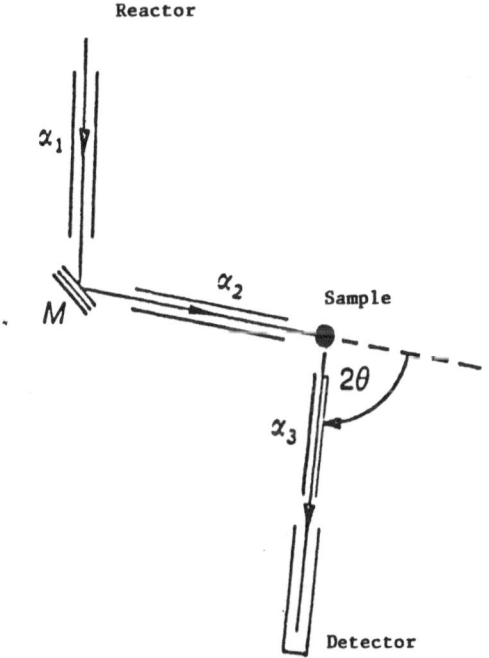

Figure 1. Diagram of a neutron powder diffractometer (λ fixed, 2θ variable); M=monochromator. $\alpha_1, \alpha_2, \alpha_3$, define the divergences of the incoming, monochromated and scattered beams respectively.

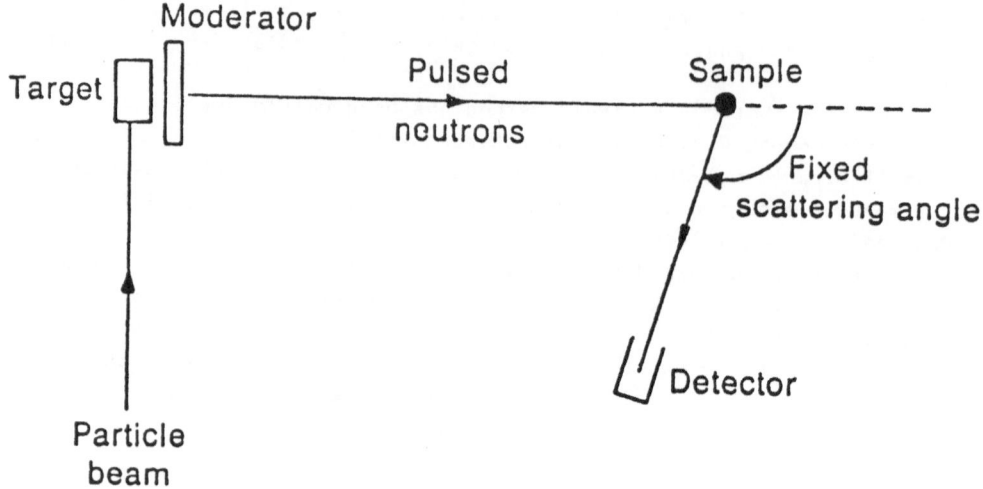

<u>Figure 2.</u> Schematic layout of a Time Of Flight diffractometer.

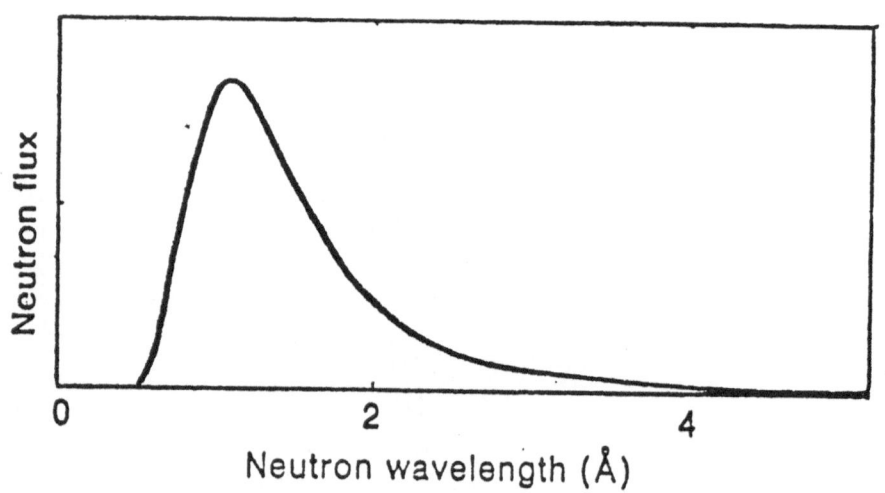

<u>Figure 3.</u> The flux spectrum $\varphi(\lambda)$ from a reactor with a moderator temperature T = 300 K.

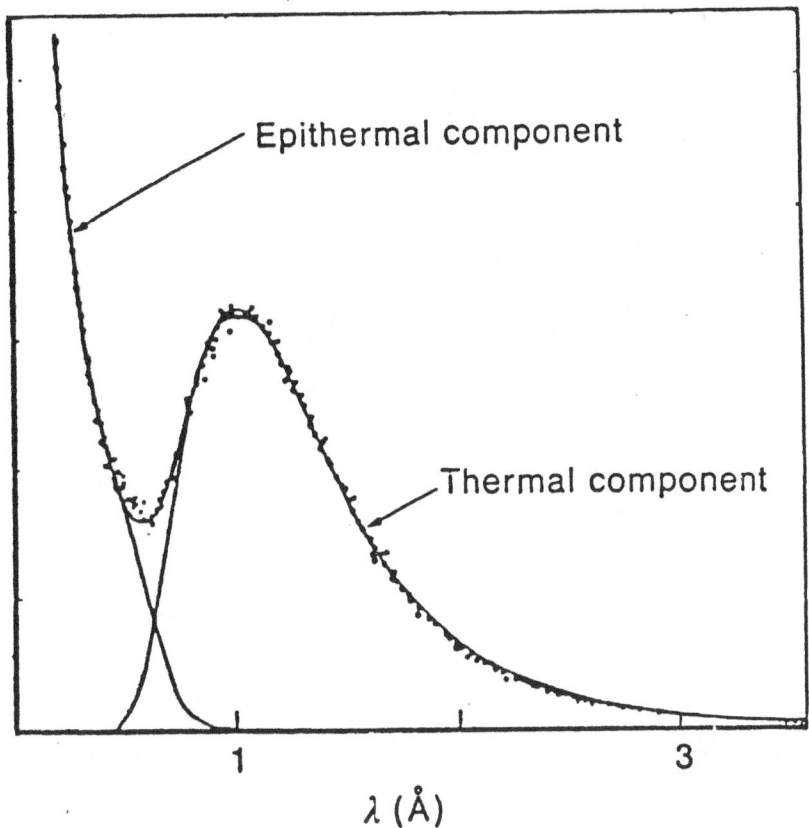

Figure 4. Measured (dotted) and fitted (line) spectrum for neutrons produced by a LINAC (see Ref.11).

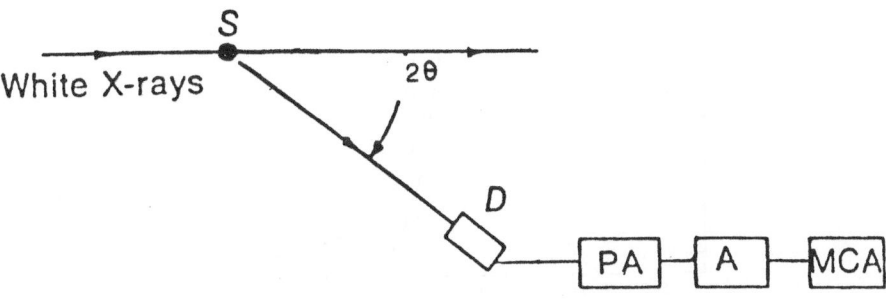

Figure 5. Diagram of an energy dispersive diffractometer; S=sample,D=solid state detector,MCA=multi-channel analyzer.

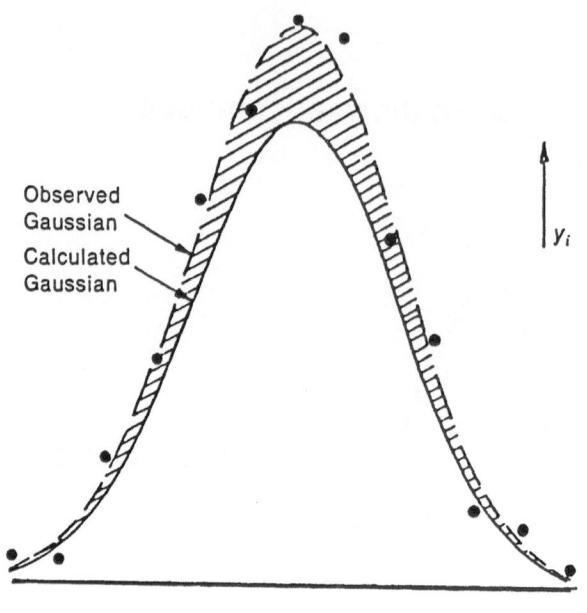

Observed
Gaussian

Calculated
Gaussian

y_i

$2\theta \longrightarrow$

Figure 6. Diagram illustrating correlation between the residuals for pair of points belonging to the same Bragg peak.

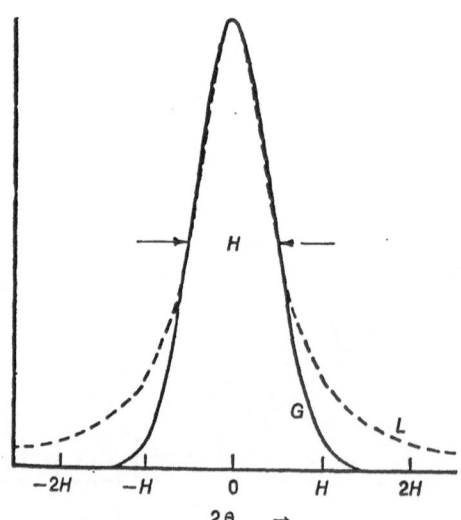

Figure 7. Lorentzian peak (L) and Gaussian peak (G) with same HWFM (H) and same heigths.

8.References

1) H.M.Rietveld Acta Cryst., **22**, 151 (1967).

2) H.M.Rietveld J.Appl.Cryst., **2**, 65 (1969).

3) A.K.Cheetham,J.C.Taylor J.Solid State Chem., **21**, 253 (1977).

4) A.Albinati,B.T.M.Willis J.Appl.Cryst., **15**, 361 (1982).

5) J.C.Taylor Aust.J.Phys., **38**, 519 (1985).

6) A.W.Hewat Chem.Scripta, **26A**, 119 (1986).

7) Ch.Baerlocher Zeolites, **6**, 325 (1986).

8) G.E.Bacon "Neutron Diffraction", 3rd Ed.Oxford University Press (1975).

9) P.Suortti NBS Special Publ.No.567 (1980).

10) C.G.Windsor "Pulsed Neutron Scattering", Taylor and Francis (1981).

11) R.B.Von Dreele,J.D.Jorgensen,C.G.Windsor J.Appl.Cryst.,**15**,582 (1982).

12) B.Buras Nucleonika, **8**, 259 (1963).

13) J.O.Thomas Chem.Scripta, **A26**, 7 (1986).

14) C.P.Khattak, D.E.Cox J.Appl.Cryst., **10**, 405 (1977).

15) W.Parrish,J.C.Huang NBS Special Publication No.567,pg.95 (1980).

16) W.Parrish,M.Hart,C.G.Erickson,N.Masciocchi Adv.Xray Anal.,**29**,243 (1986).

17) J.B.Hastings,W.Thomlinson,D.E.Cox J.Appl.Cryst., **17**, 85 (1984).

18) W.Parrish,M.Hart,T.C.Huang J.Appl.Cryst., **19**, 92 (1986).

19) G.Will,N.Masciocchi,W.Parrish,M.Hart J.Appl.Cryst., **20**,in press (1987).

20) B.Buras,J.S.Olsen,L.Gerward Nucl.Instr.Meth., **135**, 193 (1976).

21) B.Buras,L.Gerward,A.M.Glazer,M.Hidaka,J.Staun Olsen J.Appl.Cryst.,**12**,
                                                           531 (1979).

22) G.Caglioti,A.Paoletti,F.P.Ricci Nucl.Instr., **3**, 223 (1958).

23) C.J.Howard J.Appl.Cryst., **17**, 482 (1984).

24) T.M.Sabine,P.J.Clarke J.Appl.Cryst., **10**, 277 (1977).

25) M.Sakata,M.J.Cooper J.Appl.Cryst., **12**, 554 (1979).

26) M.J.Cooper Acta Cryst., **A38**, 264 (1982).

27) M.J.Cooper,K.D.Rouse,M.Sakata Zeit.Krist., **157**, 101 (1981).

28) A.Albinati,M.J.Cooper,K.D.Rouse,B.T.M.Willis Acta Cryst.,**A36**,265 (1980).

29) G.A.Lager,F.K.Ross,F.J.Rotella,J.D.Jorgensen J.Appl.Cryst.,**14**,137 (1981).

30) J.J.Didisheim,K.Yvon,P.Fischer,D.Shaltiel J.Less Comm Met.,**73**,335 (1980).

31) A.G.Nord,T.Stefanidis Acta Chem.Scand., **A37**, 715 (1983).

32) M.Marezio Chemica Scripta, **26A**, 91 (1986).

33) E.Prince J.Appl.Cryst., **14**, 157 (1981).

34) A.W.Hewat,T.M.Sabine Aust.J.Phys., **34**, 707 (1981).

35) H.G.Scott J.Appl.Cryst., **16**, 159 (1983).

36) E.Baharie,G.S.Pawley J.Appl.Cryst., **16**, 404 (1983).

37) J.Hill,I.C.Madsen J.Appl.Cryst., **17**, 297 (1984).

38) R.A.Young,D.B.Wiles J.Appl.Cryst., **15**, 430 (1982).

39) P.Suortti,M.Ahtee,L.Unonius J.Appl.Cryst., **12**, 365 (1979).

40) M.Ahtee,L.Unionus,M.Nurmela,P.Suortti J.Appl.Cryst.,**17**,352 (1984).

41) W.I.F.David,J.C.Mathewman J.Appl.Cryst., **18**, 461 (1985).

42) W.I.F.David J.Appl.Cryst., **19**, 63 (1986).

43) J.Hill,C.J.Howard J.Appl.Cryst., **18**, 173 (1985).

44) N.P.Pyrros,C.R.Hubbard J.Appl.Cryst.,**16**, 289 (1983).

45) T.C.Huang,W.Parrish Adv.X-ray Anal., **21**, pg 275 (1978).

46) G.Will,W.Parrish,T.C.Huang J.Appl.Cryst., **16**, 611 (1983).

47) C.Baerlocher Proc.6th Int.Zeolite Conf. pg 823 (1984).

48) M.Jarvinen,M.Merisalo,A.Pesonen,O.Inkinen J.Appl.Cryst.,**3**,313 (1970).

49) A.Pesonen J.Appl.Cryst., **12**, 460 (1979).

50) Th.H.De Keijser,E.J.Mittemeijer,M.C.F.Rozendaal J.Appl.Cryst.,**16**, 309 (1983).

51) C.Greaves J.Appl.Cryst., **18**, 48 (1985).

52) D.Louer Chem.Scripta **26A**, 17 (1986).

53) J.W.Wisser J.Appl.Cryst. **2**, 89 (1969).

54) D.Taupin J.Appl.Cryst. **6**, 380 (1973).

55) R.Shirley NBS Spec.Publ. No.567, pg 361 (1980).

56) E.Baharie,G.S.Pawley Acta Cryst. **A35**, 323 (1979).

57) G.S.Pawley,G.A.Mackenzie,O.W.Dietrich Acta Cryst. **A33**,142 (1977).

58) P.E.Werner Chem.Scripta **26A**, 57 (1986).

59) G.S.Pawley J.Appl.Cryst. **14**, 357 (1981).

60) P.R.Rudolf,A.Clearfield Acta Cryst. **B41**, 418 (1985).

61) A.Nørlund Christensen,M.S.Lehmann,M.Nielsen Aust.J.Phys. **38**,497 (1985).

62) P.R.Rudolf,C.S.Molina,A.Clearfield J.Phys.Chem. **90**, 6122 (1986).

63) A.K.Cheetham,W.I.F.David,M.M.Eddy,R.J.B.Jakeman,M.W.Johnson Nature **320**, 46 (1986).

64) J.P.Attfield,A.W.Sleight,A.K.Cheetham Nature **322**, 620 (1986).

65) G.A.Lager,F.K.Ross,F.J.Rotella,J.D.Jorgensen J.Appl.Cryst. **14**, 137 (1981).

66) G.E.Bacon,E.J.Lisher,G.S.Pawley Acta Cryst. **B35**, 1400 (1979).

67) A.Albinati,M.J.Cooper,K.D.Rouse,B.T.M.Willis NBS Spec.Publ.No.567, pg 203 (1980).

68) G.S.Pawley J.Appl.Cryst. **13**, 630 (1980).

69) G.S.Pawley,A.W.Hewat Acta Cryst. **B41**, 136 (1985).

70) M.L.Putkonen,R.Feld,C.Vettier,M.S.Lehmann Acta Cryst. **B41**,77 (1985).

71) G.S.Pawley Acta Cryst. **B34**, 523 (1978).

72) A.Gibaud,A.LeBail,A.Bulou J.Phys. C **19**,4623 (1986).

73) K.Refson,G.S.Pawley Acta Cryst. **B42**, 402 (1986).

DISCUSSION

ERRORS IN METHODS OF STRUCTURE DETERMINATION

It is important to distinguish between precision as the internal consistency of an experiment and accuracy as the extent to which the results represent reality.

Precision is represented by the standard deviation of a least squares analysis of the errors in a given experiment. Only if all the parameters capable of correlating with the structural variables are appropriately included in the least squares matrix will the analysis be correct.

There is much evidence that the standard deviations involved in structure determination by diffraction can be defined in a statistically correct way. Parameters measured at the Argonne National Laboratory as a function of variables such as temperature, pressure or composition give statistically satisfactory plots against the variables, and similar conclusions have been reached in Australia when a given experiment has been replicated many times (Hill and Masden, J. Appl. Cryst., 17, 297 (1984)).

Temperature factors are the parameters which have most often been miscalculated. The original Rietveld code and the one in use for many years at ILL failed in this respect by not including the background in the least squares matrix as one of the parameters correlating with the structural variables. In careful work with powders, however, temperature factors agree with the single crystal values. Particular care is needed with anisotropic temperature factors. If the Hamilton ratio test is correctly used to decide whether or not each atom should be treated anisotropically, refinement gives components of the anisotropic temperature factors which agree with the single crystal values to within statistics for all hard materials. For soft materials such as organic compounds, where there is much thermal diffuse scattering under the peaks, the process is more tedious and is not usually attempted. Further

312

complications can be caused by any preferred orientation in powder samples.

Accuracy can best be tested by comparing the results obtained for a given structure by different methods, when variations in systematic error will be revealed by discrepancies exceeding 3 in correctly estimated standard deviations. An example of that approach is the comparison in Prof Cheetham's paper (table 4) of the structural parameters for α-$CrPO_4$ derived from synchrotron X-ray powder, neutron powder and single crystal data.

In the Rietveld method errors are thought to have decreased in recent years as a result of improvements in instrumentation and in data analysis, and are now approaching those in single crystal work. Choice of an appropriate model for refinement is very important in the Rietveld method.

REAL-TIME NEUTRON POWDER DIFFRACTION

J.PANNETIER
Institut Laue Langevin
156X 38042- Grenoble (France)

ABSTRACT

The use of a high-flux neutron source together with large curved one-dimensional position sensitive detectors (PSD's) allows a complete powder diffraction pattern to be recorded simultaneously on a time scale of a few minutes or less so that transient phenomena such as chemical reactions in the solid state or first-order phase transitions can now be studied in real-time. Such measurements provide direct information on the progress of a reaction (kinetic law), but may also shed light on the transformations of structure and/or morphology which occur in the reactants during the course of a reaction. The instrumental parameters relevant to the design of time-dependent experiments are discussed. The applications are illustrated by four kinds of experiments : kinetic studies, thermodiffractometry, texture and stroboscopic measurements.

1 - INTRODUCTION

Static powder diffraction (either by X-Ray or neutrons) is a standard technique for the detection and identification of crystalline phases and for the quantitative determination of their volume fraction. Indeed, a diffraction pattern is determined by the exact atomic arrangement in a material and is like a fingerprint in that no two compounds give rise to exactly identical diagrams. In principle a proper measurement of a powder diffraction pattern affords the possibility to characterize the composition, structural arrangement (line positions and intensities) and morphology (line breaths and shapes) of any crystalline material.

Although X-ray diffraction is by far the simpler and less expensive method of powder diffraction (at least with traditional sources), neutrons may provide otherwise inaccessible information and some examples will be considered later in this paper. Detailed comparison between X-ray and neutron diffraction is outside the scope of this paper and can be found in most textbooks on diffraction methods (see, for example, G.E. BACON [1]). It is worthwhile, however, to draw attention to advantages of the low absorption cross-section of most elements for neutrons; this is obviously useful when the sample has to be contained in a controlled environment such as furnace, cryostat or reaction cell but, principally, means that neutron beams probe the bulk sample whereas X-rays often see only a thin layer at the surface. This (together with the almost random dependence of the neutron scattering amplitude on atomic number) clearly confers some advantages in the study of heterogeneous chemical reactions involving mixtures of heavy and light atoms reactants.

M. A. Carrondo and G. A. Jeffrey (eds.), Chemical Crystallography with Pulsed Neutrons and Synchrotron X-Rays, 313–355.
© *1988 by D. Reidel Publishing Company.*

The use of neutron powder diffraction has increased rapidly over the last decade. This renewed interest is the result of the construction of high-resolution, high intensity powder diffractometers and of the development of data analysis methods such as Rietveld profile refinement [2], which allows precise structural information to be obtained from powder data. As a consequence, neutron powder diffraction is now often replacing single crystal methods and is expected to play a major role in various fields of chemistry and materials science; it is already the preferred method for studying the structure of materials which cannot easily be prepared as single crystals (for example catalysts, fast-ion conductors, zeolites etc.). However, it must be reminded that the intensity of the neutron sources, even at high-flux reactors, is weak compared to the intensity available with synchrotron radiation or even conventional X-ray sources; this implies that the data acquisition rates on the best neutron powder high-resolution diffractometers are still of the order of a few hours which clearly precludes their use in most studies of time-dependent chemical or physical phenomena.

This article describes an alternative approach of neutron diffraction based on the use of high flux/medium resolution diffractometers equipped with position sensitive detectors (PSD's). This approach is believed to be of fundamental interest to investigate simultaneously the kinetic, mechanistic and structural features of transformations in the solid state.

The paper is organized as follows : in the next section, we shall present the general features of real-time diffraction. Section 3 discuss some technical aspects of real-time neutron powder diffraction (RTNPD) and try to answer the beginner's question "How do I prepare and perform a time-resolved neutron powder diffraction experiment ?". The next section is a brief survey of the data analysis; in the fifth section, a number of experiments are presented in an attempt to illustrate the various aspects of RTNPD. The last part of the paper is a brief discussion of the respective advantages of steady and pulsed neutron sources.

2 - REAL-TIME DIFFRACTION

New generation of powerful radiation sources and development of fast detection systems make it possible nowadays to perform diffraction measurements in times so short that rapid changes of diffraction patterns can now be observed in real-time. The main advantage of time-resolved diffraction compared to its static counterpart is to provide temporal information about structural/morphological changes occuring in a sample during a chemical or physical process. Such information can be extremely valuable in studies of chemical reactions (for instance to identify intermediate species or to decipher the mechanism of transformations), crystallizations and phase transformations; of particular importance is the ability of the method to detect transient intermediates that might otherwise go unnoticed in static measurements.

From a general point of view, the common feature of most real-time diffraction measurements consists first in bringing a sample into a non-equilibrium state by applying an external perturbation and second in monitoring the sample behavior at different subsequent time intervals to follow its relaxation towards equilibrium. Obviously, such studies will be feasible only if the time scale of a single measurement (acquisition of a complete diffraction pattern with adequate statistical accuracy) is much shorter than the relaxation time of the process (at least in case of non-reversible processes). Therefore, owing to the intense beams of X-ray available from conventional and synchrotron radiation sources, it may seem that X-ray diffraction is

1 - Attempt to illustrate real-time experiments

316

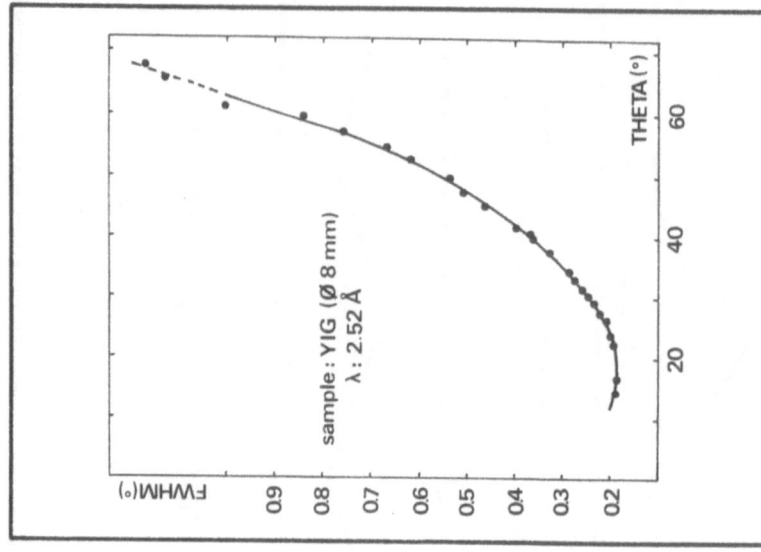

Resolution curve of D1B.

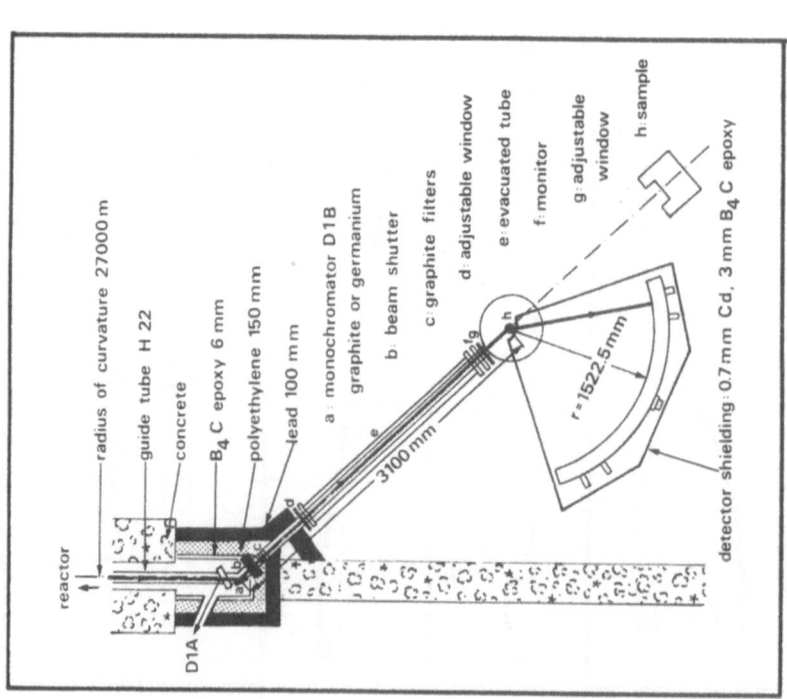

Schematic view of D1B.

2 - Schematic drawing and resolution curve of the PSD diffractometer D1B at the ILL (adapted from ref. 46)

particularly suited to time-resolved experiments; indeed, the applicability of real-time X-ray diffraction using synchrotron radiation in various fields of condensed matter research as already been demonstrated [3,4,5] and measurements with resolution in the nanosecond range have already been performed [6,7]. In contrast, real-time uses of neutron powder diffraction are not so well documented yet (8,9,10). However, its time resolution, now in the range 0.5 to 10 minutes, is well adapted to many solid state processes which often have characteristic times of the order of a few hours.

The system under investigation can be brought to a metastable state in many various ways that are schematically illustrated in Fig. 1 ; this can be obtained for instance :

- by mixing reactants to initiate a chemical reaction (a)
- by an abrupt change (a) of a thermodynamic variable (temperature, pressure, magnetic field etc ...)
- by applying a short pulse (b,e) of temperature, pressure, electric or magnetic field or light to trigger a transformation : e.g. light-induced polymerization, dielectric or magnetic relaxation.
- by a continuous (c) or cyclic (d,f) variation of an external parameter (temperature or field sweep for instance) .

The experimental procedure will not, however, be the same for reversible and non-reversible processes, and this will lead to rather different time-scales of measurement :

- in the case of non-reversible processes, time runs only in one direction and the time dependence is followed simply by recording data sequentially : the sampling time t_s is then the shortest time period in which one pattern with sufficient statistics is acquired and it has to be much shorter than the characteristic time of the process to investigate ; this time resolution t_s depends only on the beam flux, the scattering power of the sample and the efficiency of the detector . With the best reactor-based PSD's diffractometers, t_s is presently of the order of a minute. Although this time resolution may sound extremely poor when compared to the values currently achieved by X-ray diffractometers operating at synchrotron sources [6,7], one must remember that this is a convenient time scale for many transformations in the solid state (first-order transformations, chemical reactions , etc) ; moreover, in this field, the specific properties of the neutron makes neutron powder diffraction (NPD) a unique tool to tackle problems involving magnetic phenomena, hydration/dehydration reactions, X-ray sensitive processes, complex sample environments etc...
- the conditions for reversible time-dependent phenomena (by this we only mean a process which can be repeated many times after a cyclical start signal) are less critical ; indeed, if the relaxation time of the process is shorter than the time needed to gather sufficient statistics, the process can be repeated and the data accumulation continued until adequate statistics are accumulated. In a typical experiment, each cycle of the perturbation t_c is divided into an integer number of time slices t_s and the corresponding data are recorded separetely and accumulated over many periods (Fig. 1) . The time resolution is therefore not limited by the beam flux nor by the sample scattering power but rather by the electronics of data acquisition (dead time, storage capacity) or the pulsing frequency of the neutron source.

3 - EXPERIMENTAL ASPECTS OF A REAL-TIME NEUTRON POWDER DIFFRACTION MEASUREMENT

3-1- The diffractometer

Real-time neutron powder diffraction has been made possible by the development of curved one-dimensional PSD's with wide angular coverage [11] : these detectors allow simultaneous collection of data for multiple scattering angles. Although the primary motivation for using PSD's is to obtain a more efficient use of the scattered neutrons, one may also argue that their use finally opens up the field of real-time crystallography. For the technical point of view, four aspects are crucial for RTNPD :

- the **PSD** which is the "heart" of the machine. For real-time application, one requires this detector to be :
  - *stationary* : this is obviously needed to achieve reasonable temporal resolution but, in addition, allows minute modifications of a diffraction pattern to be observed.
  - *wide* : to survey simultaneously a large part of reciprocal space. Angular coverages of 80° [12] and 130° [13] are currently available
  - *curved, continuous & homogeneous* : to minimize instrumental corrections (positional and efficiency calibration). Uniformity of a few percent is now common for most gas PSD's (14).
  - *efficient* : any loss of efficiency reduces the time resolution. For all detectors, the neutron efficiency is inversely proportional to the neutron velocity. A 70 % detector efficiency at 2.5 Å can be achieved, for instance, by a ^{3}He detector operating under 4 atm.

Additional requirements, though not specific to real-time experiments, are also demanded : long-time stability ($\approx$ 1 part in 10^3), high counting rates ($\approx 10^5$ counts/cell.sec.) with small dead-time losses and low g sensitivity. The spatial resolution of the PSD, i.e. the distance between the cells, is also an important parameter in that it influences the resolution of the diffractometer (see below).

Various technologies are available to develop large thermal neutron PSD's (11) and a number of laboratories around the world have development programs underway. For the time being, most (if not all) real-time powder diffractometers are equiped with high-pressure gas detectors (BF_3, ^{3}He + Xe or ^{3}He + propane) which give a good compromise between all the above requirements.

- the **flux** at sample position : all neutron scattering research is intensity limited but this is particularly relevant to real-time scattering because increases in intensity do not simply mean that the same experiments are done faster but that totally new kinds of investigation can become possible. For a given primary flux, increases in monochromatic flux can only be achieved by moving the diffractometer as close as possible to the core and/or by using a vertically focusing, highly efficient monochromator. One of the most efficient monochromator materials is highly oriented pyrolytic graphite (H0PG) with typical mosaicity between 0.5 and 1° ; it is particularly useful at wavelength of about 2.5 Å. For shorter wavelengths, common choices are germanium or copper crystals with a properly adapted mosaic spread.

- the **angular resolution**, i.e. the ability of the diffractometer to separate neighbouring reflections. The resolution of an angle dispersive two-axis spectrometer is essentially determined by the mosaïc spread β of the monochromator and by geometrical parameters [15] :
  - the primary beam collimation (α_1)
  - the collimation α_2 of the monochromatic beam
  - the collimation α_3 of the diffracted beam.

Optimization of a conventional diffractometer (16) requires

$$\alpha_2 = 2\beta > \alpha_1 \approx \alpha_3$$

Then, at the maximum of resolution $(\theta \approx \theta_M)$ the full width at half maximum (FWHM) of a diffraction line (ignoring sample broadening) is given as

$$(FWHM)^2 \approx \alpha_1^2 + \alpha_3^2$$

However this analysis does not apply directly to PSD's : it is not possible to collimate the diffracted beam with Soller slits and α_3 is dependent on the spatial resolution of the detector, on the sample diameter and on the sample to detector distance [12]. As the spatial resolution of PSD's is usually of the order of a few millimeters, α_3 is essentially determined by the sample size ; it can be improved by reducing the sample diameter but the improvement is less than linear while the intensity is reduced by a factor proportional to the square of the diameter. This compels longer wavelengths to be used to achieve a better intensity vs. resolution compromise. As a consequence, real-time neutron diffraction with PSD's is restricted to medium resolution problems.

- the **data acquisition system** must provide for the accumulation of the neutron counts detected by the PSD, their storage, display and retrievial. The high neutron counting rates· $(\approx 10^5$ counts/sec. in the whole detector) and high speed of acquisition put severe constraints on the design of a data acquisition system : the most appropriate system is one based on individual scalers (see, for instance, chapter 4 of ref. 11).

3-2- The sample. Its transformation.

Owing to the weak interaction of the neutron with matter, neutron diffraction requires rather large amounts of sample material; a typical sample for the diffractometer D1B at the ILL has the shape of a cylinder with a height of about 50mm and a diameter of 10mm.

Almost all elements are compatible with neutron powder diffraction; the only restrictions arise from the high absorption cross section of a few isotopes (^{10}B, ^{113}Cd, ^{151}Eu or ^{157}Gd for instance). A further difficulty, at least in NPD, may sometimes result from the high incoherent scattering from hydrogenous samples (see §5-2-2). In addition, a few isotopes may become activated after some time in a high neutron beam; such safety-related problems are handled by the radio-protection teams which supervise all experiments at neutron beam facilities. In the worst cases, users may not be allowed to bring their samples back to their home laboratory.

The diffracted pattern of sample must be simple enough so that most of the lines in the angular range of interest are resolved by the instrument: orthorhombic cells with a cell volume of about 1000Å^3 are a reasonable estimate of the kind of problem which can be handled at a PSD diffractometer; lower symmetry or larger cells can also be studied if the information on the process under investigation is contained in the part of the pattern where the instrument resolution is best (see [39,40] for instance). The transformation of the sample must obviously modify in some way the diffraction pattern and has also to take place over a period of time which allows both a reasonable statistics and a proper number of samplings (see exemples in §5).

3-3- The sample environment.

The penetrating power of the neutron allows diffraction studies of samples in very complex environments. Most of these environments take advantage of the following materials :

- *Vanadium* : its low coherent cross section ($b \approx -0.038.10^{-12}\text{cm}$) means that in most experiments, it does not give rise to any significant Bragg diffraction, hence its use in making sample cans and, at least for temperature up to about 800°C, heating elements.
- *Aluminum*: : its low absorption cross section ($\mu = 0.10\text{cm}^{-1}$) makes it extremely useful to build most of the mechanical environment of the samples (heat shields, windows).
- *Cadmium*: : the total path length of thermal neutrons in this metal is only a fraction of a millimeter, making it ideal for shielding purposes.

In addition, amorphous silica (for instance in the form of quartz tubes) may sometimes provide useful containers for sensitive or agressive materials: the diffusion ring from amorphous SiO_2 is usually weak as compared to the Bragg diffraction from most crystalline samples.

Low temperature experiments can be performed routinely between room temperature and 10K with a displex refrigerator and down to about 1.5K in a pumped helium cryostat. Above room temperature, various kinds of furnaces operating either under vacuum or under controlled atmosphere are available; neutron powder diffraction experiments above 2500°C have already been reported. In the case of RTNPD measurements it is sometimes necessary to change the temperature of the sample as rapidly as possible over a wide temperature range; a furnace designed for this purpose has recently been described [47].

Most cryostats and furnaces are "clean" environments in that they dont contribute diffraction lines to the recorded pattern; with this respect, high-pressure environments bring more difficult problems which are most easily solved by time-of-fligth, fixed-angle diffraction (90° geometry).

Detailed information about the ancillary equipment available at ILL instruments can be found in [46].

4 - DATA ANALYSIS

A common feature to all real-time diffraction experiments is the huge amount of raw data collected in a single experiment : accumulation of several hundred complete

diffraction patterns per day is now quite usual for most kinetic experiments but it may sometimes exceed a few thousands . This problem of data analysis is then twofold :
• during the experiment, on-line data reduction is advantageous in that it provides diagnostic feedback on the advancement of the process under investigation. This analysis can however be restricted to limited parts of the patterns and therefore performed directly on the instrument computer ; for instance the intensity of a few Bragg reflections can be calculated after each pattern has been recorded in order to assess the progress of a transformation
• at the end of the experiment, all data can be transfered to a central computer where more sophisticated programs for data analysis are available. A convenient first step is usually to prepare a pseudo 3D or contour plot of the data which provides , in a single frame, a complete survey of the experiment. Afterwards, the diffraction patterns can be analysed by one of the following two method
- the conventional Rietveld method : the limited resolution and angle coverage of state-of-the-art PSD's restrict the use of this method to simple systems, either single-phase systems with only a limited number of structural parameters to refine or mixtures of a few structurally simple components [17]. When applicable, this approach provides the time (or any time dependent external parameter) evolution of the structural parameters (atomic positions).
- In most cases, the approach to the analysis of the pattern is pragmatic and assumes no a-priori knowledge of the data : the position of the Bragg peaks is first determined by visual inspection of the first pattern and adjusted through a least squares or maximum of likelyhood fitting ; then the program works sequentially on the different patterns from successive time slices . The line shape is normally considered as gaussian in form and the background is calculated as a first or second order polynomial. The output of such an analysis is a (long !) list of line positions, intensities and FWHM's. These data can then be used to calculate the temporal evolution of the cell parameters (e.g. thermal expansion), transformed fractions (kinetic studies), lattice strain, particle size etc.

5 - EXAMPLES OF APPLICATION

Real-time applications of neutron powder diffraction are still relatively scarce and the number of publications on the matter rather limited (see Appendix). Considering that both high-flux sources and PSD's have been in use for more than a decade, it is not altogether clear why as yet so few experiments have been performed in this field. One may expect, however, that, as neutron diffraction techniques become more familiar to a wider community of users (physical chemists, materials scientists), the emphasis on time-resolved experiments will grow rapidly. A few examples, mostly in the field of solid state inorganic chemistry, are presented below to illustrate possible applications. All these measurements have been performed at the ILL with the diffractometer D1B (fig. 2) and a wavelength of 2.51 Å provided by a focusing HOPG monochromator.

5-1- Kinetic studies (isothermal)

The use of neutron powder diffraction with PSD's to investigate chemical reaction kinetics was pionnered by C. Riekel at the ILL : most of the references to this early work, mainly concerned with chemical or electrochemical intercalation reactions, can be found in his review articles [8,18] and references therein . The examples quoted in the following have been chosen to show the broad range of application of real time neutron powder diffraction.

- **The reaction of THF vapour** on the second stage graphite intercalation compound CsC_{24} yields a ternary first stage phase of composition $CsC_{24}(THF)_x$ with $x \approx 1.7$, where the plane of the THF molecule lies down parallel to the graphite layers ; with a powder sample this process takes place over a few hours at room temperature. The corresponding structural transformations were investigated on D1B [19], using deuteriated THF, with a resolution $\Delta t = 1$min. in order to identify the intermediate stages and compositions and their kinetics of formation. An overwiew of the reaction is given in fig. 3. This picture clearly shows the three steps of the process. The binary phase CsC_{24} vanishes rapidly as an intermediate compound builds up; this material was identified as a second stage $CsC_{24}(THF)_x$ ($x \approx 1$). A few minutes later, peaks corresponding to the final first stage ternary compound show up in the diffraction pattern. Nine minutes after the beginning of the reaction the lines of the intermediate phase reach a maximum intensity then start to decrease. After five hours the binary compound CsC_{24} has completely disappeared whilst very weak peaks of the intermediate phase still exists; this transient phase can be isolated by desintercalating the THF (for instance by pumping) and was already known; however its role in the intercalation process which points to the formation of the first stage compound $CsC_{24}(THF)_{1.8}$ via a Daumas-Herold mechanism had not been identified before. As the life-time of this phase is of the order of minutes it would have probably gone unnoticed in an investigation with poorer time resolution.

- **Reactions of Portland cement components with water** [20] : the hydration of Portland cement mortars (a mixture of oxides whose main components are calcium silicates and aluminates), which provokes solidification after a few hours, is a complex chemical process which involves a large number of simultaneous reactions; for sake of convenience, Christensen and his coworkers first investigated the reactions between water and some pure oxides which are the main components of cement clinkers. Their results on the reaction of calcium silicate Ca_3SiO_5 with heavy water are illustrated in Fig. 4 and 5 : the hydration of the silicate can be divided into three steps denoted initialization, induction and reaction ranges. During the first step it is believed that there is a rapid dissolution of a small amount of Ca_3SiO_5 immediately after the solid is brought in contact with water; however the amount of silicate consumed in this reaction was too small (1-2%) to be resolved in the experiment. The following induction period results either from a reduced nucleation rate of hydration products or from the formation of a protective layer surrounding the grains for a certain time. The onset of the reaction period is characterised by the precipitation of $Ca(OD)_2$ which is the only crystalline product formed in the reaction. All silicon from the dissolution of calcium silicate is found in the amorphous gel formed during hydration. Analysis of the time dependence of the degree of hydration suggests a reaction controlled kinetics.
Additional studies of the hydration of other calcium silicates and aluminates show that the silicates have a reduced reactivity relative to the aluminates; this is reflected in time dependent mechanical properties of the end products of hydration of the Portland cement clinker.

3 - Kinetic study of the intercalation of deuterated THF in CsC_{24} (from ref. 19) : $\Delta t = 1$ min.

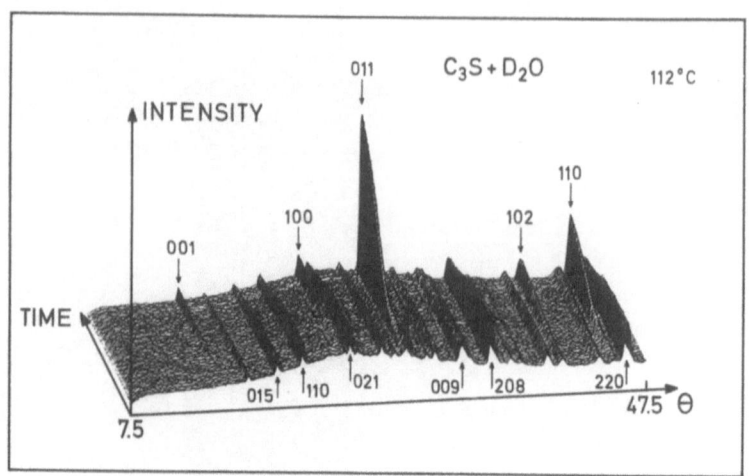

4 - Time-dependence of the neutron diffraction pattern of a mixture $D_2O + Ca_3SiO_5$ at 112°C. Note the diffusion ring
(at $\Theta \approx 24°$) from excess D_2O : $\Delta t = 20$ min. Miller indices refer to Ca_3SiO_5 pseudocell (↑) and $Ca(OD)_2$ cell (↓) (from Ref. 20).

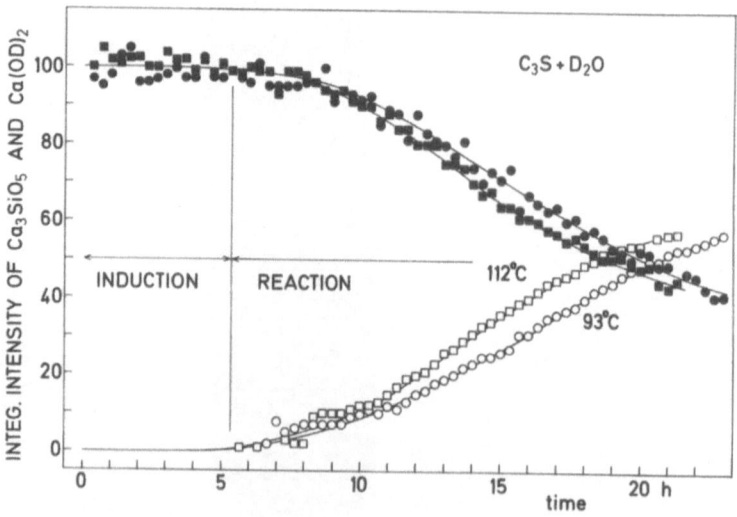

5 - Integrated intensities of Ca_3SiO_5 (●,■) and $Ca(OD)_2$ (○,□) vs. time for the reaction of Ca_3SiO_5 with D_2O at 93 and 112°C (from Ref. 20)

325

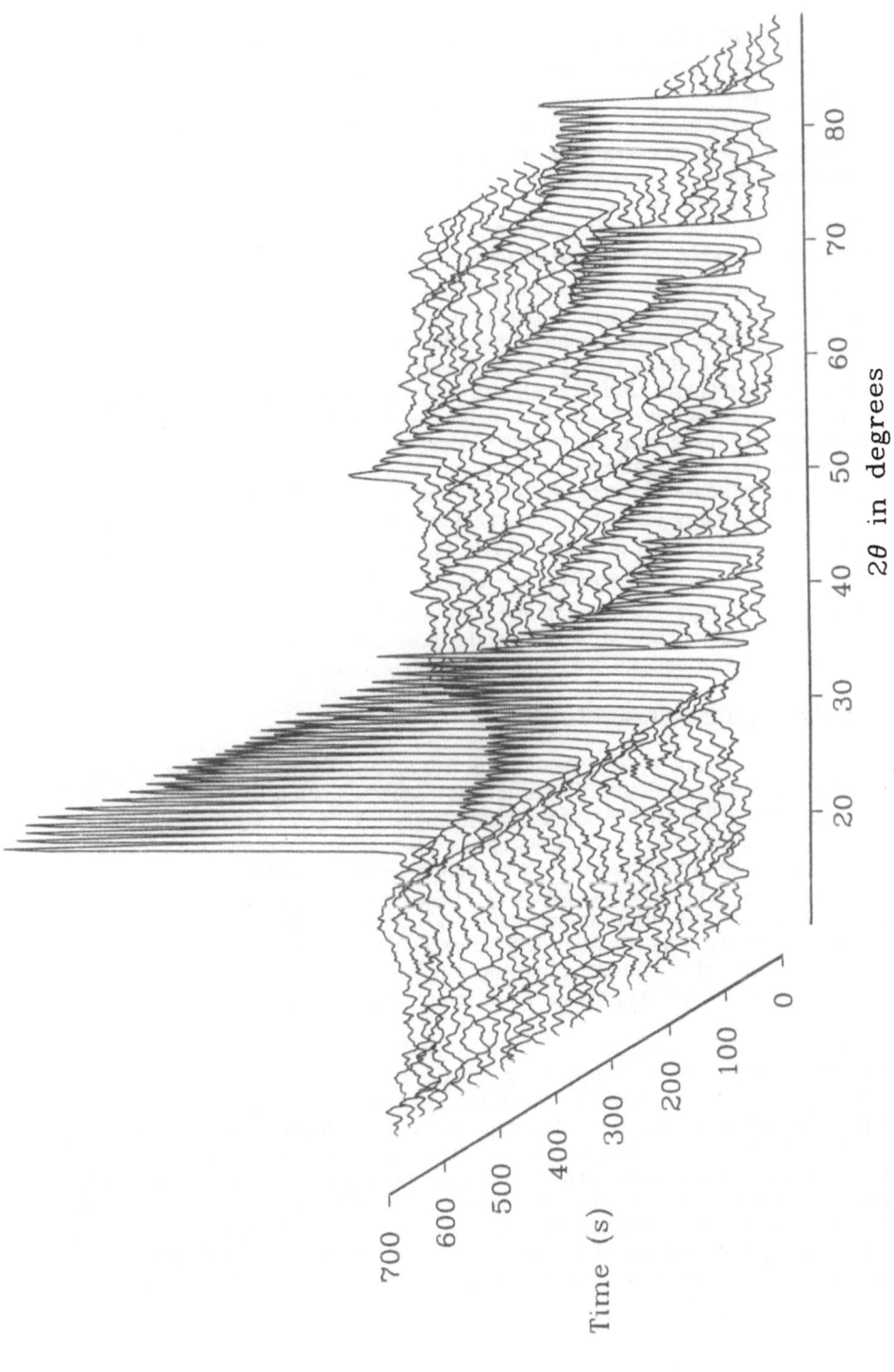

6 - Kinetics of the rhombohedral to cubic phase transition in KPF$_6$ (from Ref.21)

2θ in degrees

Time (s)

0
100
200
300
400
500
600
700

20
30
40
50
60
70
80

• **Nucleation kinetics** in the pressure-induced first-order phase transformation in KPF$_6$[21] : structural phase transformations in solids are often understood in terms of nucleation and growth processes where nuclei of stable phase form and subsequently propagate through the structure. Nucleation and growth phenomena can be studied by measuring the development of the transformation with time after the system has been perturbed from a stable equilibrium by a sudden change in temperature or pressure.

The kinetics of the phase transformation from rhombohedral (ordered CsCl-type structure) to cubic (disordered NaCl-type structure) in KPF$_6$ was investigated with a time resolution of 20 seconds (Fig.6) by monitoring the time-dependent changes in neutron powder diffraction peaks. An hydrostatic pressure-jump method was used, the equivalent temperature studies being impossible due to the large thermal capacity of powdered samples. This study shows that the transformation curves change shape from sigmoidal near the transition pressure to exponential at pressures well away from the transition pressure; this reflects a change from a nucleation and growth process to a transformation where the growth process is rapid as compared to the nucleation rate. This approach enables rate constants for the transition to be obtained and hence the thermodynamic quantities of activation energy and volume may be determined. The results help to explain the observation that freshly precipitated KPF$_6$ transforms to a rhombohedral phase on cooling whilst dried thermally cycled KPF$_6$ transforms to a monoclinically distorted NaCl-type modification under the same conditions.

A similar neutron investigation of the pressure-induced NaCl-to-CsCl type structural transformations in RbI has also been reported [22].

5-2- Neutron Thermodiffractometry (NTD)

Thermodiffractometry can be considered as one of the many experimental techniques of studying thermally stimulated processes, such as DTA (differential thermal analysis), TGA (thermogravimetric analysis), EGA (evolved gas analysis), TPR (temperature programmed reduction) etc . It has actually been in use for a while in X-ray diffraction (e.g. with high-temperature Guinier cameras) to identify phase transitions ; however the natural advantages of neutrons and, in particular, the ease of controlling the sample environment make NTD a most valuable tool to investigate thermally stimulated reactions in the solid state. Owing to its simplicity and sensitivity, it is probably one of the quickest and most efficient ways to search for crystalline phase transformations and to detect the different stages of thermal decomposition of solids.

The experimental arrangement is the same as for a conventional diffraction experiment except that the temperature of the sample, in the furnace or cryostat, is varied continuously at a rate of a few degrees per hour in a predetermined range of temperature. Diffraction patterns are extracted at constant time intervals Δt, every pattern thus providing an average picture of the sample within the temperature interval ΔT. The data collection rate depends upon the scattering power of the sample but the temperature resolution ΔT is controlled by the experimentalist; it is finally limited by the temperature gradient in the sample (and, of course, by the beam-time available !).

The range of phenomena that could be studied by NTD in condensed matter is fairly broad. A few examples, based on our recent experience with D1B at the ILL, are briefly presented below.

5-2-1- <u>Polymorphic transformations</u>

This is a straightforward application of NTD which compares directly with its X-ray counterpart. However, the use of a large detector which does not move yields a high accuracy in the determination of the (relative) Bragg angles ; this allows minute modification of the cell parameters to be observed and leads, for instance, to accurate measurements of thermal expansivity. This method has also proved to be extremely valuable to investigate the low temperature magnetic behavior of complex materials : for example, many changes of magnetic structure, unnoticed in previous fixed-temperature neutron diffraction studies, have been shown by this technique [23,24]. Further studies include temperature-induced first order phase transitions ; although it is usually difficult to extract valid kinetic parameters only from non-isothermal experiments, NTD can nevertheless provide information about temporal aspects of structural changes as they occur. Three examples are presented below :

- **Reconstructive phase transitions** of lithium iodate [25,26] : the stable modification of $LiIO_3$ undergoes two first-order phase transitions in the temperature range 200°

  to 300° ($P6_3 \rightarrow Pna2_1 \rightarrow P4_2/n$) but there are unexpectedly large discrepancies in the literature concerning both the transition temperatures and the range of stability of each phase. Extensive NTD experiments were performed with powdered samples of known particle size prepared from single crystals grown from aqueous solutions under carefully controlled conditions (temperature and pH); two typical thermodiffractograms are given in fig. 7 : the first pattern (a) corresponds to a sample obtained by grinding large crystals and the second one (b) to a "natural" powder obtained by fast evaporation. The sequences of phase transitions are clearly different: the "natural" powder which did not undergo any mechanical treatment transforms directly from hexagonal to tetragonal whereas all other samples transform through an intermediate phase γ-$LiIO_3$. It was observed that the

  intermediate phase always coexists either with α- or β-$LiIO_3$; in addition, its range of existence was found to be roughly constant ($\approx 45°C$) while the temperatures of onset of the phase transitions depend on the particule size (Fig.8). The hysteresis of the transformation could also be easily studied by using different temperature programs. These results allows to explain the apparent discrepancies between previously published results and to set up a detailed mechanism for the α

  to γ transformation [26].

- **Order / disorder transitions** will be illustrated by the case of $Sr_2Co_2O_5$ [27]. $SrCoO_{3-x}$ is the general formula of a complex perovskite related system in which various phases with different compositions have been reported. The composition $Sr_2Co_2O_5$ with the Brownmillerite-type structure is obtained by fast quenching in air from 900°C : it is stabilized by the high spin state of Co^{3+} in tetrahedral coordination. Upon heating at 750°C under nitrogen, a new phase related to the $BaNiO_3$ structure is formed; this phase is paramagnetic at room-temperature (which suggests a change of spin state of Co^{3+}) and its volume per formula unit is

328

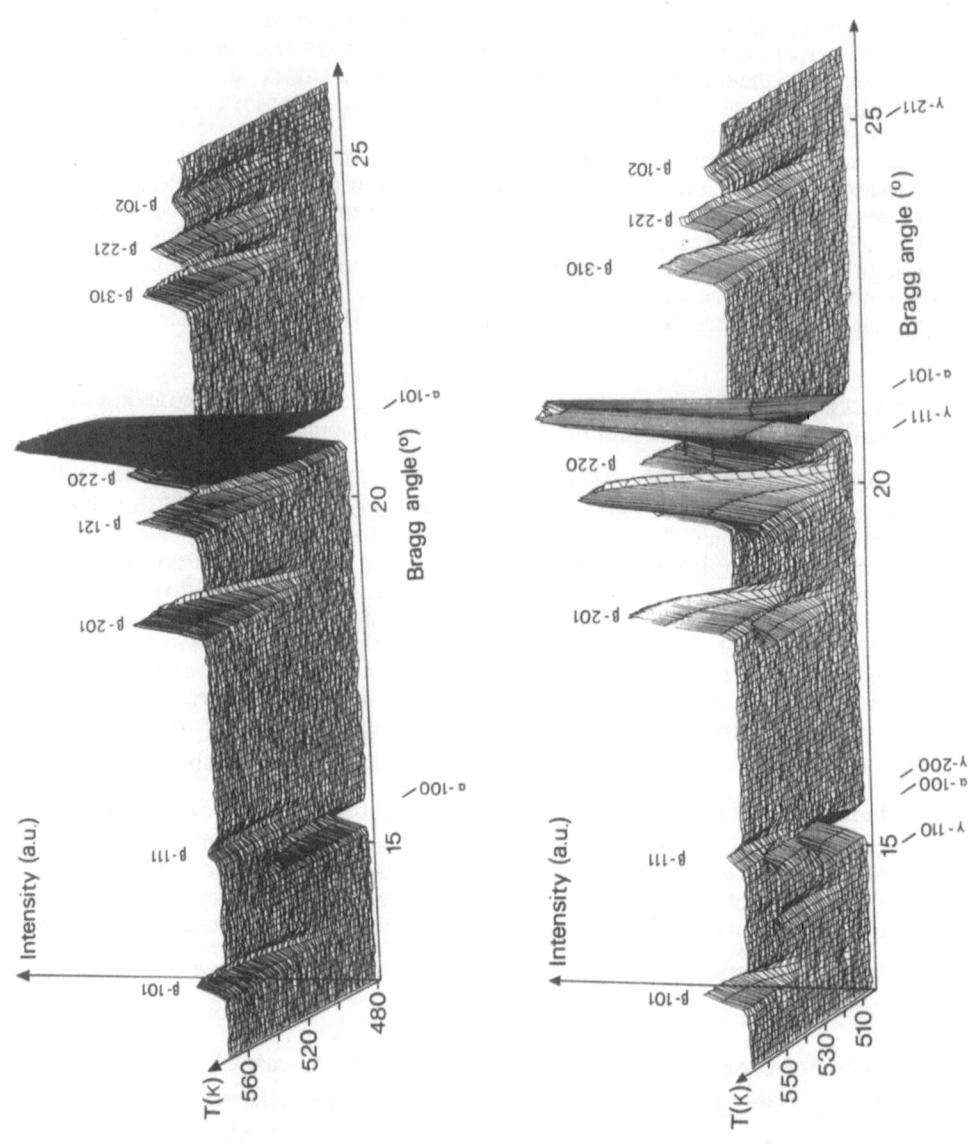

7 - Phase transition of a Lithium Iodate sample (from ref.26) :

   a- "natural" powder (average grain size 55µ)

   b- ground powder (average grain size 130µ)
$\Delta T = 1°C$, $\Delta t = 6$ min.

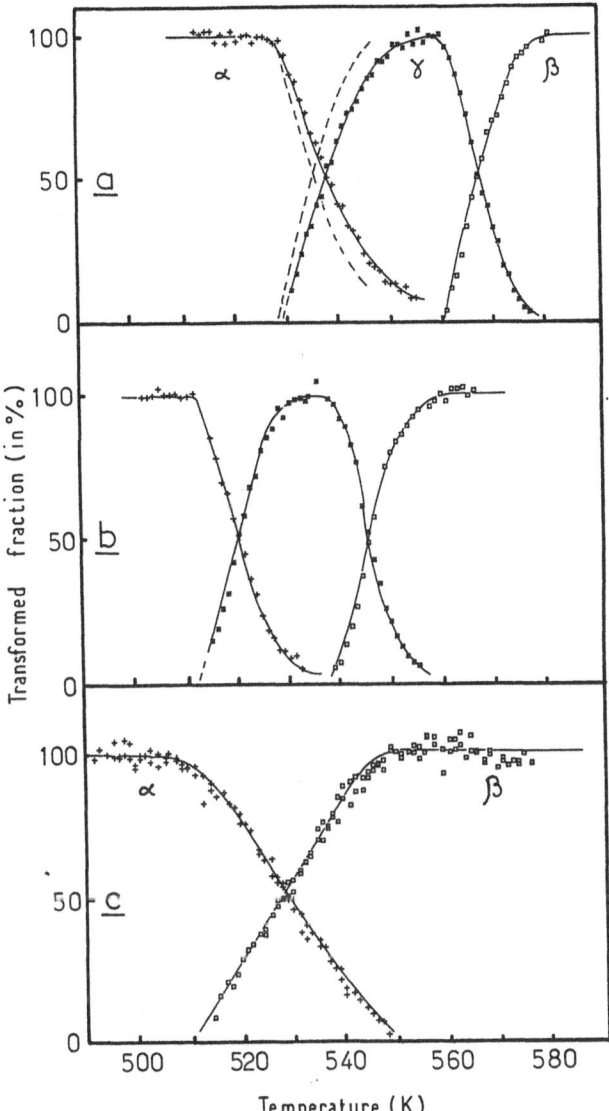

8 - Temperature evolution of the percentage of α–, γ- and β- phases during the first heating of different lithium iodate samples (from ref. 26) :

a)- ground powder (average grain size 15μ)

b)- ground powder (average grain size 130μ)

c)- "natural" powder (average grain size 55μ)

The dotted curves in a) represent the true equilibrium curves calculated by taking into account the kinetics of transformation (see Ref.26)

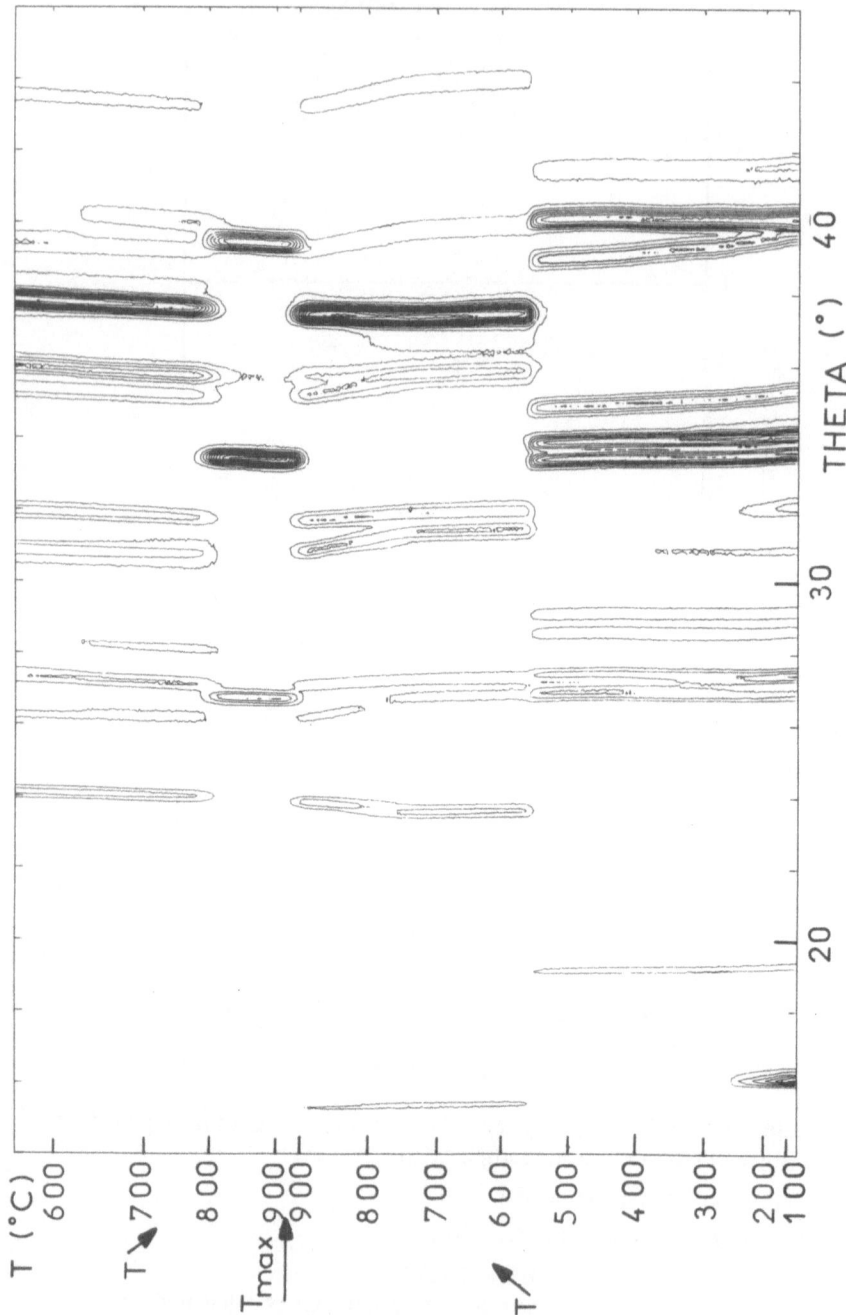

9 - Contour plot (intensity) of the neutron powder diffraction pattern of $Sr_2Co_2O_5$ as a function of temperature (from ref. 27). $\Delta T = 1.2°C$, $\Delta t = 2min$.

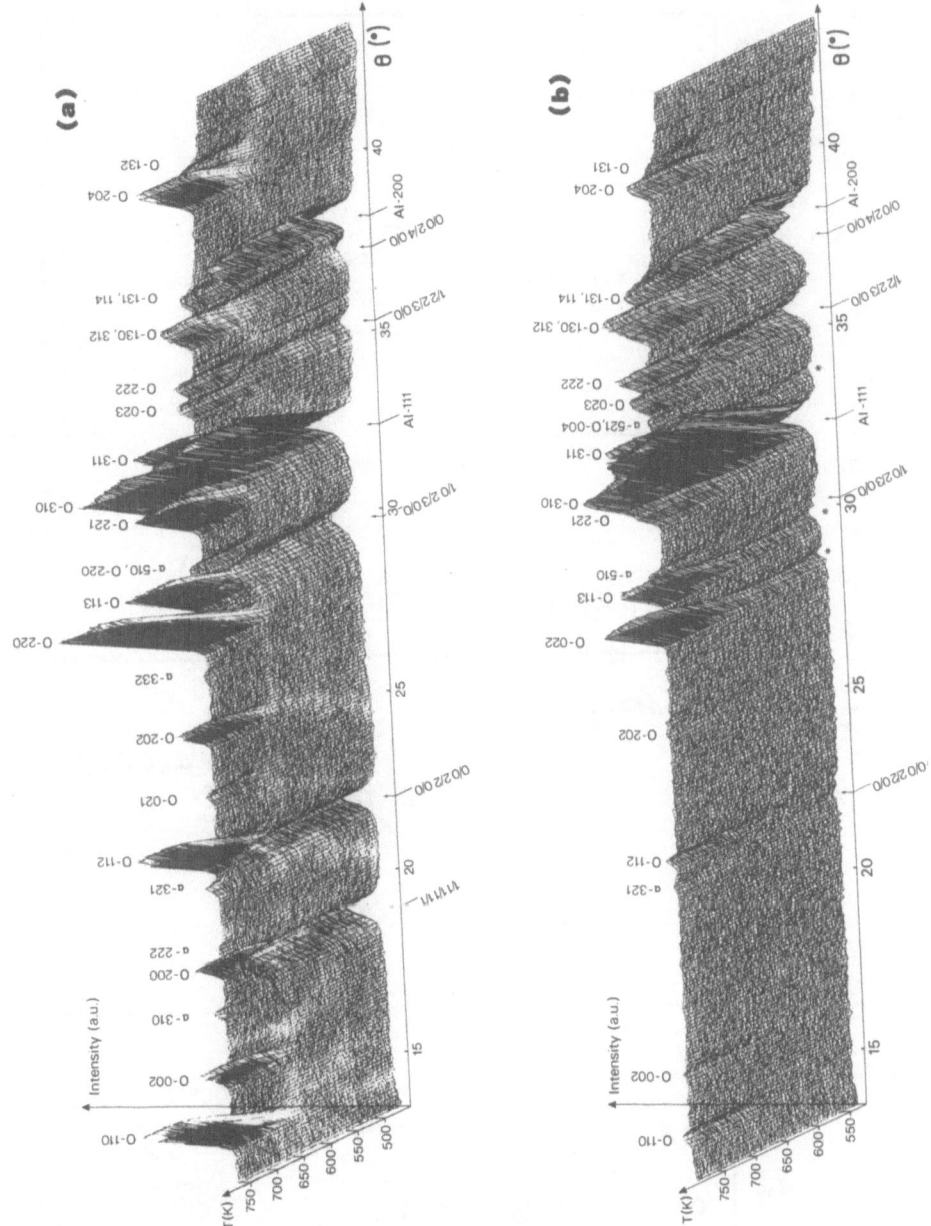

10 - Thermodiffractometry patterns of the i-$Al_{85}SiMn_{14}$ (a) and
i-$Al_{85}Si(Mn_{0.72}Fe_{0.28})_{14}$ (b) phases [28]. Six-index notation is used for the
i-phases while the Bragg peaks of the crystalline phases are labeled with the usual
three Miller indices. The star (*) indicates traces of 0-Al_6Mn in the as-quenched
sample. $\Delta T = 1.2°C$, $\Delta t = 3min$.

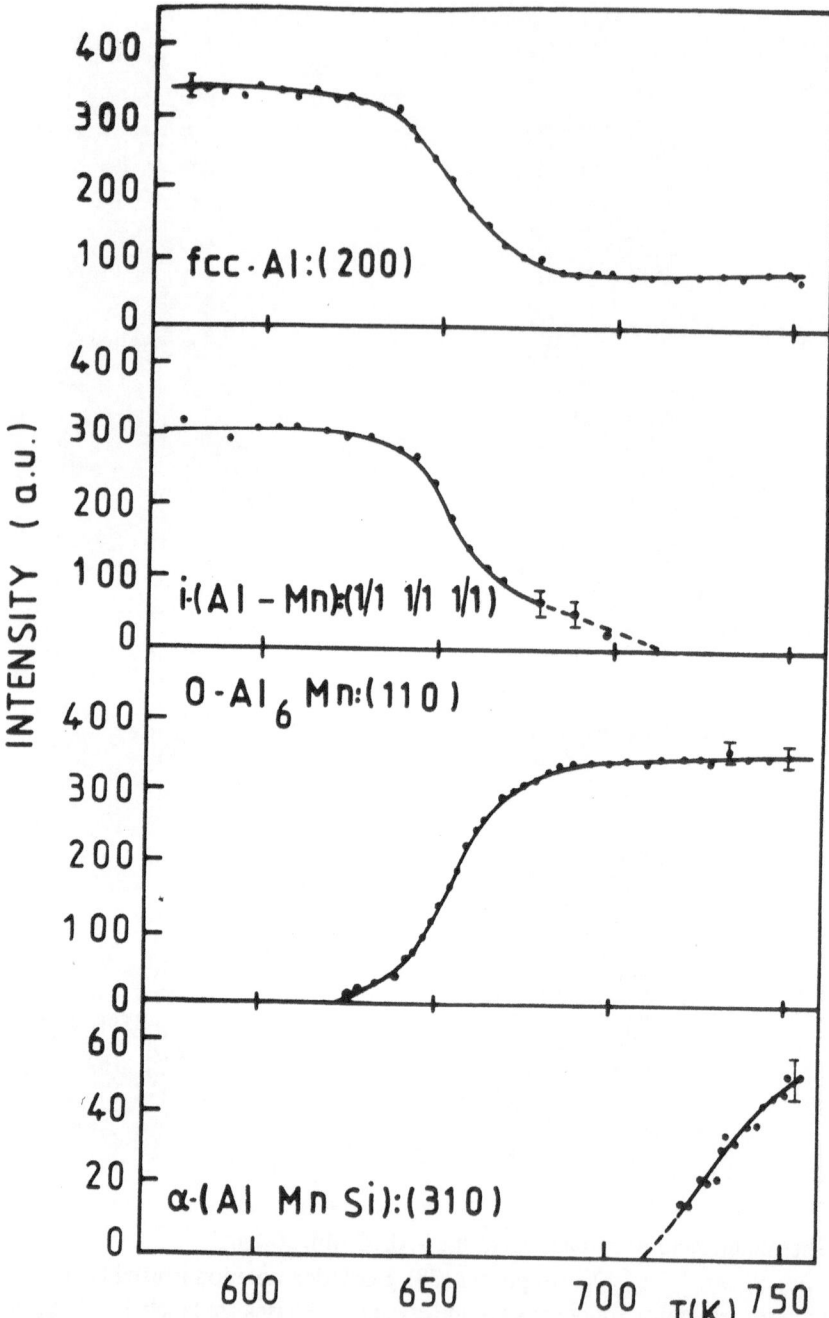

11 - Thermal variation of the diffracted intensities of the phases observed during the crystallization of i-$Al_{85}SiMn_{14}$ [28].

~11% lower than that of the Brownmillerite. The study of this phase transition was carried out by NTD in the range RT to 920°C both upon heating and cooling. The results are illustrated in Fig.9 and the different phases of $Sr_2Co_2O_5$ are summarized below :

Transition		Temperature(°C)	
B_m →	B_p	278	$T_{Néel}$
B_p →	R	530-588	Reconstructive
R →	R'	747	Displacive
R' →	P	882-915	Order-Disorder
P →	R'(?)	840-772	Ordering

The phase noted R has a rhombohedral cell related to $BaNiO_3$ while R' is probably an incommensurate distortion of R; the upper temperature phase P has an average cubic perovskite structure indicating a dynamic disorder of the oxygen atoms and vacancies.

Owing to the moderate crystallographic complexity of most of the above phases, a modified (multi-pattern) Rietveld profile refinement method could be used to refine the structural parameters of each phase [17] : this is probably the first exemple of real-time powder crystallography !

• **Thermal transformations of icosahedral quasiperiodic crystals** of the Al-Mn system [28] : the discovery by Schechtman et al. [29] of Al-Mn quasi-periodic alloys with long-range icosahedral symmetry and experimentally discrete diffraction patterns has brought to evidence a new class of materials which apparently fits to the concept of 3D Penrose tiling and generated extensive theoretical and experimental studies. These rapidly quenched alloys are metastable phases which can be prepared in the form of poly(quasi)crystalline samples mixed with various amounts of crystalline aluminum. Neutron powder diffraction is well adapted to the study of these phases because the only stable manganese isotope ^{55}Mn has a negative coherent scattering length ($b_{Mn} = -0.373.10^{-12}$cm.) and can be replaced by various amounts of transition metals such as Fe or Cr which have positive scattering length ($b_{Fe} = 0.954.10^{-12}$; $b_{Cr} = 0.363.10^{-12}$). Therefore, by modifying the amount of substitution, the average contrast of the 3d metal species in the alloy may be negative, null or positive; of special interest is the so-called "zero scattering" composition (for instance $Mn_{0.72}Fe_{0.28}$) which allows to study the sole aluminum network.

The crystallisation of a quasi-crystalline phase has been investigated on two alloys with nominal composition $Al_{85}SiMn_{14}$ and $Al_{85}Si(Mn_{0.72}Fe_{0.28})_{14}$: the results are summarized in Fig.10 which shows that the icosahedral phase transforms into orthorhombic Al_6Mn (or $Al_6Mn_{0.72}Fe_{0.28}$) at a temperature of about 630K. Due to the presence of 1 atomic % Si in the sample, a small amount of α-AlMnSi phase is also formed but at higher temperature (≈710K) and, most interestingly, well after all traces of icosahedral phase have vanished (fig.11); this suggests that the silicon

334

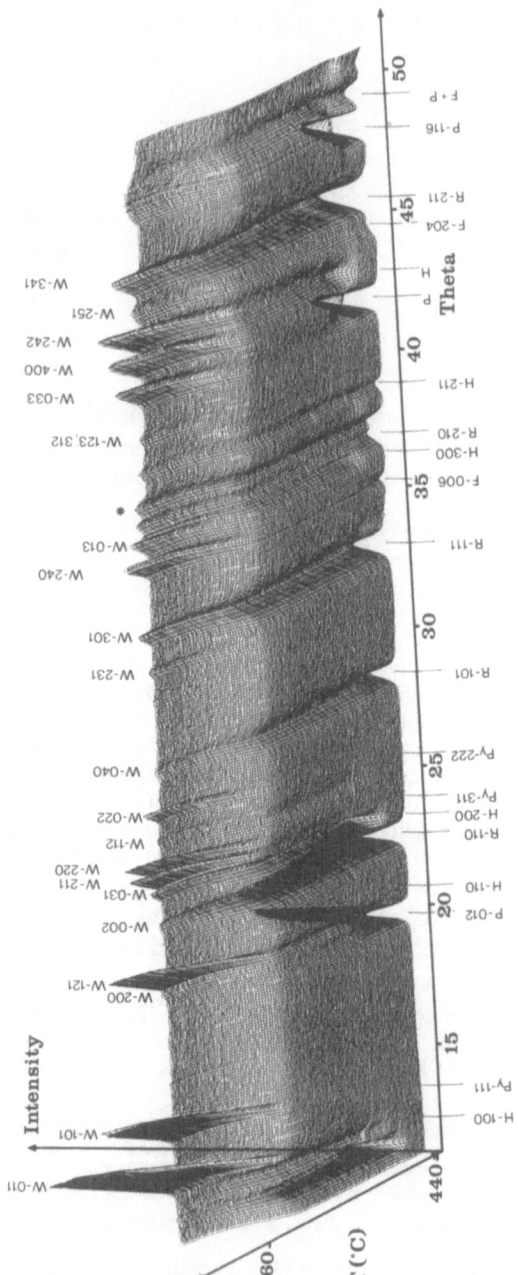

12 - Thermal evolution of the powder pattern of the inverse Weberite $Fe_2F_5,2H_2O$ (from ref. 23). Note the decrease of background intensity corresponding to the evolution of water

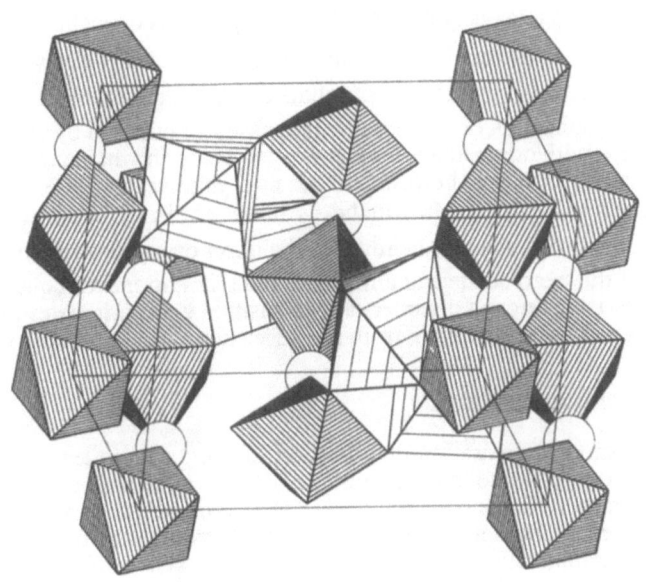

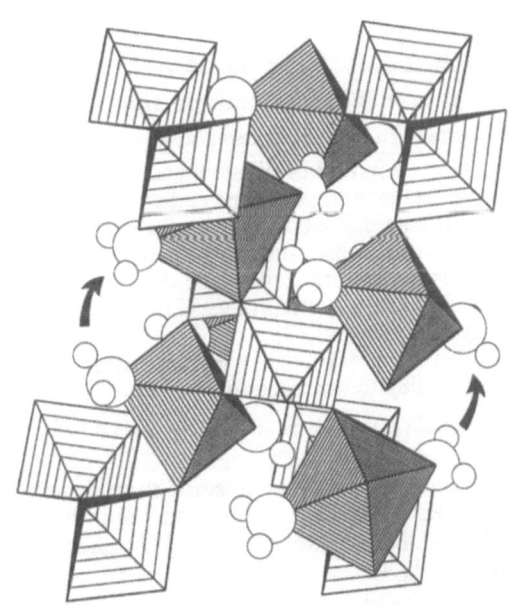

13 - Schematic representation (using STRUPLO (32)) of the inverse Weberite
$Fe_2F_5(H_2O)_2$ and pyrochlore $Fe_2F_5(H_2O)$ structures. Iron coordination octahedra
are hatched and the protonic species are represented as open circles.

atoms are homogeneously dispersed in the i-phase as quenched and have to form clusters in the o-Al_6Mn phase for the α-phase to nucleate. The volume fraction of the residual f.c.c. Al disappears simultaneously to the growth of Al_6Mn, showing that the crystallization of the icosahedral phase is monitored by atomic transport at its interface with crystalline aluminum; analysis of the diffracted intensities from the crystalline phases before and after crystallization yields an estimated formulation $Al_{4.98}Si_{0.07}Mn$ for the icosahedral phase. In addition, the diffuse scattering was observed to decrease at the onset of crystallization of $Al_{85}SiMn_{14}$ but not of the zero-scattering alloy : the transformation from quasiperiodicity to periodicity is therefore a disorder-order transition but it originates mainly from the transition metal atoms and not from the Al subnetwork.

5-2-2- Thermal decompositions

The investigation of hydration/dehydration of solids provides probably the best examples to demonstrate the usefulness of NTD. Indeed, the high incoherent background from hydrogenous samples is usually considered as very inconvenient in neutron powder diffraction in that it severely decreases the quality of the pattern (peak-to-background ratio) . This inconvenience still exists in NTD experiments but it can sometimes be turned to advantage because this incoherent scattering provides a straightforward measure of the proton content of the material under investigation. This affords the possibility of investigating **simultaneously** the composition (proton content) and structural characteristics of a sample. The following experiments illustrate the usefulness of the method.

- **Dehydration** of $Fe_2F_5,2H_2O$: this mixed valence fluoride (Fe^{2+}/Fe^{3+}) can be described as an inverse Weberite built up from corner-sharing $[Fe^{3+}F_6]$ and $[Fe^{2+}F_4(H_2O)_2]$ octahedra. It dehydrates above $\approx 130°C$ under vacuum and finally leads above 380°C to a mixture of divalent and trivalent iron fluorides. However, the dehydration process is rather complex and was believed to involve a pyrochlore phase as well as various ferric fluoride hydrates as intermediate species [30,31]. A NTD investigation in the temperature range 100° to 400°C with a resolution $\Delta T \approx$ 1°C provided a detailed description of the dehydration by allowing to identify quantitatively all the intermediate phases [23]. The sequence of reactions can be summarized as follows (see fig. 12) :
  - between 130° and $\approx 160°C$, the Weberite transforms to a pyrochlore structure according to the reaction :

$$Fe_2F_5(H_2O)_2 \;\rightarrow\; Fe_2F_5(H_2O) \;+\; H_2O\,\nearrow$$
Weberite   Pyrochlore

This reaction can be easily understood from the close similarities between the two structures : apart from the removal of half the water molecules, it only implies a small rotation of the Fe^{2+} octahedra around the **a** axis of the Weberite cell (fig.13).

- in the range 160°-170°C, the dehydration scheme changes progressively from the above equation to the following one :

$$Fe_2F_5(H_2O)_2 \rightarrow Fe_2F_5(OH) + H_2O\nearrow + 1/2\,H_2\nearrow$$
Weberite   Pyrochlore

which implies a partial oxidization of Fe^{2+} into Fe^{3+}.
Although the difference between the above two decomposition equations is rather small, the temperature evolution of the background intensity and transformed fraction of the Weberite unambiguously points to this mechanism of internal oxidization which could actually be confirmed by Mössbauer spectroscopy [23].

- upon further heating (between 170° and 215°), the latter mechanism competes with a third (and last!) decomposition reaction :

$$Fe_2F_5(H_2O)_2 \rightarrow FeF_2 + FeF_3 + H_2O\nearrow$$
Weberite   Rutile   HTB

where $HTB-FeF_3$ stands for hexagonal tungsten bronze FeF_3, a new form of iron trifluoride which is usually obtained by hydrothermal synthesis [33].
- above 335°C, $HTB-FeF_3$ undergoes an irreversible transition into the thermodynamically stable form of iron fluoride, $rh-FeF_3$ [34].

In spite of many independent studies [30,31] making use of conventional diffractometric and thermoanalytical methods, this complex decomposition scheme was not understood before this investigation ; it could be solved from a single NTD experiment only because this technique provides simultaneously a measure of the proton content (from the incoherent background intensity) and a quantitative estimate of the various phases existing in the sample (from the Bragg peak intensities).

- **Dehydration** of $WO_3,1/3H_2O$: this new hydrate (hereafter noted HYD) of tungsten trioxide has an orthorhombic structure closely related to the hexagonal tungsten bronze (HTB) structure [35] ; upon heating, it transforms first into a new form of tungsten trioxide $HTB-WO_3$ and, at higher temperature, into perovskite type (PTB) WO_3 [36]. A NTD investigation [9] of this sequence of dehydration / transition shows (fig. 14) that the transformation HYD - HTB actually occurs in two steps :
  - first a dehydration reaction which preserves the HYD framework (this implies that half of the W atoms are now in a five-fold coordination),
  - then a reconstructive phase transition from the "dehydrated HYD structure" to $HTB-WO_3$.

Interestingly, the FWHM of the $0k0$ reflections of HYD (and only these reflections) is strongly affected by the latter transition ; in the HTB phase they can actually be represented by the sum of two gaussian peaks centered at the same Bragg angle and with intensities in the ratio of 1/2. The stronger reflection has almost the FWHM expected from the instrumental resolution of the diffractometer whereas the weaker one has a width corresponding to an average crystal dimension of about 200Å along this direction (**b** axis of HYD structure). This observation has an obvious bearing on the mechanism of the transformation HYD - HTB and suggests that it proceeds through a shear of the HYD structure perpendicular to its b axis.

This last example demonstrates that, in additon to the possibility of detecting intermediate stages and providing kinetic information about chemical processes in the solid state, NTD can also bring information about the structural mechanism of transformations.

5-2-3- Crystallization processes

The study of crystallization of amorphous materials provide information on the stability of the amorphous phase with respect to the crystalline state but can also shed light on the mecanism of nucleation and growth in a topologically disordered structure. However, conventional methods (electrical resistivity, calorimetry, Mossbauer spectroscopy) are usually not well adapted to such investigations ; high-speed neutron powder diffraction which probes the bulk sample, even through complex environment, is a convenient and fast method to obtain quantitative information on the successive stages of the transformation. An example concerning the crystallization of the amorphous alloy $Fe_{77}P_{11.5}C_{11.5}$ [37] is presented in fig. 15. The crystallization is seen to proceed in two stages :

- below $T_p \approx 320°C$, the material is amorphous
- between T_p and $T_c \approx 400°C$, there is a mixture of crystalline Fe and new amorphous phase
- above T_c, all the sample is in a crystalline state : $Fe + Fe_3P + Fe_3C$

The mixture of these phases corresponds to the crystalline equilibrium state. This experiment was carried out with alloys of different compositions and with several heating rates [37] ; it was concluded that the initial alloy evolves towards an "ideal" amorphous alloy with composition $\approx Fe_{75}(PC)_{25}$; the excess of iron is rejected as crystalline Fe.

5-3- Texture measurements

The texture of a polycrystalline material is the orientation distribution of its crystallites with respect to a sample fixed reference system e.g. rolling plane and rolling direction in metallic materials or mesoscopic and macroscopic fabric coordinates in geological samples. The texture determines the orientation dependent mean values of anisotropic properties in polycrystalline samples (for instance the strength of a material in various directions) and depends on the history of the material such as plastic deformation, phase transformations and crystallization. This latter point is especially relevant in geology where it provides information about processes which have taken place millions of years ago.

Texture is quantified by the crystalline orientation distribution function (CODF) [38]. The experimental starting point of this calculation are quantitatively determined pole figures (a stereographic projections of the diffraction intensity of a given Bragg reflection as the sample is rotated) the number of which must be the larger the lower the crystal symmetry. Pole figures are usually measured, by X-ray or neutron diffraction, by using a standard 4-circles crystal diffractometer. Neutron diffraction allows much bigger samples to be used (typically a $1cm^3$ volume) which gives smooth pole figures and consequently a precise CODF. As a pole figure determination can be

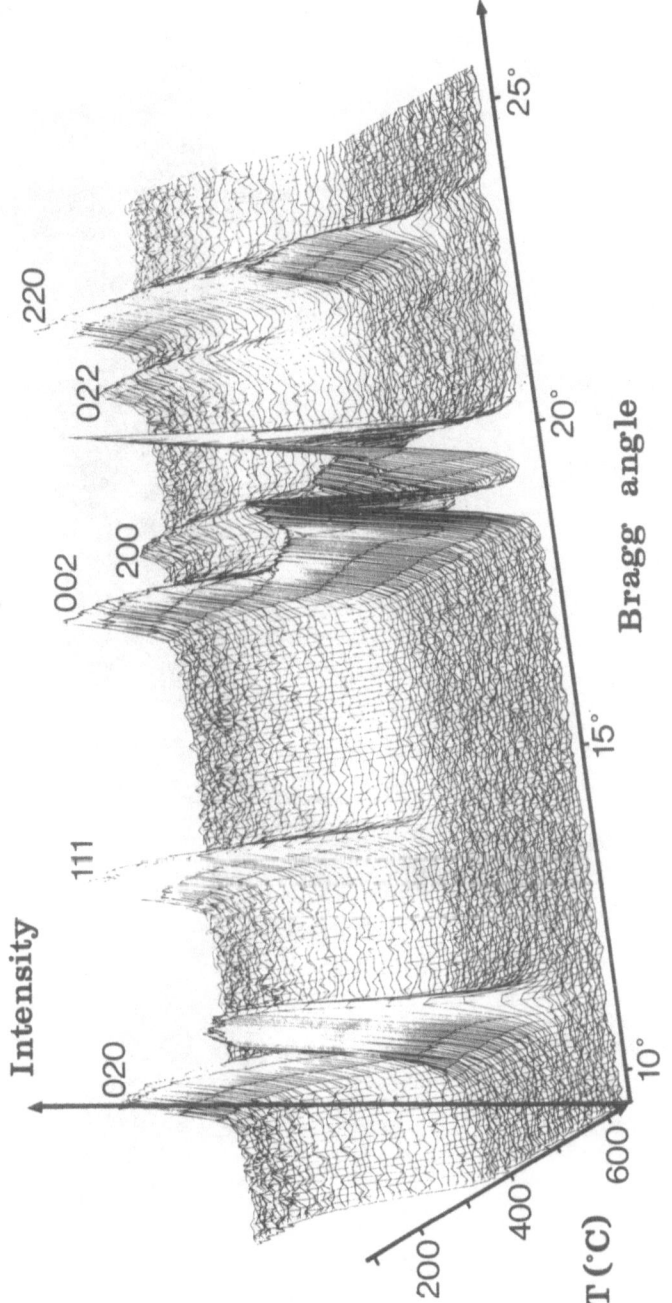

14 - Temperature variation of the neutron powder diffraction pattern of $WO_3, 1/3 H_2O$ (from ref. 9)

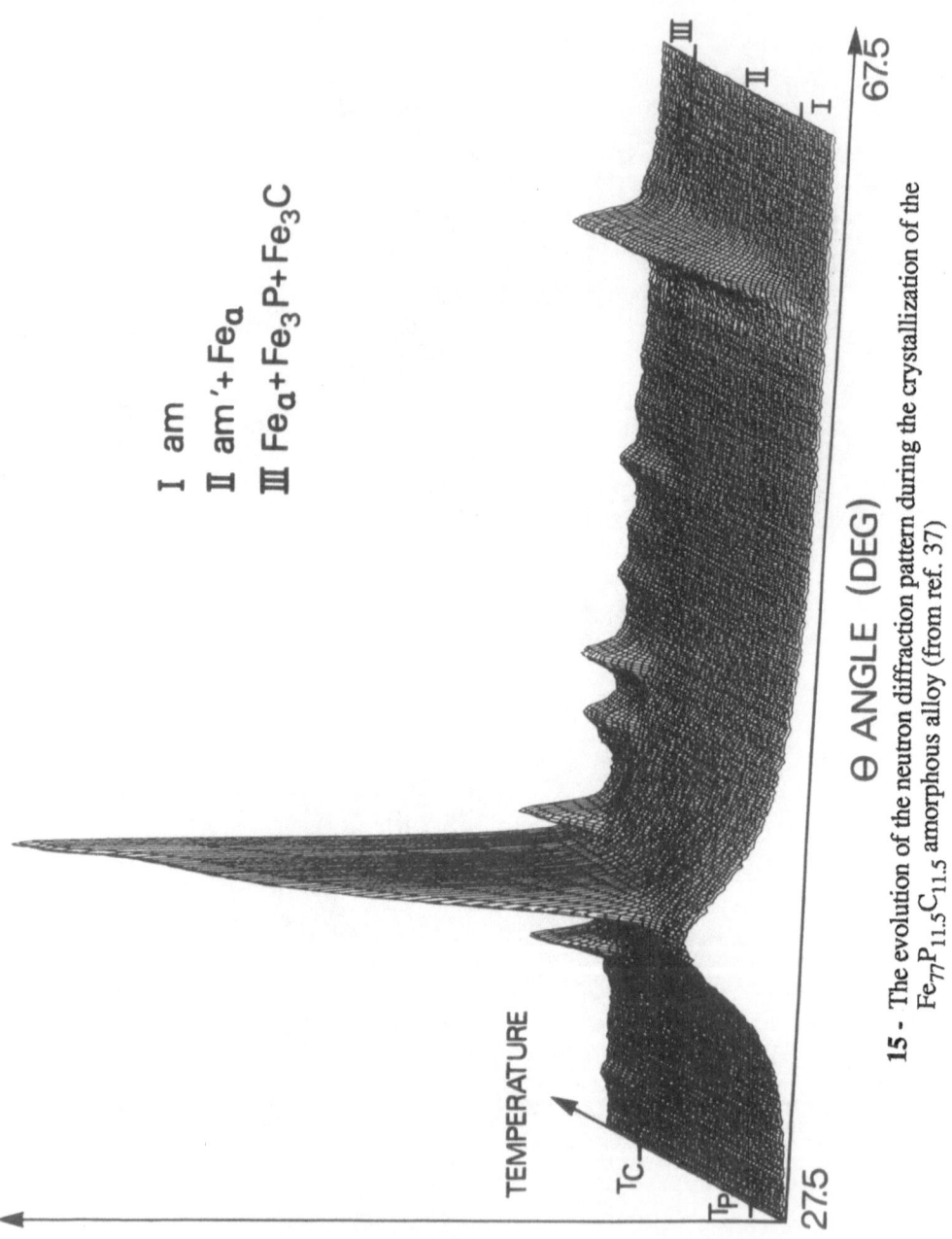

15 - The evolution of the neutron diffraction pattern during the crystallization of the $Fe_{77}P_{11.5}C_{11.5}$ amorphous alloy (from ref. 37)

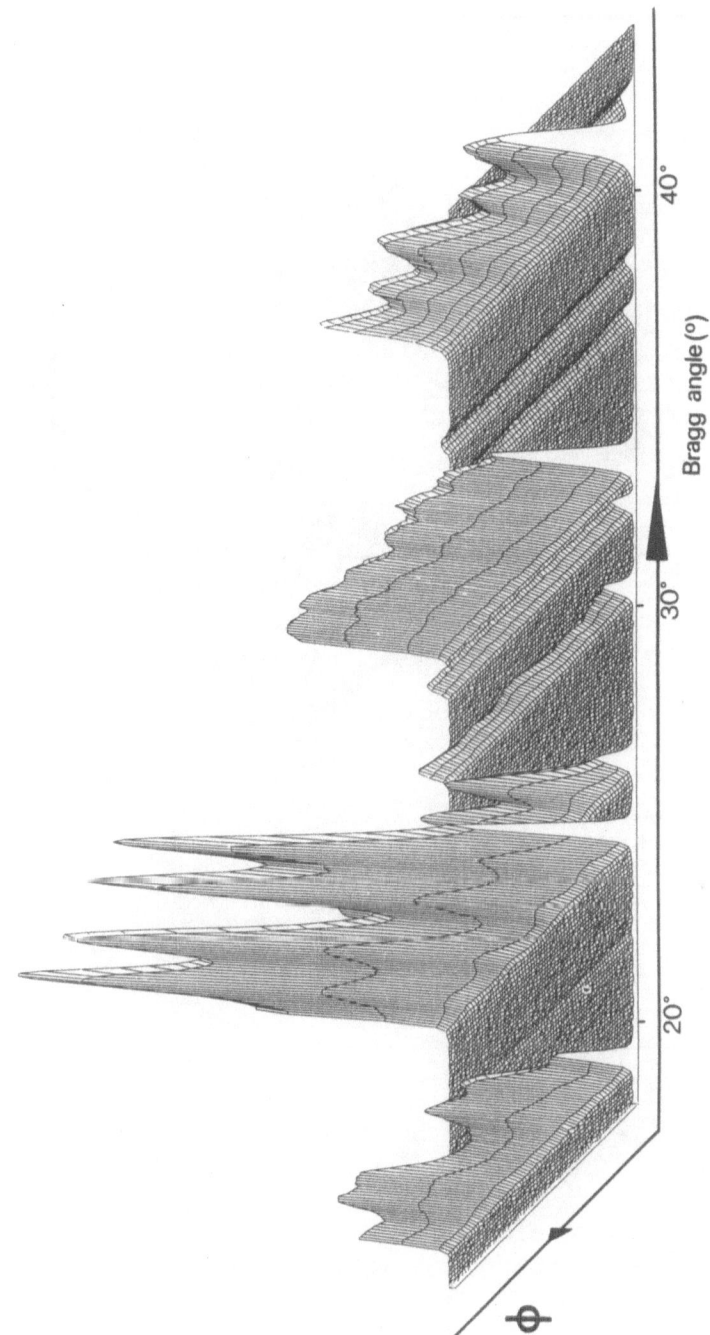

16 - Measurement of a pole figure from a calcite sample [41]

342

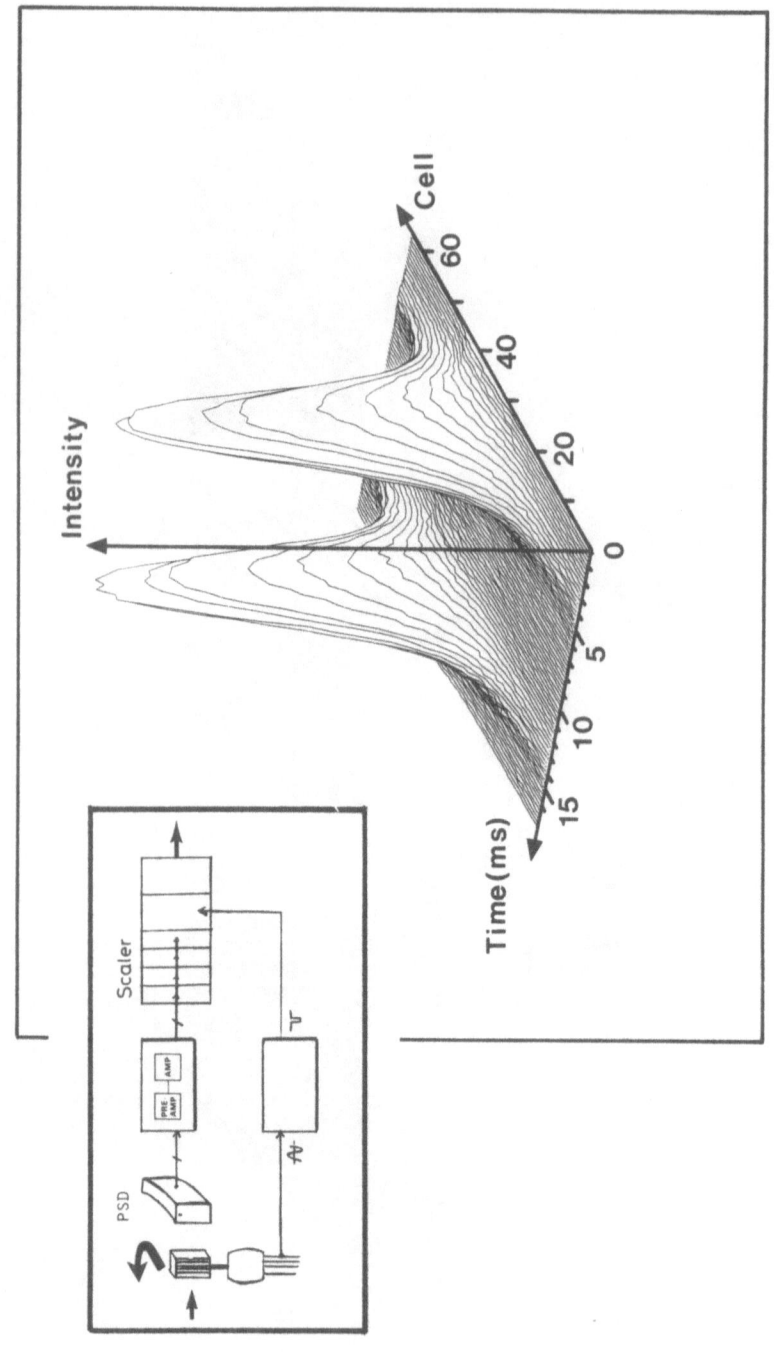

17 - An example of stroboscopic diffraction experiment (from ref. 44) : dynamical measurement of the rocking curve of a grafoil sample

performed on a medium flux neutron source on a conventional diffractometer, it is not strictly speaking a real-time measurement; however with a single detector the measuring time for low crystal symmetry materials would be prohibitive. More rapid data collection is possible if use is made of a PSD because a complete pattern is measured at the same time. The method have been successfully used on D1B to investigate the texture of a triclinic plagioclase with large cell parameters [39,40]. An exemple of raw data taken by the same method from a calcite sample is given in Fig.16.

A different technique based on the use of a linear PSD has been developed at Risø National Laboratory [42]; a variable incident wavelength is used to give a scattering angle for the desired reflection of 90°, so that the Debye-Scherrer cone becomes a plane. Part of the pole figure is thus measured at just one sample setting. This method is rapid enough to enable the real-time evolution of texture (e.g. during annealing) to be studied.

5-4- Stroboscopic measurements

There are great but still largely unrealized opportunities of new applications in this area; basically this concerns all real-time studies of structural responses of condensed matter to external perturbations such as temperature, stress, electric fields, magnetic fields, etc... ; either periodic or stepwise perturbations can be applied. As discussed previously (§ 2), the time resolution for these periodic processes is not limited by the scattering power of the sample but rather by the instrumentation.

Short time measurements can be achieved rather easily by synchronizing the data acquisition and the perturbation (steady source) or, alternatively, the stimulation and the burst of neutrons (pulsed source) . For example, resolutions of about 10 msec. have already been obtained at pulsed sources [43] and the data acquisition of the two-axis diffractometer D20 under development at the ILL achieves a resolution of about 2.5 msec. with a dead-time of 30 µsec. [44].

In principle, the temporal resolution could be extended down to times of about 10 µsec. by tuning **both** the frequency and the phase of the perturbation with respect to the pulsed measurement (multiscaling measurement) or the pulsed source : the method is to initiate the process at the same frequency as the pulsed measurement (or source), but to displace the phase of the perturbation with respect to the data acquisition (or the time of arrival of the neutrons at the sample). No experiment in this time domain has been reported yet but feasibility tests have been performed by P. Convert on the D20 prototype diffractometer at the ILL [44] : the complete rocking curve (fig. 17) of a grafoil sample (oriented graphite with large mosaic spread) was measured by rotating the sample at 50 Hz ; data were accumulated over several periods, each period being divided in 80 time-intervals, thus bringing the time resolution to 250 µsec.

It is probably worth to mention, at this point, that such measurements are getting close to the theoretical limits of time-resolution which can be estimated from the neutron velocity :

$$v \; (\text{cm } \mu\text{sec.}^{-1}) = 0.3955 / \lambda \; (\text{Å})$$

for instance, when $\lambda = 2.5$ Å, it takes more than 6 µsec. for a beam of neutrons to travel through a 1 cm sample. A time resolution of about 100 µsec. can therefore be considered as a reasonable goal for the near future ; although research has still been very limited in this domain, one may expect it to grow rapidly as new instrumentation is made available.

6 - CONTINUOUS VS. PULSED SOURCES

Accelerator-based pulsed neutron sources have been developped rapidly during the last decade and time-of-flight (TOF) diffractometers installed at such sources are now competitive with the best constant wavelength high resolution diffractometers (for a review of instrumentation at pulsed neutron sources, see for instance ref. 45). At the present time, the time-averaged flux of neutron from pulsed sources does not exceed that in reactors, but it has been claimed that, owing to their temporal structure, pulsed sources are best suited to time-resolved experiments. This is an argument which certainly deserves further discussion, first of all because most of the time-resolved neutron powder diffraction experiments published up to now (see appendix) have been performed on conventional reactor-based diffractometers ! Once again, it is convenient to distinguish between irreversible (one-shot) and periodic measurements :

- for **irreversible processes**, the only relevent parameter is the ratio of the time of interest of the phenomena t_R to the shortest period of time t_s in which one diffraction pattern is collected. Therefore, as long as t_s is longer than the repetition rate of the pulsed source (typically $t_R > 10$ msec.) and assuming the time-averaged flux is the same, TOF diffractometers are almost equivalent to reactor-based instruments : they just sample the time interval t_s in a different way. Below this limit, the pulsed source instrument would in principle perform better than a reactor diffractometer but this would imply that a single burst of neutrons is bright enough to provide adequate statistical accuracy : this would require beam intensities much higher than is presently available at existing neutron diffractometers.

  Incidentally, the good resolution of TOF diffractometers at large d-spacings should in principle confer some advantages in studies involving long-range ordering periods like staging processes during intercalation processes ; unfortunately, this is also the wavelength range where the beam intensity is the weakest, thus making the instrument less powerful.

- **periodic measurements** are employed to study time-dependent reversible phenomena stimulated by pulsed environments. The temporal structure of the pulsed sources thereby seems to be well adapted to such studies and, indeed, the only experiment yet published in this field (a study of electric-fields induced domain motion in ferroelectric $NaNO_2$ [43]) was performed at a pulsed neutron source. However, this does not mean that "pulsed sources are best" in this field ; for sake of clarity, we can consider three domains of time resolution :

  - Time-resolution of about 10^{-2} sec. can be achieved by matching the pulsed perturbation to the pulsed source ; the lower limit of resolution is then given by the pulse repetition rate of the source (typically about 10 msec. or more on current machines) ; but a serious inconvenience arises from the fact that the frequency of the perturbation (i.e. a parameter of the process to investigate) **must** be tuned to the frequency to which the accelerator is run. This of course does not occur for steady source ! In this case, the sample is bathed in a constant flux of neutrons and it is the rate of data acquisition which is tuned to the frequency of the cyclic perturbation, whatever it is ; the time resolution is then limited only by the speed of the data acquisition system. In this time domain, both pulsed sources and reactor diffractometers can compete, but steady sources certainly provide more flexibility.

  - Higher time resolution of about 10^{-4} sec. can be obtained on both sources (see 5-4). However, in the case of TOF diffractometers an additional difficulty arises from the use of a white beam : different wavelength neutrons travel at different speeds and thus probe the sample at different times ; for instance, with a source-to-sample distance of 20 m the difference of time of flight between 1 and 5 Å is more than 600 µsec.. Therefore, at these time resolutions only a limited part of the diffraction data can be used.

- For resolutions of the order of 10^{-5} sec., pulsed source methods have a unique advantage over steady methods ; this occurs, for instance, in the case where the fraction of time that the sample is in its interesting state is small (very high magnetic fields for example). In such case, it is not the temporal evolution which is investigated but a status which can be maintained only for a short period of time.

In conclusion, it is to early to see clearly an answer as to "which type of source is more suited to real-time diffraction" ; it can be argued however that, as long as the time-averaged flux is about the same, pulsed source diffractometers present no significant advantage over reactor instruments, except when the time resolution to achieve is of the order of the pulsed width.

7 - CONCLUSION

Real-time neutron diffraction is a relatively new field of research but one can expect wider applications to develop as more instrumental facilities become available and as the imagination of experimenters is bringing new exciting problems. We have tried in this article to present the current possibilities and limitations of time-resolution experiments using neutron and to show how materials science may benefit from dynamical studies using diffraction techniques. We can look forward to seeing the list of applications expand in the next future.

8 - ACKNOWLEDGMENTS

I would like to thank my collegues of the I.L.L. who contributed over the years to my education in neutron diffraction and the many others who initiated or contributed to the experiments described in this article. Particular thanks are due to P. Convert for introducing me to the field of stroboscopic measurements and to G. Ferey for many stimulating discussions on the applications of NTD in solid state chemistry.

LIST OF REFERENCES

1 - G.E. BACON
Neutron Diffraction, Clarendon Press, Oxford (1975)

2 - H.M. RIETVELD
Acta Cryst. 22, 151 (1967)

3 - M. CAFFREY & D.H. BILDERBACK
Nucl. Inst. & Meth. **208**, 495 (1983)

4 - K. KOSTEN & H. ARNOLD
J. Applied Crystallogr. **17**, 206 (1984)

5 - H.D. BARTUNIK
Rev. Physique Appliquée **19**, 671 (1984)

6 - B.C. LARSON, C.W. WHITE, T.S. NOGGLE & D.M. MILLS
Phys. Rev. Lett. **48**, 337 (1982)

7 - D.M. MILLS , B.C. LARSON , C.W. WHITE & T.S. NOGGLE
Nucl. Inst. & Meth. **208** , 511 (1983)

8 - C. RIEKEL , in *Position-sensitive Detection of Thermal Neutrons*   (ed. P.
Convert and J.B. Forsyth) , 267 . Academic Press (1983)

9 - J. PANNETIER
Chemica Scripta **26A**, 131 (1986)

10 - J. PANNETIER
Ber. Bunsenges. Phys. Chem. **90**, 634 (1986)

11 - For a recent discussion of developments in this field, see :   *Position-sensitive
Detection of Thermal Neutrons* (ed. P. Convert and J.B. Forsyth), Academic
Press (1983)

12 - P. CONVERT, D. FRUCHART, E. ROUDAUT & P. WOLFERS
in *Position-sensitive Detection of Thermal Neutrons*
(ed. P. Convert and J.B. Forsyth), 302,   Academic Press (1983)

13 - M. IIZUMI
Physica **136B** , 36 (1986)

14 - E. ROUDAUT in *Position-sensitive Detection of Thermal Neutrons* (ed. P.
Convert and J.B. Forsyth), 294, Academic Press (1983)

15 - G. CAGLIOTTI, A. PAOLETTI & F.P. RICCI
Nucl. Inst. & Meth. 3, 223 (1958)

16 - A.W. HEWAT
Nucl. Inst. & Meth. **127** , 361 (1975)

17 - J. RODRIGUEZ & M. ANNE
Private communication

18 - C. RIEKEL
Progr. Solid State Chem. **13** , 89 (1980)

19 - M. GOLDMANN, F.BEGUIN & J. PANNETIER
unpublished results

20 - A.N. CHRISTENSEN,H. FJELLVAG & M.S. LEHMANN
Acta Chem. Scand **A39**, 593 (1985)

21 - L. BRAGANZA, A. FITCH & J. COCKROFT
in preparation

22 - Y YAMADA, N.HAMAYA, J.D. AXE & S.M. SHAPIRO
Phys. Rev. Lett. **53**, 1665 (1984)

23 - Y. LALIGANT
These, Université du Maine - Le Mans (1986)

24 - Y. LALIGANT, M. LEBLANC, J. PANNETIER & G. FEREY
J. Physics **C 19**, 1081 (1986)

25 - J. PANNETIER, E. COQUET, J. BOUILLOT & J.M. CRETTEZ
C.R. Hebd. Acad. Sci. Paris **209**, 541 (1984)

26 - J.M. CRETTEZ, E. COQUET, B. MICHAUX, J. PANNETIER, J. BOUILLOT, P. ORLANS, A. NONAT & J.C. MUTIN
Physica *in press*

27 - J. RODRIGUEZ, J. PANNETIER & M. ANNE
Solid State Comm. *in press*

28 - J. PANNETIER, J.M. DUBOIS, C. JANOT & A. BILDE
Phil. Mag. *in press*

29 - D. SHECHTMAN, I. BLECH, D. GRATIAS & J.W. CAHN
Phys. Rev. Lett. **53**, 1951 (1984)

30 - P. CHARPIN & Y. MACHETEAU
C.R. Hebd. Acad. Sci. Paris, **280C**, 61 (1975)

31 - K.J. GALLARGHER & M.R. OTTAWAY
J. Chem. Soc., Dalton Trans. 978 (1975)

32 - R.X. FISCHER
J. Applied Crystallogr. **18**, 258 (1985)

33 - M. LEBLANC, G. FEREY, P. CHEVALLIER, Y. CALAGE & R. DE PAPE
J. Solid State Chem. **47**, 53 (1983)

34 - M. LEBLANC, J. PANNETIER, G. FEREY & R. DE PAPE
Rev. Chimie Minérale **22**, 107 (1985)

35 - B. GERAND, G. NOWOGROCKI & M. FIGLARZ
J. Solid State Chem. **38**, 312 (1981)

36 - B. GERAND, G. NOWOGROCKI, J. GUENOT & M. FIGLARZ
J. Solid State Chem. **29**, 429 (1979)

37 - C. TETE, M. VERGNAT, G. MARCHAL & P. MANGIN
Solid State Comm. **53**, 191 (1985)

38 - H.J. BUNGE
Quantitative Texture Analysis,H.J. Bunge & C. Esling Ed.
Deutsche Gesellschaft für Metallbunde;Oberursel,Germany (1981)

39 - H.J. BUNGE, H.R. WENK & J. PANNETIER
Textures and Microstructures **5**,153 (1982)

40 - H.R. WENK, H.J. BUNGE, E. JANSEN & J. PANNETIER
Tectonophysics **126**,271 (1986)

41 - H.R. WENK
Private communication

42 - N. HAUSEN, T. LEFFERS & J.K. KJEMS
Acta Metall. 29,1523 (1981)

43 - N. NIIMURA & M. MUTO
J. Phys. Soc. Japan **35**, 628 (1973)

44 - P. CONVERT
Workshop on high resolution neutron powder diffraction
ILL, 1-3 August 1984
and private communication

45 - J.M. CARPENTER, G.H. LANDER & C.G. WINDSOR
Rev. Sci. Instrum. **55**, 1019 (1984)

46 - Neutron beam facilities at the HFR. This publication is subject to periodical
revision and copies may be obtained from the Scientific Secretary (ILL).

47 - S. KATANO, Y. MORII & M. IIZUMI
ORNL Solid State Division Progress Report - 6306 (1986)

Appendix - A list of references on real-time neutron powder diffraction experiments.

1- ISOTHERMAL EXPERIMENTS

1-1- Phase transformations   (Ref.)

NaCl - CsCl	RbI	3.5Kbars/RT	1
β - α	Sn	213°-253°C	2
Disorder / Order	Ni_3Mn	470°C	2
trigonal - cubic	KPF_6	RT	3
Crystallization	FeF_3,xHF	280°-300°C	4
Crystallization	FeOOH	RT - 100°C	5
Crystallization	$Fe_{79}B_{13}Si_9$	450°-520°C	6

1-2- Chemical reactions (gas-solid)

NH_3/ND_3 exchange	$TaS_2.NH_3$	215K-223K	7
C_5D_5N intercalation	TaS_2	RT	8
Intercalation of H_2	KC_8	300K-400K	9
AsF_5 doping	Polyacetylene	RT	10
AsF_5 doping	Polyparaphenylene	RT	11
SbF_5, I_2, IBr doping	Polyacetylene	RT	12,13
C_6D_6 intercalation	RbC_{24}	RT	14
K intercalation	Graphite	280°C	15
THF intercalation	RbC_{24}	RT	16
THF intercalation	CsC_{24}	290K	17
Intercalation of H_2	CeNi	RT	18

1-3- Chemical reactions (liquid-solid)

H/D exchange	γ-AlOOH	RT - 100°C	19
Hydration	Cements	RT - 120°C	20,21
Hydration	Gypsum	RT - 75°C	22

1-4- Electrochemical reactions

Intercalation of H_2	TaS_2	RT	23
Topotactic reduction	TaS_2	RT	24
Intercalation of D_2	TiNi	RT	25

2- THERMODIFFRACTOMETRY

2-1- Phase transformations

Reconstructive	$LiIO_3$	200° - 300°C	26
Magnetic transition	$Fe_2F_5,2H_2O$	2K - 60K	27
Order-Disorder	$PbSnF_4$	RT - 412°C	28
Structural transitions	KPF_6	160K - RT	29
Structural transitions	FeF_3	RT - 400°C	30,35
Structural transitions	$Sr_2Co_2O_5$	RT - 920°C	31

2-2- Chemical reactions

Solid - Solid Reaction	$Sb_2O_5 + Ta_2O_5$	RT - 600°C	32

2-3- Thermal decompositions

Dehydration	$WO_3,1/3H_2O$	RT - 640°C	33
Dehydration	$Fe_2F_5,2H_2O$	RT - 440°C	30,36
Decomposition	$H_{1/3}Li_{2/3}IO_3$	RT - 300°C	34
Deammoniation	$(NH_3)_xFe_2F_6$	RT - 400°C	35
Dehydration .	$ZnFeF_5,2H_2$	RT - 500°C	36
Dehydration	$MoO_3,2H_2O$	RT - 800°C	37

2-4- Crystallization processes

FePC	310° - 440°C	38
FeF_3, xHF	RT - 420°C	39
i-Al_6Mn	RT - 760K	40
$(Fe,Mn)_{80}P_{20}$	RT - 740K	41

3- TEXTURE MEASUREMENTS

Plagioclase	RT	42
Calcite	RT	43
Copper	RT	44

4- STROBOSCOPIC MEASUREMENTS

Domain motion	$NaNO_2$	RT	**45**
Relaxation (SANS)	Spin-glass	RT	**46**
Stretching (SANS)	Polymers	RT	**47**

References of Appendix

1 - Y. YAMADA, N. HAMAYA, J.D. AXE & S.M. SHAPIRO
Phys. Rev. Lett. **53**, 1665 (1984)

2 - M. IIZUMI
Physica **136B**, 36 (1986)

3 - L. BRAGANZA, A.N. FITCH & J.K. COCKROFT
to be published

4 - M. LEBLANC, G. FEREY & J. PANNETIER
unpublished results

5 - A.N. CHRISTENSEN & M.S. LEHMANN
Acta Chem. Scand. **A34,** 771 (1980)

6 - S. KATANO, Y. MORII & M. IIZUMI
ORNL Solid State Division Progress Report - 6306 (1986)

7 - C. RIEKEL
Solid State Comm. **28**, 385 (1978)

8 - C. RIEKEL & C.O. FISHER
J. Solid State Chem. **29**, 181 (1979)

9 - J.P. BEAUFILS, T. TREWERN, R.K. THOMAS & J.W. WHITE
J. Chem. Soc. , Faraday Trans. I, **78**, 2387 (1982)

10 - C. RIEKEL, H.W. HAESSLIN, K. MENKE & S. ROTH
J. Chem. Phys. **77**, 4254 (1982)

11 - H.W. HAESSLIN & C. RIEKEL
Synthetic Metals **5**, 37 (1982)

12 - C. RIEKEL, H.W. HAESSLIN, K. MENKE & S. ROTH
Synthetic Metals **10**, 31 (1984)

13 - C. RIEKEL, H.W. HAESSLIN, K. MENKE & S. ROTH
Mol. Cryst. Liq. Cryst. *in press*

14 - A. HAMWI, P. TOUZAIN & C. RIEKEL
Synthetic Metals **2**, 153 (1980)

15 - C. RIEKEL, A. HAMWI & P. TOUZAIN
Synthetic Metals **15**,345 (1986)

16 - B. MARCUS, P. TOUZAIN & A. HAMWI
Carbon **24**, 403 (1986)

17 -   M. GOLDMANN, F. BEGUIN & J. PANNETIER
       unpublished results

18 -   A. PERCHERON-GUEGAN, C.LARTIGUE & J.C. ACHARD
       unpublished results

19 -   A.N. CHRISTENSEN, M.S. LEHMANN & P. CONVERT
       Acta Chem. Scand. **A36**, 303 (1982)

20 -   A.N. CHRISTENSEN & M.S. LEHMANN
       J. Solid State Chem. **51**, 196 (1984)

21 -   A.N. CHRISTENSEN, H.FJELLVAG & M.S. LEHMANN
       Acta Chem. Scand. **A39**, 593(1985)

22 -   A.N. CHRISTENSEN, M.S. LEHMANN & J. PANNETIER
       J. Applied Crystallogr. **18**, 170 (1985)

23 -   C. RIEKEL, H.G. REZNIK, R. SCHOLLORN & C.J. WRIGHT
       J. Chem. Phys. **70,** 5203 (1979)

24 -   C. RIEKEL, H.G. REZNIK & R. SCHOLLORN
       J. Solid State Chem. **34,** 253 (1980)

25 -   C. POINSIGNON, M. ANNE & J. PANNETIER
       unpublished results

26 -   J.M. CRETTEZ, E.COQUET, B. MICHAUX, J. PANNETIER,
       J.BOUILLOT, P. ORLANS, A.NONAT & J.C.MUTIN
       Physica *in press*

27 -   Y. LALIGANT, M. LEBLANC, J. PANNETIER & G. FEREY
       J. Phys. C **19,** 1081 (1986)

28 -   T. BIRCHALL, G. DENES, K. RUEBENBAUER & J. PANNETIER
       Hyperfine Interactions **29,**1331 (1986)

29 -   J.K. COCKROFT, B.E.F. FENDER, A.N. FITCH & G.J. KEARLEY
       ILL-Experimental Report 5.22.270 (1985)

30 -   Y. LALIGANT
       These, Universite du Maine - Le Mans (1986)

31 -   J. RODRIGUEZ, J.M. GONZALEZ-CABRET, J.C. GRENIER,
       J. PANNETIER & M. ANNE
       Solid State Comm. *submitted*

32 -   A.N. CHRISTENSEN & H. FJELLVAG
       ILL-Report 5.25.187 (1985)

33 -   J. PANNETIER
       Chemica Scripta **26**A,131 (1986)

354

34 - J. BOUILLOT, E. COQUET, J. PANNETIER & J-M. CRETTEZ
Physica **B136**, 493 (1986)

35 - G. FEREY, J. PANNETIER, H. DUROY, R. DE PAPE
unpublished results

36 - Y. LALIGANT, G. FEREY & J. PANNETIER
unpublished results

37 - J. PANNETIER
unpublished results

38 - C. TETE, M. VERGNAT, G. MARCHAL & P. MANGIN
Solid State Comm. **53,** 191 (1985)

39 - M. LEBLANC, G. FEREY, J-M. GRENECHE, A. LEBAIL, F. VARRET,
R. DE PAPE & J. PANNETIER
J. Physique **46,** C8-175 (1985)

40 - J. PANNETIER, J.M. DUBOIS, C. JANOT & A. BILDE
Phil. Mag. *in press*

41 - C. JANOT, B. GEORGES, J. PANNETIER & D. BOUMAZOUZA
International Conf. MGS-1986 *in press*

42 - H.R. WENK, H.J. BUNGE, E. JANSEN & J. PANNETIER
Tectonophysics **126**,271 (1986)

43 - H.R. WENK
Private communication

44 - D. JUUL JENSEN & J.K. KJEMS
Textures and Microstructures Vol.5,239 (1983)
Gordon & Breach, London

45 - N. NIIMURA & M. MUTO
J. Phys. Soc. Japan **35,** 628 (1973)

46 - Y. ISHIKAWA, M. ARAI, M. FURUSAKA & M. MERA
KENS report IV, KEK internal 83-4

47 - R.C. OBERTHUR
Rev. Physique Appliquée **19**, 663 (1984)

DISCUSSION

QUASI-CRYSTALS

It is hoped that the preparation of a single quasi-crystal
from the Li-Al-Cu system will soon resolve the controversy about
the existence of such materials.

REAL TIME STUDIES

The adequacy of the total flux is the main factor in deciding which
technique or techniques can be applied to any particular problem.
For the smallest samples and the shortest time scales only the very
high flux of synchrotron sources will be sufficient, but there are
many studies for which neutron beams are appropriate, and sometimes
advantageous. Neutron instruments are convenient in flux, resolution
and data acquisition time for many thermally induced phase changes
and chemical reactions, especially when sensitivity to hydrogen atoms
is important. The greater penetrating power of neutrons compared
with X-rays facilitates the use of complex environments, as in a
recent study by Dr Pannetier of the charge and discharge of a battery,
when the whole of a specially constructed cell was placed within
the neutron beam.

It should be possible to study chemical reactions between gas and
solid phases by either neutrons or X-rays, perhaps with some degree
of surface sensitivity in the latter case.

INTERMEDIATE STAGES IN PHASE TRANSFORMATIONS

The extent to which diffraction gives information about the
intermediate stages of the transformation of one phase into
another depends on the example chosen. In the dehydration of the
WO_3 hydrate described above line broadening indicates the mechanism
of change, but in other neutron work nothing is seen, perhaps
because the resolution is insufficient. In synchrotron studies of
two protein transformations, Dr. Harding reported that the protein
crystal first appeared to become rather disordered, then the
ordering improved as the change was completed (time scale
20-60 min).

CHEMICAL CRYSTALLOGRAPHY WITH PULSED NEUTRONS AND SYNCHROTRON RADIATION

H.B. Stuhrmann,
GKSS - Research Centre
Department of Macromolecular Structure Research
Max - Planck - Straße
D - 2054 Geesthacht
F. R. Germany

ABSTRACT. The pulsed structure of synchrotron radiation and thermal neutrons enters into chemical crystallography by two ways. Nuclear resonant scattering may profit from the time structure of synchrotron radiation yielding more detailed information about molecular structure, whereas time resolved experiments reveal their dynamics. As the time of flight techniques of neutron scattering and time resolved scattering experiments are covered elsewhere, the impact of Mößbauer methods will be discussed here in more detail. Nuclear resonant scattering may gain considerable importance in macromolecular structure research, as some nuclei have large resonant scattering lengths. The potential of nuclear resonant scattering is compared with some recent developements of anomalous X-ray scattering and nuclear spin dependent neutron scattering.

1. PULSED SYNCHROTRON RADIATION

For fast time domain measurements it is important to know the detailed bunch shape during an experiment either as a function calculable from measured machine parameters or directly from diagnostic devices.

The pulse structure of synchrotron radiation is a copy of the closed sequence of electron "bunches" stored in a high energy ring accelerator. The radio frequency (rf) fields used to replenish the emitted synchrotron radiation energy have an accelerating action on the charged particles only during half of their period and deaccelerate during the other. In this way "phase focussing" of the electrons occurs so that they are gathered in "bunches" which run in phase with the accelerating part of the field. The length of this bunch is typically 1/10 of the rf-wavelength. The number of bunches which are distributed around the whole circumference is an integer, because the rf-period has to be a higher harmonic of the fundamental orbital period. Not all of these "buckets" have to be filled with bunches of electrons. DORIS, for example, has 480 buckets, SPEAR has 280. The total current in a machine is , of course, smaller if only one bucket is filled. It can increase at most with the square root of the number of bunches.

M. A. Carrondo and G. A. Jeffrey (eds.), Chemical Crystallography with Pulsed Neutrons and Synchrotron X-Rays, 357–378.
© 1988 by D. Reidel Publishing Company.

358

Fig 1 shows the pulse structure of DORIS on various time scales: 120
bunches are filled and the repetition rate of one bunch at a fixed point
of the orbit is 1 µs, each with a width of 0.15 ns. It is important to
mention that the pulse length can further be decreased to the ps range.
This feature is taken into account as an option for the beam optic when
storage rings are designed as synchrotron radiation sources.

If a storage ring runs in the single bunch mode, or if in general the
bunch occupation number is smaller than the maximum possible value, one
should for time dependent studies, always keep in mind that some of the
"empty" buckets might have restfills of 10^{-3} of the maximum. Leakage of
particles during injection and from one bunch into others can be
responsible for it.

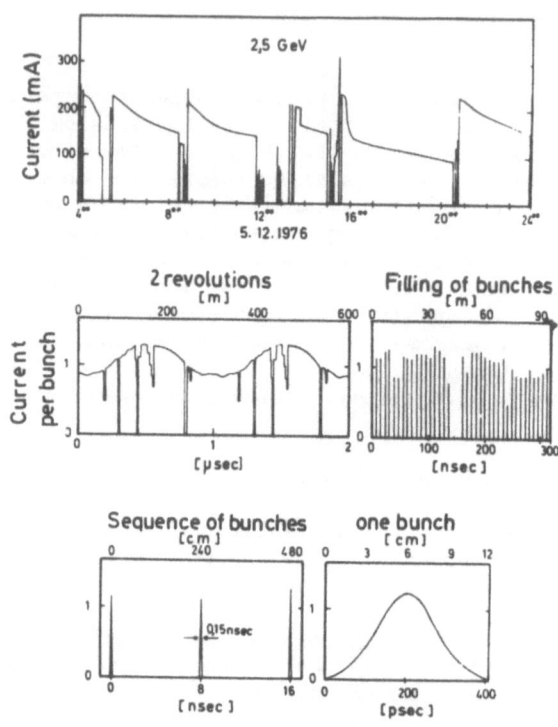

Fig. 1 Time structure of DORIS on various time scales.

In a synchrotron the time structure due to the rf-field and orbital
frequency is roughly the same as in a storage ring. Added is the
injection frequency (50 Hz at DESY). During the first 10 ms of one cycle
the particles are speeded up to maximum energy. Then they are ejected
from the ring into an experiment or into a storage ring. The spectrum
which is emitted at a certain time of the acceleration process is a
function of the momentary energy. If an experiment views the optical
part of the spectrum, it receives photons for a longer time than another

instrument which detects X-ray photons. - In conclusion the duty factor should be mentioned. The peak intensity from one bunch is much higher than the average intensity. If t_b is the width of one bunch and t_0 the repetition time of one pulse, average intensity I_a and peak intensity I_p are related for n bunches by

$$I_p = I_a \; t_0 \, / \, t_b \qquad\qquad (1)$$

For DORIS the peak intensity may be higher by 1000 / .15 = 6600

2. FREE ELECTRON LASER IN THE X-RAY REGION

The coherence properties of the eleectromagnetic radiation emitted by electrons in an undulator agnetic field can be exploited to obtain laser action, by reflecting the bremsstrahlung light rays back into the elec- tron beam by means of mirrors (1). The most important distinction between a Free Electron Laser (FEL) and an ordinary laser is the fact that the former does not depend on the principle of optical pumping of atomic levels, and subsequent stimulated emission, which is the basis of ordinary lasers. So far lasers have not been demonstrated to be feasable in the X-ray region because the lifetimes and cross sections of transi- tions involving K atomic levels are not favorable for laser action.

On the other hand a free electron laser is not limited by this kind of restriction: it would appear that if the parameters(magnetic field, radii of curvature, elctron energy, etc.) are suitable chosen, and mirrors with high reflectivities (say, > 0.9) can be built, nothing prevents the realisation of a laser in the X-ray region (λ = 2 - 3 Å).

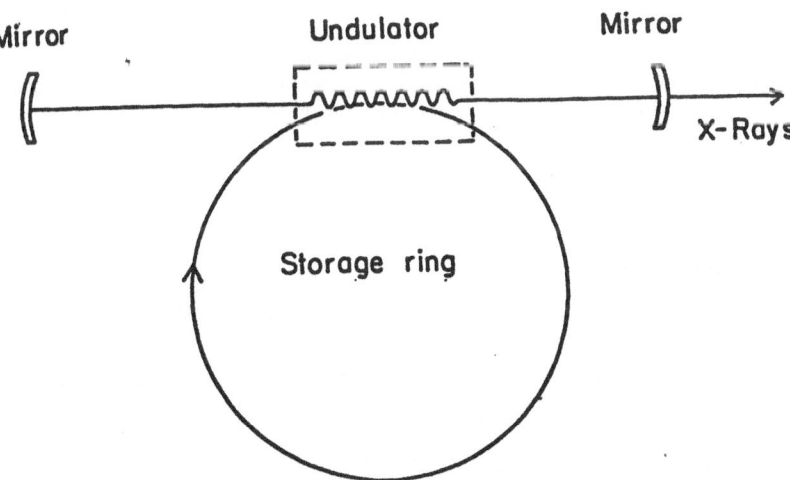

Fig. 2 . Schematic view of a Free Electron Laser (see text).

A common scheme for FEL makes use use of a straight section of an
electron storage ring (Fig. 2). The laser action is confined within the
cavity delineated with a dotted line, and the two spherical mirrors
reflect back into the cavity the light beams emitted in the forward
direction by the elcetrons. For continuous laser amplifiction up to
saturation it is necessary to keep the gain in the lasing free electron
"medium" and the reflectivity of the mirrors as high as possible.
Starting from not too unrealistic assumptions Collela and Luccio (2)
show that the power available from an X-ray FEL is 250 watts on average,
with a peak value of 4 10^6 watts. The intensity of synchrotron radiation
pulses emitted by an X-ray laser would be stronger by several orders of
magnitude.

3. PULSED THERMAL NEUTRONS

Pulsed neutron beams are produced either by
-      chopping a continous neutron beam from a reactor, or by
-      spallation of very heavy nuclei (e.g. uranium) by pulsed high
        energy (1 GeV) protons (3)
Spallation neutron sources are at Rutherford Appleton Laboratory
(England) and at Los Alamos (USA). The plan of a very powerful pulsed
neutron source has been worked out by the SNQ project group at KFA
Jülich (F.R. Germany)(4). We refer to these results in Tab 1 and compare
them with the pulse characteristics of synchrotron radiation from DORIS
at DESY Hamburg

The pulse length of proton bunches emerging from the linear accelerator
may be of the order of some tenth of a second. A 5 mA proton current (50
Hz, 0.35 s pulse duration) would produce a neutron flux comparable to
that produced by the high flux reactor of the Institut Laue Langevin
(ILL) at Grenoble. The difference however is that this mean flux would
be produced by the SNQ in bursts with a peak-to-average ratio of 40. The
moderation of the high energy neutrons leaving the target stretches the
pulse of thermal neutrons to 0.5 ms and that of cold neutrons to about
0.7 ms. (Fig. 3a)

Fig. 3 a
Time dependence and
relative intensities
expected from the
neutron pulses of the
room temperature H_{2O}
moderator and the cold
D_2 source with
graphite reflector.
The peak to average
flux ratio is 35 and 20.
From (4).

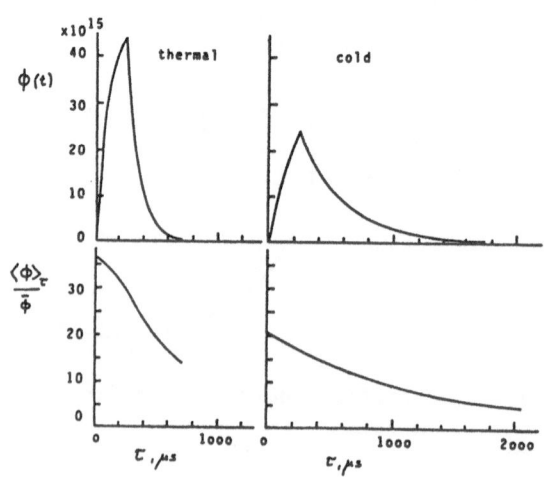

Tab. 1 . Pulsed neutrons and synchrotron radiation

	synchrotron radiation from DORIS at DESY, Hamburg	thermal neutrons from SNQ (planned)
pulse length (mm)	50	500 - 2000
pulse duration	140 ps	0.3 - 1 ms
photons (neutrons) per pulse ($\Delta\lambda / \lambda$) = 1.5 A	10^6 (.001)	10^6 (.02)
repetition rate single bunch: 480 bunches :	1.04 MHz 500 MHz	50 Hz
typical flux at the sample (cps) ($\Delta\lambda / \lambda$) = 1.5 A	10^{12} (.001)	$5\ 10^7$ (.02)

The pulse characteristics of DORIS compare to those encountered at other
electron storage rings (see also (5))
- VEPP 2M and VEPP 4 at Novosibirsk, USSR
- SRS Daresbury (England)
- Photon factory at Tsukuba (Japan)
- SPEAR II at Stanford and CESR at Cornell (USA)

Slightly longer pulses (10 to 50 cm length) are produced by the storage
rings - Adone at Frascati (Italy)
- DCI at Orsay (France)
- VEPP 3 at Novosibirsk (USSR)
- NLS at BNL Brookhaven (USA)

4. HOW TO USE PULSED RADIATION IN DIFFRACTION EXPERIMENTS

An important difference in the use of the pulsed nature of synchrotron
radiation and neutrons is due to the fact that the latter have a mass.
The momentum of a particle travelling at a speed v behaves as if it were
a wave with the wavelength λ :

$$m \vec{v} = h \vec{k} \qquad \text{with} \quad k = 2\pi/\lambda \qquad (2)$$

$\vec{k}$ is the wave vector. For neutrons we have

$$\lambda(A) = 3.96 / v\ (km/s) \qquad (3)$$

Neutrons emerging from a moderator at room temperature have a nearly
Maxwellian velocity distribution centered at v = 3.6 km/s which
corresponds to λ = 1.1 A. As neutrons of different wavelength travel at

different speeds an originally compact bunch will be stretched out in
the flight direction. Short wave neutrons will reach the detector
earlier than the long wavelenth neutrons. Taking diffraction patterns at
short enough intervalls will allow to use the whole spectrum of thermal
neutrons for diffraction studies. The gain of a pulsed neutron source
with respect to a continuous source (averge intensity being equal) is
between one and two orders of magnitude. This aspect is discussed in
more detail by H. Fueß (6) (this course).

Synchrotron radiation is propagating at the speed c of light

$$c = \nu \lambda \qquad\qquad (4)$$

ν is the frequency.
There is no dispersion of c in the vacuum. The time-of-flight method
described above cannot be applied with pulsed synchrotron radiation.

However the pulsed structure of synchrotron radiation offers unique
advantages for running drift chambers as position sensitive detectors.
It is the time-of-flight of photoelectrons which can be used to locate
diffracted photons (last section of this talk).

Common aspects of the use of pulsed neutrons and synchrotron radiation –
at least in principle – are encountered in time resolved experiments for
the study of structural and spectroscopic changes in molecules. A
technical difference arises from the time scale (see Tab. 1). A major
impact van be expected from X-ray Free Electron Lasers (XFEL) once they
will become available.

5. TIME RESOLVED EXPERIMENTS

The amplitude A (or geometrical structure factor) of a molecular
structure is given by

$$A(Q) = \sum_n b_n \exp(i\, \vec{Q} \cdot \vec{r}_n) \qquad\qquad (4)$$

The intensity S(Q) is given by

$$S(\vec{Q}) = |A|^2 \qquad\qquad (5)$$

For single crystals S(Q) appears in the points of the reciprocal
lattice. For crystalline sheets and fibres the reciprocal lattice
degenerates to a regular arrangments of rods and planes respectively. In
the case of random distribution of molecules (e.g. dilute solution) the
averaged intensity I(Q)

$$I(Q) = \frac{1}{4\pi} \int S(\vec{Q})\, d\Omega \qquad\qquad (6)$$

Time resolved scattering experiments may aim at
– structural changes; these will influence the atomic coordinates $\vec{r}_n$

- spectroscopic changes with concomitant change of the scattering
  length b_n of a selected species of atoms.

Both synchrotron radiation and pulsed neutron sources deliver pulses of similar strength. In diffraction experiments not more than one out of a hundred incident photons will be scattered. The number of scattered photons from a puls crossing the crystalline sample is of the order of 10 000 at most. This is usually not sufficient to derive any structural information from the scattering pattern. The experiment would have to be repeated several times in order to obtain a well-defined diffraction pattern. The quality of the data must be good enough in order to monitor changes of the scattering pattern which then can be analysed in terms of structural changes.

What is the aim of time resolved scattering experiments and how could it profit from the pulse structure ?

Time resolved X-ray scattering is preferred if the structural relaxation (e.g. macromolecular rearrangement, self assembly, etc.) cannot be monitored by the easier spectroscopic methods in the region of visible light or near UV.

The creation of a non-equilibrium state should be very fast. It should need less time than the rate of reaction or relaxation.
It should also occur within the duration of the pulse. Major differences between the application of synchrotron radiation and neutron scattering are to be expected at this point (see Tab. 1)
The creation of a non-equilibrium state should be efficient. That is the density of excited molecules should be high.
The creation of a non-eqilibrium state should be homogeneous in the whole sample volume.

There are various ways to meet the above conditions:

Mixing of two reactants. - This is a most conveniently done in solution studies. The mixing time can be as short as some milliseconds. - For crystals of biological structure the high water content allows the penetration of the reactant into the crystal. The creation of the non-equilibrium state then is diffusion controlled. This process is much slower than direct mixing methods.
Other perturbation methods:
Temperatures can be changed over several degrees and stabilised within one second. - The propagation of a pressure jump is very much faster. High repetition rates could be achieved by sound waves.

Flash light. - This method is most favorable for the use of pulsed synchrotron radiation. The flash light source could be (Tab. 2) (5)
- synchrotron radiation
- lasers
- other (incoherent) light sources

Tab. 2   Comparison of pulsed sources (5)

	synchrotron radiation	lasers	incoherent sources
wavelength range	.03 nm – 1 cm	tunable over narrow ranges in ultraviolet and visible; some lines below 200 nm and many in the infrared	wide variety of sources are needed to cover the range from 15 nm to radio frequencies
intensity (number of photons) per pulse within a 0.1% wavelength band	10^9	10^{10} in 1 ps pulse of width 10 cm^{-1}	10^6
minimum approximate pulse duration	100 ps	0.2 ps	1 ns
pulse repetition	1 – 500 MHz	dc to 100 MHz	dc to 100 MHz
source	incoherent	coherent	incoherent
source size	1 mm^2	1 mm^2	few mm^2
beam divergence	10 mrad	5 mrad	isotropic

For the observation of structural changes by diffraction at a repetition rate of 1 MHz as it is given by synchrotron radiation see H. Bartunik (this course)

Many of the above mentioned methods of creating non-equilibrium states of the sample which have become very popular in the millisecond region of time resolved scattering experiments (compare lecture of C. Riekel) match the time scale of pulsed neutron sources. They are not convenient for experiments in the submicrosecond region. For the use of pulsed synchrotron radiation pulsed light sources (Tab. 2) will probably offer the only way of efficient sample perturbation.

6. MÖSSBAUER DIFFRACTION.

This method has been proposed by F. Parak and it had been started at the EMBL Outstation at DESY, Hamburg in the late seventies. The preparative steps are documented in the thesis of C. Hermes (1981)(8). The aim is the direct measurement of the iron diffraction pattern in host structures, e.g. iron in hemoglobin by recoil-less nuclear resonance scattering.

This method in many respects is very similar to nuclear resonance scattering of neutrons and to anomalous scattering of X-rays by atomic electrons. The scattering length b described by

$$b(\lambda) = b_0 + b'(\lambda) + i\,b''(\lambda) \qquad (8)$$

where b" is the imaginary part of the scattering length (a wave with a phase shift of 90°). The real resonant part b' can be obtained from the wavelength dependence (or dispersion) of b" by the Kramers-Kronig relation

$$b''(\omega) = \frac{2}{\pi} \int \frac{\omega'\,b''(\omega)}{\omega^2 - \omega'^2}\,d\omega' \qquad (9)$$

The dispersion b" is the same as that of the absorption in the case of X-ray scattering. The optical theorem tells us

$$\acute{6} = 2\,\lambda\,b'' \qquad (10)$$

The result of an interaction of a photon with atomic electrons very much favours photoelectric absorption. Only a fraction of around 0.005 of the incident X-rays is scattered. With Mößbauer scattering from ^{57}Fe this fraction is much higher (about 0.1, see Table 3). This is also the reason why the amplitude of nuclear resonance scattering of ^{57}Fe is higher than in any other resonance scattering process of interest for chemical crystallography (9).

Given a structure $\rho(\vec{r}) = u(\vec{r}) + (b' + ib'')\,v(\vec{r})$ then the dispersion of resonant diffraction is described by four terms which differ in their dispersion (10):

$$
\begin{aligned}
S(\vec{Q}) = & & \int\int & u(\vec{r})\,u(\vec{r}')\,\cos[\vec{Q}(\vec{r}-\vec{r}')]\,dv\,dv' \\
& + 2b' & \int\int & u(\vec{r})\,v(\vec{r}')\,\cos[\vec{Q}(\vec{r}-\vec{r}')]\,dv\,dv' \\
& +(b' + b'') & \int\int & v(\vec{r})\,v(\vec{r}')\,\cos[\vec{Q}(\vec{r}-\vec{r}')]\,dv\,dv' \\
& b'' & \int\int & [u(\vec{r})v(\vec{r}')-v(\vec{r})u(\vec{r}')]\,\sin[\vec{Q}(\vec{r}-\vec{r}')]\,dv\,dv'
\end{aligned}
$$

$$= S_0(\vec{Q}) + b'S_{uv}(\vec{Q}) + (b' + b'')S_v(\vec{Q}) + b''\psi(\vec{Q}) \qquad (11)$$

The overall effect of resonant scattering is to cause the break down of Friedel's law, so that the Bijvoet pairs of reflections $S(\vec{Q})$ and $S(-\vec{Q})$ are unequal (11). The difference

$$S(\vec{Q}) - S(-\vec{Q}) = 2b''\psi(\vec{Q}) \qquad (12)$$

The sum of the Bijvoet pairs (as it appears with twinning in single crystals or crystalline powders) is

$$\frac{1}{2}[S(\vec{Q}) + S(-\vec{Q})] = S_0(\vec{Q}) + b'\,S_{uv}(\vec{Q}) + (b' + b'')\,S_v(\vec{Q}) \qquad (13)$$

This is also the dispersion of randomly oriented maromolecules in solution and of gas molecules. As we are going to compare the efficency of resonant labels with those which rely on nuclear spin ordering, we

just show the result of non-resonant labels. The imaginary part b" then
has to be omitted in Eq. 13. b' becomes the change of scattering length
due to isotopic subsitution or change of spin density in neutron
scattering (see Eq. 13a)

Tab. 3. Mößbauer data of the first excited state of ^{57}Fe.

energy	E (keV)	14.41303
width	(ev)	4.67 10^{-9}
halftime	$t_{1/2}$ (ns)	97.81
conversion coefficient δ		8.21
scattering factor ratio	$(f_n/f_e)^2$	410

(from C. Hermes, 1981) (8)

In the very near vicinity of the recoil-less nuclear resonance energy of
^{57}Fe resonance absorption of X-ray photons leads to an excitation of the
^{57}Fe nucleus and subsequent recoil-less nuclear resonance scattering or
Mößbauer scattering (12). Contrary to all other X-ray scattering processes the emission of the γ -ray is slightly delayed due to the halftime
of nearly 0.1 μs of the excited state of the iron nucleus. The decay
time of the excited iron nucleus falls well within the repetition time
of pulsed synchrotron radiation of typiclly 1 μs. The direct observation of the delayed photons would be possible if the detector could be
conveniently gated. The scattering of the delayed photons would be
entirely due to (b' + b") $S_v(Q)$ in Eq. 13

Considering the intensity of synchrotron radiation of 10^{11} photons /s/eV
only 10 to 100 photons would have a chance to be seen as delayed photons
diffracted by the iron atoms in a (macro)molecular structure. From a
hemoglobin crystal perhaps one count per minute could be expected. - The
simple experiment mentioned above is not feasable because
- the sensitivity of the detector cannot be changed by 14 orders of
magnitude a million times per second, and even if so
- the distribution of electrons in the storage ring cannot be completely
restricted to one 'bucket' of the storage ring. About 10^{-3} of the
electrons are circulating in other parts of the orbit contaminating the
pulse structure of synchrotron radiation with many very small, unequal
intermediate pulses.

In a first step of monochromatisation a double crystal monochromator is
used. The energy width of the monochromatic radiation is about 10 eV
which is 9 orders of magnitude more than the width of the Mößbauer line
of ^{57}Fe. The use of the pulsed structure for separating the Mößbauer
photons so far was not yet succesful for the reasons mentioned above.

The ultramonochromatisation was finally achieved by using pure nuclear
reflections of ^{57}Fe-YIG by E. Gerdau in 1984. (13).

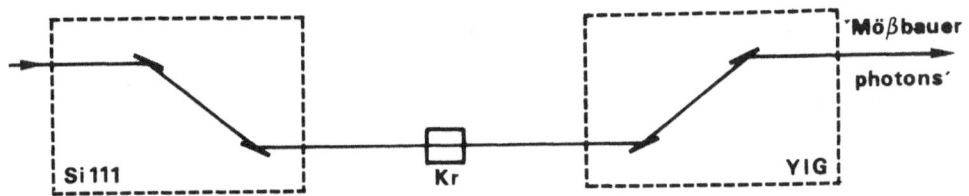

Fig.3 b. Extraction of the Mößbauer photons out of the 'white' synchrotron radiation spectrum. The K-absorption edge of Kr is used for calibration.

The usable countrate of Mößbauer diffraction is about 1 cps. The background amounts to about 0.3 to 0.5 cps. The time resolution of 1 to 2 ns is sufficient to resolve 'quantum beats' due to excitation of nuclear levels in YIG (Fig. 4). In this case the wavevector of the incident γ-radiation is nearly parallel to the internal magnetic field B and the Bragg angle was $\Theta_b = 4°$. Gerdau et al. (14) show that an explanation can be given in the kinematic approximation. The fast quantum beats with a period of 11.5 ns are due to magnetic hyperfine splitting of the 14.4 keV energy level of the ^{57}Fe nucleus by the d-levels of YIG. The slow modulation comes from electric hyperfine splitting of the d-levels which contribute to scattering. The collective width Γ leads to damping by $1 / \Gamma = \tau = 45$ ns, which is considerably smaller than $\tau = 140$ ns of the unperturbed nuclear level. An exact interpretation can only be given by the dynamic diffraction theory.

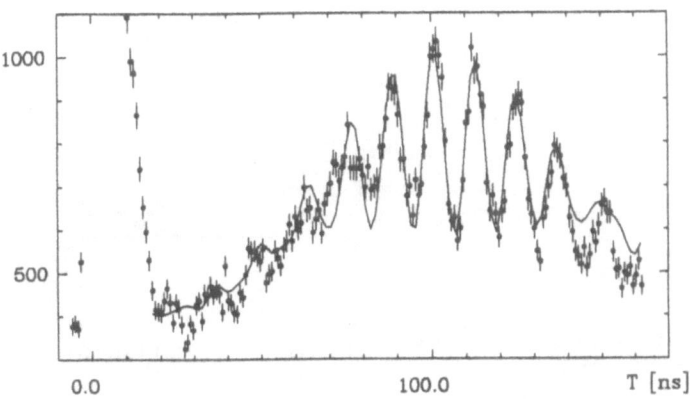

Fig. 4. Time structure after two YIG crystals (from E. Gerdau(14))

368

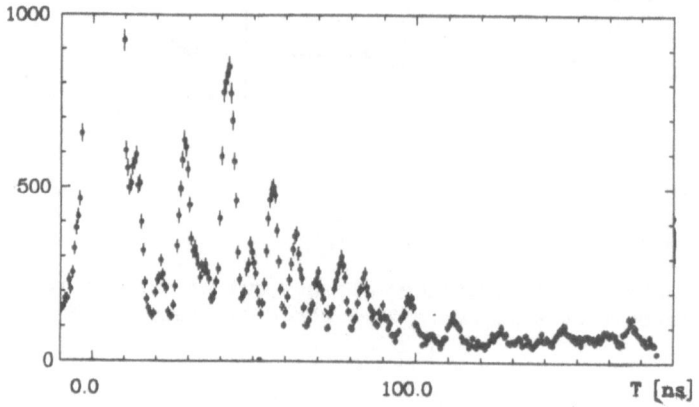

Fig. 5. Time structure after one YIG crystal (from E. Gerdau (14))

a) in Bragg: $\theta = \theta_B$

$T_{172} = 12$ ns

b) above Bragg: $\theta > \theta_B$

$T_{1/2} = 18$ ns

c) below Bragg: $0 < 0_b$

$T_{1/2} = 18$ ns

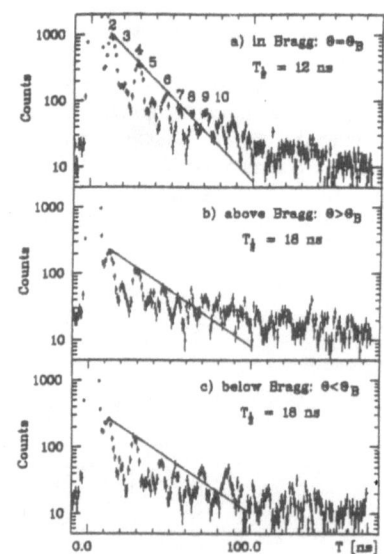

Fig 6. Time spectra after one $^{57}FeBO_3$ crystal (111 reflection) at different angles 0 [15].

The precision in the determination of hyperfine parameters could be increased by one to two orders of magnitude compared to conventional

Mößbauer spectroscopy (14). The speed up of the coherent decay in
nuclear resonant diffraction of the synchrotron radiation could be shown
more clearly in the case of $^{57}FeBO_3$. This Mößbauer filter has been
studied in order to find better conditions for the study of decay times.
Further interest in an $^{57}FeBO_3$ filter arises from the possible use as a
broadband filter at high angular collimation and a single-line filter at
the Néel temperature. The nuclear resonant diffraction of this compound
has been studied for many years using conventional ^{57}Co sources (15).

The time spectra in Fig. 6 show a nearly exponential decay of the
reflected intensity which is modulated be fast quantum beats. The decay
has a halftime of 12 ns at the Bragg position and of 18 ns at 7" above
and below the Bragg position. These half times have to be compared with
the half life time of the individual nucleus , which is 98 ns. Earlier
experiments gave a rather indirect indication by the observation of an
angular dependent broadening of the Mößbauer line. This is the first
direct observation of the speed up of coherent decay in resonant
diffraction of $^{57}FeBO_3$ (15).

7. COMPARISON WITH ANOMALOUS SCATTERING OF X-RAYS

Although the relative contribution to the total scattering intensity is
smaller than with nuclear scattering anomalous dispersion of X-ray
scattering near the ionisation energies of inner shell atomic electrons
is receiving more attention. This is so as the required accuracy of the
data is readily obtained with electronic position sensitive counters,
and this method can be applied to the vast majority of all chemical
elements. The technical difficulties with lighter elements which have
their absorption edges in the soft X-ray region are more and more
overcome by the construction of new types of diffractometers (Fig. 7).

The diffractometer shown in Fig. 7 provides rapid tunability over a wide
range of wavelengths. With θ = 7 mrad the first gold coated mirror
reflect wavelenghts from about 1.2 Å onwards. The two halves of the
second mirror surface have very different electron densities due to gold
and quartz. The reflectivity of the latter starts at wavelenghts around
3 Å. Lateral displacement of this mirror will taylor the spectrum to the
needs of monochromatisation by the single crystal monochromator over a
wide range of wavelengths extending from 1.2 to 8 Å. As the peak
reflectivity of the crystals decreases significantly at longer X-ray
wavelengths − it drops to about 0.15 at λ = 6 Å from 111 plane of Ge −
only one crystal is used. This means that the diffractometer has to
turned around the axis of the crystal monochromator. The rotation in the
vertical plane is preferred in order to avoid losses due to polarisation
of synchrotron radiation from a bending magnet section of the electron
storage ring. This instrument extends the use of resonant X-ray
scattering to the near soft X-ray scattering experiments to the near
soft X-ray spectrum (Fig. 8).

370

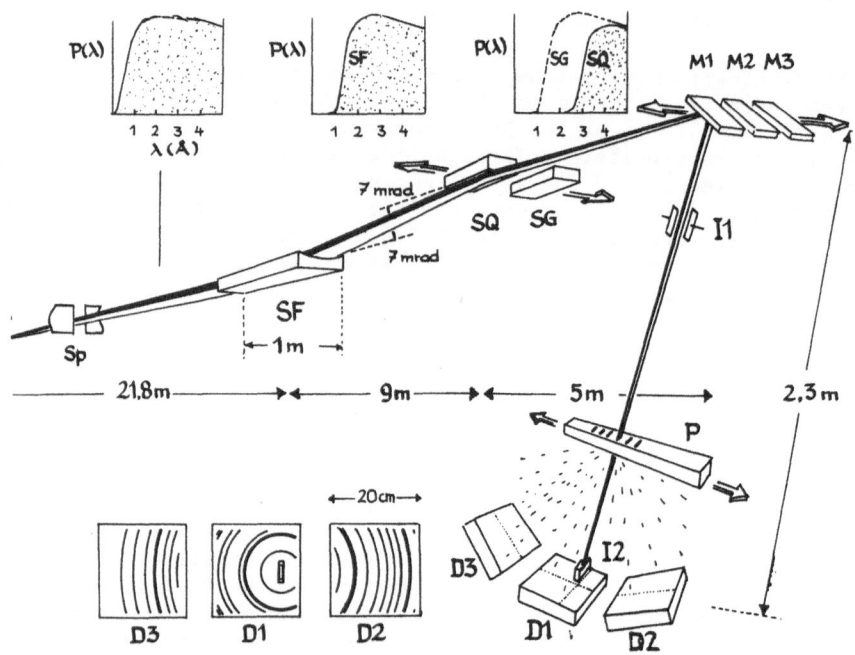

Fig. 7. Double mirror – single crystal optics for soft X-rays. SF = gold
coated double focussing mirror. SQ = plane quartz mirror. One half of it
is coated with gold (SG). Lateral displacement of the second mirror
changes the transmitted spectrum P(λ) as indicated in the third upper
insert. The incident spectrum and the spectrum after reflection by the
gold surface of the second mirror are shown in the first and the second
upper insert respectively. M1, M2, M3 are the crystal monochromators
which can be used alternatively. The following monochromator crystals
are used   111 plane of Ge   2 d = 6.53 A
                    111 plane of Si   2 d = 6.28 A
                    111 plane of InSb   2 d = 7.48 A
P = sample exchanger (driven by a stepping motor over an interval of 12
cm). I1, I2 = ionisation chambers (reduced pressure of air). D1, D2, D3
are position sensitive counters from A. Gabriel and F. Dauvergne
(European Molecular Biology Laboratory , Outstation at Grenoble). The
lower inserts show the diffraction pattern of bacteriorhodopsin as it
would appear on the three detectors using 5 A photons. The instrument is
installed at beam A1 of HASYLAB. Part of the data aquisition system is
due to C. Boulin (EMBL, Heidelberg)

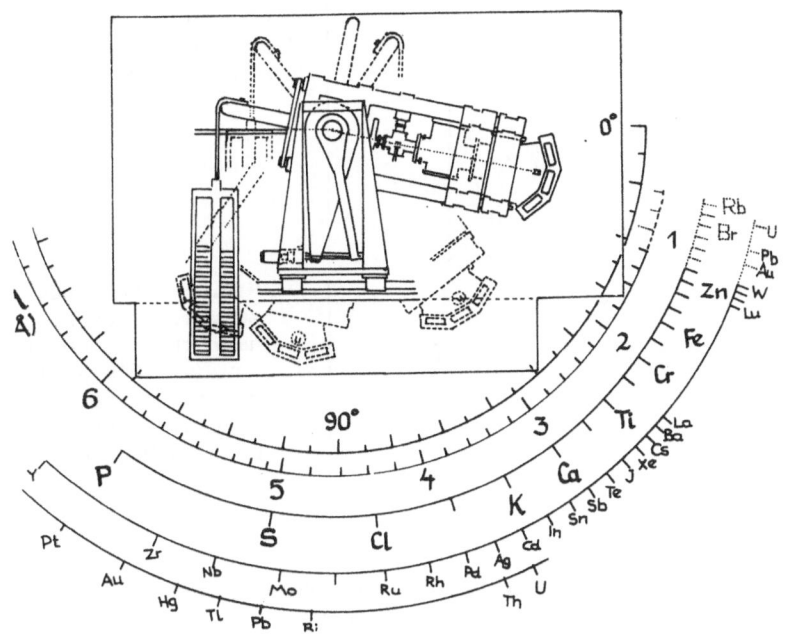

Fig. 8. The wavelength range of the diffractometer at the beam line A1
of HASYLAB. Using the 111 plane of a germanium single crytal wavelengths
up to 6.4 Å can be reached by rotating the crystal by θ = 80° and the
camera by 2 θ = ·160°. The transmitted wavelength is given by Bragg's
equation

$$n \lambda = 2 d \sin \theta$$

Some 2d values of frequently used reflecting lattice planes of
monochromator crystals are given in Fig. 7. Thus, the K-absorption edges
down to phosphorus (Z = 15), the L_3 absorption edges down to Yttrium (Z
= 39) and the M_5 absorption edges down to iridium (Z = 78) can be used.
- Replacing the germanium crystal by indium antimonide as monochromator
extends the wavelength range to 7.3 A including the K absorption edge of
silicon (Z = 14) as well.

The distance between sample and detector may vary from 30 cm to 125 cm.
At short wavelengths and at very long wavelengths, when the camera is
not too far from the horizontal direction, the distance between sample
and detector may be enlarged to 185 cm. Taking into account the vertical
height of the beam stop of 12 mm the accessible range of scattering
angles extends from .3° to 60°. The counterbalance as shown on the left
side is needed to rotate the camera safely in steps if 1.4 ".

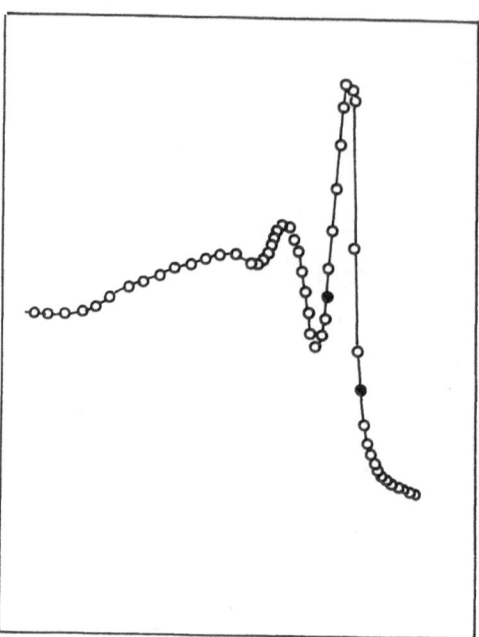

As a recent result we mention the diffraction studies of bacteriorhodopsin at wavelengths near the K-absorption edge of sulfur.

Fig. 9: The X-ray absorption edge of sulfur near the K-edge at λ = 5.018 A. The full points indicate the two wavelelengths which were used by Munk et al (16)

The seven sulfur atoms of this membrane protein give a 8% decrease of diffraction in the 1,1 reflection (Fig. 10)

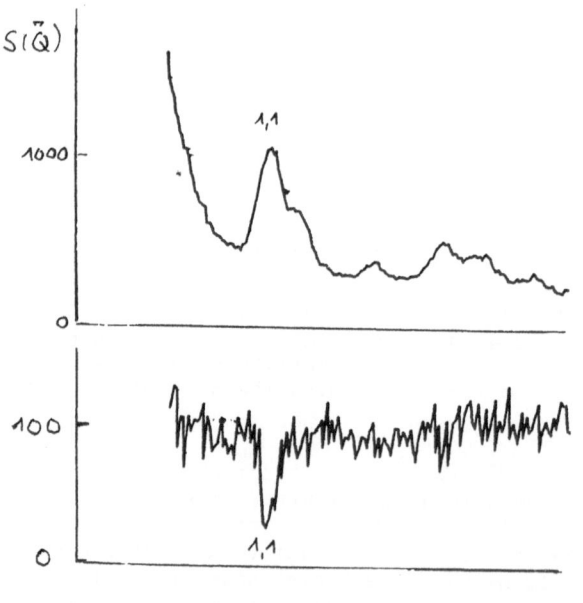

Fig. 10. Averaged diffraction pattern of purple membrane (16)

a) diffraction of bacteriorhodopsin at λ = 5 Å

b) difference diffraction intensity taken at wavelengths slightly off the main peak of K absorption 'edge' of sulfur as indicated in Fig. 9

The contribution of anomalous scattering to the intensity pattern of purple membrane is most clearly seen in the 1,1 reflection (16). More recently the same group could measure the anomalous diffraction of purple membrane to higher resolution. – In addition, the discrimination between anomalous diffraction of sulfur in methionin of the protein and sulfur in sulfonic acids of the lipid matrix has been achieved as well. This became possible as there is a substantial shift of the absorption edge of sulfur depending on wether it is encountered in $-C-S-C-$ or $-SO_3H$ (unpublished result of B. Munk et al.)

The data have been taken with a multiwire proportional counter which has been built by A. Gabriel and F. Dauvergne at EMBL.

8. COMPARISON WITH NUCLEAR SPIN DEPENDENT NEUTRON SCATTERING.

This method relies on the spin dependence of the interaction of neutrons with hydrogen nuclei and with protons in particular. As no energy transition is involved the imaginary part in Eq. 10 has to be skipped. A change of the coherent scattering length b is brought about by changing the polarisation $\vec{n}$ of the neutron spins and the polarisation of the proton spins $\vec{P}$. Eq. 13 then reduces to (17)

$$S(Q) = S_u(Q) + \vec{n}\cdot\vec{P}\, S_{uv}(Q) + P^2\, S_v(Q) \qquad (13)$$

Changing the direction of the neutron spins inverts the sign of $\vec{n}\cdot\vec{P}$ $S_{uv}(Q)$. The difference between measurements taken at opposite neutron spin direction yields

$$S(\uparrow\uparrow) - S(\downarrow\uparrow) = 2\; \vec{n}\cdot\vec{P}\, S_{uv}(Q) \qquad (14)$$

The sum of these data corresponds to unpolarised neutron scattering from a polarised target:

$$\frac{1}{2}[\; S(\uparrow\uparrow) + S(\downarrow\uparrow)\;] = S_u(Q) + P^2\, S_v(Q) \qquad (15)$$

Due to the progress of proton spin polarisation made in high energy physics laboratories , especially at CERN (Geneva), this method of scattering length variation has become feasable in macromolecular structure research. Recently a series of polarised neutron scattering experiments have been carried out at the neutron reactor of GKSS Research Centre at Geesthacht in collaboration with ILL and CERN. So far the targets were frozen solutions of proteins, tRNA and the large subunit of ribosomes in deuterated solvent. Proton spin polarisations up to 77 % could be obtained by dynamic nuclear polarisation in the presence of an organic Cr(V) compound at T = 0.4 K and H = 2.5 Tesla. As an example we show the data obtained from bovine serum albumin at a proton polarisation of 63% (Fig. 11).

The drastic changes in small angle scattering demonstrate the power of this method. The inversion of the spin direction of the incident

374

polarised neutron corresponds to a change of the scattering length of
the protons at 63 % polarisation by $2.8 * 0.63 * 10^{-12}$ cm $= 1.76 * 10^{-12}$
cm which corresponds to the scattering length of 6.5 electrons in X-ray
scattering. Isotopic substitution of ^{1}H by ^{2}H (=D) leads to a change of
the scattering length by $1.04 * 10^{-12}$ cm only. Nuclear spin dependent
scattering is the most sensitive way to look at hydrogen in
(macro)molecular structures. This method is not only important for the
in situ structure determination of ribosomal proteins where the contrast
can be enhanced by a deuterated matrix, but it also can be applied to
any other system where the proton density distribution is sufficiently
inhomogeneous.

Fig. 11
Polarised neutron scattering
by the dynamic polarised
protons of bovine serum
albumin dissolved in a
mixture of heavy water and
deuterated glycol.

A proton polarisation of
63 % was achieved after
2 hours 4 mm microwave
irradiation in the
presence of Cr(V) at
T = 0.35 K and a
magnetic field H = 2.5 T

The upper curve is ob-
tained when the spin
direction of the incident
polarised neutron is anti-
parallel to the target polarisation. The lower curve
reflects the decreased
contrast between the protein
and the solvent when the
spins of protons and neu-
trons are parallel. (18)

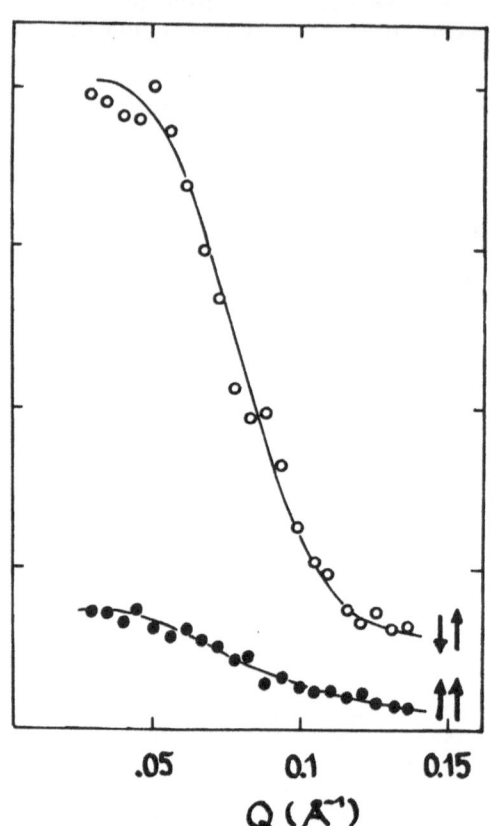

9. A DRIFT CHAMBER TYPE DETECTOR FOR SYNCHROTRON RADIATION

Finally we shall turn to the use of the pulsed structure of synchrotron
radiation in position sensitive counters. It is another example of the
efforts of EMBL to make use of the single bunch mode operation of the
storage ring DORIS at EMBL in the years 1978 to 1981.

The duty cycle for the emission of X-rays at DORIS operated in the single bunch mode is 10^{-4}. With intense scatterers the probability of more than one photon being detected co-incidentally within the short duration of the bunch will become very high. Single photon detectors with processing times much shorter than the bunch period of 1 µs would not change this situation, since multiple events will occur within 100 ps.

Following an idea f J. Hendrix (EMBL)(19) , the time structure of the radiation of the storage ring operated in the single bunch mode can be turned into an advantage. It is known when photons can arrive at the detector. This additional information is used by the high energy physicists in the principle of the drift chamber. Hendrix and Fürst (19) used this principle to design a new type of position sensitive X-ray detectors (Fig. 9). The detector is of circular form, consisting of a drift space with a central cathode and anode wires placed in a concentric cylindrical configuration around it. There are 360 anode wires, one per degree with a diameter of 25 µm. The electrical field in the drift space has 1/r dependence as in a cylindrical capacitor. The field is linearized by a number of concentric circular field-shaping electrodes on the mylar front-window and on the printed circuit backplane of the detector.

A fast plastic scintillator, placed inside the hollow central cathode, will produce a light signal each time it is hit by the primary beam. A photo- multiplier generates a timing pulse which is synchronous with the passage of

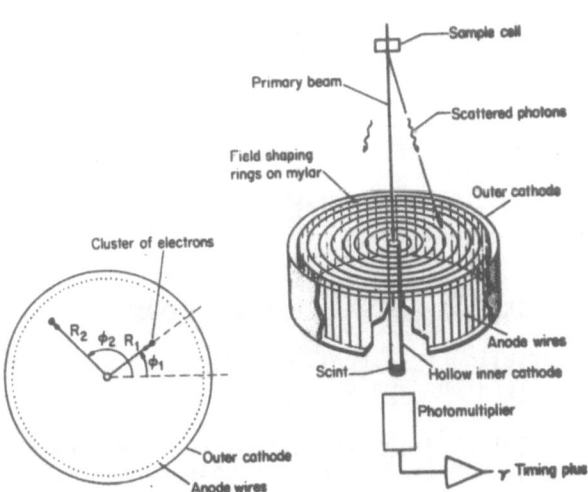

Fig. 12. Principle of a drift chamber detector.

376

the bunch in the storage ring. The signal indicates the moment at which
clusters of the primary photoelectrons can be deposited in the filling
gas of the detector by scattered photons. The clusters drift radially to
the anode elements at the circumference of the detector. At the wires
gas amplifiction takes place as in a usual proportional counter. The
time intervals elapsed between the synchronisation pulses of the
photo-multiplier and the signals from the anode wires are a measure of
the radial distribution of the scattered photons. The most important
advantage of this detector system lies in the fact that during the short
period of a flashes of X-ray radiation, several clusters of primary
electrons can be deposited in the gas. They will drift simultaneously
and independently of each other to the anode elements. The effects of
ionizing events during the 100 ps length of the bunch are stored in the
gaseous medium of the detector, retaining their positional information.
The instants of arrival of the signals are randomized due to the
differences in drift time, spreading the duty cycle for processing from
10^{-4} to one.

Part of the results presented here – anomalous X-ray scattering and
nuclear spin dependent neutron scattering – were obtained in projects of
the University of Mainz at HASYLAB (DESY) and the GKSS Research Centre
at Geesthacht. These were supported by the Bundesministerium für
Forschung und Technology at Bonn (Grant Nr. 05 353FAI and 03-I21E05).

REFERENCES

1. See e.g. 'Free Electron Generators of Coherent Radiation', S.F.
Jacobs, G.T. Moore, H.S. Pilloff, M. Sargent III, M.O. Scully and R.
Spitzer, Editors, Physics of Quantum Electronics, Addison – Wesley Publ.
Vols. 5-7-8-9 1979-82

2. R. Colella and A. Luccio, 'Proposal for a Free Electron Laser in the
X-ray Region', Brookhaven National Laboratory – 33986 (1983)

3. Neutron scattering in the Nineties IAEA Conference at KFA Jülich
(1985)

4. B.Alefeld, G. Bauer, H. Halling, H. Lang, J. Schelten, G. Thamm and
Z. Zettler, in Neutron Scattering in the Nineties, 361-367, (1985)

5. Synchrotron Radiation Research, H. Winick and S. Doniach, Editors,
Plenum Press, New York and London (1980)

6. H. Fueß, 'Application of Neutron Scattering in Chemistry– Comparison
between Continuous and Pulsed Sources ' this course (1987)

7. J. Bordas and E. Mandelkow 'Time Resolved X-ray Scattering from
Solutions Using Synchrotron Radiation', In Fast Methods of Physical
Biochemistry and Cell Biology, R.I. Shalafi and S.M. Fernandez, Editors,
Elsevier Science Publishers (1983)

8. C. Hermes, Thesis, Technical University of Munich (1981)

9. R.L. Mößbauer, 'The application of Anomalous Dispersion of Radiation to the Structure Analysis of Macromolecular Crystals' in Anomalous Scattering", S. Rameseshan and S.C. Abrahams, Editors , Munksgaard, Copenhagen (1975)

10. H.B. Stuhrmann, 'Resonance Scattering in Macromolecular Structure Research' Advances in Polymer Science 67, 123-163 (1985)

11. M.M. Woolfson : An Introduction to X-Ray Crystallography, Cambridge University Press (1970)

12. F. Parak, R.L. Mößbauer and W. Hoppe , Berichte der Bunsengesellschaft für Physikalische Chemie, 74, 1207 (1970)

13. E. Gerdau, R. Rüffer, H. Winkler, W. Tolksdorf, C.-P. Klages, J.P. Hannon , Phys. Rev. Letters 54 , 835

14. E. Gerdau, R. Rüffer, R. Hollatz, J.P. Hannon ' 'Quantum Beats' durch Anregung der Kernniveaus in YIG' HASYLAB Jahresbericht 1986, 367, DESY, Hamburg (1987)

15. U. van Bürck, R.L. Mößbauer, E. Gerdau, R. Rüffer, R. Hollatz, J.P. Hannon, G.V. Smirnov 'Speed-up of Coherent Decay in Nuclear Resonant Diffraction of the Synchrotron Radiation', HASYLAB Jahresbericht 1986, 370, DESY, Hamburg (1987)

16. B. Munk, G. Goerigk, H.B. Stuhrmann, G. Büldt, H.-J. Plöhn, 'Resonante Röntgenstreuung an der K Absorptionskante des Schwefels' HASYLAB Jahresbericht 1986, 354, DESY, Hamburg (1987)

17. H.B. Stuhrmann, O. Schärpf, M. Krumpolc, T.O. Niinikoski, M. Rieubland , A. Rijllart, 'Dynamic Nuclear Polarisation of Biological Matter', Eur. Biophys. J. 14, 1-6 (1986)

18. W. Knop, K.H. Nierhaus, V. Novotny, T.O. Niinikoski, M. Krumpolc, J.M. Rieubland, A. Rijllart, O. Schärpf, H.J. Schink, H.B. Stuhrmann, R. Wagner, 'Polarised Neutron Scattering from Dynamic Polarised Targets of Biological Origin', Helvetica Physica Acta 59 741-746 (1986)

19. J. Hendrix 'Position Sensitive Detectors', in Uses of Synchrotron Radiation in Biology, H.B. Stuhrmann editor, Academic Press , London and New York (1982)

DISCUSSION

NUCLEAR SPIN DEPENDENT NEUTRON SCATTERING IN BIOLOGICAL SAMPLES

The extent to which the paramagnetic Cr(V) complex diffuses throughout the protein crystal will determine the homogeneity of the proton polarisation. The response of the nmr signal to polarisation changes, which is the most sensitive method of checking, shows that the proton polarisation has always been fairly homogeneous.

The critical reflectometer on the pulsed neutron beam at the Rutherford Appleton Laboratory offers, in principle, a method of studying proton polarisation in a membrane. The membrane could be floated on a fluid containing the paramagnetic chromium complex, and, after polarisation, the neutron beam could be reflected from the other side of the membrane. Depth profiling by scanning the neutron wavelength at constant angle should reveal the depth of penetration into the membrane of the change in proton scattering length due to polarisation. Since the total reflection of neutrons depends on the coherent scattering length, this application should be a straightforward one, but the temperature would have to be as low as in the Hamburg experiments reported above. Work on membranes and membrane proteins may start in Hamburg in 1988 when a new cryostat should be available.

The accuracy with which the distance between ribosomal proteins can be measured in the ribosomal contrast method is about 5%. The H/D ratio in each protein is controlled by growing it in the appropriate cell culture medium. Such protons are mostly not labile, and the 20% that do subsequently exchange make only a minor contribution to the contrast. The 13 smaller proteins whose structures need to be found are rather flexible, so sometimes only fragments crystallise; the parts of the structures which have so far been determined are mainly β-sheets.

Recent Theoretical Advances in Macromolecular Structure Determination

Jerome Karle
Laboratory for the Structure of Matter
Naval Research Laboratory
Washington, D. C. 20375-5000, U.S.A.

There have been a number of recent theoretical formulations that have led to new formulas and methodologies for investigating macromolecular structure by use of the techniques of anomalous dispersion and isomorphous replacement. In this part, several of the new developments will be outlined in connection with the attached reprints. A summary of the main ideas and results will be discussed in terms of formulas and implications that appear in the reprints in greater detail.

Exact Algebraic Analysis of Multiple Wavelength Anomalous Dispersion Data

(International Journal of Quantum Biology Symposium $\underline{7}$, 357-367 (1980))

A recent algebraic analysis of multiple-wavelength anomalous dispersion data has resulted in a set of simultaneous equations that are both exact and linear. As a worthwhile complement to this development, it is expected that the collection of data for multiple-wavelength anomalous dispersion experiments will be more and more facilitated by use of the increasingly accessible synchrotron radiation facilities.

The exact algebraic analysis which is valid for any number of anomalous scatterers and any variety of types of anomalous scatterer can be found in the attached reprint from the Journal of Quantum Chemistry (JQC). The simplest case is given by atoms that scatter normally together with one type of anomalous scatterer. The appropriate equation for this case is given by Eq. (14) in JQC. The types of unknown quantities that occur in Eq. (14) are representative of the unknowns that occur in the most general case that can be formed by combining Eqs. (7), (17), (18) and (19). Evidently, if the unknown quantities in Eq. (14) are taken to be

$$|F_{1,h}^{n}|^2, \quad |F_{2,h}^{n}|^2, \quad |F_{1,h}^{n}||F_{2,h}^{n}|\cos(\phi_{1,h}^{n}-\phi_{2,h}^{n}) \text{ and}$$

$$|F_{1,h}^{n}||F_{2,h}^{n}|\sin(\phi_{1,h}^{n}-\phi_{2,h}^{n}),$$

the equations are linear in these variables and they can be readily evaluated from a set of similtaneous equations generated from measurements at various wavelengths, λ, and at h and $-h$. The analysis is facilitated by the use of the quadratic relation, $\sin^2(\phi_{1,h}^{n}-\phi_{2,h}^{n}) + \cos^2(\phi_{1,h}^{n}-\phi_{2,h}^{n})=1$ and application of the least-squares technique.

379

M. A. Carrondo and G. A. Jeffrey (eds.), Chemical Crystallography with Pulsed Neutrons and Synchrotron X-Rays, 379–418.

The unknown quantities are characterized, in general, by intensities, e.g. $|F_{1,h}^n|^2$, and phase differences, e.g. $(\phi_{1,h}^n - \phi_{2,h}^n)$ multiplied by products of structure factor magnitudes, whose values are independent of the wavelength, λ. They are associated with structure factors that represent only normal scattering from the anomalously scattering atoms represented by $|F_{2,h}^n|$ and $\phi_{2,h}^n$ in addition to $|F_{1,h}^n|$ and $\phi_{1,h}^n$ for the normally scattering atoms in Eq. 14 of JQC. These quantities do not change in value as the wavelength of the experiment changes and thus we can set up an exact set of simultaneous equations. A special characteristic of the analysis given in the JQC paper is the treatment of the intensity for the measured anomalous scattering data, $|^{\lambda}F_h|^2$, at some wavelength, λ, that separates the normal scattering from that derived from both the real and imaginary parts of the corrections to the atomic scattering factors. It is for that reason that the resulting equations (14), or, in general, (7), (17), (18) and (19) contain explicitly unknown quantities that are independent of λ and known quantities, obtained from tables of the real and imaginary corrections to anomalous scattering, that act as the coefficients of the unknown quantities and vary with λ.

The general formulation of the multiple-wavelength anomalous dispersion analysis given in the JQC paper has a number of favorable characteristics. In addition to the unknown quantities, i.e. intensities of scattering and phase differences associated with the non-anomalous scattering, the phases and magnitudes involved are those that would be obtained from individual types of atoms as if each type were present in isolation from the rest. Knowledge of the intensities for the structure formed by a particular type of atom can facilitate the determination of the structure formed by this particular type of atom. Once the structure is known for any of the types of atom present, the entire structure can be readily determined. As noted, the anomalous scattering enters the simultaneous equations as separate factors in terms of known, tabulated quantities and with appropriate definitions of unknown quantities, the simultaneous equations are linear and involve no approximations.

An additional feature of Eq. (14) concerns the fact that heavy atoms acting as anomalous scatterers are often substituted into a native substance that scatters essentially non-anomalously. Under such circumstances, intensity data for the native material correspond to the $|F_{1,h}^n|^2$ in Eq. (14), thus reducing the number of unkown quantities.

Evaluation of Triplet Phase Invariants from Exact Algebraic Analysis of

Anomalous Dispersion Data

(Acta Crystallographica A$\underline{40}$, 526-531 (1984))

As noted in the previous section, the algebraic analysis gives values for phase differences of the type $\phi_{1,h}^n - \phi_{2,h}^n$. In order to derive values for the phases of the non-anomalous scatterers, $\phi_{1,h}^n$, from the values of the $\phi_{1,h}^n - \phi_{2,h}^n$, it is evidently necessary to have values for the $\phi_{2,h}^n$. These may be obtained from a determination of the structure of the anomalous scatterers by use of a Patterson-type map with coefficients

$(|{}^\lambda F_n| - |{}^\lambda F_{-h}|)^2$ (see Eq. 20 of JQC). An alternative determination of the structure of the anomalously scattering atoms may also be obtained from phase determination applied to the values of the $|F_{2,h}^n|$ derived from the algebraic analysis described in JQC.

Structure determinations of anomalous scatterers may not always be successful. In those cases, the structures of non-anomalously scattering atoms may, perhaps, be obtainable from known values for triplet phase invariants of the type $\phi_{1,h}^n + \phi_{1,k}^n + \phi_{1,(\bar{h}+\bar{k})}^n$. Values for triplet phase invariants may be obtained from the exact algebraic analysis which gives values for the $\phi_{1,h}^n - \phi_{2,h}^n$. As seen in Acta Cryst. A40, page 58 (AC40, p. 528) in the part on "Triplet phase invariants", appropriate sets of phase differences can be combined to give

$$\phi_{1,h}^n + \phi_{1,k}^n + \phi_{1,(\bar{h}+\bar{k})}^n - \phi_{2,h}^n - \phi_{2,k}^n - \phi_{2,(\bar{h}+\bar{k})}^n = A_{hk} \qquad (6)$$

For large associated structure factor magnitudes, the triplet phase invariants for the structure of the anomalously scattering atoms, $\phi_{2,h}^n + \phi_{2,k}^n + \phi_{2,(\bar{h}+\bar{k})}^n$ can be expected to be close to zero and therefore the triplet phase invariants for the non-anomalously scattering atoms are equal, to good approximation, to

$$\phi_{1,h}^n + \phi_{1,k}^n + \phi_{1,(\bar{h}+\bar{k})}^n \simeq A_{hk} \qquad (7)$$

With exact data, the only uncertainty in this evaluation of triplet phase invariants for the structure of the non-anomalous scatterers is the deviation of the triplet phase invariants from zero. A study of this was made in reprint AC40, pp. 529-531 for a very simple structure, a 3-atom structure in space group P1, and a more complex structure, a 9-atom structure in an asymmetric unit of space group P2$_1$2$_1$2$_1$. In these studies the statistical distributions of $\cos(\phi_{2,h}^n + \phi_{2,k}^n + \phi_{2,(\bar{h}+\bar{k})}^n)$ were obtained and compared with the results of several theoretical studies. It was shown that closeness to unity for the cosine of the triplet phase invariant could be easily achieved for the 3-atom structure, AC40, Fig. 1. It could also be seen in AC40, Fig. 1, that the commonly used formula for evaluating triplet phase invariants (formula S shown with symbol X in Fig. 1) seriously underestimates their value for very simple structures. Some of the other formulas e.g., D and I, fit better. For the more complex structure, the 9-atom structure in an asymmetric unit of space group P2$_1$2$_1$2$_1$, it is possible to find acceptably large values for the cosines of the triplet phase invariants but, as expected, not as readily as for simpler structures, AC40, Fig. 2. The fit of the theoretical functions to the calculated values is better than that for the 3-atom structure. The commonly used formula, S, still underestimates somewhat the values of the cosine invariants.

The main conclusion for macromolecules from the study of AC40 is that, because the structure of the main anomalous scatterers in these molecules are usually quite simple, it is often readily possible to satisfy the condition of high accuracy for $\phi_{2,k}^n + \phi_{2,k}^n + \phi_{2,(\bar{h}+\bar{k})} \simeq 0$ and therefore achieve high accuracy for the triplet phase invariants given by (7) for the non-anomalously scattering structures.

Essentially Unique Results from One-Wavelength Anomalous Dispersion Data

(Acta Crystallographica A$\underline{41}$, 387-394 (1985))

The usual analysis of the implications of the data from a one-wavelength anomalous dispersion experiment leads to the conclusion that such data generate a twofold ambiguity in the evaluation of phase differences, e.g.$(\phi_{1,h}^{n}-\phi_{2,h}^{n})$. The purpose of this study has been to show that additional information contained in one-wavelength anomalous dispersion data and not normally used in the analysis that implies a twofold ambiguity can be used to obtain essentially unique values for the phase differences with potentially useful accuracy. The investigation was limited to the case of one predominant type of anomalous scatterer in a one-wavelength experiment.

The basis for the analysis is again the exact algebraic analysis given in JQC. In a one-wavelength experiment, there is one more unknown quantity than defining equations. The basic equation for this study is JQC, Eq. 14. As seen from Eqs. 9-11 of the reprint relevant to this study, which we label AC41, approximate values for two of the variables in Eq. 14 of JQC can be obtained from the measured quantities $|F_{\lambda h}|$ and $|F_{\lambda \bar{h}}|$.

For the test calculations with cytochrome c550·PtCl$_4^{2-}$ whose results are summarized in Tables 1 and 2 of AC41, the following procedure was followed:

1. Estimates of the various values of $|F_{2,h}^{n}|^2$ were made by use of Eq. 11 of AC41. The values of S to be used in this equation were based on a statistical analysis of examples having the same atomic composition as the substance of interest.

2. On the basis that values for the $|F_{2,h}^{n}|$ were thus approximately known, the least-squares System II was set up.

3. Three different widely spaced starting values were used for $|F_{1,h}^{n}|$. They were based on Eq. 9 of AC41.

4. With an approximate value for $|F_{2,h}^{n}|^2$, three starting values for $|F_{1,h}^{n}|^2$ and least-squares System II, the calculations converged sometimes all three times to the same results for the unknown quantities, $|F_{1,h}^{n}|$ and $(\phi_{1,h}^{n}-\phi_{2,h}^{n})$, and more often two times for one set of values and one time for the alternative. In the second case, the values that occurred twice were accepted. This was not always the best choice, but it usually was.

It is apparent from Tables 1 and 2 that in the test example, acceptable accuracy is obtained for the $\phi_{1,h}^{n}-\phi_{2,h}^{n}$ even though large errors were introduced into the values of the $|F_{\lambda h}|-|F_{\lambda \bar{h}}|$. Most of the error is ultimately seen to reside in the values of the $|F_{1,h}^{n}|$ and $|F_{2,h}^{n}|$.

A geometric analysis of the basis for obtaining unique or essentially unique values for the phase differences is given in an Appendix to AC41. It is seen, for example, that the two alternative values for the $\phi_{1,h}^{n}-\phi_{2,h}^{n}$

can be related to significantly different values for the $|F_{1,h}^n|$. This is the basis for resolving the ambiguity.

Phase Information from Single Isomorphous Replacement or One-Wavelength

Anomalous Dispersion Given Heavy-Atom Information

(Acta Crystallographica A42, 246-253 (1986))

This study, which we label AC42, is concerned with several matters, an investigation of the accuracy of a formula for computing triplet phase invariants having values anywhere between $-\pi$ and π for single isomorphous replacement or one-wavelength anomalous dispersion, the applicability of formulas for computing phase differences which, with knowledge of the structure of the heavy-atoms or of the anomalously scattering atoms, could afford a large number of phase values for the initiation of a structure determination of a native or non-anomalously scattering structure and a possible strategy for the use of triplet phase invariants.

The general formula for the calculation of the cosines of triplet phase invariants is given by Eq. 1 of AC42. This equation can be specialized for application to single isomorphous replacement, Eq. 6 of AC42, and to one-wavelength anomalous dispersion, Eq. 15 of AC42.

Formulas for phase differences are given in Eqs. 10, 11, 23 and 24 of AC42. A way to employ such equations is to set a criterion concerning how close to a magnitude of unity the right sides of these equations must be, e.g. 0.7660 or 0.9387, for the calculation to be accepted. Such values are regarded as acceptable deviations from unity. When accepted, the left sides of Eqs. 10, 11, 23 and 24 of AC42 are treated as if the right sides had a magnitude of unity. When the structures of the heavy atoms or anomalous scatterers is known, their phases can be evaluated and then Eqs. 10, 11, 23 and 24 can be used to define the phases of the native or non-anomalously scattering atoms in terms of those for the heavy atoms or anomalous scatterers. An alternative way to obtain values of phases for non-anomalous scatterers from phase differences has already been described in AC41.

Test calculations of various formulas were carried out on cytochrome c550·PtCl$_4^{2-}$ in AC42. Tables 2 and 5 concern the calculation of the cosines or sines of triplet phase invariants. Table 4 concerns the evaluation of phases in isomorphous replacement by use of Eqs. 10 and 11. The calculations imply a potential for future application. Although, single isomorphous replacement and one-wavelength anomalous dispersion are investigated separately in AC42, in order to evaluate some individual characteristics, it would be better in practice to combine them.

As seen from the Introduction and Concluding Remarks of AC42, there have been many algebraic and probabilistic investigations associated with the anomalous dispersion and isomorphous replacement techniques. Recently, there have been a number of probabilistic and algebraic investigations devoted to triplet phase invariants. As a practical matter, it may be difficult to extract phase information from the evaluations of triplet phase invariants for macromolecules. This can arise from errors in the evaluations and inherent ambiguities in the extraction of individual phase values from the triplets. Perhaps use of triplet phase invariants can be faciliated by use of initial phase evaluations from phase differences and known values of phases for the heavy atom or anomalously scattering structures, as described in AC42.

Summary Conclusions

1. The availability of a system of exact, linear simultaneous equations for the analysis of multiple-wavelength anomalous dispersion data from any number or type of anomalous scatterers, combined with the greatly enhanced experimental facilities associated with synchrotron radiation sources, suggests that the ease with which analyses of macromolecular structure can be performed will increase in the future.

2. Other theoretical developments that are potentially useful are evaluations of triplet phase invariants augmented by phase information obtained from phase differences and knowledge of the structures of the heavy atoms or anomalous scatterers.

3. The general observation is that more information is available from isomorphous replacement and anomalous dispersion experiments than is normally extracted and there are a number of theoretical tools and experimental facilities available for taking advantage of the opportunities that are afforded.

DISCUSSION

DIRECT METHODS

The use of direct methods in the small molecule field became widespread only when a few people wrote programs which many people could use. For such methods, having many data is more important than having fewer, but more accurate, data.

HEAVY ATOMS AND ANOMALOUS DISPERSION

In the methods described above the entire process depends on having reasonable data and reasonable substitution. But frequently one cannot get good isomorphous substitution or the heavy atoms do not go in well. In those cases one can try to find a molecule of known structure likely to be sufficiently similar to the unknown protein to act as a starting point for getting some phases, and then work up the structure from that stage. This approach, however, can be very difficult and is not often successful.

Dr Harding had attempted to solve the structure of a protein, cytochrome c4, twice as large as the cytochrome mentioned above. Data were obtained using synchrotron radiation at the wavelength of the iron absorption edge for the native protein, and data for another wavelength without the iron anomaly could also be fed in. The iron anomalous differences gave a Patterson map from which the iron atoms were well located. However, an attempt to phase the whole protein by the method Hendrickson used for crambin gave an electron density map which was very suggestive of a structure but not quite clear enough to be interpreted unambiguously.

Dr Karle explained that crambin is not typical. It was an excellent crystal giving excellent data and the sulphur atoms were very well frozen. Otherwise the anomalous scattering of only six sulphur atoms would not solve the structure of even a small protein. The method Dr Karle had described above was different from that used by Hendrickson for crambin.

With neutrons one can do something similar to isomorphous replacement by isotopic substitution, which can change the magnitude and sometimes the sign of the scattering length. The methods described above could in principle be used, but it must be remembered that not many atoms are substituted when isomorphous replacement is used with X-rays. If all the hydrogens, for example, were exchanged for deuterium, the resulting problem could be almost as difficult as solving the original structure. For neutrons, therefore, the methods should be applicable provided not too many atoms are exchanged.

Dr Karle thought that it was inappropriate to refer to the mathematical techniques discussed here as direct methods for protein structure determination. The only link between the methods for small molecules and for macromolecules was the occurrence of triplet phase invariants in both. But the latter never proceeded in the same way as the former because stepwise use of triplet phase invariants for macromolecules would involve so many steps that the cumulative errors would be prohibitive. The methods are, in reality, well established heavy atom anomalous dispersion and isomorphous replacement with a little more careful look at what the mathematics and the physics can contribute.

Some Developments in Anomalous Dispersion for the Structural Investigation of Macromolecular Systems in Biology

JEROME KARLE

Laboratory for the Structure of Matter, Naval Research Laboratory, Washington, D.C. 20375, U.S.A.

Abstract

A discussion is given of the anomalous dispersion technique in structure analysis. This is accompanied by a general algebraic analysis with no approximations for any number and type of anomalous scatterer. The resulting relations are largely linear with appropriate selection of unknown quantities. The quantities of interest are expressed only in terms of functions of the nonanomalous parts of the atomic scattering factors and are separate for each type of anomalous scatterer. This theory, the advent of increased ease in performing multiple-wavelength experiments, and some recent progress made in applying a one-wavelength experiment to the solution of a protein structure suggest considerable potential for the future application of the anomalous dispersion technique to structural investigations of complex macromolecular systems in biology.

Introduction

The methods that have been used effectively to determine structures of biological macromolecules have involved the introduction of heavy atoms into structures of interest with subsequent application of the isomorphous replacement and anomalous dispersion techniques in diffraction analysis. Both techniques take advantage of special diffraction characteristics associated with the presence of the heavy atoms. In the case of isomorphous replacement, advantage is taken of the fact that the introduction of heavy atoms often does not materially alter the atomic positions of the unsubstituted structure, and the heavy atoms are few enough that their positions are readily located. In anomalous dispersion experiments, advantage is taken of the fact that the atomic scattering factors of the heavy atoms undergo large changes at x-ray wavelengths that occur in the vicinity of their absorption edges. Applications have depended once again upon having few enough heavy atoms so that their positions could be readily determined.

The technique used most often to determine the structures of proteins has been multiple isomorphous replacement, i.e., the combined use of several isomorphous substitutions. For the most part, anomalous dispersion has played a secondary role, having usually been applied in such a way as to give auxiliary information to that obtained from multiple isomorphous substitution.

A number of recent developments, both experimental and theoretical, indicate that the anomalous dispersion technique has considerable untapped potential

International Journal of Quantum Chemistry: Quantum Biology Symposium 7, 357–367 (1980)
0360-8832/80/0007-0357$01.10

388

KARLE

for future application and that the required measurements will soon be more readily performed. Important use of anomalous scattering has recently been made in the direct determination of the structure of a small ($\sim$5000 daltons) protein crambin by Hendrickson and Teeter [1]. The protein molecule contains six cysteine residues in disulfide linkages, and the structure was solved by use of the anomalous scattering from the sulfur atoms measured at a single wavelength. The great facilitation of the structure determination by the few and relatively weak anomalous scatterers suggests that much greater analytical facility could accrue from the presence of stronger or more numerous anomalous scatterers. Conceivably, large numbers of strong anomalous scatterers would enter large, complex biological systems in the course of the formation of heavy atom derivatives. Some native macromolecular substances contain numerous significant anomalous scatterers, e.g., metal cluster compounds. The presence of large numbers of heavy atoms would represent a departure from customary circumstances in applications of the anomalous dispersion technique. Normally the number of heavy atoms in the structures under study is limited to a few, so that they can be readily located by simple techniques such as the calculation of a Patterson map.

The question arises concerning how effective use might be made of numerous anomalous scatterers in structures in which a large part of the total scattering may come from atoms that do not scatter anomalously. If it were possible to determine the intensities of scattering that would be obtained if only the anomalously scattering atoms comprised the structure, such intensities could be used in methods for direct structure determination [2] to locate the heavy atoms. Structures composed of as many as 100 or more anomalously scattering atoms in the asymmetric unit would be accessible to analysis.

To this end, a theory that effects the separate determination of the intensities of scattering for the anomalous and nonanomalous scattering and also gives the differences between certain phase angles is developed here. In fact, the theory is presented finally in a very general form to take into account the possibility of having numbers of anomalous scatterers of many different types. Examination of the systems of equations that arise shows that with an appropriate choice of variables, many of the equations occur in usefully linear form. Diffraction experiments performed at several different wavelengths could take advantage of the fact that the anomalous scattering varies with wavelength and thus could provide many independent bits of information.

The current development of dedicated sources of x-radiation from synchrotrons bodes well for the future opportunity to conduct multiple-wavelength experiments readily. The high intensity of the source, the tunability of wavelengths, and progress in the development of area photon detectors imply future convenience in the acquisition of data.

The use of anomalous dispersion reduces a complex problem to a much simpler one, i.e., the determination of the arrangement of the anomalously scattering atoms instead of that for the entire structure. The latter is much more readily determined, once the anomalously scattering atoms are located.

The anomalous dispersion technique was introduced into crystal structure analysis by Bijvoet and colleagues [3–6]. From its inception, it has played two important roles. One has been as an aid in helping to resolve the twofold ambiguity in phase determination that derives from single isomorphous replacement while enhancing accuracy at the same time. The second has been its important application as a tool for determining the absolute configuration of molecules. As one example of its application, Bijvoet, Peerdeman, and van Bommel determined the absolute configuration of d-tartaric acid in the form of its sodium rubidium double salt [4,5], and van Bommel determined it from the acid rubidium salt [6], confirming the chemical convention of Emil Fischer. The analysis presented here suggests a third important role for anomalous dispersion, namely, as an often unique and largely untapped resource for the direct solution of complex structures that may contain numerous anomalous scatterers.

Previous authors, Okaya and Pepinsky [7] and Mitchell [8], have discussed two-wavelength experiments with the objective of resolving ambiguities in phase information derivable from the known positions of anomalous scatterers. The possibility of using the characteristic L-multiplet of heavier atoms to excite the K-edge of lighter atoms to obtain information from several wavelengths in a single experiment had been discussed by Herzenberg and Lau [9] some time before the advent of the more convenient and intense synchrotron radiation as a tunable source. As an example, the authors made the insightful suggestion concerning the possibility of studying a small protein structure in a multiple-wavelength experiment by exciting the K-edge of sulfur atoms by means of the multiplet Mo L-radiation. It is of interest to note that the investigation of Hendrickson and Teeter [1] was performed more simply than that. CuK_α radiation was used in a single-wavelength experiment. A review of recent thoughts on a large variety of topics associated with anomalous dispersion including multiple-wavelength experiments and applications to macromolecules can be found in [10].

The theory to be presented has an earlier counterpart [11] which is generalized and reformulated in a way that separates completely the anomalous from the nonanomalous scattering. In many theoretical formulations, it has been customary to combine the contribution from the real part of the correction to the atomic scattering factor with the normal part and to treat separately the contribution from the imaginary part. The consequence of taking the path followed here is to obtain expressions for the quantities of interest, structure factor magnitudes and phase differences, which are defined solely in terms of the nonanomalous atomic scattering factors for all the atoms contained in the structure.

Theory

In this part the earlier treatment of multiple-wavelength anomalous dispersion [11] is expressed in a new form that emphasizes the largely linear features of the theory, the accessibility of separate values for intensities associated with the anomalously scattering atoms and the nonanomalously scattering atoms, and the potential of the anomalous dispersion technique for facilitating the structural investigation of complex systems. The theory is also used to discuss the interpretation of a type of Fourier map which was employed in the direct solution of the structure of a small protein [1].

Normal or nonanomalous atomic scattering factors f^n are computed on the basis that all the electrons scatter as if they were free classical electrons. This condition is fairly well approximated when the incident radiation has a frequency that is much larger than the absorption frequencies of the atom of interest. In practice, particularly in the vicinity of an absorption edge for an atom, there can be rather significant corrections for f^n which are denoted in this paper by f' and f'', the real and imaginary corrections to f^n, respectively.

When in a structure atoms are present which scatter anomalously in significant measure, the structure factor, whose magnitude is proportional to the square root of the measured intensities, is a function of the wavelength of the experiment denoted by λ. The structure factor may be written

$$^{\lambda}F_{\mathbf{h}} = F_{\mathbf{h}}^{n} + {}^{\lambda}F_{\mathbf{h}}^{a}, \tag{1}$$

where $F_{\mathbf{h}}^{n}$ is the structure factor that would be obtained if all atoms scattered nonanomalously and $^{\lambda}F_{\mathbf{h}}^{a}$ is the structure factor that derives from anomalous corrections to the normal atomic scattering. The structure factors $F_{\mathbf{h}}^{n}$ and $^{\lambda}F_{\mathbf{h}}^{a}$ are defined, ignoring vibrational effects, as

$$F_{\mathbf{h}}^{n} = |F_{\mathbf{h}}^{n}| \exp(i\phi_{\mathbf{h}}^{n}) = \sum_{j=1}^{N} f_{j\mathbf{h}}^{n} \exp(2\pi i \mathbf{h} \cdot \mathbf{r}_{j}), \tag{2}$$

and

$$^{\lambda}F_{\mathbf{h}}^{a} = |^{\lambda}F_{\mathbf{h}}^{a}| \exp(i{}^{\lambda}\phi_{\mathbf{h}}^{a}) = \sum_{j=1}^{N} ({}^{\lambda}f_{j\mathbf{h}}' + i{}^{\lambda}f_{j\mathbf{h}}'') \exp(2\pi i \mathbf{h} \cdot \mathbf{r}_{j}), \tag{3}$$

where the ϕ are the phases associated with the magnitudes $|F|$, $f_{j\mathbf{h}}^{n}$ is the normal atomic scattering factor for the jth atom in a crystal having N atoms in the unit cell, $^{\lambda}f_{j\mathbf{h}}'$ is the real part and $^{\lambda}f_{j\mathbf{h}}''$ the imaginary part of the anomalous correction to the normal atomic scattering factor for the jth atom, $\mathbf{h} \equiv (h, k, l)$, a vector whose components represent the Miller indices of the plane in the crystal associated with the particular scattered wave labeled with $\mathbf{h}$, and $\mathbf{r}_{j}$ is a vector whose components are the coordinates of the jth atom. Not all atoms need have anomalous corrections to their scattering factors. In that case some of the f' and f'' in eq. (3) could be replaced by zero. In addition, although the notation $^{\lambda}f_{j\mathbf{h}}'$ and $^{\lambda}f_{j\mathbf{h}}''$ implies that the f' and f'' are a function of $\mathbf{h}$ (or of the scattering angle), they are generally used as constants with respect to $\mathbf{h}$ [12]. The variation of f' and f'' with the scattering angle has been described by Hazell [10, p. 41]. Their important variation is with respect to the wavelength λ.

Equation (3) is rewritten as

$$^{\lambda}F_{\mathbf{h}}^{a} = \sum_{j=1}^{N} f_{\lambda j, \mathbf{h}}^{a} \exp(i\delta_{\lambda j, \mathbf{h}}) \exp(2\pi i \mathbf{h} \cdot \mathbf{r}_{j}), \tag{4}$$

where

$$f_{\lambda j,\mathbf{h}}^{a} = [(^{\lambda}f_{j,\mathbf{h}}')^2 + (^{\lambda}f_{j,\mathbf{h}}'')^2]^{1/2} \tag{5}$$

and

$$\delta_{\lambda j,\mathbf{h}} = \tan^{-1} (^{\lambda}f_{j,\mathbf{h}}''/^{\lambda}f_{j,\mathbf{h}}'). \tag{6}$$

The quantities $|^{\lambda}F_{\mathbf{h}}|^2$ are proportional to the measured scattering intensities and are readily obtained from them. They may therefore be regarded as representing the observed intensity measurements. From eq. (1) it follows that

$$|^{\lambda}F_{\mathbf{h}}|^2 = |F_{\mathbf{h}}^{n}|^2 + |^{\lambda}F_{\mathbf{h}}^{a}|^2 + 2|F_{\mathbf{h}}^{n}||^{\lambda}F_{\mathbf{h}}^{a}| \cos(\phi_{\mathbf{h}}^{n} - {}^{\lambda}\phi_{\mathbf{h}}^{a}). \tag{7}$$

Two Types of Atoms in a Structure, Both Scattering Anomalously

Equation (7) is first developed for the case that a structure contains two types of atoms, both of which scatter anomalously. In this case,

$$|F_{\mathbf{h}}^{n}|^2 = |F_{1,\mathbf{h}}^{n}|^2 + |F_{2,\mathbf{h}}^{n}|^2 + 2|F_{1,\mathbf{h}}^{n}||F_{2,\mathbf{h}}^{n}| \cos(\phi_{1,\mathbf{h}}^{n} - \phi_{2,\mathbf{h}}^{n}), \tag{8}$$

where $F_{1,\mathbf{h}}^{n}$ and $F_{2,\mathbf{h}}^{n}$ are the structure factors unaltered by anomalous scattering for each of the two types of atoms, and $\phi_{1,\mathbf{h}}^{n}$ and $\phi_{2,\mathbf{h}}^{n}$ are the corresponding phase angles.

It follows from eqs. (2) and (4) for the two types of atoms that

$$^{\lambda}F_{\mathbf{h}}^{a} = (f_{\lambda 1,\mathbf{h}}^{a}/f_{1,\mathbf{h}}^{n})|F_{1,\mathbf{h}}^{n}| \exp[i(\phi_{1,\mathbf{h}}^{n} + \delta_{\lambda 1,\mathbf{h}})]$$
$$+ (f_{\lambda 2,\mathbf{h}}^{a}/f_{2,\mathbf{h}}^{n})|F_{2,\mathbf{h}}^{n}| \exp[i(\phi_{2,\mathbf{h}}^{n} + \delta_{\lambda 2,\mathbf{h}})]. \tag{9}$$

Therefore,

$$|^{\lambda}F_{\mathbf{h}}^{a}|^2 = (f_{\lambda 1,\mathbf{h}}^{a}/f_{1,\mathbf{h}}^{n})^2|F_{1,\mathbf{h}}^{n}|^2 + (f_{\lambda 2,\mathbf{h}}^{a}/f_{2,\mathbf{h}}^{n})^2|F_{2,\mathbf{h}}^{n}|^2$$
$$+ 2(f_{\lambda 1,\mathbf{h}}^{a}/f_{1,\mathbf{h}}^{n})(f_{\lambda 2,\mathbf{h}}^{a}/f_{2,\mathbf{h}}^{n})|F_{1,\mathbf{h}}^{n}||F_{2,\mathbf{h}}^{n}|$$
$$\times \cos(\phi_{1,\mathbf{h}}^{n} - \phi_{2,\mathbf{h}}^{n} + \delta_{\lambda 1,\mathbf{h}} - \delta_{\lambda 2,\mathbf{h}}). \tag{10}$$

The third term on the right-hand side of eq. (7) is now considered and since, for example, $|F_{\mathbf{h}}^{n}| \cos\phi_{\mathbf{h}}^{n}$ is the real part of $F_{\mathbf{h}}^{n}$, $|^{\lambda}F_{\mathbf{h}}^{a}| \cos{}^{\lambda}\phi_{\mathbf{h}}^{a}$ is the real part of $^{\lambda}F_{\mathbf{h}}^{a}$, and so forth, it follows that

$$2|F_{\mathbf{h}}^{n}||^{\lambda}F_{\mathbf{h}}^{a}| \cos(\phi_{\mathbf{h}}^{n} - {}^{\lambda}\phi_{\mathbf{h}}^{a}) = 2[(f_{\lambda 1,\mathbf{h}}^{a}/f_{1,\mathbf{h}}^{n}) \cos \delta_{\lambda 1,\mathbf{h}}|F_{1,\mathbf{h}}^{n}|^2$$
$$+ (f_{\lambda 2,\mathbf{h}}^{a}/f_{2,\mathbf{h}}^{n}) \cos \delta_{\lambda 2,\mathbf{h}}|F_{2,\mathbf{h}}^{n}|^2$$
$$+ (f_{\lambda 1,\mathbf{h}}^{a}/f_{1,\mathbf{h}}^{n})|F_{1,\mathbf{h}}^{n}||F_{2,\mathbf{h}}^{n}|$$
$$\times \cos (\phi_{2,\mathbf{h}}^{n} - \phi_{1,\mathbf{h}}^{n} - \delta_{\lambda 1,\mathbf{h}})$$
$$+ (f_{\lambda 2,\mathbf{h}}^{a}/f_{2,\mathbf{h}}^{n})|F_{1,\mathbf{h}}^{n}||F_{2,\mathbf{h}}^{n}|$$
$$\times \cos (\phi_{1,\mathbf{h}}^{n} - \phi_{2,\mathbf{h}}^{n} - \delta_{\lambda 2,\mathbf{h}})]. \tag{11}$$

If eqs. (8), (10), and (11) are combined with eq. (7), it is seen that the unknown quantities $|F_{1,\mathbf{h}}^{n}|$, $|F_{2,\mathbf{h}}^{n}|$, $\cos (\phi_{1,\mathbf{h}}^{n} - \phi_{2,\mathbf{h}}^{n})$ and $\sin (\phi_{1,\mathbf{h}}^{n} - \phi_{2,\mathbf{h}}^{n})$ are expressed in terms of the observed $|^{\lambda}F_{\mathbf{h}}|$ and atomic scattering factor information that is available from tables [12]. The sine function appears from use of the addition formula for cosines applied to the cosine functions in eqs. (10) and (11). Simultaneous equations may be set up by observing $|^\lambda F_h|^2$ and $|^\lambda F_{-h}|^2$ at several wavelengths. In addition, if we choose as our variables, $|F_{1,h}^n|^2$, $|F_{2,h}^n|^2$, $|F_{1,h}^n||F_{2,h}^n|$ $\cos(\phi_{1,h}^n - \phi_{2,h}^n)$, and $|F_{1,h}^n||F_{2,h}^n| \sin(\phi_{1,h}^n - \phi_{2,h}^n)$, ignoring the relationships between them, the system of equations will be linear in these variables, permitting straightforward evaluation.

In the case of centric reflections, $|^\lambda F_h| = |^\lambda F_{-h}|$, $\sin(\phi_{1,h}^n - \phi_{2,h}^n) = 0$, and $\cos(\phi_{1,h}^n - \phi_{2,h}^n)$ assumes only the values 1 or -1. Information in this case therefore concerns the values of $|F_{1,h}^n|^2$ and $|F_{2,h}^n|^2$ and the two possible values of $\cos(\phi_{1,h}^n - \phi_{2,h}^n)$, derivable from measurements of $|^\lambda F_h|^2$ at several wavelengths, but not from duplicate measurements of $|^\lambda F_{-h}|^2$.

Atoms that Scatter Normally and One Type of Anomalous Scatterer

Anomalous dispersion experiments are commonly performed on structures that have a number of types of atoms whose anomalous scattering is quite weak and a very few atoms of a single type whose anomalous scattering is significantly stronger. The theory appropriate to this case is derivable from the previous analysis by setting $f_{\lambda 1,h}''$ equal to zero and reinterpreting $|F_{1,h}^n|^2$ as representing the intensities for all types of atoms in a structure which contribute no significant amount of anomalous scattering. Under these circumstances eq. (8) remains unchanged, except for the new interpretation of $|F_{1,h}^n|^2$, eq. (10) becomes

$$|^\lambda F_h^a|^2 = (f_{\lambda 2,h}''/f_{2,h}^n)^2 |F_{2,h}^n|^2 \tag{12}$$

and eq. (11) becomes

$$2|F_h^n||^\lambda F_h^a| \cos(\phi_h^n - {}^\lambda\phi_h^a) = 2(f_{\lambda 2,h}''/f_{2,h}^n) \cos\delta_{\lambda 2,h}|F_{2,h}^n|^2$$
$$+ 2(f_{\lambda 2,h}''/f_{2,h}^n) \cos\delta_{\lambda 2,h}|F_{1,h}^n||F_{2,h}^n| \cos(\phi_{1,h}^n - \phi_{2,h}^n)$$
$$+ 2(f_{\lambda 2,h}''/f_{2,h}^n) \sin\delta_{\lambda 2,h}|F_{1,h}^n||F_{2,h}^n| \sin(\phi_{1,h}^n - \phi_{2,h}^n). \tag{13}$$

The substitution of eqs. (8), (12), and (13) into eq. (7) gives

$$|^\lambda F_h|^2 = |F_{1,h}^n|^2 + \{1 + (f_{\lambda 2,h}''/f_{2,h}^n)[(f_{\lambda 2,h}''/f_{2,h}^n) + 2\cos\delta_{\lambda 2,h}]\}|F_{2,h}^n|^2$$
$$+ 2[1 + (f_{\lambda 2,h}''/f_{2,h}^n) \cos\delta_{\lambda 2,h}]|F_{1,h}^n||F_{2,h}^n| \cos(\phi_{1,h}^n - \phi_{2,h}^n)$$
$$+ 2(f_{\lambda 2,h}''/f_{2,h}^n) \sin\delta_{\lambda 2,h}|F_{1,h}^n||F_{2,h}^n| \sin(\phi_{1,h}^n - \phi_{2,h}^n). \tag{14}$$

As before, the variables can be considered to be $|F_{1,h}^n|^2$, $|F_{2,h}^n|^2$, $|F_{1,h}^n||F_{2,h}^n|$ $\cos(\phi_{1,h}^n - \phi_{2,h}^n)$, and $|F_{1,h}^n||F_{2,h}^n| \sin(\phi_{1,h}^n - \phi_{2,h}^n)$. They occur linearly in eq. (14) and can be evaluated from a set of simultaneous equations generated from measurements at various λ and at h and $-h$.

Special relations can be derived from eq. (14). For example,

$$|F_{1,h}^n||F_{2,h}^n| \sin(\phi_{1,h}^n - \phi_{2,h}^n) = \frac{|^\lambda F_h|^2 - |^\lambda F_{-h}|^2}{4(f_{\lambda 2,h}''/f_{2,h}^n) \sin\delta_{\lambda 2,h}} \tag{15}$$

and

$$2[1 + (f^a_{\lambda2,h}/f^n_{2,h}) \cos \delta_{\lambda2,h}]|F^n_{1,h}||F^n_{2,h}| \cos (\phi^n_{1,h} - \phi^n_{2,h})$$

$$= \frac{|^\lambda F_h|^2 + |^\lambda F_{-h}|^2}{2} - |F^n_{1,h}|^2$$

$$- \{1 + (f^a_{\lambda2,h}/f^n_{2,h})[(f^a_{\lambda2,h}/f^n_{2,h}) + 2 \cos \delta_{\lambda2,h}]\}|F^n_{2,h}|^2. \quad (16)$$

If measurements are performed at two different wavelengths the quantity $|F^n_{1,h}||F^n_{2,h}| \cos (\phi^n_{1,h} - \phi^n_{2,h})$ can be eliminated from eq. (16), leaving $|F^n_{1,h}|^2$ and $|F^n_{2,h}|^2$ as the only unknown quantities. Measurements at a third wavelength would evidently permit these quantities to be determined from the latter type of equation. Equation (14) shows that all unknowns can, in principle, be evaluated from a two-wavelength experiment. It is worth considering other forms for the equations, however, since with experimental data that contain errors, some forms may provide more accurate evaluations of the unknown quantities than others.

A General Formulation

The case of two different anomalous scatterers is easily generalized to the case of any number of types of atoms, all treated as anomalous scatterers (say q types). Equation (7) remains the same and eqs. (8), (10), and (11) are replaced, respectively, by eqs. (17), (18), and (19):

$$|F^n_h|^2 = \sum_{i=1}^{q} |F^n_{i,h}|^2 + 2 \sum_{i<j}^{q} |F^n_{i,h}||F^n_{j,h}| \cos (\phi^n_{i,h} - \phi^n_{j,h}), \quad (17)$$

$$|^\lambda F^a_h|^2 = \sum_{i=1}^{q} (f^a_{\lambda i,h}/f^n_{i,h})^2 |F^n_{i,h}|^2 + 2 \sum_{i<j}^{q} (f^a_{\lambda i,h}/f^n_{i,h})(f^a_{\lambda j,h}/f^n_{j,h})|F^n_{i,h}||F^n_{j,h}|$$

$$\times \cos (\phi^n_{i,h} - \phi^n_{j,h} + \delta_{\lambda i,h} - \delta_{\lambda j,h}), \quad (18)$$

$$2|F^n_h||^\lambda F^a_h| \cos (\phi^n_h - {}^\lambda\phi^a_h) = 2 \sum_{i=1}^{q} (f^a_{\lambda i,h}/f^n_{i,h})|F^n_{i,h}|^2 \cos \delta_{\lambda i,h}$$

$$+ 2 \sum_{i<j}^{q} |F^n_{i,h}||F^n_{j,h}| [(f^a_{\lambda i,h}/f^n_{i,h}) \cos (\phi^n_{j,h} - \phi^n_{i,h} - \delta_{\lambda i,h})$$

$$+ (f^a_{\lambda j,h}/f^n_{j,h}) \cos (\phi^n_{i,h} - \phi^n_{j,h} - \delta_{\lambda j,h})]. \quad (19)$$

A Special Type of Map

A useful type of Fourier map which can be computed from a one-wavelength experiment was suggested in 1961 by Rossmann [13]. It is defined by

$$M(\mathbf{r}) = \sum_{h} (|^\lambda F_h| - |^\lambda F_{-h}|)^2 \cos 2\pi \mathbf{h} \cdot \mathbf{r} \quad (20)$$

and provides information concerning the interatomic vectors between the atoms that scatter anomalously. When the number of anomalously scattering atoms is quite small, as has generally been the case in applications of the technique made so far, it is possible to derive the locations of the atoms in the unit cell of the crystal from the calculation given in eq. (20). An analysis of the coefficients in eq. (20) by means of eq. (14) clarifies the nature of $M(\mathbf{r})$.

Equation (14) may be written

$$|{}^{\lambda}F_{\mathbf{h}}|^2 = \alpha + \beta, \tag{21}$$

where α is comprised of the first three terms on the right of eq. (14) and β represents the last term. It follows that

$$|{}^{\lambda}F_{-\mathbf{h}}|^2 = \alpha - \beta \tag{22}$$

and

$$|{}^{\lambda}F_{\mathbf{h}}|^2|{}^{\lambda}F_{-\mathbf{h}}|^2 = \alpha^2 - \beta^2. \tag{23}$$

Therefore from eqs. (21), (22) and (23) it follows that

$$(|{}^{\lambda}F_{\mathbf{h}}| - |{}^{\lambda}F_{-\mathbf{h}}|)^2 = 2\alpha - 2(\alpha^2 - \beta^2)^{1/2}. \tag{24}$$

Often α^2 is much larger than β^2 so that

$$(|{}^{\lambda}F_{\mathbf{h}}| - |{}^{\lambda}F_{-\mathbf{h}}|)^2 \simeq \beta^2/\alpha \tag{25}$$

and when $|F_{1,\mathbf{h}}^n|^2$ is large enough so that $\alpha \sim |F_{1,\mathbf{h}}^n|^2$, eq. (22) becomes

$$(|{}^{\lambda}F_{\mathbf{h}}| - |{}^{\lambda}F_{-\mathbf{h}}|)^2 \simeq 4(f_{\lambda 2,\mathbf{h}}^n/f_{2,\mathbf{h}}^n)^2 \sin^2 \delta_{\lambda 2,\mathbf{h}}|F_{2,\mathbf{h}}^n|^2 \sin^2 (\phi_{1,\mathbf{h}}^n - \phi_{2,\mathbf{h}}^n). \tag{26}$$

If the relatively constant $f_{\lambda 2,\mathbf{h}}^n$ and $\sin \delta_{\lambda 2,\mathbf{h}}$ are replaced by a scale factor, it is seen that $M(\mathbf{r})$ can be interpreted as

$$M(\mathbf{r}) \simeq C \sum_{\mathbf{h}} (|F_{2,\mathbf{h}}^n|/f_{2,\mathbf{h}}^n)^2 \sin^2 (\phi_{1,\mathbf{h}}^n - \phi_{2,\mathbf{h}}^n) \cos 2\pi\mathbf{h} \cdot \mathbf{r}. \tag{27}$$

The coefficient in eq. (27) without the factor $\sin^2(\phi_{1,\mathbf{h}}^n - \phi_{2,\mathbf{h}}^n)$ is $(|F_{2,\mathbf{h}}^n|/f_{2,\mathbf{h}}^n)^2$, the proper coefficient for a sharpened Patterson map that represents the distribution of interatomic vectors for the atoms that scatter anomalously. Analysis shows that the effect of $\sin^2(\phi_{1,\mathbf{h}}^n - \phi_{2,\mathbf{h}}^n)$ is mainly to reduce the heights of the peaks representing interatomic vectors by a factor of about one half. A further deterioration of the signal-to-noise ratio can occur when $\alpha \sim |F_{1,\mathbf{h}}|^2$ is not a good approximation for substitution into eq. (25).

In many cases it may be possible to derive much information from a single-wavelength experiment, even when many anomalous scatterers are present. This occurs when, as a consequence of calculating $M(\mathbf{r})$ with eq. (20), the maxima associated with the anomalous scatterers are, for the most part, larger than incorrect maxima and therefore would be distinguishable. The values for the locations of the maxima, $\mathbf{r}_i - \mathbf{r}_j$ can be used to compute $|F_{2,\mathbf{h}}^n|^2$ from

$$|F_{2,\mathbf{h}}^n|^2 = p(f_{2,\mathbf{h}}^n)^2 + 2(f_{2,\mathbf{h}}^n)^2 \sum_{\substack{i<j \\ 1}}^{p} \cos 2\pi\mathbf{h} \cdot (\mathbf{r}_i - \mathbf{r}_j), \tag{28}$$

where p is the number of anomalously scattering atoms in the unit cell of the crystal. Alternatively, in complex situations the Patterson-type map can be altered by establishing a level that acts as a zero level in the map and below which all values are set equal to zero. This altered map can be inverted and the results scaled to give new values for the $|F_{2,h}^{\eta}|^2$.

The values of $|F_{2,h}^{\eta}|^2$ can be used to determine the arrangement of the anomalously scattering atoms. In favorable cases it might by possible to obtain additional information concerning the other variables by use of eqs. (14), (15), and (16).

Feasibility Indications

A Small Protein

The recent direct determination of the structure of a protein crambin, by Hendrickson and Teeter [1], by use of anomalous dispersion at one wavelength affords good insight into the potential of the anomalous dispersion technique. Crambin is a small protein composed of about 400 nonhydrogen atoms that is obtained from the seed of *Crambe abyssinica*. It contains six sulfur atoms in the form of six cysteine residues in disulfide linkages. Crambin is unusually well ordered for a protein and gives diffraction patterns that show better than 1.0 Å resolution.

The crystal structure of crambin was solved directly from the diffraction data of the native protein by use of the anomalous scattering from only six sulfur atoms. CuK_{α} diffraction data were measured to 0.95 Å spacings, including Friedel pairs ($|F_h|,|F_{-h}|$) to 1.5 Å. Maps based on eq. (20) with the Bijvoet differences squared as coefficients, $(|{}^{\lambda}F_h| - |{}^{\lambda}F_{-h}|)^2$, revealed the positions for the six sulfur atoms clearly. After refinement of the positions of the sulfur atoms by optimizing their fit to the values of the magnitudes of $(|{}^{\lambda}F_h| - |{}^{\lambda}F_{-h}|)$, the disulfide bond distances were found to be 2.02, 2.03 and 2.04 Å, in fine agreement with expected values. Phases for the nonanomalously scattering portion of the structure were derived by a probabilistic combination of the phase information from the anomalous scattering and that from the partial structure of sulfur atoms. A Fourier map was computed from the resulting phases at 1.5 Å resolution from which, by direct inspection, it was possible to locate tentative atomic positions for the entire protein and some of the solvent molecules. In the course of refinement of the structure, several corrections were made to the initial atomic arrangement. The present crystallographic R factor is 0.12.

It is most noteworthy that the anomalous scattering of only six sulfur atoms with CuK_{α} radiation sufficed to locate those atoms accurately in a structure containing about 400 C, N, and O atoms otherwise, and that it was further possible from a combination of anomalous scattering and the positions of the sulfur atoms to glean enough information to solve the entire structure of crambin.

A Model for Nucleotide Studies

To gain insight into the possible applicability of the anomalous dispersion technique to native polynucleotides, a feasibility study was performed on an intercalation complex composed of proflavine-cytidylyl-(3′,5′)-guanosine whose structure was recently investigated by Berman and co-workers [14]. The two main anomalously scattering atoms are P and S, having similar scattering characteristics. In a native polynucleotide it is expected that the main anomalous scatterers would be predominantly P atoms. With the presence of extra flavin molecules here, the effect of the S atom is to make the number of main anomalous scatterers as compared to the number of lighter atoms about the same.

Calculations of Fourier maps were made with the use of eq. (20). In order to simulate a variety of experimental situations, the quantities $|^{\lambda}F_{\mathbf{h}}|$ and $|^{\lambda}F_{-\mathbf{h}}|$ were computed from the atomic coordinates [14], and then random errors were introduced independently into each quantity at various levels of average error. In addition, data were used only to 1.5 Å resolution for the 82 nonhydrogen atom structure. With independent, random average errors of about 1.0% in the $|^{\lambda}F_{\mathbf{h}}|$ and $|^{\lambda}F_{-\mathbf{h}}|$ computed for CuK_{α} radiation, a Fourier cosine map with the $(|^{\lambda}F_{\mathbf{h}}| - |^{\lambda}F_{-\mathbf{h}}|)^2$ as coefficients was computed. All peaks except the one concerned with the P–P interatomic vector were identifiable among the highest peaks. A zero level at 0.33 of the value of the highest peak was established, and the map was inverted.

It was found that among the 20 largest intensities obtained from the Fourier inversion of the two-atom structure (1S and 1P), all but two occur among the largest 56 correct intensities. For the 35 largest intensities obtained from the Fourier inversion, all but seven occur among the largest 57 correct intensities. The number of exceptions decreases as the accuracy of $|F_{\mathbf{h}}| - |F_{-\mathbf{h}}|$ improves. Although it is not possible to be certain about new experimental problems, the results of this study suggest that anomalous dispersion be considered as a potentially valuable tool in the investigation of the structures of the more complex nucleotides.

Concluding Remarks

New theoretical developments in anomalous dispersion and some feasibility estimates based on a type of Patterson map suggested by Rossmann [13] and carried through to a new level of potential by Hendrickson and Teeter [1] suggest that the anomalous dispersion technique offers considerable future opportunity for the direct determination of the structures of macromolecular systems in biology. The theoretical developments could perhaps be most helpful in the case of complex structures in which it is not readily possible to immediately determine the atomic positions of the anomalous scatterers. They may also facilitate the determination of phases for nonanomalous scatterers from those of the anomalous ones. The advent of dedicated synchrotron sources should provide a good opportunity to pursue appropriate investigations, although there appears to be a great deal that can be done at the usual laboratory facilities.

397

Acknowledgment

The author would like to express his appreciation to Dr. Wayne Hendrickson for discussions concerning his work on the structure of the protein crambin. He also wishes to thank Mr. Stephen A. Brenner for performing the calculations on the intercalation complex described here.

Bibliography

[1] W. Hendrickson and M. Teeter, unpublished paper.
[2] F. R. Ahmed, K. Huml and B. Sedlacek, Eds., *Crystallographic Computing Techniques* (Munksgaard, Copenhagen, 1976).
[3] J. M. Bijvoet, Nature **173**, 888 (1954).
[4] J. M. Bijvoet, A. F. Peerdeman, and A. J. van Bommel, Nature **168**, 271 (1951).
[5] A. F. Peerdeman, A. J. van Bommel, and J. M. Bijvoet, Proc. R. Soc. Amsterdam B **54**, 16 (1951).
[6] A. J. van Bommel, Proc. R. Soc. Amsterdam B **56**, 268 (1953).
[7] Y. Okaya and R. Pepinsky, Phys. Rev. **103**, 1645 (1956).
[8] C. M. Mitchell, Acta Cryst. **10**, 475 (1957).
[9] A. Herzenberg and H. M. S. Lau, Acta Cryst. **22**, 24 (1967).
[10] S. Ramaseshan and S. C. Abrahams, Eds. *Anomalous Scattering* (Munksgaard, Copenhagen, 1975).
[11] J. Karle, Appl. Opt. **6**, 2132 (1967).
[12] D. T. Cromer, in *International Tables for X-Ray Crystallography*, J. A. Ibers and W. C. Hamilton, Eds. (Kynoch Press, Birmingham, England, 1974), p. 148.
[13] M. G. Rossmann, Acta Cryst. **14**, 383 (1961).
[14] H. M. Berman, W. Stallings, H. L. Carrell, J. P. Glusker, S. Neidle, G. Taylor, and A. Achari, Biopolymers **18**, 2405 (1979).

Received March 13, 1980

Triplet Phase Invariants from an Exact Algebraic Analysis of Anomalous Dispersion

By Jerome Karle

Laboratory for the Structure of Matter, Naval Research Laboratory, Washington, DC 20375, USA

(Received 11 January 1984; accepted 11 April 1984)

Abstract

In a previous investigation, a system of exact algebraic equations was derived for any number and type of anomalous scatterers. Solution of the equations provides information concerning intensities of scattering and certain phase differences. In this paper, it is shown that when appropriate combinations of the phase differences and their values are made, the result is the evaluation of the differences of pairs of triplet phase invariants, one associated with the macromolecular structure and the second associated with the structure of the anomalous scatterers. It is usually easy to satisfy the condition that the values of triplet phase invariants associated with the structures of the anomalous scatterers be close to zero. This permits the evaluation of triplet phase invariants associated with the macromolecular structure. Since the structures of the anomalous scatterers are quite simple in many of the substances of interest, a theoretical and experimental study of the distribution of values for triplet phase invariants associated with simple structures has been carried out. This has provided a quantitative insight into the distribution of values of the cosines of triplet phase invariants for such structures. It has also identified useful functions, based on knowledge of the values of normalized structure factor magnitudes, that permit a reliable prediction of those triplet phase invariants that have values close to zero. In the mathematical sense, the evaluation of the triplet phase invariants for a macromolecular structure, solely from the intensity data, is exact, except for the deviation of the triplet phase invariants for the structure of the anomalous scatterers from zero. No structural information concerning the anomalous scatterers is required. In practice, of course, experimental error will affect the accuracy of the information derived from the algebraic equations. The possibility of overdeterminacy in the equations should be beneficial in reducing the effect of experimental error.

tities in the equations are quantities that represent nonanomalous scattering and hence are independent of wavelength. This is effected by separating the contribution of the real and imaginary parts of the atomic scattering factors from that of the normal atomic scattering factors in the defining equations for the structure factors. The resulting simultaneous equations contain the quantities representing the effects of anomalous dispersion as separate factors modifying the wavelength-independent unknown quantities. Evaluation of the factors arising from anomalous dispersion is easily obtained from the tabulated values for the real and imaginary corrections to the normal atomic scattering factors.

The simultaneous equations have several favorable characteristics. With appropriate definitions for the unknown quantities, the equations are linear in the variables. The unknown quantities are comprised in part of the intensities of scattering for each of the individual types of atoms present. Their values correspond to individual structures in which each type of atom would be present in isolation from the others. Additional unknown quantities are phase differences arising from the scattering from the different types of atoms. The equations retain their favorable characteristics and exactness no matter how many types of anomalous scatterers are present.

A feature of the simultaneous equations that makes them potentially valuable is the existence of the individual intensities of scattering for the various types of atoms as unknown quantities to be evaluated by use of the equations. Once the intensities are known for the anomalous scatterers from solving the simultaneous equations, it is possible to solve for the structure of the anomalous scatterers. If this structure is too complicated to be amenable to an analysis by means of a Patterson function, it is still possible to undertake a determination of the structure by direct methods. Once the structure of just one type of the anomalous scatterers is known, the information provided by the simultaneous equations permits the solution of the entire structure.

This may well be the optimal strategy for using the simultaneous equations. However, it is also possible to use the simultaneous equations to obtain evalu-

Introduction

In a previous investigation, an algebraic analysis of multiple-wavelength anomalous dispersion data resulted in a set of simultaneous equations without approximation (Karle, 1980a). The unknown quan-

Acta Cryst. (1985). **A41**, 394–399

ations of triplet phase invariants in the absence of information concerning the structure of the anomalous scatterers. It is the purpose of this paper to show how these evaluations may be obtained. In addition, it will also be seen how information from isomorphous replacement may be introduced into the simultaneous equations.

The accuracy of the evaluations of the triplet phase invariants for a substructure consisting of non-anomalously scattering atoms, for example, depends upon how closely the values of appropriate triplet phase invariants for a substructure composed of anomalously scattering atoms approximate to zero. For the simple structures that often apply to the anomalously scattering atoms, the approximation to zero is generally easily achieved. In order to obtain a more quantitative insight into this matter, some test calculations were carried out concerning the distribution of values of the cosines of triplet phase invariants for simple structures and comparisons were made with results from theoretical formulas modified to facilitate the prediction of these distributions.

Theory

A simple result from the algebraic analysis (Karle, 1980a) that illustrates the characteristics of the simultaneous equations described above will now be presented. It concerns a structure which is composed of atoms that scatter normally and one type of atom that scatters anomalously. A representative equation is

$$|F_{\lambda h}|^2 = |F_{1,h}^n|^2 + \{1 + (f_{\lambda 2,h}^a/f_{2,h}^n)$$
$$\times [(f_{\lambda 2,h}^a/f_{2,h}^n) + 2 \cos \delta_{\lambda 2,h}] \} |F_{2,h}^n|^2$$
$$+ 2[1 + (f_{\lambda 2,h}^a/f_{2,h}^n) \cos \delta_{\lambda 2,h}]$$
$$\times |F_{1,h}^n| |F_{2,h}^n| \cos (\varphi_{1,h}^n - \varphi_{2,h}^n)$$
$$+ 2(f_{\lambda 2,h}^a/f_{2,h}^n) \sin \delta_{\lambda 2,h} |F_{1,h}^n|$$
$$\times |F_{2,h}^n| \sin (\varphi_{1,h}^n - \varphi_{2,h}^n), \quad (1)$$

where $|F_{\lambda h}|^2$ is the measured magnitude squared of the structure factor at wavelength λ, $|F_{1,h}^n|^2$ is the magnitude squared of the structure factor for the nonanomalously scattering atoms and $|F_{2,h}^n|^2$ is the magnitude squared of the structure factor for the anomalously scattering atoms, but scattering as if they were doing so normally. The measured $|F_{\lambda h}|^2$ are corrected for vibrational effects and the latter are also absent from $|F_{1,h}^n|^2$ and $|F_{2,h}^n|^2$. The quantities $f_{\lambda j,h}^a$ and $\delta_{\lambda j,h}$ are defined for a particular λ,

$$f_{\lambda j,h}^a = [(f_{\lambda j,h}')^2 + (f_{\lambda j,h}'')^2]^{1/2} \quad (2)$$

and

$$\delta_{\lambda j,h} = \tan^{-1}(f_{\lambda j,h}''/f_{\lambda j,h}'). \quad (3)$$

f' and f'' are the real and imaginary corrections, respectively, to the normal atomic scattering factor, f''. They represent the effects of anomalous dispersion and are tabulated in *International Tables for X-ray Crystallography* (Cromer, 1974). The total atomic scattering factor is then

$$f = f^n + f' + if''. \quad (4)$$

The phase angle $\varphi_{1,h}^n$ is the angle associated with the structure factor contributed by the nonanomalously scattering atoms and $\varphi_{2,h}^n$ is the angle associated with the structure factor contributed by the anomalously scattering atoms, but scattering as if they were doing so normally. The subscript 1 refers to the non-anomalously scattering atoms and the subscript 2 to the anomalously scattering ones.

Closely related equations can be formed from (1) by performing anomalous dispersion experiments at various wavelengths and noting that an equation for $\mathbf{h}$ is different from that for $-\mathbf{h}$. The equations are linear if the unknown quantities are chosen to be $|F_{1,h}^n|^2$, $|F_{2,h}^n|^2$, $|F_{1,h}^n||F_{2,h}^n| \cos (\varphi_{1,h}^n - \varphi_{2,h}^n)$ and $|F_{1,h}^n||F_{2,h}^n| \sin (\varphi_{1,h}^n - \varphi_{2,h}^n)$. Having more than the algebraic minimum of equations would permit the use of least-squares methods to help compensate for experimental error. It is also possible to take advantage of the relation $\sin^2\varphi + \cos^2\varphi = 1$ and the non-negativity of the magnitudes of the structure factors. If the anomalous scattering arises predominantly from the isomorphous addition of a heavy atom to the native structure, then measurement of the diffraction intensities for the native structure gives values for the $|F_{1,h}^n|^2$, thereby reducing the number of unknowns in the simultaneous equations. In this way, information from isomorphous replacement may be combined with that from anomalous dispersion in the equations.

An evaluation of the unknown quantities in (1) gives values for the angle differences, $\varphi_{1,h}^n - \varphi_{2,h}^n$, and for the intensities that would be obtained for the structure of the nonanomalously scattering atoms, $|F_{1,h}^n|^2$, and for the structure of the anomalously scattering atoms scattering normally, $|F_{2,h}^n|^2$. Information concerning the $|F_{2,h}^n|^2$ could be used to solve for the structure of the anomalously scattering atoms. With knowledge of this structure, values for the $\varphi_{2,h}^n$ could be computed readily, thus leading to an evaluation of the remaining unknown quantities, the $\varphi_{1,h}^n$. With this information, the computation of the structure of the nonanomalously scattering atoms is readily performed.

It is possible to generalize the result given in (1) to any number of types of atoms, all treated as anomalous scatterers (Karle, 1980a). If there are q types of atoms, the general result is

$$|F_{\lambda h}|^2 = \sum_{i=1}^{q} |F_{i,h}^n|^2 + 2 \sum_{i<j}^{q} |F_{i,h}^n||F_{j,h}^n| \cos (\varphi_{i,h}^n - \varphi_{j,h}^n)$$
$$+ \sum_{i=1}^{q} (f_{\lambda i,h}^a/f_{i,h}^n)^2 |F_{i,h}^n|^2$$
$$+ 2 \sum_{i<j}^{q} (f_{\lambda i,h}^a/f_{i,h}^n)(f_{\lambda j,h}^a/f_{j,h}^n) |F_{i,h}^n||F_{j,h}^n|$$

$$\times \cos{(\varphi_{i,h}^n - \varphi_{j,h}^n + \delta_{\lambda i,h} - \delta_{\lambda j,h})}$$

$$+ 2 \sum_{i=1}^{q} (f_{\lambda i,h}^a / f_{i,h}^n) |F_{i,h}^n|^2 \cos{\delta_{\lambda i,h}}$$

$$+ 2 \sum_{i<j}^{q} |F_{i,h}^n||F_{j,h}^n|[(f_{\lambda i,h}^a / f_{i,h}^n)$$

$$\times \cos{(\varphi_{j,h}^n - \varphi_{i,h}^n - \delta_{\lambda i,h})}$$

$$+ (f_{\lambda j,h}^a / f_{j,h}^n) \cos{(\varphi_{i,h}^n - \varphi_{j,h}^n - \delta_{\lambda j,h})}]. \quad (5)$$

The equations represented by (5) retain the features noted for the simple case represented by (1). They are exact and, if the unknown quantities are chosen in a fashion similar to the four described for (1), the equations are linear in the unknowns. Study of (5) shows that, after evaluation of the unknown quantities, the determination of only one of the substructures corresponding to one type of anomalous scatterer by use of the appropriate $|F_{i,h}^n|$ would permit the evaluation of all the phases required for the determination of all the remaining substructures.

This may well be the procedure of choice in the future. There is, however, an alternative to solving for some substructure since it is possible to obtain evaluations of triplet phase invariants by means of the algebraic equations and a simple probabilistic argument without any need to determine a substructure. This matter will now be discussed.

Triplet phase invariants

The derivation of triplet phase invariants may be illustrated by considering the case of a structure composed of essentially nonanomalous scatterers and one type of anomalous scatterer represented by (1). It is apparent that solution of the equations to evaluate the unknown quantities would give values for the phase differences, $\varphi_{1,h}^n - \varphi_{2,h}^n$. From the values of many of the phase differences, it is possible to form the sums of suitably chosen ones,

$$\varphi_{1,h}^n + \varphi_{1,k}^n + \varphi_{1,(\bar{h}+\bar{k})}^n - \varphi_{2,h}^n - \varphi_{2,k}^n - \varphi_{2,(\bar{h}+\bar{k})}^n = A_{hk}, \quad (6)$$

where A_{hk} is known from the values of the three individual phase differences comprising (6). It is seen that (6) consists of the difference of two triplet phase invariants, one for the structure consisting of non-anomalously scattering atoms, and the second for the anomalously scattering atoms (heavy-atom structure). Values for the $|F_{2,h}^n|$ indicate which of the triplet phase invariants for the heavy-atom structure are associated with large products, $|F_{2,h}^n F_{2,k}^n F_{2,(\bar{h}+\bar{k})}^n|$. Such structures are usually rather simple, so that when the products of the magnitudes of the structure factors are large the triplet phase invariants for the heavy-atom structure can quite reliably be set equal to zero. This gives, under the conditions of large products of structure factor magnitudes for the heavy-atom structure,

$$\varphi_{1,h}^n + \varphi_{1,k}^n + \varphi_{1,(\bar{h}+\bar{k})}^n \simeq A_{hk}, \quad (7)$$

an evaluation of triplet phase invariants for the unsubstituted structure.

The evaluation of a sufficient number of triplet phase invariants by use of (7) could obviate the necessity for determining the heavy-atom structure, since phase determination could, in principle, proceed from knowledge of the values of the triplet phase invariants.

Triplet phase invariant distributions

Calculations were made of the distributions of triplet phase invariants in simple equal-atom structures, three- and nine-atom structures in space group $P1$ and three- and nine-atom structures in the asymmetric unit of space group $P2_12_12_1$. The atoms were distributed at random with suitable separations between them. This is appropriate for the heavy-atom substructures that occur, for example, in heavy-atom substitution structures.

For the various structures, thousands of cosines of triplet phase invariants, $\cos{(\varphi_h + \varphi_k + \varphi_{\bar{h}+\bar{k}})} = \cos{\Phi_{hk}}$, were computed and their values were ordered according to the values of the products of associated normalized structure factors, $2N^{-1/2}|E_h E_k E_{\bar{h}+\bar{k}}|$, where N is the number of atoms in the unit cell. Groups of five thousand cosine invariants in the ordered sequence were averaged to give $\langle \cos{\Phi_{hk}} \rangle$ and plotted against the average value of the corresponding $2N^{-1/2}|E_h E_k E_{\bar{h}+\bar{k}}|$. Plots of the results for two of the structures are shown in Figs. 1 and 2.

Values for $\langle \cos{\Phi_{hk}} \rangle$ were also computed from four different theoretical formulas with the objective of determining which of the formulas can best represent the observed distributions. A general formula for the expected value of a cosine invariant is

$$\langle \cos{\Phi_{hk}} \rangle = I_1(t|E_h E_k E_{\bar{h}+\bar{k}}|) / I_0(t|E_h E_k E_{\bar{h}+\bar{k}}|). \quad (8)$$

The four different theoretical formulas are represented in (8) by different definitions for the quantity t. The definitions corresponding to the data points shown in Figs. 1 and 2 are as follows:

$S:$
$$t = 2N^{-1/2}; \quad (9)$$

$J:$
$$t = 2N^{-1/2} + 2N^{-3/2}(|E_h|^2 + |E_k|^2 + |E_{\bar{h}+\bar{k}}|^2 - 3)$$
$$+ 2N^{-5/2}[|E_h|^4 + |E_k|^4 + |E_{\bar{h}+\bar{k}}|^4$$
$$+ (11/4)(|E_h E_k|^2 + |E_h E_{\bar{h}+\bar{k}}|^2 + |E_k E_{\bar{h}+\bar{k}}|^2)$$
$$- (9/2)(|E_h|^2 + |E_k|^2 + |E_{\bar{h}+\bar{k}}|^2) + 2]; \quad (10)$$

$I:$
$$t = 2N^{-1/2}/q_1, \quad (11)$$

where

$$q_1 = 1 + 2|U_h U_k U_{\bar{h}+\bar{k}}| - |U_h|^2 - |U_k|^2 - |U_{\bar{h}+\bar{k}}|^2, \quad (12)$$
$$U = EN^{-1/2}; \quad (13)$$

$D:$
$$t = 2N^{-1/2} V_{3,p} \quad (14)$$

where

$$V_{3,p} = \frac{(1 - |U_h|^2) + (1 - |U_k|^2) + (1 - |U_{\bar{h}+\bar{k}}|^2)}{3(1 - |U_h|^2)(1 - |U_k|^2)(1 - |U_{\bar{h}+\bar{k}}|^2)}. \quad (15)$$

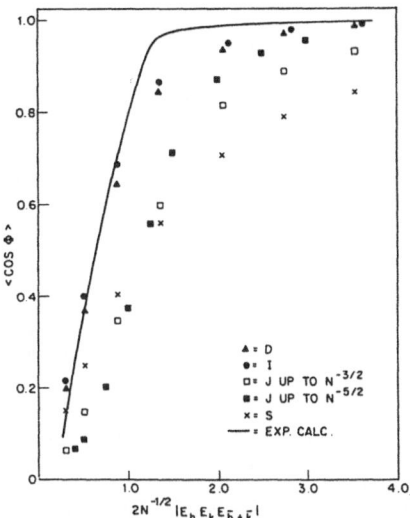

Fig. 1. Variation of the expected values of the cosines of triplet phase invariants with $2N^{-1/2}|E_hE_kE_{\bar{h}+\bar{k}}|$ for a three-atom structure in space group $P1$. The solid line is drawn among values computed experimentally from a large number of cosine invariants. Values from various theoretical formulas are plotted by means of symbols. The best fits to the larger values of $2N^{-1/2}|E_hE_kE_{\bar{h}+\bar{k}}|$ are given by functions I and D. There is a region, however, in which even these functions give estimates that are significantly too low.

In order to compute (10), (11) and (14) for given values of $|E_hE_kE_{\bar{h}+\bar{k}}|$, as required for Figs. 1 and 2, it was assumed that $|E_h| = |E_k| = |E^*_{\bar{h}+\bar{k}}| = |E_hE_kE_{\bar{h}+\bar{k}}|^{1/3}$.

The expected value formula labelled S comes from the probability distribution for a cosine invariant derived by Cochran (1955) with the use of the central limit theorem. Formula J comes from the exponential form of the conditional joint probability distribution for a cosine invariant (Karle, 1972; Karle & Gilardi, 1973), a form designed to diminish the effect of asymptotic convergence when normalized structure factors of large magnitude are present. The first term in J corresponds to S and the second and third terms are higher-order corrections. Formulas I and D have a heuristic origin and are intended to correct for the fact that formula S underestimates the distribution when the products of normalized structure factor magnitudes $|E_hE_kE_{\bar{h}+\bar{k}}|$ are large. Formula I is derived from characteristics of a third-order determinantal inequality (Karle, 1972) and formula D is derived from a joint probability distribution expressed in terms of the determinants that are associated with the non-negativity of electron density distributions in a crystal (Karle, 1978; Karle, 1980b).

The results of the computations for a three-atom structure in space group $P1$ are shown in Fig. 1. It

is seen that the distribution for the structure composed of just three atoms is not well represented by S or even by J, the joint distribution carried out to higher-order terms in the exponential form. There are actually two plots for J given in Fig. 1, one that omits the term of order $N^{-5/2}$ and one that includes it. Quite favorably for the assumption leading to (7) from (6), the distribution of the cosine invariants is much closer to unity than predicted by S or J. The theoretical formulas I and D give a better fit, but the actual distribution is still quite apparently closer to unity in the range $2N^{-1/2}|E_hE_kE_{\bar{h}+\bar{k}}| = 1-3$ and greater.

At the other extreme of complexity in the test problems is the nine-atom structure in space group $P2_12_12_1$. The results of the computations for this structure are shown in Fig. 2. It is seen that the function S again predicts values here that are too low. The functions J, I and D give a rather good fit to the experimental curve. The points for J are omitted in Fig. 2. They fit the experimental curve quite precisely except below $2N^{-1/2}|E_hE_kE_{\bar{h}+\bar{k}}| \sim 0\cdot4$ where the calculation of J gives values that are slightly high.

Experimental calculations of the averages of cosines of triplet phase invariants were made for a

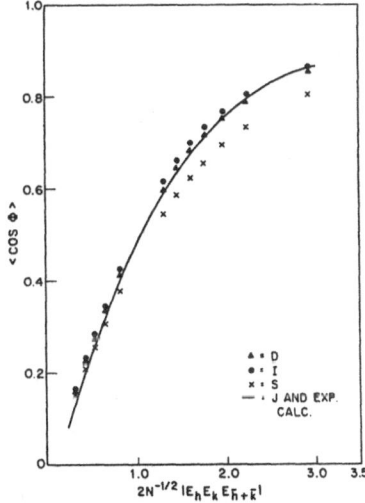

Fig. 2. Variation of the expected values of the cosines of triplet phase invariants with $2N^{-1/2}|E_hE_kE_{\bar{h}+\bar{k}}|$ for a nine-atom structure in the asymmetric unit of a unit cell in space group $P2_12_12_1$. The solid line is drawn among values computed experimentally from a large number of triplet phase invariants. Values from various theoretical formulas are plotted by means of symbols. The function J is not explicitly plotted since it coincides with the experimental curve. The function S gives estimates that are too low for the values of the cosine invariants for the larger values of $2N^{-1/2}|E_hE_kE_{\bar{h}+\bar{k}}|$. The function D is a good fit to the experimental curve in the latter region while function I has a tendency toward modest overestimation as the values of $2N^{-1/2}|E_hE_kE_{\bar{h}+\bar{k}}|$ decrease.

second three-atom structure in space group $P1$ and a second nine-atom structure in space group $P2_12_12_1$ to determine whether significant deviations from the distributions observed for the first structures would occur. None were observed.

For the nine-atom structure in $P1$ and the three-atom structure in the asymmetric unit of $P2_12_12_1$, J fitted the large values well down to a value of about 0.86 for the average of the cosines of the triplet phase invariants but S underestimated seriously the larger values of the cosines of the triplet phase invariants. J underestimated the cosine invariants between values of about 0.45 and 0.86. Below 0.45, it again continued to fit well. The possibility that the term of order $N^{-7/2}$ would improve this was not tested. The calculations I and D fitted well from the largest values for the cosine invariants down to a value of about 0.60 after which they overestimated the values.

In some additional test calculations, it was found that, for a two-atom structure in space group $P1$, the average values of experimentally calculated triplet cosine invariants were unity until the value of $2N^{-1/2}|E_h E_k E_{\bar{h}+\bar{k}}|$ fell below about 0.55, at which point the values decreased precipitously toward zero and slightly negative numbers. In the region below $2N^{-1/2}|E_h E_k E_{\bar{h}+\bar{k}}| = 2$, all the theoretical functions, D, I, J and S underestimated badly the experimental calculations. The functions I and D were better than J and S. For a one-atom structure in the asymmetric unit of space group $P2_12_12_1$, the values of the exponentially calculated cosine invariants were fairly well represented by I and D down to $2N^{-1/2}|E_h E_k E_{\bar{h}+\bar{k}}| \sim 0.6$, after which they overestimated the values of the cosine invariants. The estimates from functions S and J were systematically too low.

The calculations indicate that for the simple structures that are expected to occur in heavy-atom derivatives of macromolecules, it should be easy to satisfy the criterion for (7), namely that triplet phase invariants for the heavy-atom structure have values close to zero for the larger values of $2N^{-1/2}|E_h E_k E_{\bar{h}+\bar{k}}|$ appropriate to the heavy-atom structure. The calculations also show that the functions I and D provide a satisfactory description of the distributions except for an underestimation of the distribution for two and three-atom structures in space group $P1$ in the range $2N^{-1/2}|E_h E_k E_{\bar{h}+\bar{k}}| = 1-3$. The three-atom test example implies that, for very simple structures, the distribution of triplet invariants is closer to zero than even theoretical formulas improved to take account of the effect of large values of the magnitudes of the normalized structure factors would predict, a rather favorable circumstance. I and D can often be used as theoretical estimators of the closeness to zero of triplet phase invariants for simple structures. Alternatively, a few calculations such as the ones performed here can be used as a guide to the validity of (7). The reliability of (7) will depend, in addition, on the accuracy with which the A_{hk} can be obtained from experimental data.

Concluding remarks

This paper shows that the values of triplet phase invariants can be obtained from an exact algebraic analysis of anomalous dispersion data. This can be achieved with very high accuracy in the mathematical sense since the only uncertainty would derive from deviations from zero of triplet phase invariants associated with the structure of the anomalous scatterers. For macromolecules, the latter structures are usually quite simple and it is readily possible to satisfy the requirement for (7) to be valid, namely that the triplet phase invariants associated with the structure of the anomalous scatterers have values close to zero. The accuracy of the triplet phase invariants obtained from (7) is then mainly dependent upon the accuracy with which A_{hk} in (7) can be obtained from experimental data. It is easy to imagine circumstances in a multiwavelength experiment in which the number of equations can exceed the number of unknowns by a factor of two or more. This has the potential to reduce the effect of experimental error and thereby improve the accuracy.

It was also noted that the algebraic equations, from which the theoretical results of this paper follow, were of a form that permitted intensity information from isomorphous replacement experiments to be easily introduced into the equations, thus reducing the number of known quantities to be determined.

Use of the algebraic equations to provide the values of triplet phase invariants may not be the optimal way to derive phase information from them. In order to obtain phase information from the equations otherwise, however, it would be necessary to solve for the structure of at least one type of anomalous scatterer from knowledge of the corresponding intensities that occur as unknown quantities in the equations. If for some reason this structure determination falters, it may be possible to proceed with the use of the values of the triplet phase invariants.

There are some additional virtues of the anomalous dispersion technique that are worth noting. Essentially, the power of the method does not deteriorate with complexity. The power depends upon the contribution of the anomalous scatterers to the measured intensities relative to that of the nonanomalous scatterers. So long as a favorable ratio is maintained, complex systems remain accessible. A second point of interest concerns the fact that the number of atoms in a structure that are strong anomalous scatterers in the usual systems of interest is a rather small fraction of the total. It is therefore appropriate to expect that a rather large amount of data would be obtained from an anomalous dispersion experiment relative to the number of atoms that scatter anomalously. This facilitates the determination of the structure of the anomalous scatterers by use of the corresponding intensities that are obtained from the solution of the algebraic equations. It also facilitates the evaluation of a large number of triplet phase invariants from (7) because with many data the criterion for the validity of this equation, namely values close to zero for the triplet phase invariants associated with the structure of the anomalous scatterers, is readily satisfied in a large number of instances.

I wish to thank Mr Stephen Brenner for writing the appropriate programs and making the computations reported here.

This research was supported in part by USPHS grant GM30902.

References

COCHRAN, W. (1955). *Acta Cryst.* 8, 473–478.
CROMER, D. T. (1974). *International Tables for X-ray Crystallography*, Vol. IV, edited by J. A. IBERS & W. C. HAMILTON, pp. 148–151. Birmingham: Kynoch Press.
KARLE, J. (1972). *Acta Cryst.* B28, 3362–3369.
KARLE, J. (1978). *Proc. Natl Acad. Sci. USA*, 75, 2545–2548.
KARLE, J. (1980a). *Int. J. Quantum Chem.* 7, 357–367.
KARLE, J. (1980b). *Acta Cryst.* A36, 800–802.
KARLE, J. & GILARDI, R. D. (1973). *Acta Cryst.* A29, 401–407.

Unique or Essentially Unique Results from One-Wavelength Anomalous Dispersion Data

By J. Karle

Laboratory for the Structure of Matter, Naval Research Laboratory, Washington, DC, USA 20375-5000

(*Received* 1 *November* 1984; *accepted* 26 *February* 1985)

Abstract

An experiment that is sensitive to anomalous dispersion effects will produce at one wavelength independent intensity information at a reciprocal-lattice point and its negative. These pairs of intensities are known as Bijvoet pairs. The usual analysis of the implications of Bijvoet pairs leads to the conclusion that they generate a twofold ambiguity in the evaluation of certain phase differences. In this paper, it is shown that additional information contained in the Bijvoet pairs, and not normally used in the analysis leading to the implication of twofold ambiguity, can be used to obtain unique or essentially unique values for the phase differences of interest with potentially useful accuracy. The accuracy, of course, depends upon the accuracy of the data, but a test example has shown considerable insensitivity to such errors. The analysis presented here is based on an exact algebraic analysis of the intensity equations associated with the anomalous dispersion technique. Although the theory is quite general, applying exactly to any number or type of anomalously scattering atoms at any number of wavelengths, the application here concerns the case of one type or one predominant type of anomalously scattering atoms in a one-wavelength experiment. It is noted that in the two equations associated with the Bijvoet pairs there are three unknown quantities. It is shown, however, that the two intensity data provide enough information to evaluate the three unknown quantities to good approximation in an essentially unique fashion, which, in addition, can be effected in a least-squares calculation. The phase information of interest that is obtained concerns the values of phase differences, $\varphi_{1,\mathbf{h}}^n - \varphi_{2,\mathbf{h}}^n$, between phases associated with the structure of nonanomalously scattering atoms and those associated with the structure of the anomalously scattering atoms, respectively, with all atoms scattering as if there were no anomalous dispersion.

Introduction

An exact algebraic analysis of multiple-wavelength anomalous dispersion data resulted in an essentially linear system of simultaneous equations (Karle, 1980). The unknown quantities are intensities and phase differences that would be obtained if there were no anomalous dispersion. The equations are exact for any number or types of anomalously scattering atoms. Variation of the values of the observed intensities is expressed in the simultaneous equations by means of factors that modify the unknown quantities and can be evaluated from known values for the real and imaginary corrections to the atomic scattering factors. Isomorphous replacement information is readily incorporated into the system of equations.

An evident application of the simultaneous equations would be to measure intensity data at several wavelengths so that the number of independent data would at least equal the number of unknown quantities. Analyses should benefit in increased accuracy from an excess of independent data over unknown quantities. In this paper, a particularly simple but potentially useful case is considered, namely, a one-wavelength experiment applied to the case in which the substance of interest contains one type or one predominant type of anomalously scattering atom. An algebraic analysis of the case of one type of anomalous scatterer has recently been presented by Woolfson (1984) in terms of two sets of data, one with anomalous scattering and one without. The discussion here is restricted to only one set of data with anomalous effects.

In a one-wavelength experiment involving anomalous dispersion, values are obtained for independent intensities associated with a reciprocal vector and its negative, Bijvoet pairs. Each Bijvoet pair gives rise to two independent equations. Each of the equations contains as unknown quantities two intensities concerned with the nonanomalous scattering, and a phase difference, *i.e.* three unknown quantities. It would appear that two independent equations containing three unknown quantities would be of little use. However, the two unknown intensities bear an approximate relationship to the two measured intensities, which can be made use of. As a consequence, it will be seen that a system of equations can be generated that can be solved by a least-squares technique to yield values for the unknown intensities and the unknown phase differences in a unique or essentially unique fashion. The term 'essentially unique' implies the existence of a procedure that distinguishes between two alternative values, even when they are fairly close. In the latter case, a variety

Acta Cryst. (1985). A41, 387-394

of selections would approximate the correct answer well. Information concerning the structure of the anomalous scatterers is not required. The potential of the mathematical system to be described here may be compared to the usual analysis of a one-wavelength anomalous dispersion experiment in noncentric systems, which would, in general, imply the existence of a twofold ambiguity, even when the structure of the anomalous scatterers is known.

In the past, unique phase information has been extracted from one-wavelength anomalous dispersion experiments. A useful way for resolving the ambiguity that was normally obtained was suggested by Peerdeman & Bijvoet (1956) and by Ramachandran & Raman (1956). It was to choose, between the two alternatives found in their analyses, that phase that was closest to the phase that could be computed from known positions for the anomalously scattering atoms. This gave a correct choice in a majority of instances. This method was used by Dale, Hodgkin & Venkatesan (1963) in the study of the structure of an aquo cyanide of the natural vitamin B_{12} nucleus containing cobalt. More recently, use was made in a probabilistic fashion by Hendrickson & Teeter (1981) of the known positions of anomalously scattering sulfur atoms in the investigation of the structure of the macromolecule, crambin, to facilitate the resolution of phase ambiguities. In all these applications, the known structure of the anomalously scattering atoms was used. The analysis presented here for one-wavelength anomalous dispersion data is distinguished from the earlier ones in that it gives essentially unique phase information without the use of structural information concerning the anomalously scattering atoms. In fact, only information concerning the chemical nature of these atoms is required.

Although this paper does not concern triplet phase invariants it is noteworthy that triplet phase invariants can be evaluated unambiguously from one-wavelength anomalous dispersion data by probabilistic means (Hauptman, 1982; Giacovazzo, 1983) and by the application of certain rules (Karle, 1984b).

Theory

We consider the case of a structure composed of atoms that scatter nonanomalously and atoms, all of the same type, that scatter anomalously. The appropriate equations, which were obtained from the aforementioned analysis (Karle, 1980), may be written

$$|F_{\lambda h}|^2 = |F_{1,h}^n|^2 + \alpha_h |F_{2,h}^n|^2$$
$$+ \beta_h |F_{1,h}^n||F_{2,h}^n| \cos (\varphi_{1,h}^n - \varphi_{2,h}^n)$$
$$+ \gamma_h |F_{1,h}^n||F_{2,h}^n| \sin (\varphi_{1,h}^n - \varphi_{2,h}^n) \quad (1)$$

and

$$|F_{\lambda \bar{h}}|^2 = |F_{1,h}^n|^2 + \alpha_h |F_{2,h}^n|^2$$
$$+ \beta_h |F_{1,h}^n||F_{2,h}^n| \cos (\varphi_{1,h}^n - \varphi_{2,h}^n)$$
$$- \gamma_h |F_{1,h}^n||F_{2,h}^n| \sin (\varphi_{1,h}^n - \varphi_{2,h}^n), \quad (2)$$

where

$$\alpha_h = 1 + (f_{\lambda 2,h}^a / f_{2,h}^n)[(f_{\lambda 2,h}^a / f_{2,h}^n) + 2 \cos \delta_{\lambda 2,h}] \quad (3)$$
$$\beta_h = 2[1 + (f_{\lambda 2,h}^a / f_{2,h}^n) \cos \delta_{\lambda 2,h}] \quad (4)$$
$$\gamma_h = 2(f_{\lambda 2,h}^a / f_{2,h}^n) \sin \delta_{\lambda 2,h}. \quad (5)$$

$|F_{\lambda h}|$ is a known structure-factor magnitude whose value is obtained from a measurement of the intensity at a particular wavelength, λ, for a given reciprocal vector, h, $|F_{1,h}^n|$ is the magnitude of the corresponding structure factor for the nonanomalously scattering atoms, $|F_{2,h}^n|$ is the magnitude of the corresponding structure factor for the anomalously scattering atoms scattering as if there were no anomalous scattering and $\varphi_{1,h}^n - \varphi_{2,h}^n$ is the difference between the phases associated with $|F_{1,h}^n|$ and $|F_{2,h}^n|$, respectively. Evidently, the subscript 1 refers to the nonanomalously scattering atoms and the subscript 2 refers to the anomalously scattering ones. Further definitions of the quantities in (3)–(5) follow from the definition of the atomic scattering factor for an atom of type q,

$$f_{\lambda q,h} = f_{q,h}^n + f_{\lambda q,h}' + i f_{\lambda q,h}'', \quad (6)$$

where $f_{q,h}^n$ is the normal atomic scattering factors and $f_{\lambda q,h}'$ and $f_{\lambda q,h}''$ are the real and imaginary corrections. Equations (3)–(5) contain the following for $q = 2$:

$$f_{\lambda q,h}^a = (f_{\lambda q,h}'^2 + f_{\lambda q,h}''^2)^{1/2} \quad (7)$$
$$\delta_{\lambda q,h} = \tan^{-1} (f_{\lambda q,h}'' / f_{\lambda q,h}'). \quad (8)$$

The quantities f' and f'' are normally treated as independent of h and are tabulated so that appropriate values for α_h, β_h and γ_h may be readily computed.

The two equations representing the Bijvoet pairs are (1) and (2). It is seen that these equations involve three unknown quantities, $|F_{1,h}^n|$, $|F_{2,h}^n|$ and $\varphi_{1,h}^n - \varphi_{2,h}^n$. This apparent imbalance between the number of independent data and number of unknown quantities can be overcome to yield unique or essentially unique values for $\varphi_{1,h}^n - \varphi_{2,h}^n$ without any knowledge of the heavy-atom structure. This is accomplished by recognizing that $|F_{1,h}^n|$ and $|F_{2,h}^n|$ are approximately definable in terms of the measured Bijvoet pairs and that the values obtained for $\varphi_{1,h}^n - \varphi_{2,h}^n$ from (1) and (2) are relatively insensitive to errors in the values for $|F_{1,h}^n|$ and $|F_{2,h}^n|$.

In this section three systems of defining equations are presented that can be treated in a least-squares fashion. System I is a general system, but can be replaced for convenience with system II or III, if desired. In fact, system II has been used with the test calculations that are presented. The defining equations are treated in the usual least-squares fashion, namely, the sum of the squares of the errors in the defining equations are minimized.

Unlike most least-squares calculations, test calculations have shown that it is possible to proceed with systems I-III even though the number of independent equations does not exceed the number of unknown quantities. This is no doubt facilitated by the fact that in the systems all the defining equations are linear except for one that is quadratic, giving a broad range of convergence. Suitable starting values for the unknown quantities are thus fairly readily obtained.

In the next part, approximate statistical formulas are given for the evaluation of $|F_{1,\mathbf{h}}^n|$ and $|F_{2,\mathbf{h}}^n|$. The philosophy of their use is to recognize their approximate nature and consider a range of starting values for $|F_{1,\mathbf{h}}^n|$ and $|F_{2,\mathbf{h}}^n|$ based on the initial evaluations. The effect of the variation of starting values on the resulting values of the desired phase differences from the least-squares calculation is then observed for consistency. In some instances in the test calculations, it was found that no variation on the initial evaluation of $|F_{2,\mathbf{h}}|$ was required for the succeeding least-squares analysis. Each time new starting values are employed, a new least-squares calculation is performed. Results obtained in good agreement with each other, independently of the broad range of starting values, are a measure of the reliability of the calculation. The ultimate test is, of course, the agreement of the results with the correct values.

A geometric analysis of the interrelationships among the mathematical quantities that arise is given in the Appendix.

Estimates for $|F_{1,\mathbf{h}}^n|$ and $|F_{2,\mathbf{h}}^n|$

A statistical argument would suggest that $|F_{1,\mathbf{h}}^n|$ could be estimated from

$$|F_{1,\mathbf{h}}^n| \sim 0.5\, W_{1,\mathbf{h}}(|F_{\mathbf{Ah}}| + |F_{\mathbf{A\bar{h}}}|), \qquad (9)$$

where

$$W_{1,\mathbf{h}} = \left\{ \frac{\sum_{j=1}^{N_{non}} f_{j\mathbf{h}}^{n2}}{\sum_{j=1}^{N_{non}} f_{j\mathbf{h}}^{n2} + \sum_{j=1}^{N_{ano}} [(f_{j\mathbf{h}}^n + f_j')^2 + f_j''^2]} \right\}^{1/2} \qquad (10)$$

and N_{non} and N_{ano} are the number of nonanomalously and anomalously scattering atoms, respectively, in the unit cell. Equations (9) and (10) may be compared with similar ones (Karle, 1984a) whose purpose it is to calculate values for the $|F_{\mathbf{h}}^n|$, the structure-factor magnitudes for the total structure when all atoms scatter nonanomalously.

For $|F_{2,\mathbf{h}}^n|$, it can be shown that

$$|F_{2,\mathbf{h}}^n|^2 \simeq S\{\|F_{\mathbf{Ah}}| - |F_{\mathbf{A\bar{h}}}\| / [2(f_{A2}''/f_{2,\mathbf{h}}^n)]\}^2, \qquad (11)$$

where f_{A2}'' is treated as independent of scattering angle and S is a scale factor that is equal to 1 when the angles $\varphi_{\mathbf{Ah}}$ and $-\varphi_{\mathbf{A\bar{h}}}$ are equal. The latter circumstance gives a minimum value for $|F_{2,\mathbf{h}}^n|^2$. An estimate of values for S can be based on test examples having the same atomic composition as the substance of interest.

There are two probability theories giving identical results, except for notation (Hauptman, 1982; Giacovazzo, 1983), which afford an alternative way to estimate $|F_{2,\mathbf{h}}^n|^2$. A test calculation with exact data for cytochrome c550.PtCl$_4^{2-}$ showed these theories to underestimate the values of $|F_{2,\mathbf{h}}^n|^2$ by about 1-30% for the first 2000 differences, $\|F_{\mathbf{Ah}}| - |F_{\mathbf{A\bar{h}}}\|$, listed with the largest first. The underestimation increased roughly as the differences decreased. This may occur because heavy-atom structures in macromolecules are quite simple and therefore are not comprised of a sufficient number of atoms to have the statistical properties predicted by standard probabilistic analyses. Some special studies of this matter have appeared recently (Shmueli, Weiss, Kiefer & Wilson, 1984; Karle, 1984c).

System I

Equations (1) and (2) can be considered as linear in four variables defined as

$$x_1 = |F_{1,\mathbf{h}}^n|^2 \qquad (12)$$

$$x_2 = |F_{2,\mathbf{h}}^n|^2 \qquad (13)$$

$$x_3 = |F_{1,\mathbf{h}}^n||F_{2,\mathbf{h}}^n| \cos(\varphi_{1,\mathbf{h}}^n - \varphi_{2,\mathbf{h}}^n) \qquad (14)$$

$$x_4 = |F_{1,\mathbf{h}}^n||F_{2,\mathbf{h}}^n| \sin(\varphi_{1,\mathbf{h}}^n - \varphi_{2,\mathbf{h}}^n). \qquad (15)$$

Rewriting (1) and (2) gives

$$|F_{\mathbf{Ah}}|^2 = x_1 + \alpha_{\mathbf{h}} x_2 + \beta_{\mathbf{h}} x_3 + \gamma_{\mathbf{h}} x_4 \qquad (16)$$

and

$$|F_{\mathbf{A\bar{h}}}|^2 = x_1 + \alpha_{\mathbf{h}} x_2 + \beta_{\mathbf{h}} x_3 - \gamma_{\mathbf{h}} x_4. \qquad (17)$$

In addition, there exists a quadratic relationship among the variables

$$x_1 x_2 = x_3^2 + x_4^2. \qquad (18)$$

If (11), which defines x_2, is added to (16)-(18), there are four equations defining four unknown quantities. Three of the equations are linear in the variables and one is quadratic. This system is amenable to least-squares solution with one important caution. The predominance of (18) in the system must be tempered. This can be accomplished in many ways. Two ways that are used are to either divide (18) by x_2 or take the square root of both sides of (18). In the case of the former, the possibility exists that in some intermediate step in the least-squares process x_2 could become quite small and cause instability and, in the case of the latter, an intermediate excursion of $x_1 x_2$ into negative values would inhibit the taking of the square root. Suitable computer programming can anticipate and overcome such eventualities.

Initial values for the unknown quantities can be obtained for x_1 from (9), for x_2 from (11), for x_3 from

$$x_3 = (|F_{\lambda \mathbf{h}}|^2 + |F_{\lambda \bar{\mathbf{h}}}|^2 - 2\alpha_{\mathbf{h}} x_2 - 2x_1)/2\beta_{\mathbf{h}} \qquad (19)$$

once initial values are obtained for x_1 and x_2 and for x_4 from

$$x_4 = (|F_{\lambda \mathbf{h}}|^2 - |F_{\lambda \bar{\mathbf{h}}}|^2)/2\gamma_{\mathbf{h}}, \qquad (20)$$

where (19) and (20) are obtained by adding and subtracting (16) and (17), respectively.

In the case that the only information available is values for $|F_{\lambda \mathbf{h}}|^2$ and $|F_{\lambda \bar{\mathbf{h}}}|^2$, a suitable system of equations for a least-squares procedure involves the use of (11) and (16)-(18). Equations (19) and (20) can replace (16) and (17) in the least-squares calculation. In test examples, it has been found that convergence is enhanced by combining all the equations, giving for the defining equations (11) and (16)-(20). These six equations are called system I. In practice, it has been found useful to consider several initial values for x_1 and carry through the least-squares calculation several times. In that case, it is possible to dispense with the calculation of $W_{1,\mathbf{h}}$ in (9) and simply base the initial values on

$$|F_{1,\mathbf{h}}^n| \sim 0 \cdot 5\varepsilon(|F_{\lambda \mathbf{h}}| + |F_{\lambda \bar{\mathbf{h}}}|), \qquad (21)$$

where ε can assume, for example, the values $0 \cdot 7$, $1 \cdot 0$ and $1 \cdot 3$. It is most likely, for the simple substitution of heavy atoms in a macromolecule, that the range of values for $|F_{1,\mathbf{h}}^n|$, made accessible in the least-squares calculation by the latter initial values, would include the correct value for $|F_{1,\mathbf{h}}^n|$.

System II

The least-squares procedure can be varied in several ways. For example, an alternative procedure could be based on the use of (11) to provide a value for $|F_{2,\mathbf{h}}^n|$ that is held constant throughout the calculation. Treating $|F_{2,\mathbf{h}}^n|$ as a known constant can give rise to a somewhat different set of equations that replace (16)-(20). We define new variables

$$x_5 = |F_{1,\mathbf{h}}^n| \cos (\varphi_{1,\mathbf{h}}^n - \varphi_{2,\mathbf{h}}^n) \qquad (22)$$

$$x_6 = |F_{1,\mathbf{h}}^n| \sin (\varphi_{1,\mathbf{h}}^n - \varphi_{2,\mathbf{h}}^n) \qquad (23)$$

and have the following new set of defining equations for the least-squares system:

$$|F_{\lambda \mathbf{h}}|^2 - x_1 - \alpha_{\mathbf{h}}|F_{2,\mathbf{h}}^n|^2 - \beta_{\mathbf{h}}|F_{2,\mathbf{h}}^n|x_5 - \gamma_{\mathbf{h}}|F_{2,\mathbf{h}}^n|x_6 = 0 \qquad (24)$$

$$|F_{\lambda \bar{\mathbf{h}}}|^2 - x_1 - \alpha_{\mathbf{h}}|F_{2,\mathbf{h}}^n|^2 - \beta_{\mathbf{h}}|F_{2,\mathbf{h}}^n|x_5 + \gamma_{\mathbf{h}}|F_{2,\mathbf{h}}^n|x_6 = 0 \qquad (25)$$

$$x_1 - x_5^2 - x_6^2 = 0 \qquad (26)$$

$$|F_{\lambda \mathbf{h}}|^2 + |F_{\lambda \bar{\mathbf{h}}}|^2 - 2x_1 - 2\alpha_{\mathbf{h}}|F_{2,\mathbf{h}}^n|^2 - 2\beta_{\mathbf{h}}|F_{2,\mathbf{h}}^n|x_5 = 0 \qquad (27)$$

$$[(|F_{\lambda \mathbf{h}}|^2 - |F_{\lambda \bar{\mathbf{h}}}|^2)/(2\gamma_{\mathbf{h}}|F_{2,\mathbf{h}}^n|)] - x_6 = 0, \qquad (28)$$

where the value for $|F_{2\mathbf{h}}^n|$ is obtained from (11). Initial

values for x_1 can be obtained from (21), x_5 from (27) and x_6 from (28). The five equations (24)-(28) are called least-squares system II.

The case of isomorphous replacement (system III)

In the case that isomorphous replacement data are also available, measured values of the intensities for the native macromolecule provide values for the $|F_{1,\mathbf{h}}^n|$. It is then possible to solve for $|F_{2,\mathbf{h}}^n|$ and $\varphi_{1,\mathbf{h}}^n - \varphi_{2,\mathbf{h}}^n$. In this case, approximate information for $|F_{2,\mathbf{h}}^n|$ from (11) could distort the results and this equation is therefore either omitted from the least-squares system I (11, 16-20) or included with low weight. Values obtained for the $|F_{2,\mathbf{h}}^n|$ can be used to determine the heavy-atom structure and therefore, ultimately, values for the $\varphi_{2,\mathbf{h}}^n$. Having values for the $\varphi_{1,\mathbf{h}}^n - \varphi_{2,\mathbf{h}}^n$ and, separately, values for $\varphi_{2,\mathbf{h}}^n$ leads to values for the desired $\varphi_{1,\mathbf{h}}^n$.

An alternative set of equations to system I for the least-squares calculation when $|F_{1,\mathbf{h}}^n|$ is known can be considered. We define new variables

$$x_7 = |F_{2,\mathbf{h}}^n| \cos (\varphi_{1,\mathbf{h}}^n - \varphi_{2,\mathbf{h}}^n) \qquad (29)$$

$$x_8' = |F_{2,\mathbf{h}}^n| \sin (\varphi_{1,\mathbf{h}}^n - \varphi_{2,\mathbf{h}}^n) \qquad (30)$$

and have the following new set of defining equations for the least-squares system:

$$|F_{\lambda \mathbf{h}}|^2 - |F_{1,\mathbf{h}}^n|^2 - \alpha_{\mathbf{h}} x_2 - \beta_{\mathbf{h}}|F_{1,\mathbf{h}}^n|x_7 - \gamma_{\mathbf{h}}|F_{1,\mathbf{h}}^n|x_8 = 0 \qquad (31)$$

$$|F_{\lambda \bar{\mathbf{h}}}|^2 - |F_{1,\mathbf{h}}^n|^2 - \alpha_{\mathbf{h}} x_2 - \beta_{\mathbf{h}}|F_{1,\mathbf{h}}^n|x_7 + \gamma_{\mathbf{h}}|F_{1,\mathbf{h}}^n|x_8 = 0 \qquad (32)$$

$$x_2 - x_7^2 - x_8^2 = 0 \qquad (33)$$

$$|F_{\lambda \mathbf{h}}|^2 + |F_{\lambda \bar{\mathbf{h}}}|^2 - 2|F_{1,\mathbf{h}}^n|^2 - 2\alpha_{\mathbf{h}} x_2 - 2\beta_{\mathbf{h}}|F_{1,\mathbf{h}}^n|x_7 = 0 \qquad (34)$$

$$[(|F_{\lambda \mathbf{h}}|^2 - |F_{\lambda \bar{\mathbf{h}}}|^2)/(2\gamma_{\mathbf{h}}|F_{1,\mathbf{h}}^n|)] - x_8 = 0, \qquad (35)$$

where the value of $|F_{1,\mathbf{h}}^n|$ is obtained from experiment. Initial values for x_2 can be obtained from (11), x_7 from (34) and x_8 from (35). The five equations (31)-(35) are called least-squares system III. Equation (11) may be added to system III with a low weight. The reason for including it with a low weight is that it is possible, because of experimental error, to have inconsistent values among the three quantities $|F_{\lambda \mathbf{h}}|$, $|F_{\lambda \bar{\mathbf{h}}}|$ and $|F_{1,\mathbf{h}}^n|$. If these values are not greatly inconsistent, the least squares may be brought to reasonable convergence by including (11) with a low weight. The weight should be a compromise between dominance of the result by (11), as obtained from too large a weight, and insufficient influence to affect the convergence, as obtained from too small a weight. In one application to a problem in macromolecular structure determination, a weight of $0 \cdot 01$ for (11) was found to be suitable.

Test calculations

Test calculations were performed on exact data and also on data into which errors were introduced. The

Table 1. *Calculation of least-squares system* II *for samples of data from cytochrome* c550.PtCl$_4^{2-}$ *in which only the Pt atoms scatter anomalously*

The samples are based on **h** associated with an ordered sequence of $\||F_{\text{A}\mathbf{h}}|-|F_{\text{A}\bar{\mathbf{h}}}|\|$ in which the largest one is first. The effect of errors in the data was tested by multiplying $\||F_{\text{A}\mathbf{h}}|-|F_{\text{A}\bar{\mathbf{h}}}|\|$ by the error factor listed in the second column and effecting the error, in this case, by readjusting the value of $|F_{\text{A}\bar{\mathbf{h}}}|$. The actual distribution of the errors between $|F_{\text{A}\mathbf{h}}|$ and $|F_{\text{A}\bar{\mathbf{h}}}|$ does not materially affect the results. The average magnitudes of error for $|F_{1,\mathbf{h}}^n|$, $|F_{2,\mathbf{h}}^n|$ and $(\varphi_{1,\mathbf{h}}^n - \varphi_{2,\mathbf{h}}^n)$ are seen in columns 3, 4 and 5, respectively. The total number of independent data is 3250 at 2·5 Å resolution and the radiation is Cu $K\alpha$.

Sample	Error factor	Average % error $\|F_{1,\mathbf{h}}^n\|$	Average % error $\|F_{2,\mathbf{h}}^n\|$	Average error (rad) $\varphi_{1,\mathbf{h}}^n - \varphi_{2,\mathbf{h}}^n$
1–100	1·50	21	49	0·21
901–1000	1·50	21	50	0·35
1601–1700	1·50	27	51	0·48
1–100	1·25	14	25	0·16
901–1000	1·25	20	30	0·34
1601–1700	1·25	26	37	0·47
1–100	1·00	13	7	0·17
901–1000	1·00	19	17	0·37
1601–1700	1·00	27	28	0·49
1–100	0·75	19	25	0·29
901–1000	0·75	20	27	0·47
1601–1700	0·75	30	30	0·56
1–100	0·50	24	50	0·48
901–1000	0·50	22	51	0·59
1601–1700	0·50	33	53	0·65

data were computed at 2·5 Å resolution for Cu $K\alpha$ radiation from the coordinates for cytochrome c550.PtCl$_4^{2-}$ from *Paracoccus denitrificans* (Timkovich & Dickerson, 1976). The structure factors were computed in two ways. One calculation introduced anomalous effects from the Pt atom alone and the second included anomalous effects from the Pt, Fe, S and Cl atoms. The first calculation models the case when there would be only one type of anomalous scatterer. This calculation not only represents an important experimental case, but also provides a basis of comparison for determining the effect on the errors of including all four types of anomalous scatterers in the data while treating the data as if the Pt atoms were the one predominant type of anomalous scatterer.

Calculations based on system II are presented in Table 1. They concern only the data that contain anomalous effects from the Pt atoms alone. In system II, $|F_{2,\mathbf{h}}^n|$ is estimated from a statistical analysis and the least-squares system is solved for the values of $|F_{1,\mathbf{h}}^n|$ and $(\varphi_{1,\mathbf{h}}^n - \varphi_{2,\mathbf{h}}^n)$. The value of $|F_{2,\mathbf{h}}^n|$ was determined from (11) where the value of S was obtained from a statistical calculation based on an arbitrary structure with the same chemical composition as cytochrome c550.PtCl$_4^{2-}$. The factor S assumes that value for which the average value of $|F_{2,\mathbf{h}}^n|$ in some sample of reflections, as computed from (11), is equal to the

average value of $|F_{2,\mathbf{h}}^n|$ for the same type of sample, as obtained from the arbitrary structure.

The samples of reflections for which the statistical estimates of S were made coincide with those listed in the first column of Table 1. The numerical sequence refers to the sets of reflections for which the calculations of system II were made and are based on the sequence of values of $\||F_{\text{A}\mathbf{h}}|-|F_{\text{A}\bar{\mathbf{h}}}|\|$, with the largest first. If a value for $|F_{\text{A}\mathbf{h}}|$ is very small, it may be detrimental to the calculations. In this case, no lower limit was used, but it is worth considering in future calculations. The starting values for $|F_{1,\mathbf{h}}^n|$ were the three values described for (21). The three calculations converged either to the same three values for $|F_{1,\mathbf{h}}^n|$ and $(\varphi_{1,\mathbf{h}}^n - \varphi_{2,\mathbf{h}}^n)$ or to two the same and one different. The values accepted were either three the same or two the same. A rejection criterion could be based on the magnitude of the discrepancy when two results differ from the third. This was not done here.

Errors were introduced by taking the correct values for $|F_{\text{A}\mathbf{h}}|-|F_{\text{A}\bar{\mathbf{h}}}|$, multiplying them by the factors listed in column 2 of Table 1 and readjusting the values of the $|F_{\text{A}\bar{\mathbf{h}}}|$ to be consistent with the resulting smaller or larger differences. An even distribution of the errors among both the $|F_{\text{A}\mathbf{h}}|$ and $|F_{\text{A}\bar{\mathbf{h}}}|$ did not alter the results significantly. The introduction of rather large errors into the differences, $|F_{\text{A}\mathbf{h}}|-|F_{\text{A}\bar{\mathbf{h}}}|$, had a remarkably small effect on the average errors for $|F_{1,\mathbf{h}}^n|$ and $(\varphi_{1,\mathbf{h}} - \varphi_{2,\mathbf{h}})$.

The other type of error that could be considered is an error in the scaling of $|F_{\text{A}\mathbf{h}}|$ and $|F_{\text{A}\bar{\mathbf{h}}}|$. Because of the nature of the defining equations, this type of error affects only the accuracy of values of $|F_{1,\mathbf{h}}^n|$ and $|F_{2,\mathbf{h}}^n|$ but not those for the desired phase differences $(\varphi_{1,\mathbf{h}}^n - \varphi_{2,\mathbf{h}}^n)$. Because scaling error does not affect the values of the phase differences, it was not investigated further.

Test calculations based on data from cytochrome c550.PtCl$_4^{2-}$ in which Pt, Fe, Cl and S atoms were the anomalous scatterers are shown in Table 2. In the analysis of the data by use of system II, the Pt atoms were treated as the sole predominant type of anomalous scatterer. The nature of the computations was the same as those that contributed to Table 1 with the exception that it was often necessary to modify the value of $|F_{2,\mathbf{h}}^n|$, as obtained from (11), in order to bring the system to convergence. This was done by modifying the starting value of $|F_{2,\mathbf{h}}^n|$ by multiplying by the factors, 1·9, 1·6, 1·3, 1·0, 0·7, in succession until convergence was obtained. The results shown in Table 2 are seen to be comparable to those in Table 1 with only a modest increase in the average errors.

It is seen from Table 1 that only the largest differences, $\||F_{\text{A}\mathbf{h}}|-|F_{\text{A}\bar{\mathbf{h}}}|\|$, give reliable values for the $|F_{2,\mathbf{h}}^n|$ and then only if the errors in these differences are small. Table 2 shows that when the four types of anomalous scatterers are considered, none of the

Table 2. *Calculation of least-squares system* II *for samples of data from cytochrome c550.PtCl$_4^{2-}$ in which Pt, Fe, Cl and S atoms scatter anomalously*

The samples are based on h associated with an ordered sequence of $\|F_{\lambda b}| - |F_{\lambda \bar{b}}\|$ in which the largest one is first. The effect of errors in the data was tested by multiplying $\|F_{\lambda b}| - |F_{\lambda \bar{b}}\|$ by the error factor listed in the second column and effecting the error, in this case, by readjusting the value $|F_{\lambda \bar{b}}|$. The actual distribution of errors between $|F_{\lambda b}|$ and $|F_{\lambda \bar{b}}|$ does not materially affect the results. The average magnitudes of error for $|F_{1,b}^n|$, $|F_{2,b}^n|$ and $(\varphi_{1,b}^n - \varphi_{2,b}^n)$ are seen in columns 3, 4 and 5, respectively. The total number of independent data is 3250 at 2·5 Å resolution and the radiation is Cu $K\alpha$.

Sample	Error factor	Average % error $|F_{1,b}^n|$	Average % error $|F_{2,b}^n|$	Average error (rad) $\varphi_{1,b}^n - \varphi_{2,b}^n$
1–100	1·50	30	85	0·29
901–1000	1·50	31	79	0·40
1601–1700	1·50	27	75	0·67
1–100	1·25	24	56	0·25
901–1000	1·25	30	56	0·38
1601–1700	1·25	27	55	0·69
1–100	1·00	22	31	0·28
901–1000	1·00	30	38	0·41
1601–1700	1·00	27	43	0·72
1–100	0·75	27	23	0·41
901–1000	0·75	33	30	0·47
1601–1700	0·75	28	42	0·75
1–100	0·50	31	39	0·67
901–1000	0·50	35	45	0·59
1601–1700	0·50	29	52	0·80

estimates of $|F_{2,b}^n|$ is reliably obtained. A two-wavelength experiment, which would contribute additional defining equations, or accurate information concerning the values of $|F_{1,b}^n|$, as may be obtained from an isomorphous replacement experiment, would afford an opportunity to obtain more accurate values for the $|F_{2,b}^n|$ and the other unknown quantities as well. Reliable data for the $|F_{2,b}^n|$ would be valuable for the determination of the heavy-atom structure. Knowledge of the latter gives values for the $\varphi_{2,b}^n$. Once these are known, the desired values for the $\varphi_{1,b}^n$ may be obtained from the known values of the differences, $\varphi_{1,h}^n - \varphi_{2,b}^n$. The values for the $\varphi_{1,b}^n$ then permit the immediate calculation of a Fourier map of the electron distribution.

Summary remarks

Least-squares techniques for treating data obtained from one-wavelength anomalous dispersion experiments have been presented, which led, in test calculations, to unique evaluations of the phase differences, $\varphi_{1,b}^n - \varphi_{2,b}^n$, within acceptable ranges of accuracy, even when the errors in the differences $\|F_{\lambda b}| - |F_{\lambda \bar{b}}\|$ were rather large. The uniqueness of the results derives from inherent information in the measured intensities concerning the normal structure factors for the non-anomalously and anomalously scattering atoms. The calculations involved are very simple and can be carried out at a relatively high rate. It has been pointed

out elsewhere (Karle, 1984c) that the phase differences $(\varphi_{1,b}^n - \varphi_{2,b}^n)$ can be formed into triplet phase invariants for use in phase determination for the nonanomalously scattering atoms if difficulties should arise in the determination of the heavy-atom structure.

In order to make the calculations described in this paper, the only information required concerning the anomalously scattering atoms is the chemical nature of the predominant type of scatterer. The calculations should have fairly broad application in practice. The presence of more than one predominant anomalous scatterer of widely differing atomic number may require the use of multiple-wavelength experiments and the associated theory (Karle, 1980).

Use of multiple-wavelength data has the potential of enhancing the accuracy of the analyses in any case. The point of the presentation here, however, has been to show how much information is derivable from a one-wavelength experiment and the potential utility of such an experiment.

I wish to thank Mr Stephen Brenner for writing the appropriate programs and making the computations reported here.

This work was supported in part by USPHS grant GM30902.

APPENDIX

Geometric analysis

In this part a geometric construction is discussed that shows the relationships between the quantities that enter into the three systems of defining equations that are solved in a least-squares fashion to obtain values for the phase differences $\varphi_{1,b}^n - \varphi_{2,b}^n$. The construction not only provides a graphical understanding of the quantities involved and how they are related, but also affords a clear insight into the basis for obtaining unique or essentially unique values for the phase differences in a one-wavelength anomalous dispersion experiment when there is one type or one predominant type of anomalous scatterer present.

The equations used in the construction are

$$F_{\lambda b} = F_b^n + F_{\lambda b}^a \tag{36}$$

$$F_{\lambda \bar{b}}^* = F_{\bar{b}}^{n*} + F_{\lambda \bar{b}}^{a*} \tag{37}$$

$$F_b^n = F_{\bar{b}}^{n*} \tag{38}$$

$$F_{\lambda \bar{b}}^* = F_b^n + F_{\lambda \bar{b}}^{a*} \tag{39}$$

$$F_b^n = F_{1,b}^n + F_{2,b}^n \tag{40}$$

$$F_{\lambda b} = F_{1,b}^n + F_{2,b}^n + F_{\lambda b}^a \tag{41}$$

$$F_{\lambda \bar{b}}^* = F_{1,b}^n + F_{2,b}^n + F_{\lambda \bar{b}}^{a*} \tag{42}$$

$$F_{\lambda b}^a = \exp(i\delta_{\lambda 2})(f_{\lambda 2}^a/f_{2,b}^n)F_{2,b}^n \tag{43}$$

$$F_{\lambda \bar{b}}^{a*} = \exp(-i\delta_{\lambda 2})(f_{\lambda 2}^a/f_{2,b}^n)F_{2,b}^n \tag{44}$$

$$F_{\lambda b} = F_{\lambda \bar{b}}^* + 2i(f_{\lambda 2}''/f_{2,b}^n)F_{2,b}^n. \tag{45}$$

Equation (36) defines the structure factor for a structure that scatters anomalously, $F_{\lambda \mathbf{h}}$, in terms of a structure factor appropriate to the same structure when all atoms scatter nonanomalously, $F_{\mathbf{h}}^n$, plus a structure factor that represents the total contribution from the anomalous scattering, both real and imaginary, $F_{\lambda \mathbf{h}}^a$. Except for the quantity $F_{\mathbf{h}}^n$, all quantities appearing in (36)–(45) are defined in the body of the paper. Equation (36) is the source equation from which (1) and (2) were derived (Karle, 1980). Equation (37) follows from (36) and is written as the complex conjugate because it is convenient to use it in that form in the construction. Since (38) is valid when atoms do not scatter anomalously, (39) follows. Equation (40) is a statement that $F_{\mathbf{h}}^n$ is composed of the sum of the structure factors for the atoms that scatter nonanomalously, $F_{1,\mathbf{h}}^n$, and the structure factors for the atoms that scatter anomalously when the anomalous part of the scattering is omitted, $F_{2,\mathbf{h}}^n$. Equations (41) and (42) follow from (36)–(40). Equations (43) and (44) are valid when there is one type of anomalous scatterer. They follow from relations that arise in the derivation of (1) and (2) (Karle, 1980). Equation (45) is immediately derivable from (36), (39), (43) and (44).

The construction shown in Fig. 1 is based on (36)–(45). Equations (1) and (2) were derived from these equations without approximation and their characteristics are properly represented by the characteristics of the diagram. We now proceed to construct the diagram in Fig. 1 in a stepwise fashion. We assume

that the magnitudes $|F_{\lambda \mathbf{h}}|$, $|F_{\lambda \bar{\mathbf{h}}}|$ and $|F_{2,\mathbf{h}}^n|$ are known. In actual practice, values for $|F_{\lambda \mathbf{h}}|$ and $|F_{\lambda \bar{\mathbf{h}}}|$ are obtained from experiment and, as described in the paper, values for $|F_{2,\mathbf{h}}^n|$ may be obtained approximately from values for $|F_{\lambda \mathbf{h}}|$ and $|F_{\lambda \bar{\mathbf{h}}}|$ by use of (11). It is also assumed that the chemical identity of the anomalously scattering atoms is known. A value is selected for $\varphi_{2,\mathbf{h}}^n$, the angle associated with $|F_{2,\mathbf{h}}^n|$, to facilitate the construction of the diagram. The value of $\varphi_{2,\mathbf{h}}^n$ is arbitrary and is not determined by the mathematics. This means that the resulting diagram is not determined in orientation. Rather, it can be arbitrarily rotated about an axis perpendicular to the plane of the diagram and placed at the origin.

On the basis of the assumed information, the triangle having solid lines with sides $|F_{\lambda \mathbf{h}}|$, $|F_{\lambda \bar{\mathbf{h}}}|$ and b can be drawn, representing (45). A second triangle ambiguously placed can also be drawn, as shown by the triangle with dashed lines and the side labeled b. It can be obtained from the original triangle by rotating about the vector of length b placed at the origin. We continue with consideration of the triangle with solid lines and proceed with the construction of (41). All information is available to compute $F_{2,\mathbf{h}}^n + F_{\lambda \mathbf{h}}^a$ on the right side of (41). It is labeled with a c in Fig. 1. The head of this vector is attached to the head of the vector of magnitude $|F_{\lambda \mathbf{h}}|$. Once this is done, it is possible to complete (41) by drawing the vector $F_{1,\mathbf{h}}^n$ to close the triangle. Note that a value for $|F_{1,\mathbf{h}}^n|$ has been determined here and that a value for the associated angle $\varphi_{1,\mathbf{h}}^n$ has also been determined relative to the assumed value for $\varphi_{2,\mathbf{h}}^n$. The vector of magnitude d can now be drawn from the tip of $F_{1,\mathbf{h}}^n$ to the tip of $F_{\lambda \bar{\mathbf{h}}}^*$ to form a triangle that represents (42). The vector of magnitude d is entirely consistent with the construction that could be made from $F_{2,\mathbf{h}}^n + F_{\lambda \bar{\mathbf{h}}}^{a*}$, where the magnitude of $F_{\lambda \bar{\mathbf{h}}}^{a*}$ is labeled with a.

The triangle with sides of magnitude b, c and d is also placed at the origin in order to help illustrate some angles involved in the construction and also to indicate that for a given $\varphi_{2,\mathbf{h}}^n$ there is no other place that the construction shown in solid lines can occur. The only other position in which the triangle with sides of length $|F_{1,\mathbf{h}}^n|$, c and $|F_{\lambda \mathbf{h}}|$ can occur is indicated by the triangle having dashed lines and a side of length c. It can be placed by rotating the triangle with solid lines about the vector of length c placed at the origin. A similar circumstance applies to the triangle with sides of length $|F_{1,\mathbf{h}}^n|$, d and $|F_{\lambda \bar{\mathbf{h}}}|$. Its ambiguous alternative is shown by the dashed triangle with a side of length d. It is evident that the elements of the main construction do not combine elsewhere on the circle of rotation for a fixed value of $\varphi_{2,\mathbf{h}}^n$.

The question arises concerning the source of the alternative result often obtained in the calculations discussed in the paper, i.e. two different results were often obtained as the value of $|F_{1,\mathbf{h}}^n|$ was varied through its three assigned starting values. The expla-

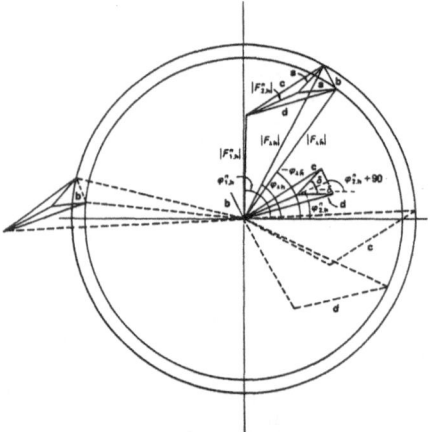

Fig. 1. A construction showing the relationships among the quantities occurring in the equations that are used to evaluate the phase differences $\varphi_{1,\mathbf{h}}^n - \varphi_{2,\mathbf{h}}^n$. Interpretation of the construction can be correlated with the experiences encountered in performing the test calculations described in the text. $\mathbf{a} = |F_{\lambda \mathbf{h}}^a| = |F_{\lambda \bar{\mathbf{h}}}^a| = (f_{\lambda 2}''/f_{2,\mathbf{h}}'')|F_{2,\mathbf{h}}^n|$; $\mathbf{b} = 2(f_{\lambda 2}''/f_{2,\mathbf{h}}'')|F_{2,\mathbf{h}}^n|$; $\mathbf{c} = |F_{2,\mathbf{h}}^n + F_{\lambda \mathbf{h}}^a|$; $\mathbf{d} = |F_{2,\mathbf{h}}^n + F_{\lambda \bar{\mathbf{h}}}^{a*}|$.

nation follows. On the basis of the information available, the triangle with sides b, c and d could just as well have been attached to the side b of the alternative dashed-line triangle. The magnitudes $|F_{Ab}|$, $|F_{A\bar{b}}|$ and $|F_{2,h}^n|$ would still be preserved as well as the assumed value for $\varphi_{2,h}^n$. As seen in Fig. 1, such a triangle with sides of length b, c and d has been attached to the dashed triangle with the common base of vectors of length c and d extending far to the left. This determines an alternative and quite different vector, $F_{1,h}^n$. It is quite long, about 2·4 times longer than the initially determined $|F_{1,h}^n|$, extending as a dashed line from the origin to the base of the vectors of length c and d. The associated angle, $\varphi_{1,h}$, is also rotated somewhat more than 90° farther than the initially determined $\varphi_{1,h}^n$. The magnitude of the initial $|F_{1,h}^n|$ is about 0·63 of the average value of $|F_{Ab}|$ and $|F_{A\bar{b}}|$ and that of the alternative is about 1·5 times larger.

It is evident now from Fig. 1 how two alternative sets of results arise and that the alternatives would be distinguishable by use of approximate knowledge of the value of $|F_{1,h}^n|$. It is of interest to review the assumptions inherent in the diagram as they relate to practical circumstances. There are experimental errors in $|F_{Ab}|$ and $|F_{A\bar{b}}|$ and $|F_{2,h}^n|$ can be obtained only approximately from the latter two intensities. The arbitrariness of $\varphi_{2,h}$ does not play a role in the calculations since the quantity evaluated is $\varphi_{1,h}^n - \varphi_{2,h}^n$, which is invariant to rotation of the diagram in Fig. 1 around the origin. The effects of the various uncertainties are illustrated in the test calculations. Because of the uncertainties, the variation of starting values for $|F_{1,h}^n|$ and, on occasion, $|F_{2,h}^n|$ was introduced into the calculations in order to explore the field of convergence. With the computer used for the test calcula-

tions, 6000 distinct least-squares computations were performed in one minute.

The diagram in Fig. 1 emphasizes the important practical significance of having additional information concerning $|F_{1,h}^n|$, $|F_{2,h}^n|$ and $\varphi_{2,h}^n$. As noted, information concerning $|F_{1,h}^n|$ is available from an isomorphous replacement experiment since $|F_{1,h}^n|$ represents the magnitude of the structure factor for the native substance. If the structure of the anomalous scatterers is determined initially, values for the $|F_{2,h}^n|$ are available to enhance the accuracy of the calculations and values for the $\varphi_{2,h}^n$ are available for the evaluation of the $\varphi_{1,h}^n$ from values of the $\varphi_{1,h}^n - \varphi_{2,h}^n$. The immediate calculation of the electron distribution of the structure of interest would follow.

References

DALE, D., HODGKIN, D. C. & VENKATESAN, K. (1963). *Crystallography and Crystal Perfection*, edited by G. N. RAMACHANDRAN, pp. 237–242. New York, London: Academic Press.
GIACOVAZZO, C. (1983). *Acta Cryst.* A39, 585–592.
HAUPTMAN, H. (1982). *Acta Cryst.* A38, 632–641.
HENDRICKSON, W. A. & TEETER, M. (1981). *Nature (London)*, 290, 107–113.
KARLE, J. (1980). *Int. J. Quantum Chem.* 7, 357–367.
KARLE, J. (1984a). *Acta Cryst.* A40, 1–4.
KARLE, J. (1984b). *Acta Cryst.* A40, 4–11.
KARLE, J. (1984c). *Acta Cryst.* A40, 526–531.
PEERDEMAN, A. F. & BIJVOET, J. M. (1956). *Acta Cryst.* 9, 1012–1015.
RAMACHANDRAN, G. N. & RAMAN, S. (1956). *Curr. Sci.* 25, 348–351.
SHMUELI, U., WEISS, G. H., KIEFER, J. E. & WILSON, A. J. C. (1984). *Acta Cryst.* A40, 651–660.
TIMKOVICH, R. & DICKERSON, R. E. (1976). *J. Biol. Chem.* 251, 4033–4046.
WOOLFSON, M. M. (1984). *Acta Cryst.* A40, 32–34.

Triplet Phase Invariants from Single Isomorphous Replacement or One-Wavelength Anomalous Dispersion Data, Given Heavy-Atom Information

By Jerome Karle

Laboratory for the Structure of Matter, Naval Research Laboratory, Washington, DC 20375-5000, USA

(*Received 17 September 1985; accepted 14 January 1986*)

Abstract

Certain general algebraic formulas for computing triplet phase invariants become accessible when structural information is available concerning the replacement atoms in isomorphous replacement or the predominant type of anomalously scattering atoms in one-wavelength anomalous dispersion experiments. The formulas of interest are presented and subjected to a number of test calculations to obtain insight into their accuracy and to determine the effects of errors in the data. The formulas are simple to calculate and some possible strategies for their use are discussed.

Introduction

On the basis of certain mathematical and physical considerations that pertain to isomorphous replacement or anomalous dispersion experiments, rules (Karle, 1983, 1984*a*, *b*) and algebraic formulas (Karle, 1984*c*, 1985*a*) were derived for the evaluation of triplet phase invariants. Among these was a general formula, applicable to both types of experimental data and not yet evaluated, that requires information concerning the structure factor magnitudes for the structure of the heavy atoms (or anomalously scattering atoms). This general formula is investigated here. Further accuracy would accrue if the phases for the heavy-atom structure were also known. This latter information is not mandatory, however, since the heavy-atom phases enter the formula as triplet phase invariants which can often be set equal to zero to good approximation. In the case of anomalous dispersion the formula is applicable when there is one type or one predominant type of anomalous scatterer. In the case of isomorphous replacement, the formula permits the calculation of the cosines of phase invariants in the range from −1 to 1 and for anomalous dispersion it permits the calculation of the sines of phase invariants in the range −1 to 1.

This investigation may be regarded as a further exploration of the mathematical tools available for application to the analysis of macromolecular structure. The particular role that triplet phase invariants may play remains to be determined. When the heavy-atom structure is known, it is possible in single isomorphous replacement to obtain numerous initial phase values that could possibly be extended and refined by use of the triplet phase invariants.

Diffraction experiments concerning isomorphous substitution of macromolecules are normally accompanied by anomalous dispersion effects. Because both types of data are readily measurable, it can be expected that data of sufficient accuracy to permit the application of both techniques will often be attainable, as the history of macromolecular structure determination has already shown (*e.g.* Adman, Sieker & Jensen, 1973). The combining of both techniques would be the optimal way to handle the data. A way to do this with the use of exact algebraic equations has already been discussed (Karle, 1984*c*). Here single isomorphous replacement and one-wavelength anomalous dispersion are treated individually as if the other did not exist.

A way to obtain essentially unique values for phase differences from one-wavelength anomalous dispersion data, *i.e.* the two intensities measured at **h** and −**h**, has been described (Karle, 1985*b*). As has been previously noted (Karle, 1985*b*), the possible existence of such a calculation was indicated by the uniqueness of values for the triplet phase invariants derived by Hauptman (1982*b*), Giacovazzo (1983) and subsequently by Karle (1984*a*). Knowledge of the structure of the predominant type of anomalous scatterers is not required to obtain the phase differences but, if it is known, phase values for the structure of the macromolecule can be obtained from the values of the phase differences. Evaluations of the triplet phase invariants of the type examined here may be useful for refining the phase values for macromolecules.

Table 1. *Quantities involved in the applications of* (1) *to single isomorphous replacement and one-wavelength anomalous dispersion data*

The quantities $_m\mathcal{F}_{1,\mathbf{h}}$, $_m\mathcal{F}_{2,\mathbf{h}}$ and $_m\mathcal{F}_{3,\mathbf{h}}$ are defined by the corresponding entries in columns 2, 3, 4, respectively.

Case m	$_m\mathcal{F}_{1,\mathbf{h}}$	$_m\mathcal{F}_{2,\mathbf{h}}$	$_m\mathcal{F}_{3,\mathbf{h}}$
i	$F_{\mathbf{h}PH}$	$F_{\mathbf{h}P}$	$F_{\mathbf{h}H}$
1	$F_{\Lambda\mathbf{h}}$	$F_{\Lambda b}^{*}$†	$F_{\Lambda\mathbf{h}}^{a} - F_{\Lambda\bar{\mathbf{h}}}^{a*}$†

† The asterisk denotes complex conjugate.

Acta Cryst. (1986). A**42**, 246–253

Theory

A general formula has been derived (Karle, 1984a, equation 31) of the form

$$\cos\left({}_m\varphi_{2,\mathbf{h}} + {}_m\varphi_{2,\mathbf{k}} + {}_m\varphi_{2,(\bar{\mathbf{h}}+\bar{\mathbf{k}})} - {}_m\varphi_{3,\mathbf{h}} - {}_m\varphi_{3,\mathbf{k}} - {}_m\varphi_{3,(\bar{\mathbf{h}}+\bar{\mathbf{k}})}\right)$$
$$\simeq T/|{}_m\mathscr{F}_{3,\mathbf{h}m}\mathscr{F}_{3,\mathbf{k}m}\mathscr{F}_{3,(\bar{\mathbf{h}}+\bar{\mathbf{k}})}|, \qquad (1)$$

where

$$T = (|{}_m\mathscr{F}_{1,\mathbf{h}}| - |{}_m\mathscr{F}_{2,\mathbf{h}}|)(|{}_m\mathscr{F}_{1,\mathbf{k}}| - |{}_m\mathscr{F}_{2,\mathbf{k}}|)$$
$$\times (|{}_m\mathscr{F}_{1,(\bar{\mathbf{h}}+\bar{\mathbf{k}})}| - |{}_m\mathscr{F}_{2,(\bar{\mathbf{h}}+\bar{\mathbf{k}})}|). \qquad (2)$$

There are numerous definitions of the ${}_m\mathscr{F}_{i,\mathbf{h}}$ ($i = 1, 2, 3$), listed in a recent publication (Karle, 1985a). When $m = i$, the $\mathscr{F}$ concern quantities associated with isomorphous replacement and when $m = 1$, the $\mathscr{F}$ concern quantities associated with anomalous dispersion, as shown in Table 1.

For isomorphous replacement,

$$F_{\mathbf{h}PH} - F_{\mathbf{h}P} = F_{\mathbf{h}H}, \qquad (3)$$

where $F_{\mathbf{h}PH}$ is the structure factor for a macromolecule substituted with heavy atoms, $F_{\mathbf{h}P}$ is the structure factor for the original macromolecule and $F_{\mathbf{h}H}$ is the structure factor for the structure formed by the heavy atoms alone.

For anomalous dispersion,

$$F_{\lambda\mathbf{h}} - F^*_{\lambda\bar{\mathbf{h}}} = F^a_{\lambda\mathbf{h}} - F^{a*}_{\lambda\bar{\mathbf{h}}}, \qquad (4)$$

where $F_{\lambda\mathbf{h}}$ is the structure factor for a macromolecule containing anomalous scatterers at some incident wavelength, λ, and $F^a_{\lambda\mathbf{h}}$ is the structure factor for the structure formed by the anomalous scatterers. For the case of one type of anomalous scatterer present (Karle, 1984a, equation 10 when $j = 2$)

$$F^a_{\lambda\mathbf{h}} - F^{a*}_{\lambda\bar{\mathbf{h}}} = 2i(f''/f^n_{2,\mathbf{h}})F^n_{2,\mathbf{h}}, \qquad (5)$$

where f'' is the imaginary correction to the normal atomic scattering factor, $f^n_{2,\mathbf{h}}$, of the anomalous scatterers and $F^n_{2,\mathbf{h}}$ is the structure factor for the anomalous scatterers, scattering as if there were no anomalous effects present.

With the definitions in Table 1 and use of (5), it is possible to calculate the right side of (1) and thereby evaluate the cosine function. The $|{}_m\mathscr{F}_{1,\mathbf{h}}|$ and $|{}_m\mathscr{F}_{2,\mathbf{h}}|$ are obtained from experimental measurement and the $|{}_m\mathscr{F}_{3,\mathbf{h}}|$ are obtained from derived information, for example, the heavy-atom structure. The angles on the left side of (1) are associated with the corresponding $|{}_m\mathscr{F}_{2,\mathbf{h}}|$ and $|{}_m\mathscr{F}_{3,\mathbf{h}}|$. It had not been brought out in the original derivation of (1) that, with the approximations used, the same function on the right side of (1) would be associated with any of the eight cosine functions obtained by substituting any or all of the ${}_m\varphi_{2,\mathbf{h}}$, ${}_m\varphi_{2,\mathbf{k}}$ and ${}_m\varphi_{2,(\bar{\mathbf{h}}+\bar{\mathbf{k}})}$ with the corresponding ${}_m\varphi_{1,\mathbf{h}}$, ${}_m\varphi_{1,\mathbf{k}}$ and ${}_m\varphi_{1,(\bar{\mathbf{h}}+\bar{\mathbf{k}})}$. In other words, the formula is insensitive to the distinction between the values of ${}_m\varphi_{1,\mathbf{h}}$ and ${}_m\varphi_{2,\mathbf{h}}$. In recognition of this and for the purpose of facilitating quantitative comparison between the cosine function and the right side of (1), the equation is rewritten

$$\cos\left({}_m\bar{\varphi}_{\mathbf{h}} + {}_m\bar{\varphi}_{\mathbf{k}} + {}_m\bar{\varphi}_{(\bar{\mathbf{h}}+\bar{\mathbf{k}})} - {}_m\varphi_{3,\mathbf{h}} - {}_m\varphi_{3,\mathbf{k}} - {}_m\varphi_{3,(\bar{\mathbf{h}}+\bar{\mathbf{k}})}\right)$$
$$\simeq T/|{}_m\mathscr{F}_{3,\mathbf{h}m}\mathscr{F}_{3,\mathbf{k}m}\mathscr{F}_{3,(\bar{\mathbf{h}}+\bar{\mathbf{k}})}|, \qquad (6)$$

where

$$_m\bar{\varphi}_{\mathbf{h}} = 0.5({}_m\varphi_{1,\mathbf{h}} + {}_m\varphi_{2,\mathbf{h}}). \qquad (7)$$

This becomes for isomorphous replacement ($m = i$)

$$_i\bar{\varphi}_{\mathbf{h}} = 0.5(\varphi_{\mathbf{h}PH} + \varphi_{\mathbf{h}P}) \qquad (8)$$

and for anomalous dispersion ($m = 1$)

$$_1\bar{\varphi}_{\mathbf{h}} = 0.5(\varphi_{\lambda\mathbf{h}} - \varphi_{\lambda\bar{\mathbf{h}}}). \qquad (9)$$

The argument of the cosine function on the left side of (6) contains a triplet phase invariant that is composed of average phases, as defined in (7), and a second triplet phase invariant composed of phases associated with the heavy-atom structure or anomalous scatterers. If the heavy-atom structure is known, the value of this second triplet phase invariant can be calculated and from knowledge of the value of the right side of (6), the value of the triplet phase invariant of interest ${}_m\bar{\varphi}_{\mathbf{h}} + {}_m\bar{\varphi}_{\mathbf{k}} + {}_m\bar{\varphi}_{\bar{\mathbf{h}}+\bar{\mathbf{k}}}$ can be evaluated, usually with a twofold ambiguity.

Under certain circumstances, the exact formulas for isomorphous replacement

$$\cos(\varphi_{\mathbf{h}P} - \varphi_{\mathbf{h}H}) = (|F_{\mathbf{h}PH}|^2 - |F_{\mathbf{h}P}|^2$$
$$- |F_{\mathbf{h}H}|^2)/2|F_{\mathbf{h}P}||F_{\mathbf{h}H}| \qquad (10)$$

and

$$\cos(\varphi_{\mathbf{h}PH} - \varphi_{\mathbf{h}H}) = (|F_{\mathbf{h}PH}|^2 - |F_{\mathbf{h}P}|^2$$
$$+ |F_{\mathbf{h}H}|^2)/2|F_{\mathbf{h}PH}||F_{\mathbf{h}H}| \qquad (11)$$

may be useful, namely, when the right sides of (10) and (11) have values in the vicinity of ± 1. In such cases, the $\varphi_{\mathbf{h}P}$ or φ_{HPH} are approximately equal to or π away from $\varphi_{\mathbf{h}H}$.

Test calculations

Test calculations were performed on exact data and also on data into which errors were introduced. The data were computed at 2·5 Å resolution for Cu $K\alpha$ radiation from the coordinates for cytochrome c550.PtCl$_4^{2-}$ from *Paracoccus denitrificans* (Timkovich & Dickerson, 1976). For the isomorphous replacement tests, the Pt atoms were regarded as comprising the heavy-atom structure. For the anomalous dispersion tests, the structure factors were computed in two ways. One calculation introduced anomalous effects from the Pt atoms alone and the second included anomalous effects from the Pt, Fe, S and Cl atoms. The first calculation models the case when there would be only one type of anomalous scatterer. This calculation not only represents an important experimental case, but also provides a basis of comparison for determining the effect on the errors of including all four types of anomalous scatterers in the data while treating the data as if the Pt atoms were the one predominant type of anomalous scatterer.

Isomorphous replacement

Calculations of values for triplet phase invariants in an isomorphous replacement experiment by use of (6) are presented in Table 2. Appropriate definitions for the use of (6) are given in Table 1. The calculations were based on 400 independent reflections except for row 6 in which 800 independent reflections were selected from among the 3252 acentric ones available from 2·5 Å data. They were chosen on the basis of the largest values for $\|F_{PH}| - |F_P\|$. The first column shows the number of invariants that were formed. When a cut-off value different from zero is given in the second column, it means that reflections were omitted from the calculations when the values for their corresponding $|F_P|$ and $|F_{PH}|$ were less than the cut-off value. For those data into which random errors were introduced (rows 3–7), an acceptance criterion was used for handling those instances in which the right side of (6) exceeded 1·0. If the right side exceeded 1·5, the calculation was rejected, otherwise the value was set back to 1·0.

In order to obtain an estimate of the accuracy of the calculations obtainable from (6), two types of error were computed, type I and type II. Type I is an average magnitude of error

$$\langle |\Phi - \cos^{-1}[\text{right side of (6)}]| \rangle, \qquad (12)$$

where Φ represents the sum of angles (of known value in the test problems). The arc cosine function is generally twofold ambiguous and may also require a shift of $\pm 2\pi$ to find the value closest to that of the

Table 2. *The evaluation of average triplet phase invariants in isomorphous replacement by use of* (6) *from* 2·5 Å *data for cytochrome* c550.PtCl$_4^{2-}$

The calculations were based on 400 independent reflections except for row 6 in which 800 independent reflections were chosen from among the 3252 acentric ones available on the basis of the largest values for $\|F_{PH}| - |F_P\|$. The last row was added to show the effect of including centric reflections in the data set and comparing the result from (6) to the correct value for a single triplet phase invariant, the one having the largest triple product of associated structure factor magnitudes, instead of the average of eight of them. For those data into which random errors were introduced (rows 3–7), an acceptance criterion was used for handling those instances in which the right side of (6) exceeded 1·0. If the right-side value exceeded 1·5, the calculation was rejected, otherwise the value was set back to 1·0.

Number of invariants	Cut-off for $|F_P|, |F_{PH}|$	Av. magnitude of error for $|F_P|$ $|F_{PH}|, |F_H|$ (%)	Av. magnitude of error type I (rad)	Av. magnitude of error type II
18 933	0	0	0·42	0·30
17 110	150	0	0·38	0·27
15 940	0	5	0·59	0·37
13 125	150	5	0·57	0·34
6453	0	10	0·81	0·49
55 880	0	10	0·82	0·54
5524	150	10	0·76	0·46
18 007	0	0	0·19	0·08

known Φ. In computing (12), the latter value of the arc cosine was used. The purpose of the computation was solely to present an estimate of the accuracy in radians. Type II is an average magnitude of error

$$\langle |\cos \Phi - \text{right side of (6)}| \rangle. \qquad (13)$$

Type I and type II errors appropriate to the calculations in Table 2 are shown in columns 4 and 5, respectively.

It is seen from Table 2 that large numbers of invariants can be computed with no apparently serious increase in the average error. This is shown by the calculations in rows 5 and 6 which differ in the use of 400 and 800 independent data, respectively, for the computation of the values of the invariants. It would appear that the number of independent data used could be safely increased and significant increases in the number of invariants computed would also accrue from the use of one- and two-dimensional data. Having a cut-off value for $|F_P|$ and $|F_{PH}|$ does not appear to have a great effect on the average error of the calculations, but it may prevent the occurrence of large errors from structure factors of small magnitude. Significant benefits would be obtained from keeping the average error in the structure factor magnitudes in the vicinity of 5% rather than 10%. The calculation of the 55 880 invariants and associated errors in row 6 of Table 2 required about 1 min and 40 s on the Cray X-MP/12 with the use of a rather simple program.

Table 3. *Average magnitude of discrepancy in isomorphous replacement between sets of eight triplet phase invariants and their averages calculated from* 2·5 Å *data for cytochrome* c550.PtCl$_4^{2-}$

In any set, the eight invariants are formed by adding φ_{hP} or φ_{hPH}, φ_{kP} or φ_{kPH} and $\varphi_{(\bar{h}+\bar{k})P}$ or $\varphi_{(\bar{h}+\bar{k})PH}$. The invariants in the first row were formed from 400 and those in the second row were formed from 800 independent reflections selected from the 3252 acentric reflections on the basis of the largest values for the $\|F_{PH}| - |F_P\|$. Since (6) is interpreted as providing the values of the average triplet invariants, this computation gives some insight into how well an average represents any member of the set of eight.

Number of invariants	Av. magnitude of error (rad)
18 933	0·34
99 367	0·37

As implied by the left side of (6), only the values of average triplet invariants, $_m\bar{\varphi}_h + {_m\bar{\varphi}_k} + {_m\bar{\varphi}_{h+k}}$, are expected to be the useful and meaningful context in which the invariants will be obtained from theory with the use of experimental data. The average invariants are the averages of eight different invariants formed by adding φ_{hP} or φ_{hPH}, φ_{kP} or φ_{kPH} and $\varphi_{(\bar{h}+\bar{k})P}$ or $\varphi_{(\bar{h}+\bar{k})PH}$. It is of interest to gain some insight into how well the value of an average invariant represents the value of any one member. Calculations of the average magnitude of error between the value of the average invariants and those of their eight members are shown in Table 3. The invariants in the

Table 4. *The evaluation of individual phases,* φ_{hP} *and* φ_{hPH}, *in isomorphous replacement by use of* (10), (11) *and* 2·5 Å *data for cytochrome* $c550.PtCl_4^{2-}$

Those values of the phases, φ_{hP} and φ_{hPH}, that were determined to be sufficiently close to the values of the corresponding φ_{hH} for the heavy-atom structure by use of (10), (11) were set equal to the φ_{hH}. This was done when the values of the right sides of (10), (11) were equal to or larger than the values in column 2, otherwise the phases were not evaluated. The subsets of 400 and 800 reflections were selected on the basis of the largest values for the differences, $\|F_{PH}| - |F_P\|$. Average percentage random errors introduced into the data are indicated in column 3.

Number of reflections	Limit for |right sides| of (10), (11)	Av. magnitude of error for $|F_P|, |F_{PH}|, |F_H|$ (%)	Number of φ_P accepted	Av. magnitude of error for φ_P (rad)	Number of φ_{PH} accepted	Av. magnitude of error for φ_{PH} (rad)
3252	0·9387	0	721	0·17	843	0·17
3252	0·7660	0	1463	0·35	1629	0·34
3252	0·9397	5	856	0·40	932	0·36
3252	0·7660	5	1403	0·47	1591	0·43
3252	0·9397	10	1094	0·57	1152	0·51
3252	0·7660	10	1522	0·59	1698	0·54
800	0·9397	10	571	0·51	598	0·45
800	0·7660	10	665	0·51	734	0·46
400	0·9397	10	333	0·46	348	0·41
400	0·7660	10	364	0·47	388	0·41

first row were formed from 400 and those in the second row from 800 independent reflections selected from the 3252 acentric reflections available on the basis of the largest values for the $\|F_{PH}| - |F_P\|$. The first column of Table 3 gives the number of average invariants and the second gives the average magnitude of error in radians for all of them.

It would appear that a large number of initial phase values may be obtained from (10) and (11) with acceptable accuracy. This is indicated by the results presented in Table 4. The reflections were ordered according to the largest values for the $\|F_{PH}| - |F_P\|$ and the number of acentric reflections available was 3252. The samples consisting of 400 and 800 reflections were ordered subsets of the latter. The second column indicates limits for the magnitude of the right sides of (10), (11). The limit of 0·9397 implies that as much as a calculated 20° difference between the two angles would be acceptable and 0·7660 implies that as much as a 40° difference would be acceptable. When there are errors in the data, as shown in the third column, the actual differences that are accepted can exceed 20 or 40°. Columns 4–7 show the numbers of phases, φ_{hP} and φ_{hPH}, that were set equal to the corresponding φ_{hH} and the average magnitude of error. It is seen that little is gained in terms of accuracy by restricting the number of reflections and, evidently, fewer phases are evaluated under such circumstances. The one- and two-dimensional data in the space group of cytochrome $c550$, $P2_12_12_1$, are centric. Application of (10), (11) to such data should give many errorless phase evaluations.

A possibly useful strategy in single isomorphous replacement would be to use (10) and (11) to obtain a large set of initial phase values and then apply the evaluations of the triplet phase invariants from (6) or from probabilistic methods (Fortier, Moore & Fraser, 1985) an extension of an analysis of Hauptman (1982a) to refine further and extend the phases by use of, for example, a least-squares procedure. The work of Fortier et al. is relevant to the

analysis in this paper because it presents a formula that can evaluate the cosines of triplet phase invariants having any value in the range -1 to $+1$.

It is possible to compare only the results in the last row of Table 2 of this paper directly with those in Table 2 of Fortier et al. The other entries in Table 2 are based on average invariants that are averages of eight of them, as noted previously. In addition, several items in Table 2 were computed from data containing random errors. The entries in Table 2 of Fortier et al. concern individual invariants computed from exact data. Comparison shows that the results of the last row of Table 2 of this paper are of an accuracy similar to the best calculations given in Table 2 of Fortier et al. The intention of the presentation in the first seven rows of Table 2 of this paper is to give an insight into what might be expected in actual experimental circumstances. The results would be improved if centric data were included in the calculations.

Anomalous dispersion

Calculations of the values for triplet phase invariants in an anomalous dispersion experiment involving one predominant type of anomalous scatterer by use of (6) are presented in Table 5. The radiation used was Cu $K\alpha$. Appropriate definitions for the use of (6) are given in Table 1. For $m = 1$, we find from (5) that

$$_m\varphi_{3,h} = \pi/2 + \varphi_{2,h}^n, \tag{14}$$

where $\varphi_{2,h}^n$ is the phase associated with $F_{2,h}^n$, the structure factor in (5) for the single type of anomalous scatterers present, scattering as if there were no anomalous effects. The same mathematics holds approximately when one predominant type of anomalous scatterer is present among others. For the case of one predominant type of anomalous scatterer present, (6) becomes from use of (9) and (14)

Table 5. *The evaluation of average triplet phase invariants in anomalous dispersion by use of* (15) *from* 2·5 Å
data for cytochrome c550.PtCl$_4^{2-}$

The calculations were based on 400 independent reflections except for row 6 in which 800 independent reflections were chosen from among the 3252 acentric ones available from 2·5 Å data on the basis of the largest values for $\|F_{\Lambda h}| - |F_{\Lambda \bar{h}}\|$. The structure factors used for the calculations in rows 1–4 were based on the Pt atoms alone as the anomalous scatterers, whereas those in rows 5 and 6 were based on the Cl, S, Fe and Pt atoms as the anomalous scatterers. In all cases, the data were analyzed as if the Pt atoms were the sole predominant anomalous scatterers. The radiation used was Cu $K\alpha$. Errors were introduced for the individual differences, $\|F_{\Lambda h}| - |F_{\Lambda \bar{h}}\|$, by multiplying randomly by values that ranged from 0·2 to 1·8, accounting for the 40% average error listed in the last row of column 3. The values of $|F_{\Lambda h}|$ and $|F_{\Lambda \bar{h}}|$ were suitably adjusted to account for the changed difference. For those data into which errors were introduced (rows 3–6), an acceptance criterion was used for handling those instances in which the right side of (15) exceeded 1·0. If the right side exceeded 1·5, the calculation was rejected, otherwise the value was set back to 1·0.

Number of invariants	Cut-off for $\|F_{\Lambda h}\|, \|F_{\Lambda \bar{h}}\|$	Av. magnitude of error for $\|F_{\Lambda h}| - |F_{\Lambda \bar{h}}\|$ (%)	Av. magnitude of error $\|F_{2,h}^n\|$ (%)	Av. magnitude of error type I (rad)	Av. magnitude of error type II
16 662	0	0	0	0·41	0·28
16 099	100	0	0	0·42	0·28
5960	0	40	3	0·80	0·47
6046	0	40	5	0·79	0·47
3141	0	40	3	0·88	0·56
41 797	0	40	3	0·83	0·55

$$\sin\left({}_1\bar{\varphi}_h + {}_1\bar{\varphi}_k + {}_1\bar{\varphi}_{h+k} - \varphi_{2,h}^n - \varphi_{2,k}^n - \varphi_{2,(h+k)}^n\right)$$
$$\simeq -T/|{}_m\mathscr{F}_{3,hm}\mathscr{F}_{3,km}\mathscr{F}_{3,(h+k)}|, \qquad (15)$$

where T is obtained from (2) and Table 1 and the denominator of the right side of (15) is obtained from taking appropriate magnitudes in (5), e.g.

$$_1\mathscr{F}_{3,h} = 2(f''/f_{2,h}^n)|F_{2,h}^n|. \qquad (16)$$

The calculations shown in Table 5 were based on 400 independent reflections except for the sixth row in which 800 independent reflections were chosen from among the 3252 centric ones available from 2·5 Å data on the basis of the largest values for $\|F_{\Lambda h}| - |F_{\Lambda \bar{h}}\|$. The first column shows the number of invariants that were formed. The structure factors used for the calculations in rows 1–4 were based on the Pt atoms alone as the anomalous scatterers. In all cases, the data were analyzed as if the Pt atoms were the sole predominant anomalous scatterers. For the second row, in which a cut-off value different from zero is given, reflections were omitted from the calculations when the values for their corresponding $|F_{\Lambda h}|$ and $|F_{\Lambda \bar{h}}|$ were less than the cut-off value. Errors were introduced for the individual differences, $|F_{\Lambda h}| - |F_{\Lambda \bar{h}}|$, by multiplying randomly by values that ranged from 0·2 to 1·8, accounting for the 40% average magnitude of error listed in the last four rows of column 3. Random errors of average magnitude 3 and 5% for $|F_{2,h}^n|$ are listed in the last four rows of column 4. For those data into which errors were introduced (rows 3–6), an acceptance criterion was used for handling those instances in which the right side of (5) exceeded 1·0. If the right side exceeded 1·5, the calculation was rejected, otherwise the value was set back to 1·0. Comparison of row 3 with row 1 indicates that the acceptance criterion caused a dramatic drop in the number of invariants accepted. There may well be a better criterion so that more invariant evaluations could be accepted without sig-

nificant loss of accuracy. It does not seem from comparison of rows 1 and 2 in columns 5 and 6 that the introduction of the cut-off had a significant effect on the resulting errors in the calculations. The cut-off value for $|F_{\Lambda h}|$ and $|F_{\Lambda \bar{h}}|$ may, however, similarly to isomorphous replacement calculations, prevent the occurrence of large errors from structure factors of small magnitude. Errors of type I and type II in columns 5 and 6 are describable by use of (12) and (13), respectively, or, equivalently, by use of

$$\langle|\Phi' - \sin^{-1}[\text{right side of (15)}]|\rangle \qquad (17)$$

and

$$\langle|\sin \Phi' - \text{right side of (15)}|\rangle, \qquad (18)$$

where Φ' represents the sum of angles on the left side of (15). The calculation of (17) was carried out in a fashion comparable to that for (12). A comparison of rows 3 and 4 in which the calculations were the same except for the different average magnitudes of error for the $|F_{2,h}^n|$ indicates that even greater average errors in $|F_{2,h}^n|$ could be tolerated.

It is seen from Table 5 that large numbers of invariants can be computed without reaching unacceptable average errors. This is indicated by the calculations in rows 5 and 6 which differ in the use of 400 and 800 independent reflections, respectively, for the computation of the values of the invariants. The calculations for row 5 required 75 s on the Cray X-MP/12. It would appear from the indicated accuracies and the time involved that the number of independent reflections used could be readily increased beyond 800 with a consequent further increase in the number of invariants evaluated.

The triplet phase invariants composed of average phases, ${}_1\bar{\varphi}_h + {}_1\bar{\varphi}_k + {}_1\bar{\varphi}_{h+k}$, that are evaluated by use of (15) are the averages of eight different invariants formed by adding $\varphi_{\Lambda h}$ or $-\varphi_{\Lambda \bar{h}}$, $\varphi_{\Lambda k}$ or $-\varphi_{\Lambda \bar{k}}$ and $\varphi_{\lambda,(h+k)}$ or $-\varphi_{\lambda,(h+k)}$. The values for the average invariants represent the eight different invariants rather accurately for cytochrome c550.PtCl$_4^{2-}$. A calculation of the average magnitude of difference between $\varphi_{\lambda h}$ and $-\varphi_{\lambda \bar{h}}$ (Karle, 1985a, Table 3, column 5, $m = 1$) for 2900 reflections and Cu $K\alpha$ radiation gave 0·07 rad. In these calculations data for which $|F_{\lambda h}|$ and $|F_{\lambda \bar{h}}| < 100$ were not included, thus eliminating the smallest magnitudes.

It is possible that the measured data would be accurate enough to permit the evaluation of individual phases in a fashion similar to that for (10) and (11). The appropriate general equations are

$$\cos\left(_m\varphi_{2,h} - {}_m\varphi_{3,h}\right) = (|_m\mathscr{F}_{1,h}|^2 - |_m\mathscr{F}_{2,h}|^2$$
$$- |_m\mathscr{F}_{3,h}|^2)/2|_m\mathscr{F}_{2,h}||_m\mathscr{F}_{3,h}| \quad (19)$$

$$\cos\left(_m\varphi_{1,h} - {}_m\varphi_{3,h}\right) = (|_m\mathscr{F}_{1,h}|^2 - |_m\mathscr{F}_{2,h}|^2$$
$$+ |_m\mathscr{F}_{3,h}|^2)/2|_m\mathscr{F}_{1,h}||_m\mathscr{F}_{3,h}|, \quad (20)$$

where, for anomalous dispersion, case 1, $_1\varphi_{3,h}$ is given by (14),

$$_1\varphi_{1,h} = \varphi_{\lambda h} \quad (21)$$

$$_1\varphi_{2,h} = -\varphi_{\lambda \bar{h}} \quad (22)$$

and $|_1\mathscr{F}_{1,h}|$, $|_1\mathscr{F}_{2,h}|$ and $|_1\mathscr{F}_{3,h}|$ are defined in Table 1 and by taking the magnitude of (5). This gives

$$\sin\left(-\varphi_{\lambda \bar{h}} - \varphi_{2,h}^n\right) = (|F_{\lambda h}|^2 - |F_{\lambda \bar{h}}|^2 - 4(f''/f_{2,h}^n)^2|F_{2,h}^n|^2)$$
$$\times [4(f''/f_{2,h}^n)|F_{\lambda \bar{h}}||F_{2,h}^n|]^{-1} \quad (23)$$

$$\sin\left(\varphi_{\lambda h} - \varphi_{2,h}^n\right) = (|F_{\lambda h}|^2 - |F_{\lambda \bar{h}}|^2 + 4(f''/f_{2,h}^n)^2|F_{2,h}^n|^2)$$
$$\times [4(f''/f_{2,h}^n)|F_{\lambda h}||F_{2,h}^n|]^{-1}. \quad (24)$$

Comparable equations for isomorphous replacement, (10) and (11), also follow from (19) and (20) and the use of Table 1. In a manner similar to that described for isomorphous replacement, (23) and (24) can be applied by specifying a magnitude somewhat less than unity that the right sides of (23) and (24) must attain. When this value or greater is attained, the sine functions may be set approximately equal to $+1$ or -1, as appropriate, and then the $-\varphi_{\lambda \bar{h}}$ or $\varphi_{\lambda h}$ may be set approximately equal to $\varphi_{2,h}^n + \pi/2$ or $\varphi_{2,h}^n - \pi/2$, corresponding to the values of $+1$ and -1, respectively. This could give an initial set of values for a number of phases which could perhaps be refined and extended by use of computed values of triplet phase invariants from application of (15).

There is an alternative way to obtain individual phase values by algebraic means. This has been described (Karle, 1985b). The phases evaluated by these means are the $\varphi_{1,h}^n$, the phases corresponding to the structure factors for the structure of the non-anomalously scattering atoms. The formation and evaluation of triplet phase invariants consisting of the $\varphi_{1,h}^n$ has also been discussed (Karle, 1984c).

Concluding remarks

Formulas (6) and (15) provide values for triplet phase invariants in single isomorphous replacement and one-wavelength anomalous dispersion experiments when the structure of the replacement or that for a predominant anomalous scatterer is known. Actually, only the magnitudes of the structure factors for the replacement structure or the structure of the predominant anomalous scatterer are required since the triplet phase invariants for the heavy-atom phases may be set equal to zero to good approximation when the associated products of structure factor magnitudes are large. Such information is obtainable, for example, from application of the exact linear theory (Karle, 1980). When the latter theory is applied, there are also alternative ways to evaluate the triplet phase invariants (Karle, 1984c).

The potential utility of (6) and (15) is indicated by the accuracy obtained in the test calculations when reasonable random errors are introduced into the data. The use of the formulas may be facilitated by the determination of initial values for a number of phases by application of (10) and (11) for isomorphous replacement or (23) and (24) for anomalous dispersion when the phases associated with the replacement or the predominant anomalous scatterer structure are known. When (10), (11), (23) and (24) are used in the manner described, they give essentially unambiguous phase values. In general, however, they give values with a twofold ambiguity that may also be useful if, in the course of the application of the triplet phase invariants, the ambiguity could be resolved by use of phase values developed by the triplet phase invariants combined with the initially determined essentially unambiguous phase values. Throughout the history of the application of isomorphous replacement and anomalous dispersion, many ways have been developed to overcome the twofold ambiguity (see, for example, Fan Hai-fu, Han Fu-son, Qian Jin-zi & Yao Jia-xing, 1984).

The use of (10), (11), (23) and (24) has the potential to establish the appropriate enantiomorph for a macromolecule. If the structure of the replacement atoms in single isomorphous replacement is centrosymmetric and the method is used independently of anomalous dispersion information, the enantiomorph will have to be established in some other way, e.g. by the use of the triplet phase invariants in some manner comparable to that used for small-molecule structure determination. One approach to this matter has been presented by Fan Hai-fu & Gu Yuan-xin (1985) and Yao Jia-xing & Fan Hai-fu (1985).

It has been the intention in this paper to illustrate the potential information available in the individual techniques of single isomorphous replacement and one-wavelength anomalous dispersion from the use of certain algebraic formulas that can be computed quite rapidly. In actual practice, the potential would be enhanced if the techniques were combined. In addition to the formulas presented here, there are numerous algebraic (Ramaseshan & Abrahams, 1975) and probabilistic (Hauptman, 1982a, b; Giacovazzo, 1983; Pontenagel, Krabbendam, Peerdeman &

418

Kroon, 1983; Fortier, Moore & Fraser, 1985) formulas available for fashioning strategies for the analysis of experimental data. There are also relations from an exact algebraic analysis (Karle, 1980, 1984c, 1985b) which give an essentially unique result with one-wavelength anomalous dispersion data and, when used with exact data, give exact phase values. It is likely that optimal strategies will depend upon the quality and character of the data. The formulas and procedures considered here should, in any case, enhance the variety of options that can be considered.

I wish to thank Mr Stephen Brenner for writing the programs and making the computations reported here.

This work was supported in part by USPHS grant GM 30902.

References

ADMAN, E. T., SIEKER, L. C. & JENSEN, L. H. (1973). *J. Biol. Chem.* **248**, 3987–3996.
FAN HAI-FU & GU YUAN-XIN (1985). *Acta Cryst.* A41, 280–284.
FAN HAI-FU, HAN FU-SON, QIAN JIN-ZI & YAO JIA-XING (1984). *Acta Cryst.* A40, 489–495.
FORTIER, S., MOORE, N. J. & FRASER, M. E. (1985). *Acta Cryst.* A41, 571–577.
GIACOVAZZO, C. (1983). *Acta Cryst* A39, 585–592.
HAUPTMAN, H. (1982a). *Acta Cryst.* A38, 289–294.
HAUPTMAN, H. (1982b). *Acta Cryst.* A38, 632–641.
KARLE, J. (1980). *Int. J. Quantum Chem. Symp.* 7, 357–367.
KARLE, J. (1983). *Acta Cryst.* A39, 800–805.
KARLE, J. (1984a). *Acta Cryst.* A40, 4–11.
KARLE, J. (1984b). *Acta Cryst.* A40, 366–373, 374–379.
KARLE, J. (1984c). *Acta Cryst.* A40, 526–531.
KARLE, J. (1985a). *Acta Cryst.* A41, 182–189.
KARLE, J. (1985b). *Acta Cryst.* A41, 387–394.
PONTENAGEL, W. M. G. F., KRABBENDAM, H., PEERDEMAN, A. F. & KROON, J. (1983). Eighth Eur. Crystallogr. Meet., Liège, Belgium, 8–12 August 1983. Abstract 4.01-P.
RAMASESHAN, S. & ABRAHAMS, S. C. (1975). Editors. *Anomalous Scattering.* Copenhagen: Munksgaard.
TIMKOVICH, R. & DICKERSON, R. E. (1976). *J. Biol. Chem.* **251**, 4033–4046.
YAO JIA-XING & FAN HAI-FU (1985). *Acta Cryst.* A41, 284–285.

SOLVING DIFFICULT STRUCTURES

Isabella L. Karle
Laboratory for the Structure of Matter
Naval Research Laboratory
Washington, D. C. 20375-5000, U.S.A.

ABSTRACT. Common difficulties encountered in direct phase determination procedures, particularly for structures containing only light atoms, include: (1) Fragment of structure; (2) Fragment or molecule misplaced with respect to origin; (3) Elements of pseudo-symmetry; (4) Many indistinguishable phase sets; (5) Limited data; (6) Large cavities with disordered solvent; (7) Flexible molecules with disordered side chains; (8) Small values of $1/N^{1/2}$ and, consequently, low probabilities for triplets; (9) No structure or recognizable fragment resulting from strong incorrect indications from triplets and quartets.

Approaches that have been successful, at least a number of times, for overcoming some of these difficulties will be illustrated by examples from structure analyses of crystals containing a large number of atoms per asymmetric unit.

1. INTRODUCTION

This paper will address various types of difficulties encountered in solving crystal structures from diffraction data by direct phase determination procedures. A major source of difficulty is the occasional failure of the Σ_2 phase relationship for centrosymmetric space groups (Hauptman and Karle, 1953) or the similar triplet phase relationship for noncentrosymmetric space groups (Karle and Hauptman, 1950; Karle and Karle, 1964; 1966) where

$$\phi_h \sim \phi_k + \phi_{h-k}, \tag{1}$$

fails to give sufficiently good phase values for ϕ_h. The reliability of (1) is directly dependent upon the magnitude of the normalized reflections, E_h, E_k and E_{h-k} and inversely proportional to the squareroot of the number of atoms in the unit cell. In structures containing many atoms, the reliability of individual

419

M. A. Carrondo and G. A. Jeffrey (eds.), Chemical Crystallography with Pulsed Neutrons and Synchrotron X-Rays, 419–441.
© 1988 by D. Reidel Publishing Company.

triplets defining a particular ϕ_h is not as good as in smaller structures. Even in smaller structures there are very occasional problems with incorrect phase indications.

An example taken from the structure of guaianolide in $P2_1/n$ with two molecules/asymmetric unit (Posner et al,1980) shows the following contradiction in triplets with very large $|E|$ values and large probability values for the $\phi\bar{3}\,1\,1$:

$$|E| = 2.4 \qquad \bar{2}\ 2\ \overline{20} \qquad \Pi$$

$$2.9 \qquad \bar{1}\ \bar{1}\ 21 \qquad 0 \qquad \text{WRONG}$$

$$5.2 \qquad \bar{3}\ 1\ 1 \qquad 0$$

$$P_+ = 0.998 \tag{2}$$

$$\text{and}\,|E| = 3.0 \qquad \bar{2}\ 1\ \bar{2} \qquad \Pi$$

$$2.4 \qquad \bar{1}\ 0\ 3 \qquad \Pi \qquad \text{CORRECT}$$

$$5.2 \qquad \bar{3}\ 1\ 1 \qquad 2\,\Pi = 0$$

$$P_+ = 0.998$$

The phases shown in (2) are the correct values from the structure.

The incorrect indication in the upper triplet in (2) could not be discovered by the use of quartets or quintets. As shown in the quartet in (3), the same reflections as used in (2) make the quartet negative. To discover experimentally that a quartet is negative, i.e. $\phi_h + \phi_k + \phi_l + \phi_{-h-k-l} = \Pi$, the $|E|$ magnitudes for the three cross-terms, E_{h+k}, E_{h+l} and E_{k+l}, must be small. For (3),

NEGATIVE QUARTET

$$(\bar{2}\ 2\ \overline{20}) + (\bar{1}\ \bar{1}\ 21) + (2\ \bar{1}\ 2) + (1\ 0\ \bar{3}) = \Pi$$

Negative triple from (2) →

$$\Pi \qquad\qquad 0 \qquad\qquad \Pi \qquad\qquad \Pi$$

$$E = 2.4 \qquad 2.9 \qquad\qquad 3.0 \qquad 2.4$$

$$\tag{3}$$

$$E_{h+k} = +5.2 \quad (\bar{3}\ 1\ 1)$$

$$E_{h+l} = 1.3 \quad (0\ 1\ \overline{18})$$

$$E_{k+l} = 2.5 \quad (1\ \bar{2}\ 23)$$

all large magnitudes:

    do not indicate negative quartet.

the three cross-terms have large magnitudes, thus giving <u>no</u> indication that the quartet is negative, and consequently, that one of the triplets is negative.

    In examining the quintet (4) containing the negative triplet in (2), it is seen that all the cross-terms have large $|E|$ magnitudes and thus <u>do not</u> indicate a negative value for the quintet. Only after the phases are known from the determined structure, can it be ascertained that the quintet is negative and thus the triplet contained in this special kind of quintet is negative.

NEGATIVE QUINTET

$$(\bar{2}\ 2\ \overline{20}) + (\bar{1}\ \bar{1}\ 21) + (3\ \bar{1}\ \bar{1}) + (2\ \bar{1}\ 2) + (\bar{2}\ 1\ \bar{2})$$

Π	0	0	x	$-x$ = Π
$\|E\|$ = 2.4	2.9	5.2	3.0	3.0

Negative Triple   Any R

CROSS-TERMS: (4)

$|E|$

H + R	$0\ 1\ \overline{18}$	1.3	
H − R	$\bar{4}\ 3\ \overline{22}$	---	DO <u>NOT</u> INDICATE NEGATIVE QUINTET.
K + R	$1\ \bar{2}\ 23$	2.5	SIMILAR RESULTS WITH
K − R	$\bar{3}\ 0\ 19$	1.1	OTHER SELECTED
L + R	$5\ \bar{2}\ 1$	1.3	VALUES OF R.
L − R	$1\ 0\ \bar{3}$	2.4	

    The above example illustrates how incorrect values for phases can arise. A sufficient number of incorrect phases leads to (1) a fragment of the structure, or (2) a misplaced fragment in the cell (although always correctly oriented with respect to the axes), or (3) no structure.

2. PARTIAL STRUCTURES AND THE USE OF THE TANGENT FORMULA

Phase determination procedures often result in partial structures, rather than complete structures. This is particularly the case for noncentrosymmetric crystals. The partial structures are properly oriented and usually correctly located with respect to a proper origin for the space group. At times the partial structure is properly oriented but not correctly placed. One of the ways that the correct location can be found is by use of translation functions. In view of the common occurrence of properly oriented structural fragments, a number of procedures have been proposed for readily developing the partial structure into a complete one. The tangent formula offers the basis for such a procedure (Karle, 1968). The partial structure can consist of a light atom fragment or it can involve a heavy atom which may have been located by use of the Patterson function. In the case of a fragment containing a heavy atom, special precautions are required to avoid obtaining phases which correspond only to the heavy atom position.

The tangent formula (Karle and Hauptman, 1956) is

$$\tan \phi_h \approx \frac{\sum\limits_{k} |E_k E_{h-k}| \sin(\phi_k + \phi_{h-k})}{\sum\limits_{k} |E_k E_{h-k}| \cos(\phi_k + \phi_{h-k})} \tag{5}$$

where ϕ_h is the phase associated with the normalized structure factor E_h. Use of Eq. (5) in developing a complete structure is implemented by the computation of structure factors from the partial structure and the selection of an appropriate subset for extension by Eq. (5). The extended set of phases so obtained is employed in the computation of a Fourier map of the structure. When the partial structure contains a heavy atom, this procedure is clearly distinguished from the usual heavy atom method by the interposition of a calculation with the tangent formula before proceeding with the Fourier series.

A relatively small structural fragment can generate phases which are sufficiently accurate for use with Eq. (5). However, only certain of the phases computed from the partial structure are suitable for use. These phases are selected by means of an acceptance criterion which is based on the amplitudes of the diffracted rays and the corresponding amplitudes computed from the structural fragment. A phase computed from a partial structure is accepted if $|F_h|_{calc} > p|F_h|_{obs}$ where p is the fraction of the total scattering power contained in the fragment and where $|F_h|$ is associated with an $|E_h|_{obs} > 1.5$. The quantity $|F_h|_{calc}$ is the value of the structure factor amplitude computed from the partial structure and $|F_h|_{obs}$ is the experimentally observed amplitude. As a working rule, if $p < 0.25$, it is replaced by 0.25 or, at least, a number somewhat larger than p. If $p > 0.6$, it is replaced by 0.6.

Phases obtained by use of the acceptance criterion do not change very much throughout the structure development procedure as more and more atoms are added to the initial partial structure (Karle, 1968). It was found in studies starting with about 20% of the structure that the average change in the values of the phases was about 0.7 radians. It was also found that atomic positions for the partial structure changed very little, generally much less than 0.02 of a cell edge, in the course of the development of the complete structure. Incorrect atomic locations will disappear in the successive cycles. The limited variation of the initial phases and atomic positions as the structure development progresses and the disappearance of incorrect atomic positions account for the usefulness of the tangent formula in the procedure. A resume of the partial structure procedure is shown in (6).

1. Accept phases if

$$|F_h|_{calc} > p|F_h|_{obs} \quad \underline{and} \quad |E_h| > 1.5$$

   $p \approx$ fraction of known atoms

2. a. <u>Refine</u> accepted phases with tangent formula

      for 1-2 cycles.   (6)

    b. <u>DO NOT</u> refine if heavier atoms are present.

3. <u>Extend</u> phase determination with tangent formula

   to $|E|_{min} = 1.0$

4. Compute E-map.

5. Repeat, if necessary.

6. Relax space group to P1, if necessary.

A number of examples of the development of partial structures into complete ones by means of the tangent formula have been reviewed previously (Karle, 1968, 1970). Several examples will be discussed in some detail here. One is valinomycin (Karle, 1975) which is composed of atoms having almost equal atomic numbers, except for the hydrogen atoms, the second is a lithium antamanide complex for which the counter ion is bromine (Karle, 1974), and the third is a complex of Mg^{++} with two cyclic hexapeptides in which the structure was developed based on the position of the Mg^{++} ion alone (Karle and Karle, 1981).

2.1. Valinomycin, an Equal Atom Structure

424

The form of valinomycin to be discussed crystallizes in space group
P1 with two molecules (156 nonhydrogen atoms) in the unit cell. To
illustrate the capabilities of the method, the recycling procedure
was initiated with only 9 atomic positions, although about half the
atoms were actually available initially from the phase determination by the symbolic addition procedure. The 9 atoms chosen form a
chemically sensible cluster of atoms. It is worthwhile to start
with a good chemical unit even though this may mean using fewer
atoms than are indicated by a phase determination. Although 9
atoms represented about 0.06 of the structure, the value of p in
the acceptance criterion was set at 0.10. As the number of atoms
increased, p was correspondingly increased to a maximum value
of 0.6. With a start based on only 0.06 of the structure, it
required 9 cycles and a final difference map to complete the
development. The number of atoms employed at each step of the
application of the tangent formula were 9, 13, 20, 25, 29, 36, 81
and 99. The last cycle produced 141 atoms. The remaining 15 atoms
atoms can be readily obtained from a difference map. This process
is illustrated in Fig. 1 where the numbers next to the atoms
represent the cycle in which they were employed together with
those obtained previously. Atoms labeled 1,2,..., n were used in
the nth cycle to produce atoms labeled n + 1.

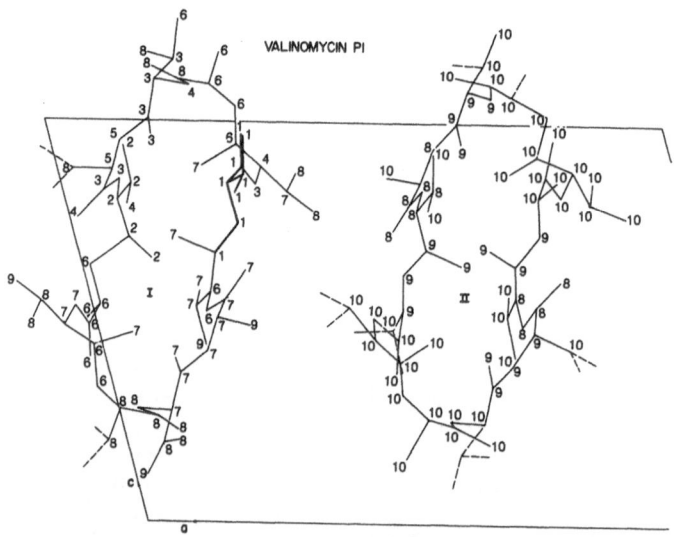

Fig. 1. The development of the structure of valinomycin from a
a structural fragment by means of the tangent formula. Atoms
labeled 1,2,..., n were used in the nth cycle to produce atoms
labeled n + 1.

Phases were selected for $|E| > 1.5$, were refined twice using the tangent formula and were then used as a basic set for proceeding with the determination of new ones. In our tangent formula program new phases are added in a stepwise fashion, each time roughly doubling the number of phases which were used as the known set in each pass of calculation. Once new phases are determined, they are used in the succeeding passes without further refinement. This process is continued with sets of unknown phases corresponding to $|E|$ values of continually decreasing magnitude until some lower limit is reached. In this calculation the lowest value considered was $|E| = 1.3$. Ordinarily the lowest value would be about 1.0, but computing capacity was a factor to be considered for this large structure. The 9 cycles of partial structure refinement are more than have been otherwise required for any structure in our laboratory by more than a factor of 2. This number might well have been cut down had the lower limit on $|E|$ been set at 1.0 instead of 1.3.

2.2. Structure Based on Heavy Atom in $P2_1$

The lithium ion complex of antamanide associated with a bromide ion crystallizes in space Group $P2_1$ with 95 atoms in the asymmetric unit, including three molecules of acetonitrile, (Karle, 1974). The development of the structure, which was initiated with the location of the bromide ion with the aid of the Patterson function, is shown in Fig. 2. The bromide ion represents 0.21 of the total scattering and this number was used for p in using the acceptance criterion for phases based on this ion. The phases were used as input to the tangent formula for the computation of additional phases in the stepwise fashion described above for the remaining $|E| > 1.1$. In space group $P2_1$, phases based on one atom result in a map which contains atoms for both enantiomorphs. Fourteen atomic positions labeled 2 in Fig. 2 were chosen which seemed to form a good fragment of one polypeptide chain and were used along with the bromide ion for the next cycle of development. With a heavy atom present it is most important not to recycle with the tangent formula for the purpose of refinement of the initial phases selected by means of the acceptance criterion. These phases would, in fact, not refine but instead, with a heavy atom present, the contribution from the light atoms would be lost and the phases would converge to those which would be obtained from the heavy atom alone. The tangent formula is used solely for phase extension when a heavy atom is present. Of the 14 light atom positions initially selected, two were incorrect in that their mirror images across a $y = 1/4$ plane should have been selected instead. This was corrected in the next cycle when 14 more light atoms were obtained to make 28 in all. Pairs of atoms corresponding to both enantiomorphs still appeared in the resulting E map, but they were of unequal

weight with the correct one usually the stronger. Choosing only
the stronger of the pair and making sure that a good structural
fragment ensued avoided errors of selection. At the next stage a
total of 54 light atoms were located and finally in the last stage,
the E map revealed all 95 atoms in the asymmetric unit including
the three molecules of the solvent, acetonitrile. At all stages,
the tangent formula (5) was used for phase extension but <u>not for
recyling</u> because of the presence of the heavy atom.

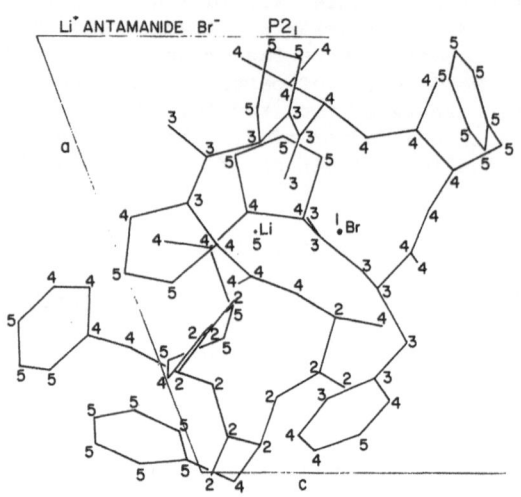

Fig. 2. The development of the structure of Li+Antamanide-Br−
from a structural fragment by means of the tangent formula. Atoms
labeled 1,2,...,n were used in the nth cycle to produce atoms
labeled n + 1.

2.3. Mg^{++} as the "Heavy Atom"

Crystals grown from a CH_3CN solution of cyclo(Gly-L-Pro-L-Pro)$_2$ and
$Mg(ClO_4)_2$ have the space group $P3_1$ with dimensions a = b =
15.744(4)Å, c = 24.002(6)Å, γ = 120°, and V = 5153.5 Å^3,
and there are three formula units in the unit cell, with the compo-
sition $C_{48}H_{68}N_{12}O_{12}Mg \cdot 2ClO_4 \cdot 4CH_3CN$. An $(|E_h|^2 - 1)$ Patterson
map showed one major peak in the Harker section. It proved to be
the Mg-Mg vector from which the entire structure was derived, even
though Mg represents only 3.0% of the total scattering. In retro-
spect, it is clear why the Mg-Mg vector was prominent although
the Cl-Cl vectors were not apparent. The analysis showed that the

thermal factor for the Mg atom, B = 2.7 Å2, implies a rigidly held atom of concentrated electron density, whereas the thermal factors for the Cl atoms, B = 7 and 10 Å^2, imply considerable positional disorder.

Two hundred fifty-five phases based on the coordinates of the Mg atom alone that satisfied the criteria that $|E_h| > 1.5$ and $|F_h|_{calc} > p|F_h|_{obs}$, where p = 0.20, were used as input to the tangent formula (5) to compute the phases for the remaining $|E_h| > 1.1$. Only those phases for which $|E_h|_{calc} > 0.5$ in the tangent formula extension were used in the initial E map, Fig. 3, which showed the positions of four O atoms at ≈ 2.1 Å from the Mg++ and arranged in a square. In four additional rounds of phase extension by the tangent formula, in which the input phases were based on the Mg atom plus any additional atoms located in the intervening E maps and the value of p was raised successively to a maximum of 0.5, the coordinates of 75 atoms were determined. The tangent formula was used only for phase extension, not phase refinement. A series of cycles of least-squares refinement on the coordinates and thermal factors of the known atoms interspersed with difference maps yielded approximate coordinates for eight highly disordered O atoms in the two ClO_4^- anions and for 12 atoms in four CH_3CN solvent molecules, also highly disordered (Karle and Karle, 1981). The structure of the complex is shown in Fig. 4.

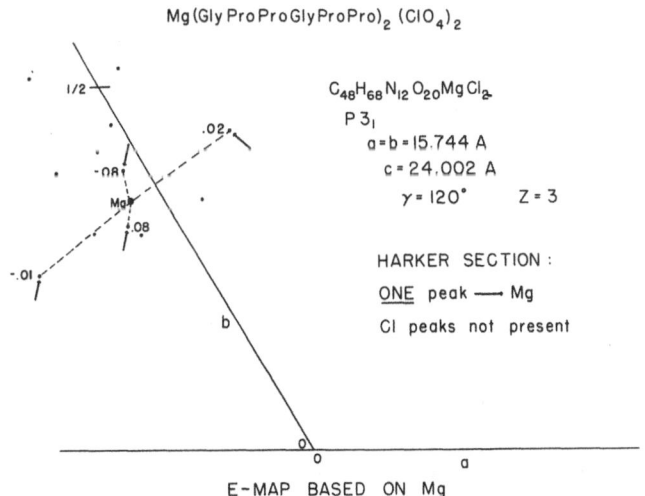

Mg(GlyProProGlyProPro)$_2$ (ClO$_4$)$_2$

$C_{48}H_{68}N_{12}O_{20}MgCl_2$
P 3$_1$
a = b = 15.744 A
c = 24.002 A
γ = 120°   Z = 3

HARKER SECTION :

ONE peak —— Mg

Cl peaks not present

E-MAP BASED ON Mg

Fig. 3 continued on next page.

428

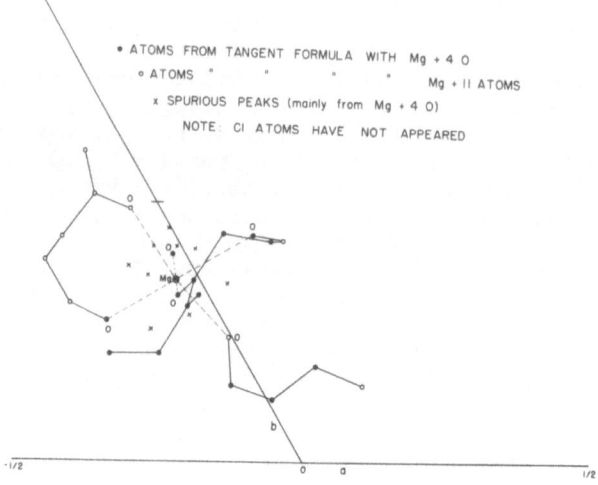

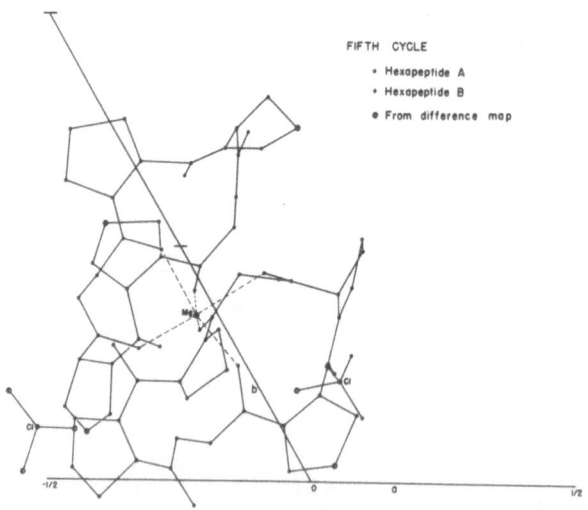

Fig. 3. Development of the Mg·2 hexapeptide structure with the use of the tangent formula based on the position of the Mg^{++} ion alone.

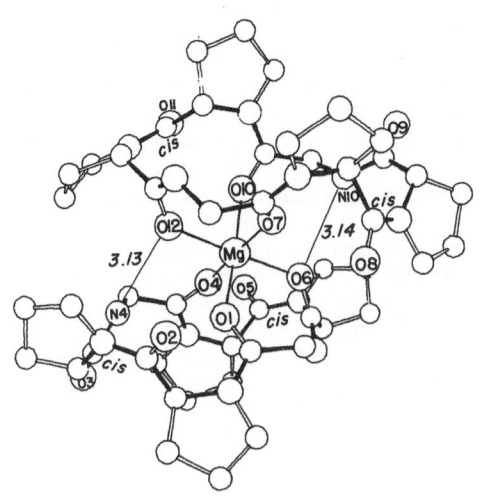

Fig. 4. View of the cyclo(Gly-Pro-Pro-Gly-Pro-Pro)$_2$Mg^{2+} complex.
Two hydrogen bonds between the upper and lower peptide molecules,
N(4)•••O(12) and N(10)•••O(6), are shown as light lines. The
six Mg--O ligands are also denoted by light lines. Four <u>cis</u>
peptide bonds occurring in the four Pro-Pro sequences are indicated. (Karle and Karle, 1981).

2.4. Other Remarks on Partial Structures

Valinomycin, Li$^+$antamanide, and Mg^{++}•2(Gly-Pro-Pro-Gly-Pro-Pro)
are examples of some of the more complicated partial structure developments. With equal atom structures in which the starting fragment consists of 25% or more of the total atoms, the complete
structure is often obtained in one or two cycles of partial structure development. The same has been found to hold for heavy atom
structures. In structures containing side chains, or solvent
molecules with large thermal parameters, B = 7 or larger, difference maps are more efficient than phase extension with the tangent
formula in locating the atoms with high thermal values. Tests have
shown that it is possible at times to start with a fragment posses-
sing only two light atoms.

3. MISPLACED MOLECULES OR FRAGMENTS

Incorrect values for some of the phases, particularly for those
introduced near the beginning of the phase determination can result
in an essentially correct depiction of the molecule, or significant
fragment thereof, but with all the atoms shifted with respect to
a correct origin. The misplaced fragment is always in the correct
orientation but merely translated with respect to one or more of
the cell axes. A number of different procedures have been used to
shift the molecular fragment to the correct position in the unit
cell, including translation functions. For space groups with two
asymmetric units, e.g. P$\bar{1}$, P2_1, etc. the simplest procedure is to
reduce the space group to P1. With the P1 procedure, the problem
of multiple minima, often encountered with translation functions,
is avoided.

   Two examples will be shown in which misplaced molecular fragments were shifted to their correct positions by means of reducing
the space group to P1. One example concerns sialic acid, a small
structure of 23 atoms (Flippen,1973), and the other Leu-enkephalin,
a large structure with more than 210 atoms (Karle et al, 1983).
In each case, the data from P2_1 were expanded to P1 by allowing

$$|F(hkl)| = |F(h\bar{k}l)|$$

$$\text{and} \quad |F(\bar{h}kl)| = |F(\overline{hkl})| . \tag{7}$$

The magnitudes of the new data remain the same, but the phases will
be determined independently.

3.1. Sialic Acid Reduced to P1.

Phases for sialic acid, $C_{11}H_{19}NO_9 \cdot 2H_2O$, space group P$2_1$, were
determined by the symbolic addition procedure for non-centrosymmet-
ric crystals (Karle and Karle, 1966) and phase determining formulas
which are based on inequality (34) of Karle and Hauptman (1950),
and its probability implications. One E map led to a partial structure which was expanded to a reasonable molecule, in view of the
expected structure, by recycling with the tangent formula (Karle,
1968). However, the molecule would not refine properly and it
soon became apparent that it was misplaced with respect to a true
origin for space group P2_1 The data set was then expanded from a
monoclinic set to a triclinic set by generating $|F_{hkl}| = |F_{h\bar{k}l}|$ and
$|F_{\bar{h}kl}| = |F_{\bar{h}\bar{k}l}|$. It was then possible to use the molecule obtained
in the original tangent formula recycling procedure as a partial
structure for the same procedure in space group P1. In this way,
the second molecule, in addition to the first molecule, and the

431

four H_2O molecules appeared in the Pl cell. The shift of the origin necessary to relate the two molecules by a two-fold screw and to place them correctly in space group $P2_1$ was easily determined. Two of the 21 atoms used as a partial structure in Pl were incorrect (see Fig. 5), but they disappeared in the process of tangent formula refinement.

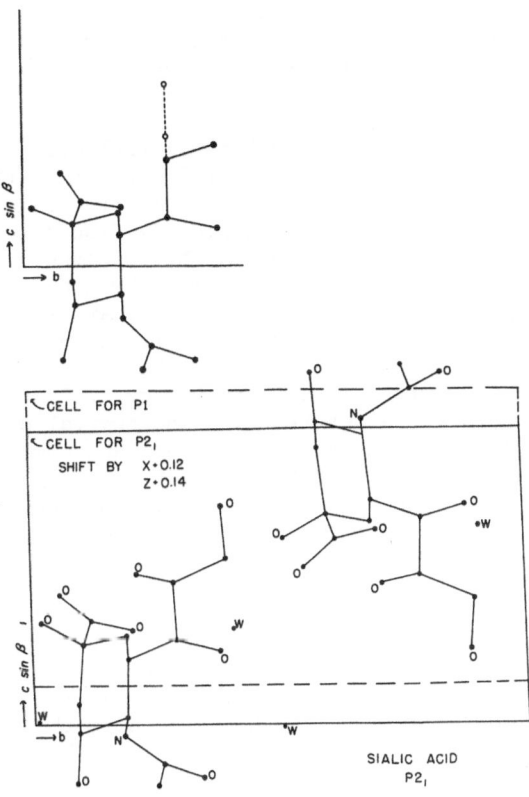

Fig. 5 (a) Misplaced molecule obtained from original phase determination which was used as a partial structure in Pl. Atoms connected by dotted lines were incorrect. (b) Final structure showing correct placement in the $P2_1$ cell (solid lines) as well as position with respect to Pl cell (dashed lines).

3.2. Leu-enkephalin, 210 Atom Structure reduced to P1.

The structure of Leu-enkephalin (Tyr-Gly-Gly-Phe-Leu) contains four conformers of the peptide plus 40-50 solvent atoms for a total of more than 230 independent C, N and O atoms in space group P2$_1$ (Karle et al, 1983). It provides an excellent vehicle for the application of formulas and techniques for direct phase determination and demonstrates that the practical limit with respect to size of a structure containing only light atoms has not been reached. This particular structure lies in an area intermediate between small-molecule crystallography and protein analysis. Small structures are normally considered to be those that have about 100 or fewer non-hydrogen atoms in the asymmetric unit. The intermediate range is an area in which few structure analyses have been reported.

The solvent, consisting of H$_2$O and N,N-dimethylformamide molecules (DMFA), comprises about 25% of the total content of the cell. Many of the solvent molecules are moderately well-ordered while others, occurring in special regions, are grossly disordered. It is conceivable that some of the disorder may be resolved by further study of the data. The low thermal factors for the atoms in the peptide molecules indicate that the peptide conformations are well established. The R-factor is 11.9% for 8155 reflection with $|F|>0$.

Since the asymmetric unit contained more than 200 C, O and N atoms with no heavier atom present, solving the crystal structure was not a routine procedure. Attempts utilizing several computerized direct-phasing multiple-solution methods were uniformly unsuccessful. The successful structure determination, although somewhat circuitous, made use of a combination of several approaches.

An immediately obvious departure from a usual distribution of $|E|$ values was the large number of reflections, 83 with unusually high $|E|$ values ranging from 3.0 to 8.3, that indicated a considerable regularity in the structure. In addition to the direct phase determination employing the Symbolic Addition Procedure, a Patterson map using ($|E|^2 - 1$) values for coefficients was calculated employing a special set of only 55 reflections with high $|E|$ values. The particular reflections chosen were those that occurred in sets with varying k indices such as $\overline{17}$,k,8(0 < k < 7) and $\overline{16}$,k,4(0 < k < 6, all with $|E| > 2.4$, in order to determine the implications of these sets concerning the structure. The peaks in this vector map were consistent with a β-sheet structure. When all the observed data were included in the ($|E|^2 - 1$) map, the vectors defining the β-sheet were sufficiently masked by vectors from other parts of the structure to make the analysis of the implications of the map much more uncertain. It is possible that in other structures such an abbreviated ($|E|^2 - 1$) map derived from only special sets of reflections may give indications of characteristic features of the structure.

Direct phase determination by the Symbolic Addition Procedure (Karle and Karle, 1966, 1968) proceeded in the usual fashion for space group $P2_1$. An origin was selected by assigning a phase value of zero to reflections $\overline{10},0,19$, $\overline{17},0,12$ and 014. The phases of reflections 020 and 060 were readily determined to be π in the initial stages. Four other symbolic assignments, a, b, c, and d, were required for the phase determination to proceed. They were assigned to $\overline{16},0,4$, $7,1,13$, 170 and $\overline{335}$, respectively. All assignments were made to reflections with $|E| > 4.0$. From the applications of the Σ_2 relationship and examination of the values from multiple phase indications, it was not possible to make an unambiguous evaluation of any of the symbols. At this point, the Σ_3 relationship for non-centrosymmetric crystals (8) (Karle, 1969), proved to be useful in evaluating symbols a, b and c as π, π and 0, while symbol d remained indeterminate.

$$\Sigma_3 \text{ FORMULA FOR } P2_1$$

$$s\, E_{h_1 0 1_1} = s\, E_{h_2 0 1_2} \frac{\Sigma}{k} \; M \left(\left| E_{\frac{h_1 - h_2}{2}, k, \frac{1_1 - 1_2}{2}} \right|^2 - 1 \right) \qquad (8)$$

where $M = (-1)^k$

Examples

$\overline{10}\ \ 0\ \ 19$	$8\ \ 0\ \ \overline{9}$	$\overline{9}\ \ k\ \ 14$	
$0*$	$a + \pi *$	$\underset{k}{\Sigma} = 7.7$	$P_+ = 0.71$
$a = \pi$			

$\overline{10}\ \ 0\ \ 19$	$\overline{10}\ \ 0\ \ 17$	$0\ \ k\ \ 1$	
$0*$	$a + b + c + \pi *$	$\underset{k}{\Sigma} = -12.9$	$P_+ = 0.24$

$\overline{10}\ \ 0\ \ 19$	$10\ \ 0\ \ \overline{17}$	$\overline{10}\ \ k\ \ 18$	
$0*$	$a + b + c + \pi *$	$\underset{k}{\Sigma} = -8.9$	$P_+ = 0.31$
$a + b = c$			

* Phases from Σ_2

Accordingly, since up to this point all phases would have values of 0 or π, as in a centrosymmetric crystal, symbol d was chosen to be +π/2, a choice that effectively selected the enantio-morph. In the resulting E map the strongest peaks represented segments of four parallel chains. The atoms, although near y ~ 1/4, were not coplanar but showed the characteristic pleats of a pleated sheet. With the aid of the indications from the restricted $(|E|^2 1)$ map, forty peaks in the E map, calculated with ~1800 reflections, were selected consistent with an antiparallel β-sheet. Phases for reflections with $|E| > 1.5$ calculated from this partial structure, and accepted if $|F_{calc}| > 0.18 \ |F_{obs}|$, were used for phase extension by the tangent formula for reflections with $|E| > 1.2$ The phase values were <u>not</u> recycled in the attempt to refine their values, since in space group $P2_1$, particularly if a heavier atom is present, or, as in this example, where many atoms lie near a plane parallel to (010), recycling has the tendency to shift the phases toward centrosymmetric values (Karle, Gilardi, Fratini and Karle, 1969; Karle, 1974). The 84 atoms in the backbones of the four conformers were found in two stages of partial structure development by use of the tangent formula and calculation of E maps.

The satisfactory appearance of the β-sheet, the many hydro-gen bonds that were indicated and the relatively low R factor of 46% for the initial stages of phase extension by the tangent for-mula suggested that the structure could be correct up to this point. The location of the 76 atoms in the side chains was very difficult, however, since the partial-structure technique did not give any further information. From the known sequence of the [Leu[5]]enkephalin, it was possible to place the twelve C^β atoms fairly accurately. From difference maps and models, the positions of most of the atoms in the side chains were eked out. Cycles of restrained least-squares refinement (Konnert, 1976; Konnert and Hendrickson, 1980) were interspersed with difference maps that showed possible water molecules separated by distances appropriate to hydrogen-bond separations. The R factor, however, could not be reduced below 29% for the 190 to 210 atoms that were included in the refinement process and this high value, in fact, indicated that the structure was incorrectly placed with respect to a proper origin.

At this point it was decided to relax the space group to P1 (Karle and Karle, 1971). The data base was doubled with $|E_{hkl}| = |E_{h\bar{k}l}|$ and $|E_{\bar{h}kl}| = |E_{\bar{h}\bar{k}l}|$. About 150 atoms comprising the back-bones and most of the side chains in the apparently misplaced structure determined above were used as the partial structure for computation of the initial phases in P1. Refinement and extensions of phases with the tangent formula and calculation of an E map based on the derived phases, with $|E|_{min} = 1.2$, yielded peaks

that reproduced the original backbones but not the side chains.
Thirty peaks in the other half of the cell, i.e. near y = 3/4,
were assigned to the backbones of additional enkephalin molecules
potentially related to the original molecules by a symmetry operation. With these additional atoms, the twofold screw operation,
for the original space group $P2_1$, could be satisfied either by
the original position of the origin, or by a new position shifted
by x - 0.068 and z - 0,135, or by another new position shifted
by x - 0.096 and z - 0.257, or even largely satisfied by origins in
in several other locations. Another round of the partial structure
development in P1, with the addition of the 30 new peaks, yielded
another 30 or so new atomic positions that could be more definitely
related to the original backbone atoms by a twofold screw placed
at x - 0.096 and z - 0.257 with respect to the original location.

The next step involved the return to space group $P2_1$ with a
shift to the new origin and a partial structure development based
initially only on the backbone atoms. Since some of the side-chain
atoms in the original structure may have been in error, they were
omitted at this stage so as not to prejudice the calculation. This
time, with the correct placement of the atoms for the backbone with
respect to an appropriate origin, use of the tangent formula only
for phase extension and not for phase refinement produced atomic
positions for the side chains quite rapidly. After three cycles
of partial structure development, positions were obtained for 156
of the 160 atoms in the four peptide molecules and seven atoms in
solvent molecules.

The initial incorrect placement of the enkephalin chains with
respect to an origin may have been caused by a subset of phases
derived from reflection $\overline{16},3,4$ ($|E| = 3.4$). The phase of $\overline{16},3,4$
was accepted from two strong Σ_2 triplets, each of which indicated
an erroneous phase displaced by nearly π from the correct value.
Under such circumstances, when a large subset of phases is incon-
sistent with the remainder of the phases, there is no combination
of values for the symbolic phases that will produce a correct E
map. The best map may be displaced from a correct origin.

The structure is shown in Fig. 6.

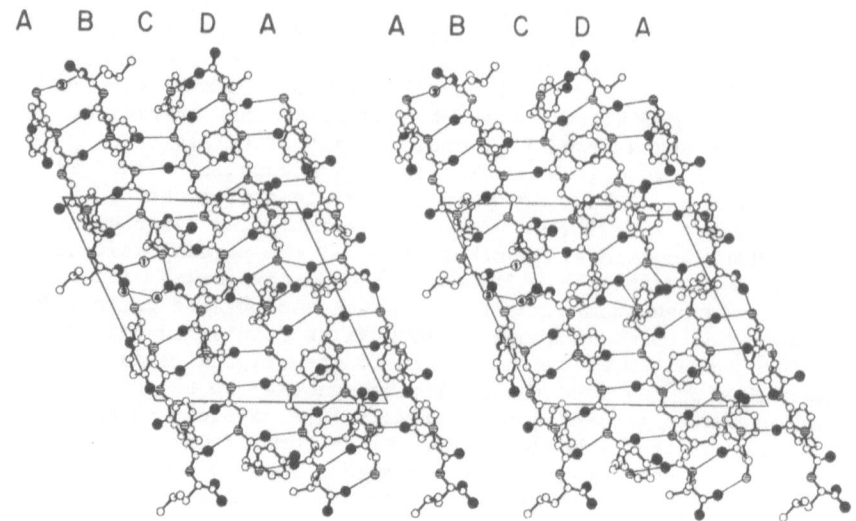

Fig. 6. Four independent molecules of enkephalin labelled A, B, C and D, each with a different conformation, form an infinite anti-parallel β-sheet in the crystal. The layer of hydrogen-bonded molecules near y ~ 1/4 is shown in the stereodiagram. The axial directions are a ↑ and c →. Four water molecules are labelled 1-4, O atoms are indicated by filled circles and N atoms by hatched circles. (Karle et al, 1983).

4. VECTOR SEARCH PROCEDURE

Recently the structure of a sixteen residue peptide containing 135 C, N and O atoms in space group P1 has been solved, Fig. 7 (Karle et al, 1987). Since the structure of a helical decapeptide composed of the same first ten residues was already known, Fig. 8 (Karle et al, 1986), the 36 backbone atoms of residues 1-9 were used as a model in the rotation function contained in the computer program DIRDIF (Beurskens et al, 1982). The vector set of the model was rotated until a best match was obtained with the vectors in the Patterson functon of the 16-residue peptide with the unknown structure. The next step usually would be to perform a translation search of the model fragment with the orientation that resulted from the rotation search with respect to the cell axes of the unknown structure. Since the unknown structure was in space group P1, no translation search was needed because the origin of a P1 cell can be chosen arbitrarily.

The remainder of the structure was derived in a number of cycles by phase extension with the use of the tangent formula. In the DIRDIF program, the tangent formula is applied to the differences between the observed structure factors and those calculated for the known fragment, whereas the observed structure factors phased by the known fragment are used in the partial structure development described in Section 2. In this particular example, the two methods of extending phases appear to be equivalent in developing the structure.

The close fit of the model to the structure of the unknown, where the r.m.s. deviation was 0.25 Å for the 36 atoms in common, no doubt facilitated greatly the search procedure.

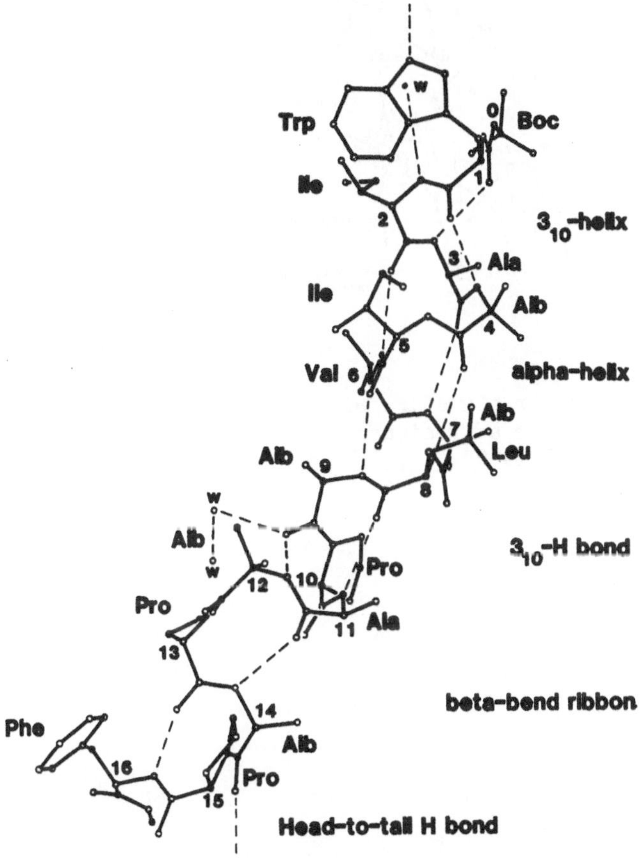

Fig. 7. The molecular structure of a helical 16-residue peptide (Karle et al, 1987).

438

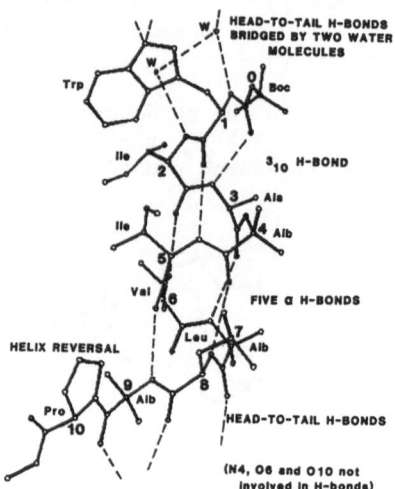

Fig. 8. The molecular structure of a helical 10-residue peptide
with the same sequence as residues 1-10 in the peptide shown in
Fig. 7 (Karle et al, 1986).

5. REFERENCES

Beurskens, P. T., Bosman, W. P., Doesburg, H. M., van den Hark, Th., Prick, P. A. J., Noordik, J. H., Beurskens, G., Gould, R. O. and Parthasarathi, V. (1982). Conformation in Biology, Eds. R. Srinivasan and R. H. Sarma. Adenine Press, New York, pp. 389-406.

Flippen, J. L. (1973). Acta Cryst. B29, 1881-1886.

Flippen-Anderson, J., Gilardi, R. and Konnert, J. H. (1983). Program RESLSQ. NRL Memorandum Report 5042, Naval Research Laboratory, Washington, D. C. 20375, USA.

Hauptman, H. and Karle, J. (1953). Solution of the Phase Problem. I. The Centrosymmetric Crystal. A.C.A. Monograph No. 3, Western Springs, Illinois: Polycrystal Book Service.

Karle, I. L. and Karle, J. (1964). Acta Cryst. 17, 835-841.

Karle, I. L. and Karle, J. (1968). Acta Cryst. B24, 81-91.

Karle, I. L., Gilardi, R. D., Fratini, A. V. and Karle, J. (1969). Acta Cryst. B25, 1469-1479.

Karle, I. L. and Karle, J. (1971). Acta Cryst. B27, 1891-1898.

Karle, I. L. (1974). J. Amer. Chem. Soc. 96, 4000-4006.

Karle, I. L. (1975). J. Amer. Chem. Soc. 97, 4397-4386.

Karle, I. L. and Karle, J. (1981). Proc. Natl. Acad. Sci. USA 78, 681-685.

Karle, I. L. and Karle, J., Mastropaolo, D., Camerman, A. and Camerman, N. (1983). Acta Cryst. B39, 625-637.

Karle, I. L., Sukumar, M. and Balaram, P. (1986). Proc. Natl. Acad. Sci. USA, 83, 9284-9288.

Karle, I. L., Flippen-Anderson, J., Sukumar, M. and Balaram, P. (1987). To be published.

Karle, J. and Hauptman, H. (1950). Acta Cryst. 17, 835-841.

Karle, J. and Hauptman, H. (1956). Acta Cryst. 9, 635-651.

Karle, J. and Karle, I. L. (1966). Acta Cryst. 21, 849-859.

Karle, J. (1968). Acta Cryst. B24, 182–186.

Karle, J. (1969). Advances in Chemical Physics, Vol. XVI, edited by I. Prigogine and S. A. Rice, pp. 131–222, New York: Interscience.

Karle, J. (1970). Partial Structures and the Tangent Formula, in Crystallographic Computing, ed. F. R. Ahmed, S. R. Hall and C. P. Huber, pp. 37–40, Copenhagen: Munksgaard.

Konnert, J. H. (1976). Acta Cryst. A32, 614–617.

Konnert, J. H. and Hendrickson, W. A. (1980). Acta Cryst. A36, 344–350.

Posner, G. H., Babiak, K. A., Loomis, G. L., Frazee, W. J., Mittal, R. D. and Karle, I. L. (1980). J. Amer. Chem. Soc. 102, 7498–7505.

DISCUSSION

SOME ASPECTS OF STRUCTURE DETERMINATION

In the example of guaianolide, all the structure factors which gave the incorrect phase values were themselves correct, and there were no unusual structural features such as pseudo-symmetry. For valinomycin, however, the pseudo-centre of symmetry of the molecule might have contributed to the initial phase determination giving only a fragment of the total structure.

When the phases based on the bromide ion in the lithium bromide complex of antamanide gave a map containing atoms for both enantiomorphs, the structure determination was not affected by which enantiomorph was chosen at that stage.

In the magnesium perchlorate peptide complex, the one major peak in the Harker section was attributed to the Mg-Mg, rather than the Cl-Cl, vector on chemical grounds. The near-neighbours of the atom concerned could be fitted to an octahedral co-ordination acceptable only for the magnesium ion and not to the tetrahedral structure required for the perchlorate ion.

There are also automated procedures, using conditional probabilities based on the observed and calculated magnitudes of the normalised reflections, for predicting tentative atomic positions. Dr Beursken's program DIRDIF can be used in that way for structure development once at least 5% of the structure is securely known.

After a starting fragment has been found, there are three ways of developing a structure: (i) by phase extension using the tangent formula; (ii) by DIRDIF; (iii) by difference Fourier, with or without prerefinement. If the fragment is large, it does not make much difference which method is used; if the fragment is small, it is not yet possible to predict which method is likely to be the best in any particular case, although for equal-atom structures, procedure (i) has been most successful, even for very small fragments.

Real Time Synchrotron Radiation Diffraction Experiments on Polymers

C. Riekel
European Synchrotron Radiation Facility
BP 220
F-38043 Grenoble Cedex
France

M. A. Carrondo and G. A. Jeffrey (eds.), Chemical Crystallography with Pulsed Neutrons and Synchrotron X-Rays, 443–485.

444

Abstract

The article reviews realtime synchrotron radiation diffraction experiments in polymer science performed at Hasylab in Hamburg. The properties of the source, camera, data collection systems and perturbation methods are briefly described. Selected examples are taken from experiments on polymer crystallization and melting (polyphosphazene, polyethylene, polyethyleneterephthalate), solid state phase transformations (polyacetylene) and heterogeneous chemical reactions (polyacetylene/I_2). Experimental possibilities at the future European Synchrotron Radiation Facility (ESRF) are discussed for two beamlines, proposed principally for small angle scattering experiments.

I.   Introduction

Synthetic, solid polymers have become increasingly important in modern society as they offer a broader spectrum of properties than natural products like wood, rubber or metals. A considerable effort is made to improve the properties of known polymers or to create new polymers with special properties. Thus solution spinning or ultradrawing of **polyethylene** (PE) (Fig.1) has dramatically improved its mechanical properties with Young moduli up to ≈ 300 GPa (theoretical limit: 324 GPa) instead of ≈ 1 GPa for low density PE./1/ Electrically conducting polymers is another rapidly evolving research topic. Thus the conductivity of the prototype polymer -**polyacetylene** (Fig.1)- can be controlled over a large range, as shown in Fig.2./2/

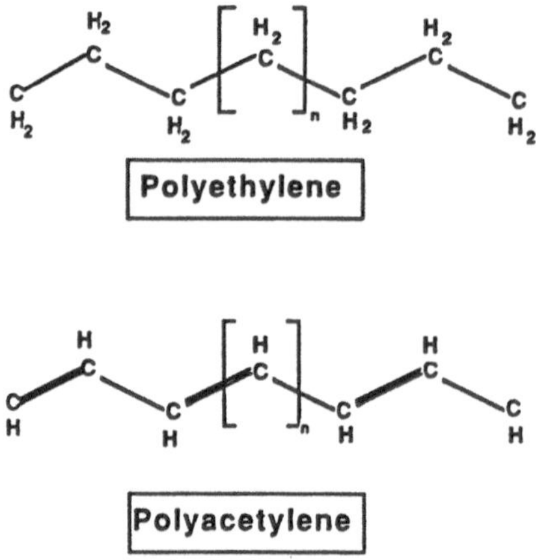

Polyethylene

Polyacetylene

Fig.1 Chain structures of polyethylene (PE) and polyacetylene (PA). The monomer units are indicated by brackets [ ].

The search for polymers with special properties profits from a better understanding of the structure/property relationship. Although a number of such problems can be studied by diffraction methods on stable or metastable polymers, the ultimate aim is to develop methods for the analysis of dynamic processes, e.g. during polymer crystallization, stretching or chemical reactions. This has led to the development of **diffraction methods** by which structural changes can be studied in **real time**./3,4/ **Real time diffraction** experiments give information on changes in **long range** electron density variations and are complimentary to **real time spectroscopic** experiments which give information on **local changes** in electron density.

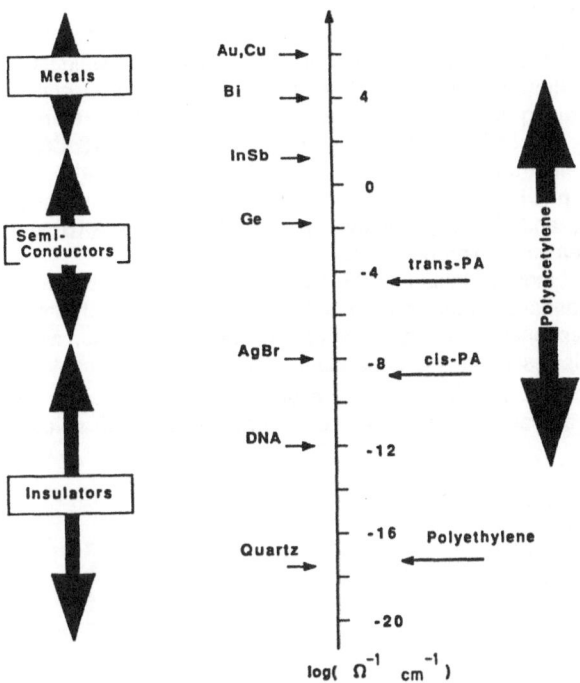

Fig.2 Comparison of the range of specific conductivities of a number of materials with those which may be reached by polyacetylene or chemically modified PA./2/ Cis- and trans-PA are the two known polyacetylene modifications.

Fig.3 shows schematically the components of a real time diffraction setup using an angular dispersive geometry. The basic difference as compared to a classical diffraction experiment is the use of an in-situ cell and the possibility to perturb the sample in order to start a structural process. In most cases, a position sensitive detector (PSD) will be necessary in order to speed up the rate of data collection.

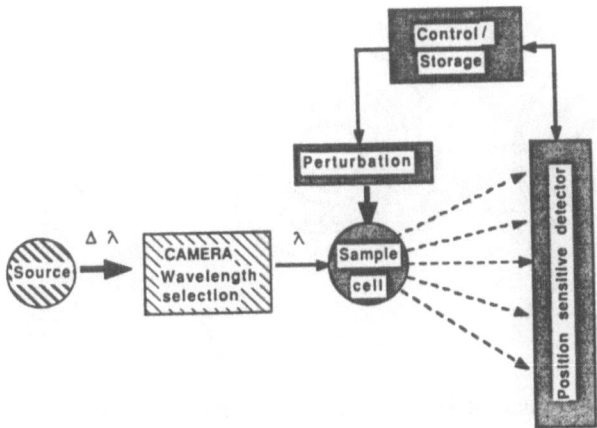

Fig.3 Schematic design of a real time diffraction instrument using an angular dispersive setup

II.   Polymer structure and morphology

Polymers are long chain molecules which generally have molecular weights between 10^4 to 10^7. Diffraction methods show that crystallized polymers usually are semicrystalline materials./5/ **Wide angle scattering** methods (**WAXS**) will therefore give information on the **structure** on the level of the unit cell. **Small angle scattering** methods (**SAXS**) ,on the other hand, are sensitive to   electron density fluctuations on a larger scale, which are generalized under the term **morphology**. Evidently the macroscopic properties of polymers are related both to the structure and the morphology. This is why both WAXS and SAXS experiments are used to study semicrystalline polymers.

Fig.4 shows three idealized morphologies and the corresponding SAXS patterns recorded by a 2D-detector. Case **A** corresponds to an lattice of ideal lamellae packets which are randomly oriented in space. Case **B** corresponds again to a lattice of lamellae packets, which show, however, thickness variations. This results in a Θ-dependant broadening of the peaks. /6,7,8/ A fibre diagram with discrete reflections, where lamellae are oriented under an angle relative to the preferred direction, is shown as case **C**. The distance **L** corresponds to the so called **long period.**

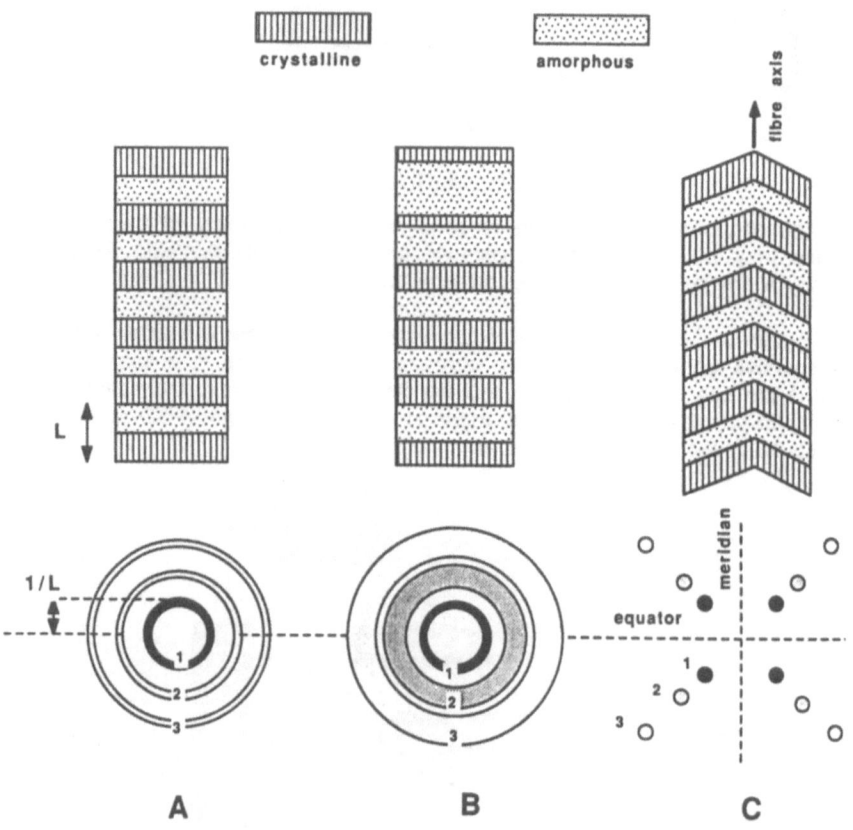

Fig.4 Schematic representation of selected polymer morphologies and of the corresponding 2D-SAXS patterns.

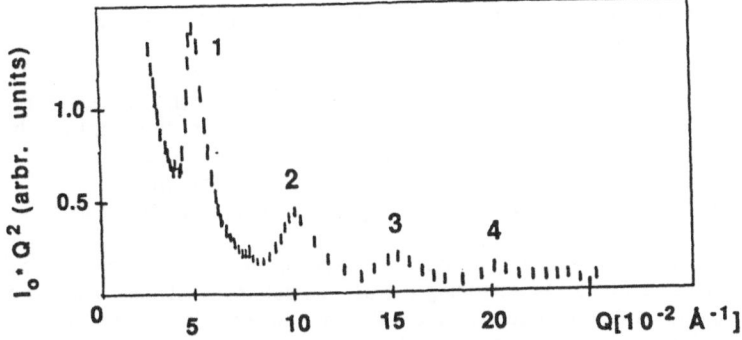

Fig.5 SAXS pattern of polyethylene measured by a 0-dimensional detector. The intensity -I-
has been multiplied by Q^2. $(Q=(4\pi \sin\theta)/\lambda)$. I_o is the primary beam intensity. The
pattern shows the n=1,2,3,4 harmonics./9/

Such models are highly idealized and can be further refined./8/ They are, however, often
sufficient in order to establish a mechanistic model of a morphological transformation.
Note furthermore that such morphologies are part of a more complex superstructures -such as
the formation of spherulites- which may reach μm-sizes./5/

As an example of a lamellar morphology, the SAXS-pattern of polyethylene -crystallized
from solution- is shown in Fig.5 /9/ The Q-dependant width of the peakprofiles suggest that the
lattice corresponds to case B. (Fig.4) Here the the average separation of the crystalline zones is
≈120 Å.

III.   Instrumentation

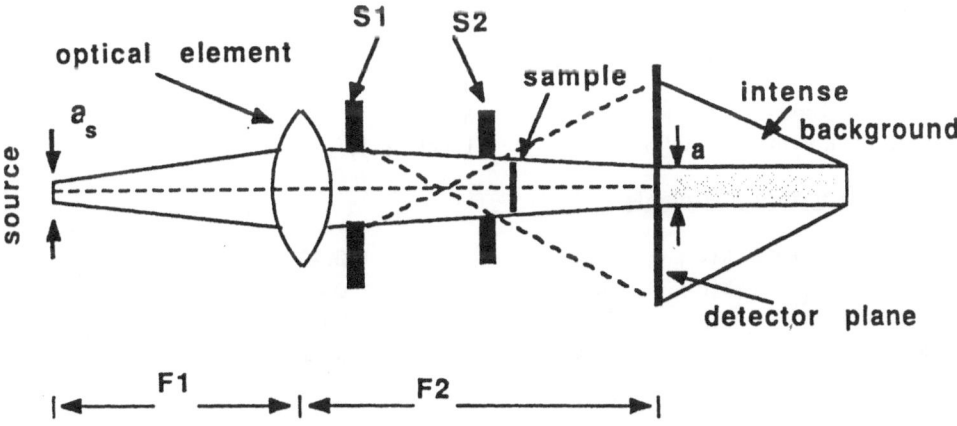

Fig.6 Schematic design of a pinhole camera with a focusing element. S1 is the aperture and
S2 the guard slit

The classical camera used to perform SAXS-experiments is the **Kratky camera.**/8/ This camera has a line focus and allows to record structures with dimensions up to ≈ 20000 Å. It has, however, a number of disadvantages:

(i) the rather long exposure time for the generally weakly scattering polymers. Real time experiments are feasible only in special cases, when the contrast between the crystalline and amorphous zones is sufficiently high./10/ (section IV)

(ii) the line focus which results in a smearing of the small angle peak

(iii) combined small- and wide-angle scattering experiments are not possible

Pinhole cameras (Fig.6) do not have the problem of peak smearing. WAXS and SAXS experiments can be performed by changing the sample/detector distance. The intensity of such cameras at X-ray generators is, however, often too low, for real time SAXS experiments. Synchrotron radiation sources -on the other hand- allow to construct high flux point focusing, pinhole cameras which can be used for real time WAXS and SAXS studies./11/ In order to obtain a maximum flux at the sample a focusing geometry has to be used, which is shown schematically in Fig.6. The **temporal resolution** reached in SAXS at present is in the second-range for irreversible processes at the polymer beamline in Hamburg./3/ Some of the instrumental aspects of the presently available instrumentation will be discussed below.

III.1 Source

The storage ring **DORIS** at **DESY** in Hamburg will serve as an example for other sources. This ring is operating at ≈ 5 GeV -with both electrons and positrons circulating- for high energy physics experiments and can then only be used parasitically. In the dedicated mode (3.7 GeV; about 1/3 of the total beamtime) only electrons are circulating.

The parameters of DORIS and of the photon beam emitted from the A1-dipole magnet are indicated in Table 1. The coordinate system used to describe the photon beam is based on the one used to describe the electron beam./12/ For a given point near the ideal electron orbit, x is the radial coordinate and z the vertical coordinate (normal to the plane of the ring).

Table 1.   Parameters of the storage ring DORIS.

Energy	(GeV)	3.7 (dedicated)	5 GeV (parasitic)
Bunch length	(ps)	140	
Revolution time	(μs)	1	
Magnet of polymer beamline		Dipole (A1)	

Photon beam from A1-dipole

Source point size	$x(\sigma)$	1.70	2.40
(mm)	$z(\sigma)$	0.30	0.47
Divergence	$x'(\sigma)$	0.31	0.47
(mrad)	$z'(\sigma)$	0.05	0.06

III.2 Camera

The camera constructed in Hamburg for the polymer community is similar to one developed for scattering on biological materials. /13/ (Fig.7) It has a focusing monochromator

[Ge(111)] and a 1.6 m long, segmented focusing mirror, made out of quartz. Instead of pinholes, pairs of orthogonal slits are used. The wavelength is 1.5 Å and the mirror serves also to cut the higher harmonics.

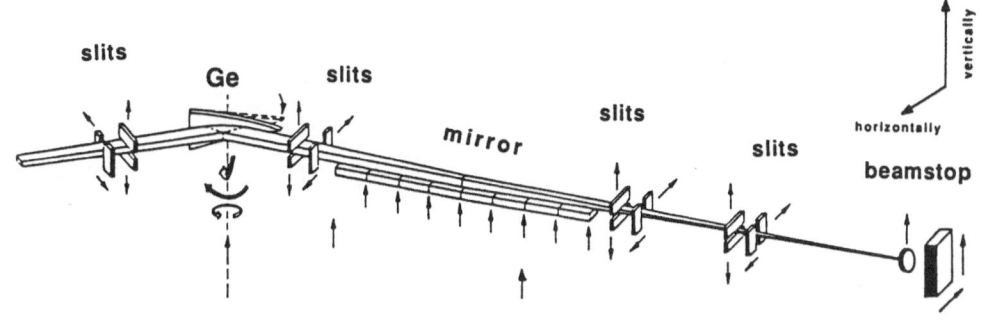

Fig.7 Double focusing camera operating at DORIS for polymer research. The arrows indicate the individual stepping motors which can be adressed via a CAMAC module. (Fig.8)

The size of the focus -a- is ideally given by $a=(F2/F1)^{*}a_s$ where a_s is the size of the source point ($a_s^x=2^{*}x; a_s^z=2^{*}z$), F1 the distance source point/optical element and F2 the distance optical element/detector plane. For a value of F1=20 m -fixed by the shielding of the ring- and a max. value of F2=9m, one calculates a vertical focus size of ≈0.3 mm at 3.7 GeV. The actual value is ,however, rather ≈1 mm due to the use of a segmented mirror, made out of eight plane 20 cm pieces./3/ In the horizontal direction, beam compression by an asymmetric cut monochromator crystal is used in order to cope with the larger horizontal divergence (Table 1) and to reduce the horizontal focus size./13/

The lower limit in **Q-range** is defined by the extension of the zone of diffuse scattering in the detector plane. $(Q=(4\pi\sin\theta)/\lambda)./3/$(Fig.6) This zone depends on the size of the beam at the aperture slits and the length of the camera. In the present case the limit in Q is ≈$6^{*}10^{-3}$ Å^{-1}. Note that this value corresponds to a linear extrapolation of the diffuse background to zero intensity. Experience on polymer spectra shows that SAXS-peaks can only be observed up to d-values of 500-600 Å. ($Q\approx10^{-2}$ Å^{-1}).

III.3 Detectors and data collection

For data collection in SAXS and WAXS experiments, gas filled and solid-state position sensitive detectors are used./3/ A complete data collection system employing a one dimensional gas filled detector, operating according to the delay line principle/14/, is shown in Fig.8. Up to 256 frames of 256 channels each can be recorded in a 64 K-word CAMAC memory. Detector waiting -t_w- and measuring -t_m- times are determined by a **time frame generator**./3,15/ The system is also used to control the stepping motors of the slits and the optical elements, as well as external perturbations (e.g. valves, stepping motor of stretching machine, pneumatic piston etc).

For anisotropic scattering patterns, a commercial Vidicon-camera with a converter screen and an on-line digitization is used. /3,16/ The minimum transfer time of 2D-frames of 256*256

450

pixels to disc is 2-3 s. No general use has been made of this detector in real time SAXS due to the limited data storage capacity of the present computer.

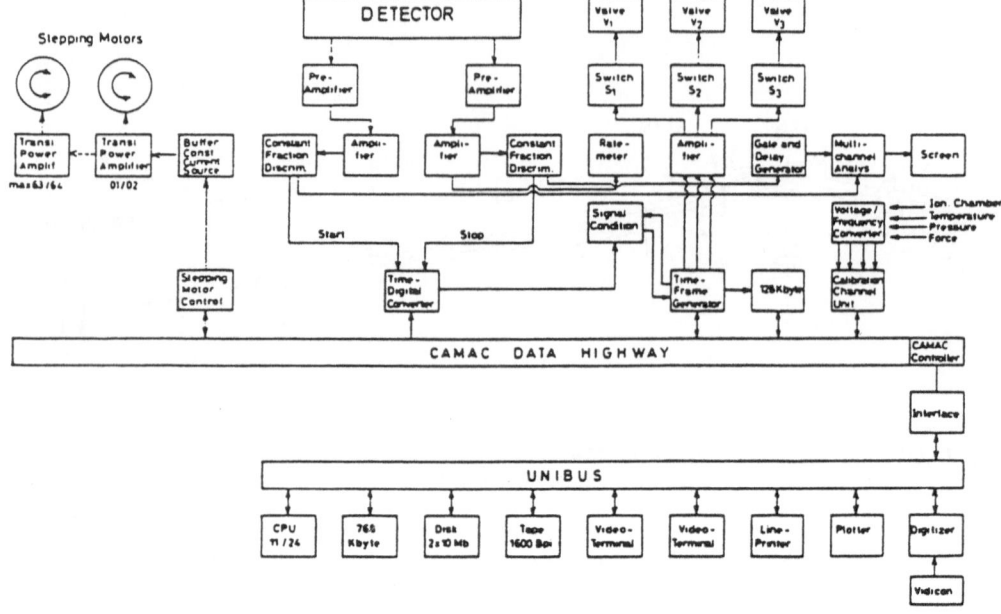

Fig.8 Data collection and instrumental control system used at the polymer beamline. A reaction cell where the valves V_1 and V_2 are used is shown in Fig.38.

III.4 Perturbation methods

A perturbation invoking a structural change can be applied to a solid only once for an irreversible process (*single shot*) or repetitive (*stroboscopic*) for reversible processes. For liquids and solutions stopped flow experiments allow to study irreversible processes in a repetitive way. (Fig.9)

The corresponding structural change in a solid can be characterized by a variable $\alpha(t)$ $(0<\alpha<1)$. The half time of conversion $-\tau-$ is defined for $\alpha=0.5$. (Fig.9)

In order to couple a pulsed source to a structural process, the repetition time of the photon puls $-t_r-$ and the width (FWHM) of the photon puls $-t_p-$ have also to be considered. (Fig.9)

Three different modes of realtime experiments are feasible depending on the timescale of the structural process.(Fig.10a,b,c)

(i) $\tau \gg t_r$ $\tau \gg t_p$

In this case the source will appear as a steady state source. A sampling of the structural process is feasible for $\tau > t_m$. Repetitive processes cannot be triggered by source pulses.

(ii) $\tau > t_r$ $\tau \gg t_p$

In this case the perturbation has to be synchronized with the photon pulses. The structural process is sampled by the photon pulses between two perturbations. The signal to noise ratio is optimized for $t_p=t_m$

(iii) $\tau<t_r$ $\tau>t_p$

Here a sampling of the structural process is possible by changing the temporal difference between the photon puls and the perturbation puls. The signal to noise ratio is optimized for $t_p=t_m$

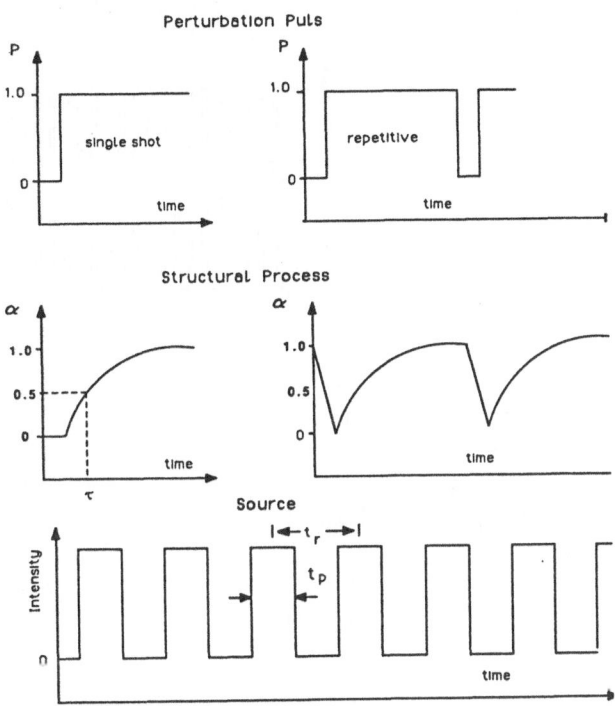

Fig.9 Schematic design of characteristic times of source, sample, perturbation puls and detector involved in real time experiments

Fig.11 shows the timescale of temporal resolutions (-t_m-) reached in a number of experiments on biological and polymeric samples. Specifically experiments on biological solutions using temperature jump /17/, stopped flow techniques/18/ or voltage jumps /19/ are indicated. For polymeric samples heating pulses /3,20/ and stretching experiments /21,22/ are indicated. The polymer experiments all belong to case (i).
The temporal resolution is limitated by the timescales of the structural processes investigated and the perturbation pulses used. Thus stepping motors used for stretching, valves used for stopped flow experiments or furnaces limit the timescale to the ms range.
The time structure of the source has practically not been used as yet. Thus a temporal resolution of ≈ 20 ns was reached for laser induced surface melting of Si./23/ This experiment belongs to case (iii). A zero dimensional detector (NaI) was used. More general use of this technique will demand the development of fast 1D- or 2D-detection systems and broader wavelength bands in order to obtain integrated reflection intensities.

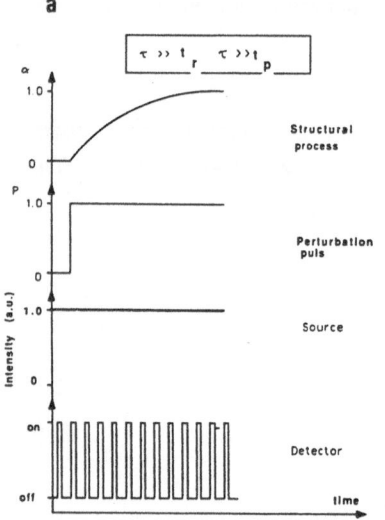

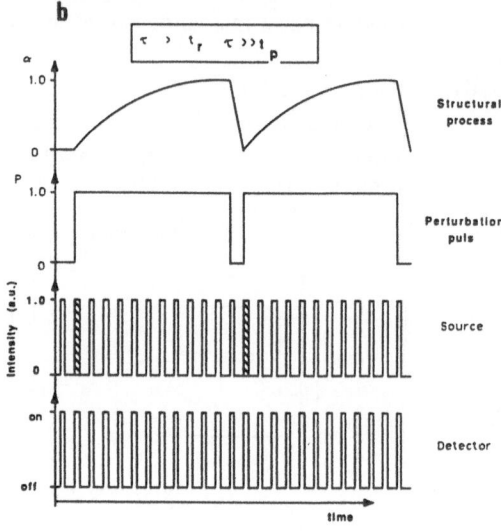

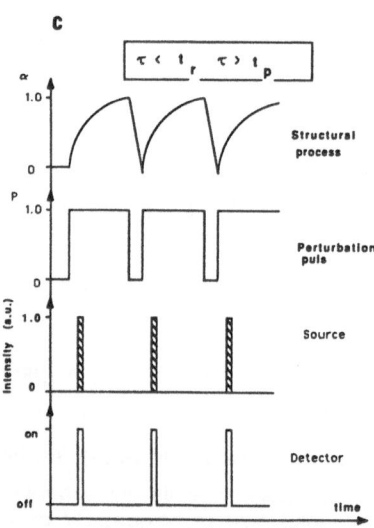

Fig.10a-c Three different possibilities to couple the timescales of structural process and the temporal variation of the photon beam intensity. For repetitive structural processes with sufficiently short half times, the perturbation puls must be triggered by source pulses. (barred rectangles)

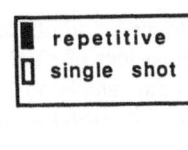

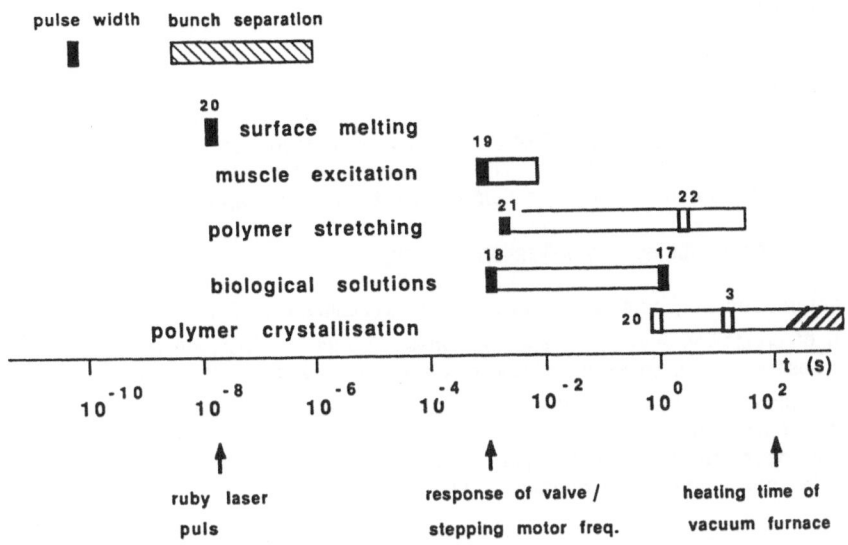

Fig.11 Characteristic timescales reached for real time experiments. The black and white bars represent typical experiments. (see text)

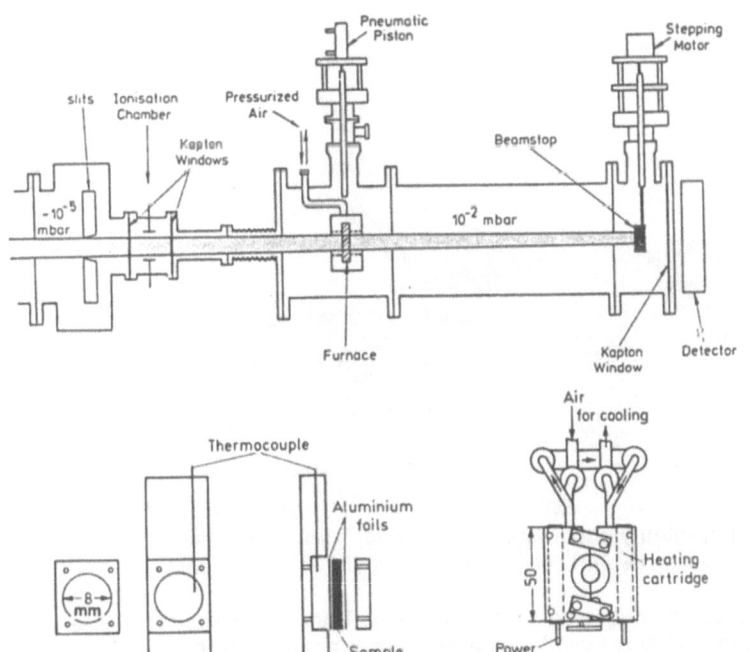

Fig.12 Schematic design of a vacuum furnace incorporated into the beamline. The heating block and the sample holder are also shown./3/

As an example for a perturbation device used in polymer science, a vacuum furnace is shown in Fig.12 . Polymer foils of <100 μm thickness can be heated up to ≈ 300°C in ≈ 120 s. The size of the beam at the sample position is ≈ 1-3 mm². Note that heating rates of a fraction of a second can be obtained by using a stream of hot gas./20/ This is possible, however, only for less sensible samples.

IV.   Examples

An increasing amount of real time diffraction experiments on polymers is performed at national synchrotron radiation facilities. The examples selected do not cover the totality of research topics initiated but will serve to demonstrate the present capabilities of the method.

<u>Morphological changes during crystallization and melting</u>

Amorphous polymers ,which have been crystallized at T_c, show a nonequilibrium distribution of crystalline and noncrystalline zones. /5,8,24,25/ This can be deduced from a number of observations such as the shape of the melting curve./5/ Upon heating above T_c a change in position of the **long period** (L) has been observed for a number of cases by real time SAXS./3/ This has also been observed for solution crystallized **poly[bis(p-methylphenoxy)phosphazene]** (**PBMPP**) which transforms from a semicrystalline phase into a mesophase at T(1)≈ 130 °C. /26/(Fig.13)

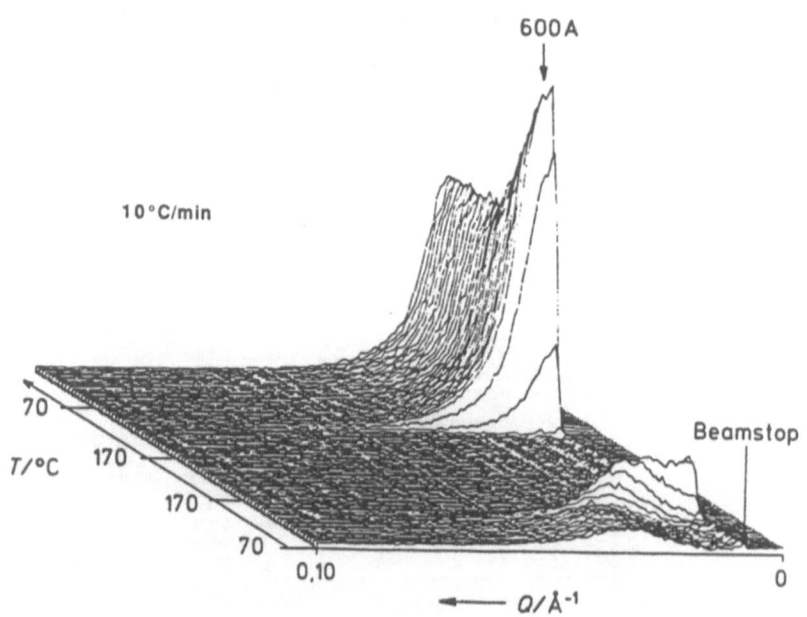

Fig.13 Change   of   SAXS   pattern   of   solution   crystallized poly[bis(p-methylphenoxy)phosphazene] upon heating. (8s/frame)./26/ The background at T=170°C has been subtracted from all patterns

These examples can be discussed in the same way. Thus the change in position of PBMPP (room temperature: L≈180 Å) can be understood by a disappearance of smaller crystallites and therefore an increase of the **long period.** No intensity is observed in the mesophase but a long period of ≈ 600 Å appears upon cooling down the material into the semicrystalline phase. Electron microscopy suggests that this is related to chain unfolding.

The plausible assumption of a preferable disappearance of smaller crystallites does not, however, give further information on the the process. This can be obtained via an analysis of the scattering power $-I_{nv}-$ (**Invariant**) which is determined by integration of the scattering intensity over all angles Θ./8/

$$I_{nv} = \int I(Q)Q^2 dQ \qquad (1)$$

I_{nv} is related to the structure of the material according to:/8/

$$I_{nv} = c^* \, w_{ce} \, (1-w_{ce})(\rho_c - \rho_a)^2 \qquad (2)$$

where ρ_c is the density of the crystals, ρ_a that of the noncrystalline regions and c is a constant depending on the intensity of the primary beam, the volume of the sample, and the geometry of the instrument. w_{ce} is the volume degree of crystallinity.

In case one stays below T_c only the term $\Delta \rho^2 =(\rho_c - \rho_a)^2$ will influence I_{nv}.

Fig.14 shows the development of I_{nv} during several temperature cycles of PBMPP upon cycling the temperature below T(1). The **Invariant** follows also in this case closely the change in temperature, which suggests a dominating influence of $\Delta \rho^2$.

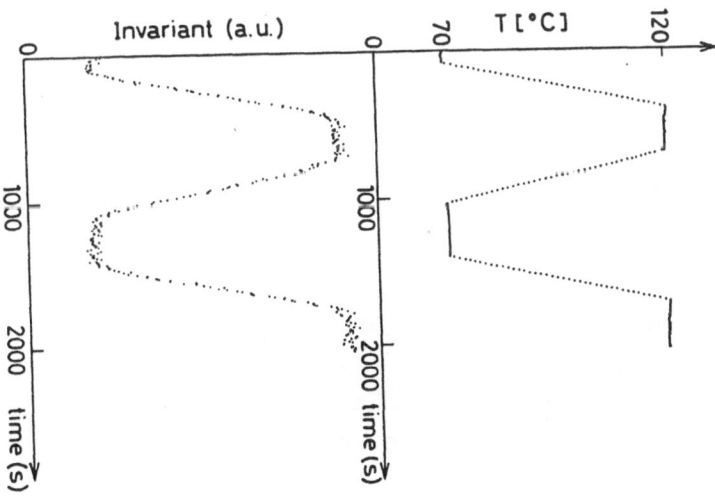

Fig.14 Change of the **Invariant** during several temperature cycles of polyphosphazene./26/

Fig.15a shows a sequence of SAXS spectra of **high density polyethylene** **(HDPE)**-isothermally crystallized at 105°C and then cooled slowly down to r.t.- upon ramp heating with 5°C/min. /27/ The corresponding **DSC (Differential Scanning Calorimetry)** traces, the variation of the **Invariant** and of the long period are shown in Fig.15b. One observes an increase of **Invariant** up to 105°C while both **Invariant** and the long period increase above 105°C until the melting point T_m. This can be interpreted as an increase of the density difference $\Delta \rho^2$ until ≈105°C, while a partial melting -which decreases w_{ce} (the crystallinity)- occurs above 105°C. Evidently the smaller crystalline lamellae melt first which implies an increase in the average separation of the crystalline lamellae. As the crystallinity of HDPE at r.t. is quite high ($w_{ce}≈0.77$), an increase of the **Invariant** is observed well beyond the start of melting above T_c. (Eq.2)

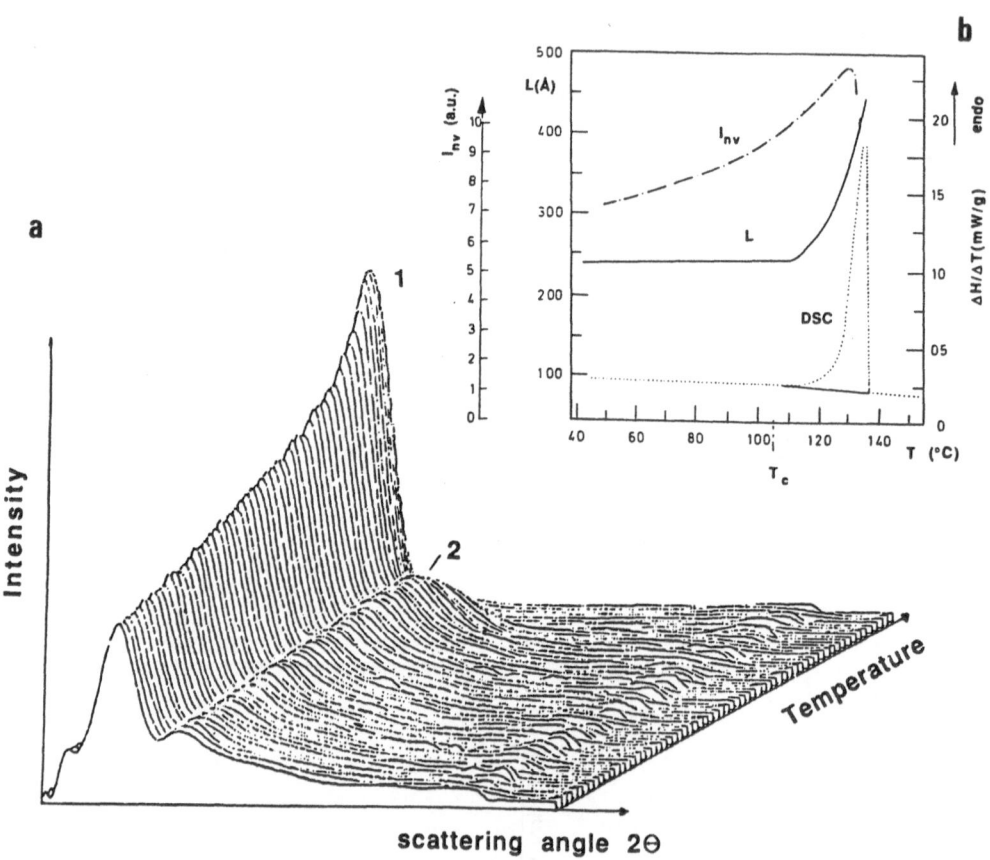

Fig.15a Sequence of SAXS patterns of high density polyethylene (HDPE) upon ramp heating with 5°C/min. The first and second order peaks can be seen./27/
Fig.15b Differential scanning (DSC) trace, variation of the **Invariant** and long period of HDPE upon heating with 5°C/min./27/

In comparison copolymerers of **linear low density polyetheylene** (LLDPE)with 1-octene show a much more complicated DSC trace.(Fig.16a)/27/ The long period increases below T_c (Fig.16b) which can be interpreted as a partial melting of zones which are richer in 1-octene, while zones which are richer in pure polyethylene remain. This is also supported by fractionation experiments./28/ The instrumental resolution was not sufficient to determine the change of **L** and I_{nv} above T_c.

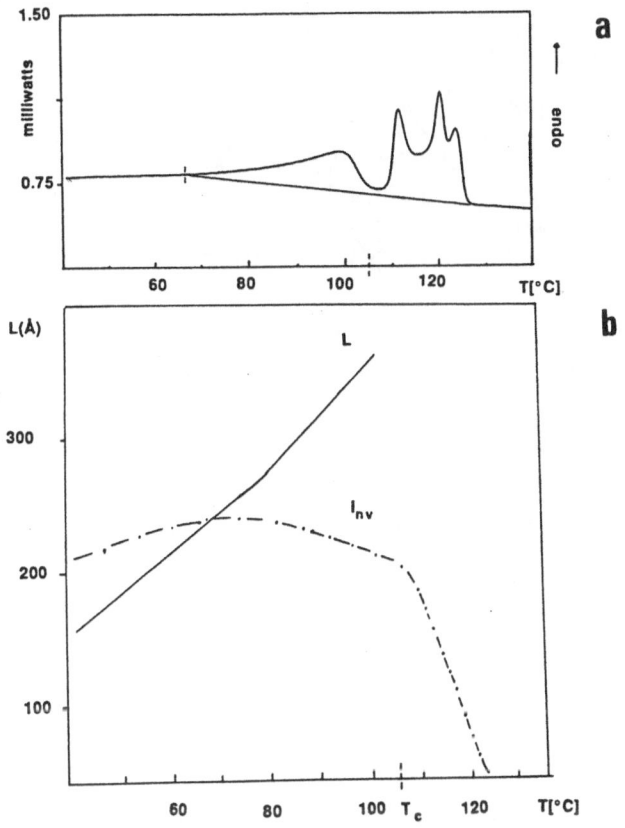

Fig.16a DSC-trace of an 1-octene copolymer of polyethylene./27/
Fig.16b Variation of **Invariant** and long period upon ramp heating of an 1-octene copolymer of polyethylene./27/

It is interesting to note that the increase of the long period at the onset of heating suggests that X-ray diffraction is in this case a more sensitive method to study partial melting than DSC. The different behaviour of the **Invariant** -as compared to HDPE (Fig.15b) is explained by the lower crystallinity of the starting material. ($w_{ce} \approx 0.44$).

Polyethyleneterephthalate   (PET) has been studied particularly in detail./3,29,30,31/ Thus the question of the onset of crystallization has been adressed by a combined WAXS and SAXS experiment on isotropic PET.(Fig.17) The WAXS patterns were recorded on films, which were changed every minute and afterwards digitized. SAXS patterns were recorded at the same time with a linear PSD./31/ A more refined experimental technique will demand the replacement of the film by an electronic detector.

458

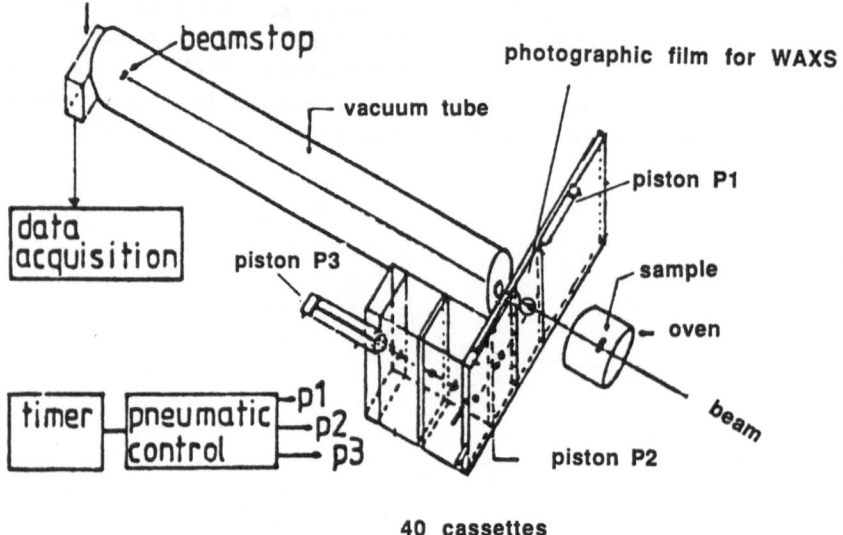

Fig.17   Schematic setup for combined WAXS and SAXS experiments./31/

Fig.18 shows the variation of the **Invariant** and the volume crystalline fraction (determined from WAXS-experiments), both scaled to 1.0 at maximum crystallization. As both curves behave similarly it was concluded that the crystallization occurs in this case by a simple growth of the phase -shown in terms of the growth of spherulites- and not by a spinodal decomposition mechanism, which would demand a change in the defect concentration and hence $\Delta \rho^2$.

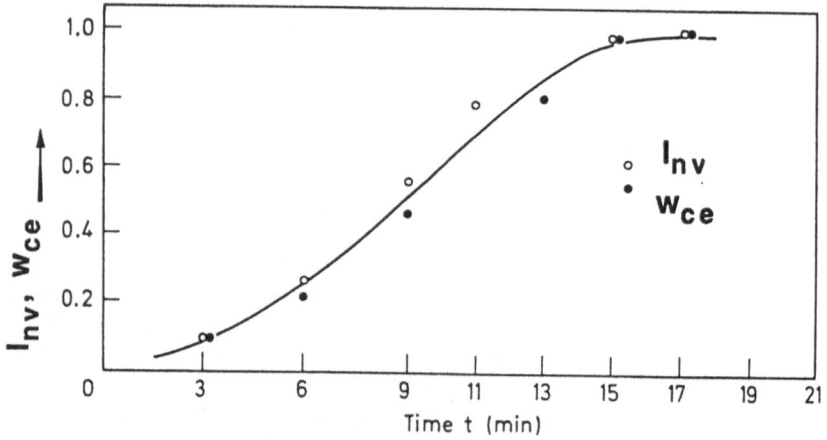

Fig.18   Change of **Invariant** and volume crystallinity during crystallisation of isotropic polyethyleneterephthalate(PET)./31/

Isothermal annealing experiments of PET samples crystallized at T_c have suggested the existence of partial melting and recrystallization for samples annealed above the T_c./3,30/ Thus Fig.19a shows the result of an annealing experiment for a material which had previously been crystallized during several temperature cycles at 120°,230°,120°,235°,120°,245°,120°C./29,30/ Crystallite sizes due to the highest crystallisation temperature are hence present.

Upon heating to 250°C one observes first an increase of the **Invariant** due to the increase in $\Delta\rho^2$. This is followed by a decrease due to a partial melting. The slight recovery is due to a recrystallization.(Fig.19b) After cooling to 125°C, the **Invariant** increases due to further crystallization.

The melting-recrystallization scheme becomes also evident from the behaviour of the small angle peak which disappears completely and is formed again at smaller angles. (Fig.20a) The sample had a similar crystallization history as the one shown in Fig.19a,b.

The result of stepwise heating with **larger temperature jumps** is shown in Fig.20b. Here the sample had been crystallized at 130°C, while annealing was performed at 235°, 240° and 245°C. The influence of $\Delta\rho^2$,melting and recrystallization are no longer clearly separable. An inflexion of the **Invariant** due to partial melting is only observed at the highest annealing temperatures.

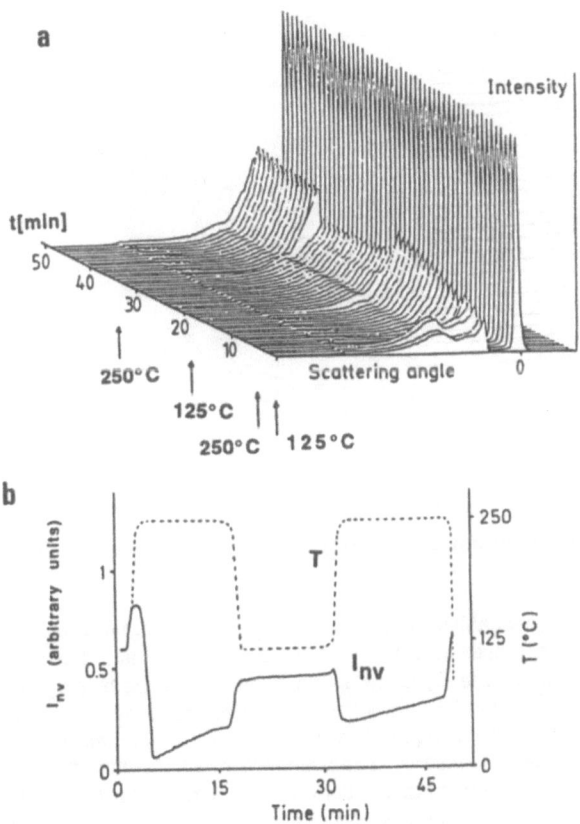

Fig.19a Change of PET SAXS-patterns during annealing./3,29,30/
Fig.19b Corresponding change of **Invariant** and temperature./3,29,30/

A mechanism has been proposed where some lamellae completely melt and a new crystallization with larger lamellae thickness occurs./29,30/ Below T_c -where a gradual peakshift is observed- there seems to be rather a gradual thickening by chain diffusion.

Such morphological processes can be reversible. Thus oriented **low density polyethylene** (LDPE) -annealed at 103°C for about a day- shows a four-point pattern. (Fig.4, case C). Temperature cycles below the annealing temperature result in a reversible variations of the intensity, separation and width of the peaks.

Fig.21a shows an intensity plot parallel to the meridian, as determined by a 1D-PSD./32/ The change of temperature, distance of the two peaks and their width is shown in Fig.21b. The long period varies between 217 Å (30°C) and 252 Å (100°C). This can be explained by a change in crystal size distribution inside the fibrils as a function of the temperature. Upon increasing the temperature, the small crystals become instable and larger crystals are built up in the course of a dynamical rearrangement of the superstructure. The change in size distribution and superstructure implies the variation of the width and position.

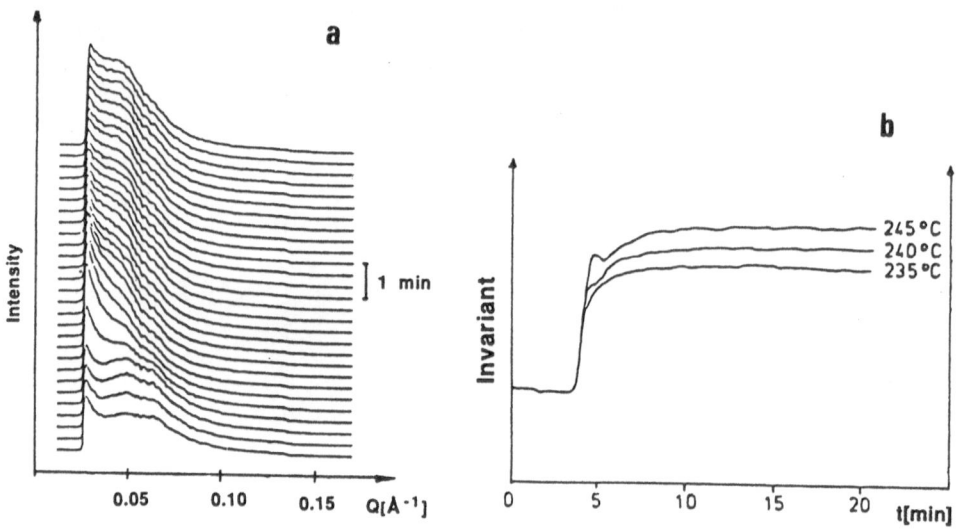

Fig.20a Change of PET SAXS-patterns upon anealing at 245°C./30/ ($Q=(4\pi\sin\theta)/\lambda$)

Fig.20b Change of **Invariant** of PET -crystallized at 130°C upon annealing at 235°,240° or 245°C./29,30/

A different mechanism seems to operate when highly oriented PET yarn is rapidly stretched (<0.1 s) to a tensile strain of ≈ 10% by a pneumatic driving device./22/ Here the four point diagram (Fig.4; case C) is transformed into a two point diagram upon stretching in a fraction of a second. Thus Fig.22 shows the change of the pattern -recorded with a 1D-position sensitive detector parallel to the equator- and a temporal resolution of 0.5 s/frame. The change in SAXS pattern involves necessarily a rearrangement of the lamellae. This can be explained either by a mechanical deformation scheme or via a partial melting process at the crystalline-amorphous interfaces./33/ As no change in the transformation rate was observed at 21°C (≈ 50°C below the glass transition temperature T_g) and ≈160 °C, it was concluded that the mechanical scheme applied in this case.

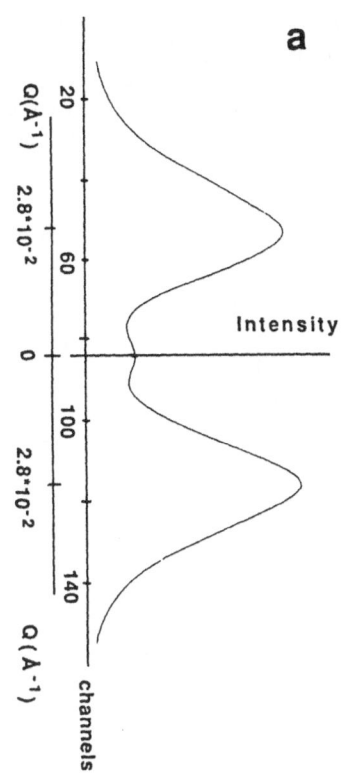

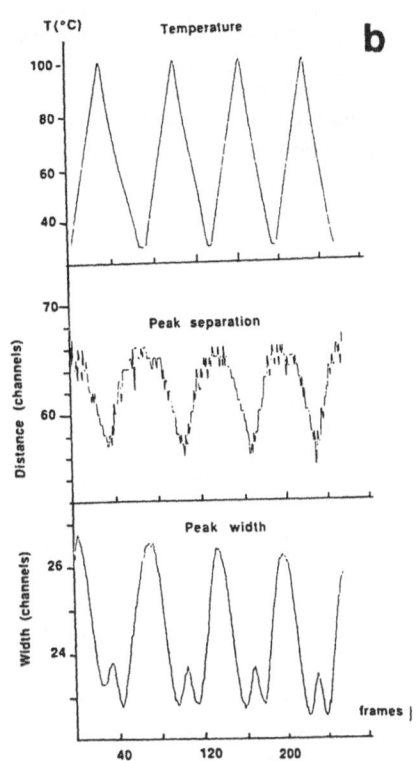

Fig. 21a SAXS-pattern of low density polyethylene -crystallized at $T_c=103°C$- measured parallel to the meridian. /32/(15s per frame)

Fig. 21b Change in peak separation and width upon cycling the temperature./32/

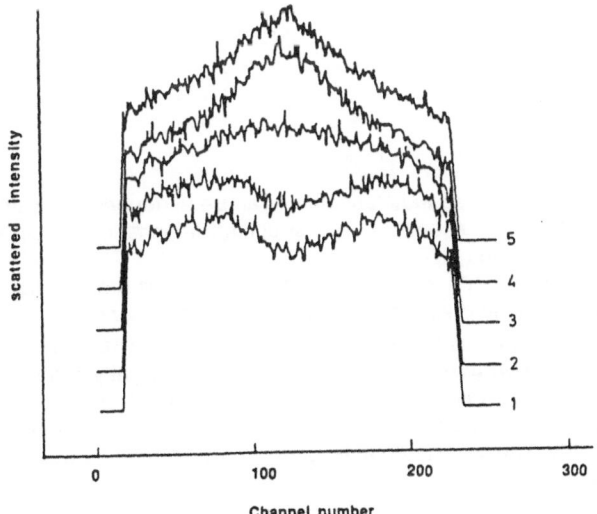

Fig. 22 Change of the SAXS pattern of PET-yarn, measured with a 1D-PSD parallel to the equator, upon stretching (0.5 s /pattern)./22/

462

Instrumental progress now allows to couple directly a DSC cell with the X-ray setup. /34,35/ Thus Fig.23 shows a correlation of the DSC trace with the change in **Invariant** for a liquid crystalline polymer **polyethylenenaphthaline-2,6-dicarboxylate** (PEN)./35/ The results can be interpreted with a crystallization in the range 165<T<200°C and an increase of the **Invariant** above 200°C due to the density difference. A further crystallization occurs at 235<T<262°C while melting occurs at >262°C

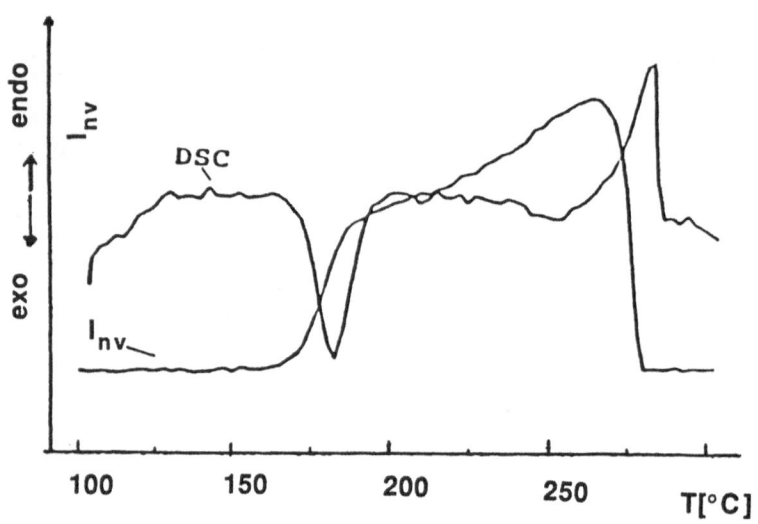

Fig.23 Change of **Invariant** and DSC-trace upon heating of polyethylenenaphthaline-2,6-dicarboxylate./35/

Solid state phase transformations

Mechanistic studies of phase transformations can be performed by realtime WAXS on a timescale of several seconds per spectrum. The structural information obtained in this way compliments again other methods such as DSC.

Thus **polyacetylene (PA)** exists in a cis- and a trans-modification. /2/ The unit cell projections of both modifications onto the chain axes and the π-bonding system are shown in Fig.24. The settings of the unit cells correspond to that of Ref.36. The fibrillar material prepared according to the **Shirakawa-method** /37/ has a crystallinity >90%. /38/ The DSC trace -recorded with a heating rate of 2.5°C/min (Fig.26)- shows two exothermic peaks. X-ray experiments indicate that peak 1 is due to the cis/trans transformation while peak 2 due to decomposition./39/ These experiments give, however, no information on possible intermediate structural steps.

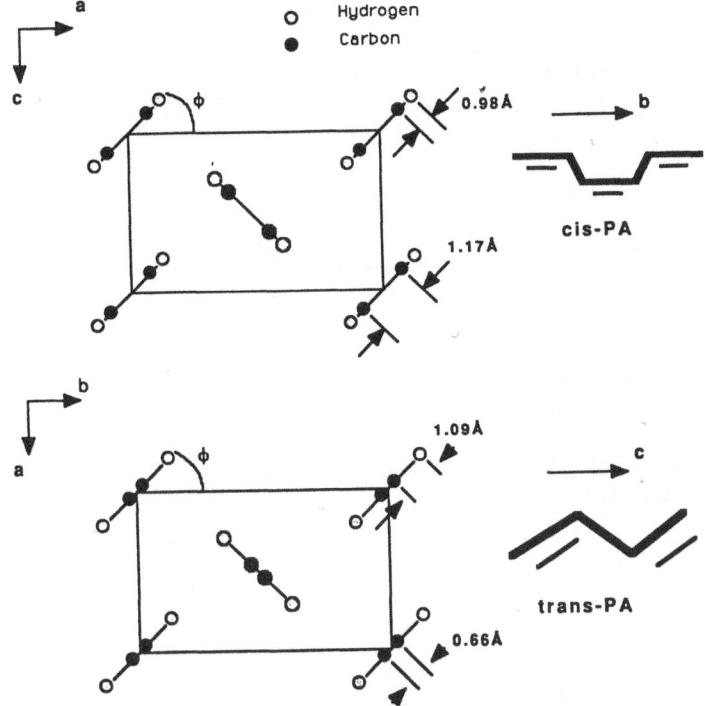

Fig. 24 Chain projections of cis- and trans-PA and π-bonding systems. Note that the C-C projection distances in the a/c(cis) and a/b(trans) planes are different. The setting angle is called ϕ

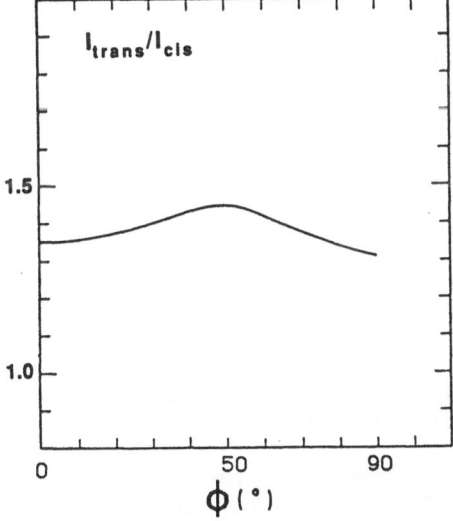

Fig. 25 Calculated intensity ratio trans-(110/020)/ cis-(101/200) as a function of the setting angle ϕ

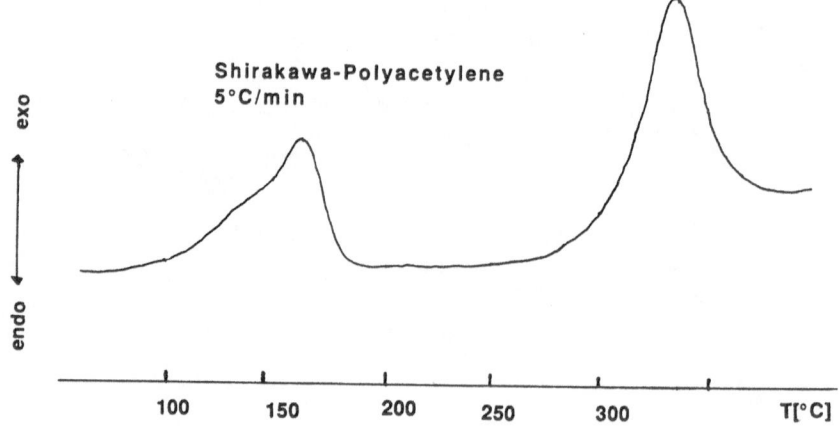

Fig.26 DSC trace of Shirakawa polyacetylene (heating rate: 5°C/min; unoriented material)./42/

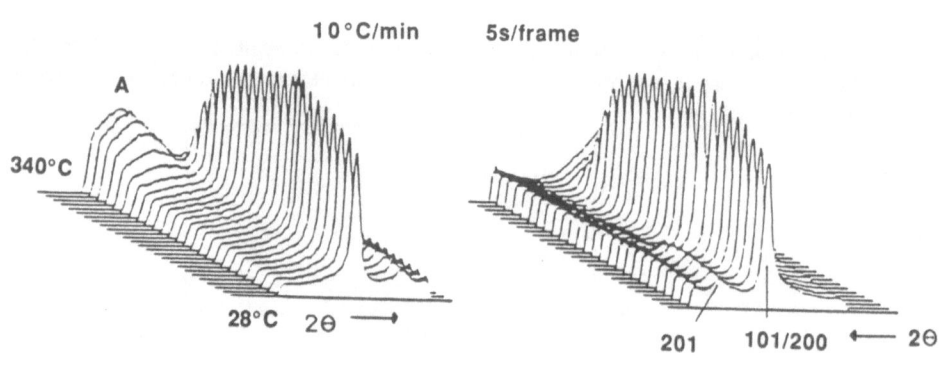

Fig.27 Sequence of WAXS patterns recorded upon ramp heating of cis-polyacetylene (10°C/min; unoriented material)./41/

In order to study this question one can either monitor the change in position of selected reflections or the change in reflection intensities. In the latter case, model calculations are feasible. Thus, for the model shown in Fig.24, an increase of the intensity of the coinciding 200/101 reflections of cis-PA upon transformation into the 020/110 reflections of trans-PA is expected. As shown in Fig.25, this increase depends somewhat on the setting angle ϕ. Structural data suggests that $\phi \approx 51°$./40/

Real time WAXS spectra, recorded at a heating rate of 10°C/min with several seconds per pattern, are shown in Figs.27,28,29 for unoriented and oriented material /41,42/ The results are basically identical. The overall crystallographic changes can be seen from the disappearance of the 201 reflection above ≈ 120°C. At ≈ 300°C a broad peak (A), due to the destruction of the crystalline lattice, can be clearly seen. These results are identical to the DSC/X-ray results./39/

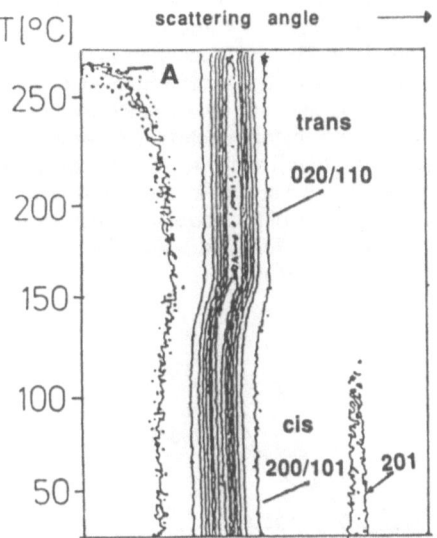

Fig.28   Change of 200/101 reflection of cis-PA upon ramp heating with 5°C/min(unoriented material)./42/

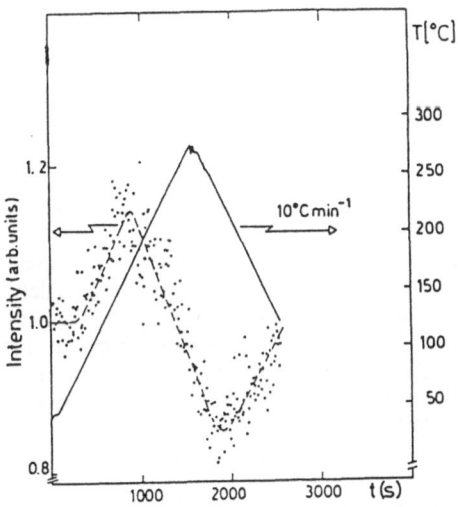

Fig.29   Change of integrated intensity of the cis-PA 200/101 reflection upon ramp heating with 10°/min (oriented material; solid line: thermocouple, see Fig.12)

The real time results show, however, a continuous transformation of the cis- into the trans structure which becomes evident from Fig.28 where the intensity is projected onto the T/Θ-plane./42/ This corresponds to X-ray results obtained from rapidly cooled down samples./43/ Further spectroscopic experiments like Raman, IR or NMR have also not given evidence for an intermediate phase./2/

The intensity of the cis-200/101 increases as expected from Fig.25 but decreases again above T≈150°C. A quantitative analysis of this increase has not been tried in view of problems with sample expansion during the transformation. (Fig.32)The absolute increase of the peak appears, however, to be smaller than expected from Fig.25 which could indicate a beginning disordering. The onset of the intensity decrease seems also to be correlated with an increase in diffuse scattering (Fig.28) which eventually develops into the broad peak A. (Fig.27) The DSC-trace suggests that the destruction of the crystalline matrix reaches a maximum at ≈340°C. (Fig.26) At these temperatures, macroscopic properties, such as the electrical resistivity, are also affected by the loss of order. Thus Fig.30 shows that in the range 150<T<260°C an initial decrease of the resistivity -as expected from the difference in specific conductivities of cis and trans-PA (Fig.2) is followed by a re-increase./44/ The value for trans-PA given in Fig.2 suggests, however, that it is also affected by disorder.

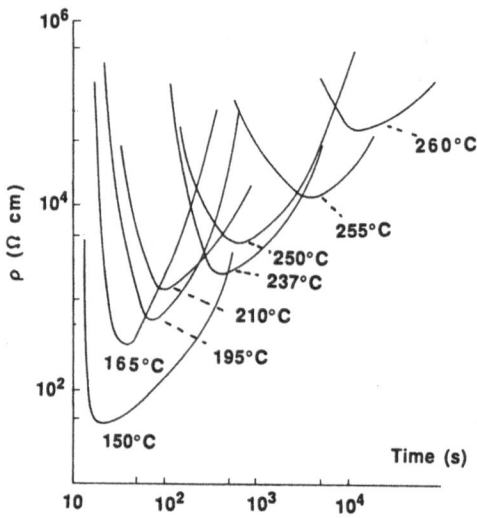

Fig.30   Change of specific resistivity of Shirakawa-PA upon annealing./44/

Although weight-loss experiments indicate an onset of decomposition only at ≈ 400°C /39/, gaschromatography shows the formation of aromatic decomposition products already after prolonged heating at ≈ 195°C /45/, which implies the development of crossbridging resulting in chain fusion and aromatic decomposition products. Part of the intensity-loss seems furthermore to be reversible (Fig.29), which is not understood at present. (a Debye-Waller factor term can be excluded) From these data it is not possible to decide whether disordering and decomposition are directly linked. A coupling of a method like mass-spectrometry -which is sensitive to gaseous decomposition products- with WAXS-experiments would therefore be highly desirable.

A more detailed examination of the intensity change suggests a somewhat more complicated picture of the cis/trans isomerization. Thus the intensity of the strongest peak increases earlier than the shift in peakposition. (Fig.31a,b) This discrepancy can be explained by a local isomerization of cis-PA chain segments without a change in unit cell and which

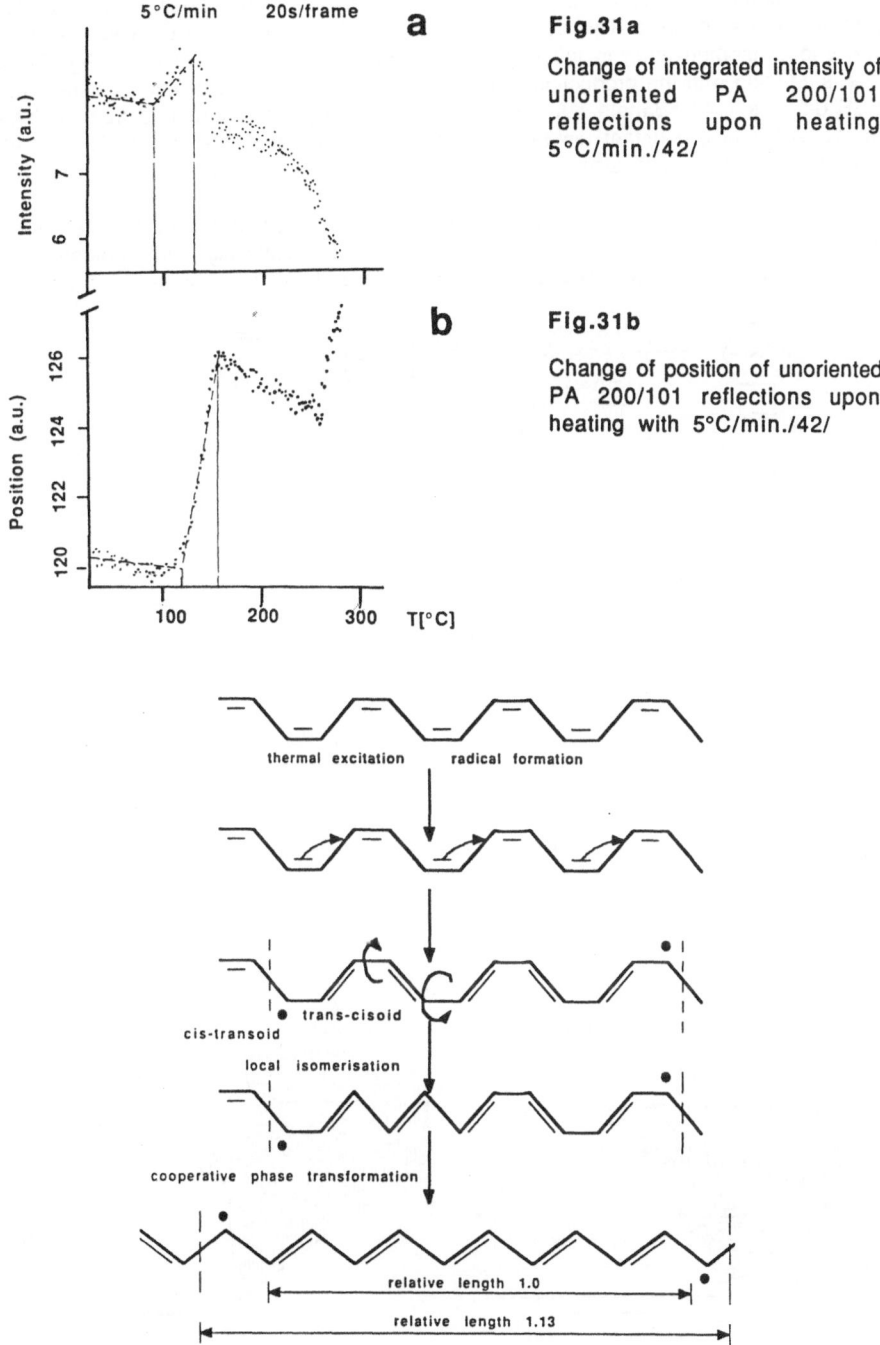

Fig.31a

Change of integrated intensity of unoriented PA 200/101 reflections upon heating 5°C/min./42/

Fig.31b

Change of position of unoriented PA 200/101 reflections upon heating with 5°C/min./42/

Fig.32 Model for the cis/trans isomerization of PA which can explain the appearance of unpaired spins, observed by ESR./46/ Note that the chain length increases by ≈13% upon isomerisation.

468

produces only a change in intensity. At a sufficiently high concentration of chain segments, the unit cell parameters also change. This can be correlated with a mechanism derived from ESR data, where a local isomerization is followed by transformation of whole chains. (Fig.32)/46/ This process is coupled with the production of unpaired spins. The activation energy determined from the ESR data is ≈ 42 KJ/mole. A similar activation energy is determined from realtime WAXS data, recorded at a constant temperature, which corroborates this assumption. /41,42/ This two step transformation can also explain the asymmetric shape of first peak in the DSC trace.(Fig.26)

Polyacetylene can also be prepared by thermal decomposition of a prepolymer by the so called **Durham route.**/47,48/ This PA variety has a nonfibrillar morphology. The decomposition reaction and the corresponding DSC-curve are shown in Fig.33./42/

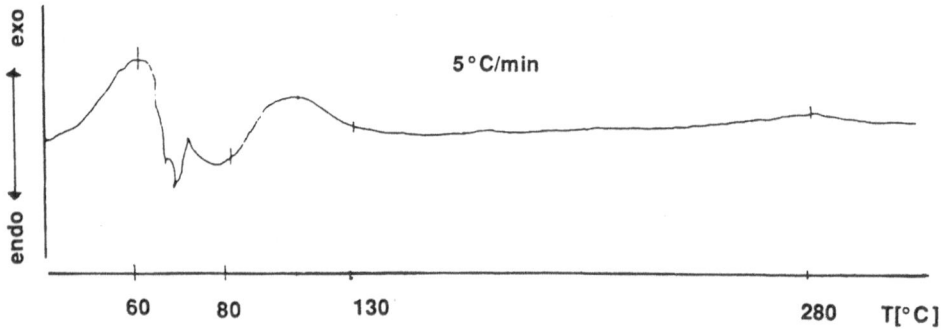

Fig.33   Decomposition reaction and DSC trace for the formation of PA according to the Durham route./47,48,42/

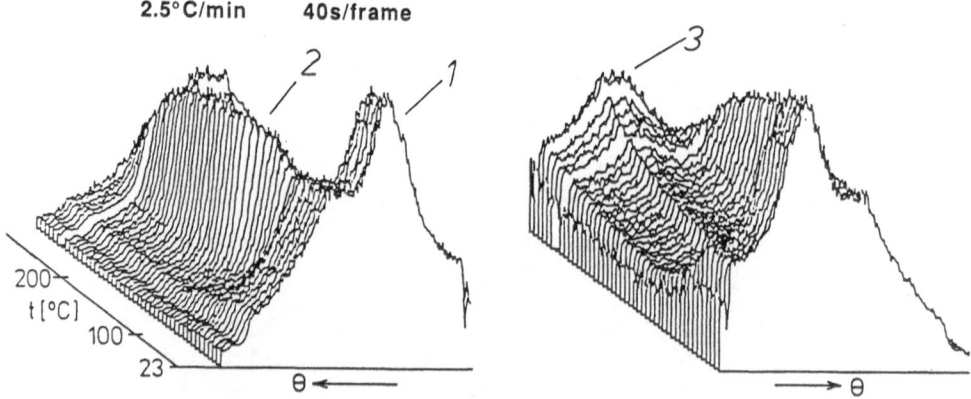

Fig.34   Corresponding sequence of WAXS patterns./42/ The small, narrow peak in the right sequence is an artifact.

A sequence of corresponding WAXS-patterns is shown in Fig.34. Note that the peaks are much broader than for the Shirakawa material which indicates less order. The mechanism of the transformation can again be determined from the change in integrated peak intensity. Thus in Fig.35, these changes are shown for the peaks 1, 2 and 3. The destruction of the prepolymer (peak 1) is obviously correlated with the appearance of the PA-peak (peak 2).

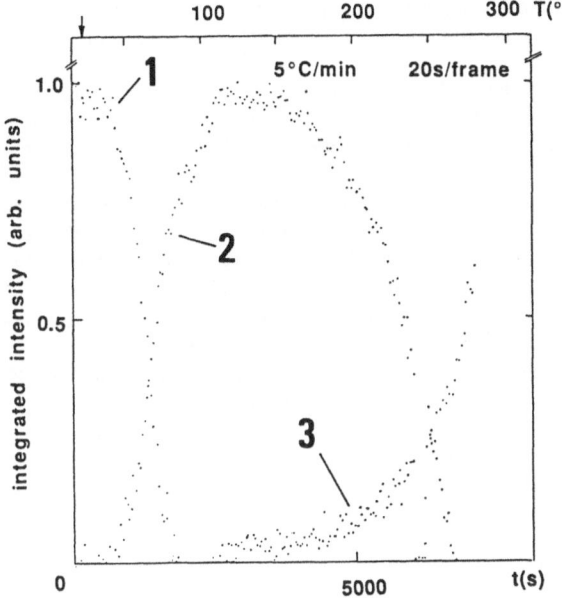

Fig.35 Change of integrated intensity of peaks 1,2 and 3 shown in Fig.34./42/ The arrow marks the onset of heating.

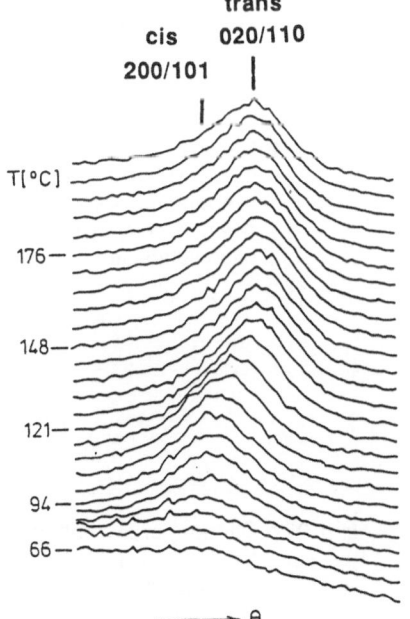

Fig.36 Change of position of peak 2 (Fig.34)./42/

470

A more detailed examination of peak 2 shows again a shift due to the cis/trans isomerisation.(Fig.36) This conclusion agrees with a laboratory X-ray study at a constant temperature of 30°C which allowed to slow down the decomposition to about 3 days./49/ At higher temperatures, decomposition of PA appears to start, as evidenced by a decrease of peak 2 and the appearance of a broader peak 3. Although the DSC-maximum appears to be at ≈ 280°C, both the decrease of peak N°2 and the increase of peak N°3 suggests an onset of the disordering of trans-PA at ≈ 100°C which is significantly lower than for trans-PA prepared from Shirakawa material. (see above) This may be related to the more disordered Durham-PA , as evidenced by the broader peaks./42,49/

The WAXS spectra appear to be much more sensitive to the destruction of the PA-lattice than the DSC data.(Fig.33) As for the Shirakawa material the available weight-loss data /50/ do not allow to decide whether the onset of disordering is also related to a decomposition of the PA-lattice.

In contrast to Shirakawa material which has due to its fibrillar morphology a dominationg hole-scattering background, SAXS-spectra can be obtained for Durham-PA. Fig.37a,b shows a sequence of such spectra obtained upon heating with 2.5°/min and $\int I(Q)./42/$ The background due to decomposed material has been subtracted.

No discrete maximum was observed for the prepolymer or PA. Although the WAXS results suggest that Durham-PA is more disordered than Shirakawa-PA, there is therefore no evidence for a semicrystalline structure. This corroborates the assumption that the formation of a semicrystalline morphology demands a flexible polymer chain as found for PE./5/

One observes ,however, a decrease of the integrated SAXS intensity during the decomposition, which is apparantly correlated with the production of cis-PA. There is no evidence that the evaporating aromatic rings create defects in the material.

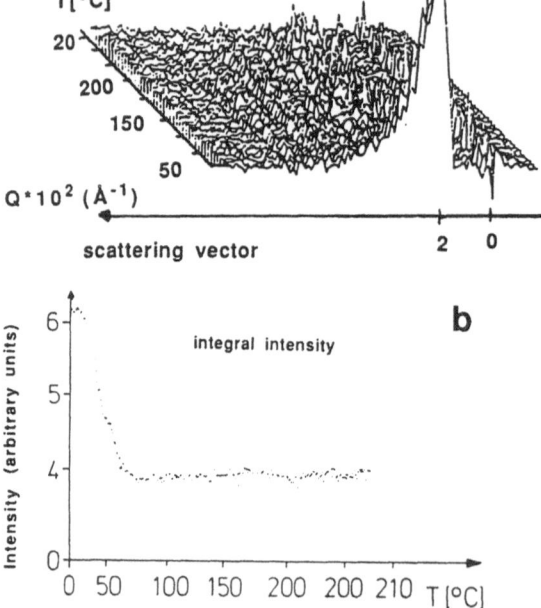

Fig.37a Change of SAXS patterns upon decomposition of Durham prepolymer./42/ Background due to decomposed material has been subtracted

Fig.37b Change of overall integrated SAXS intensity upon decomposition of Durham PA./42/

Solid/gas reactions

Wide angle diffraction experiments on the uptake of I_2 by polyacetylene have been performed in the reaction cell shown in Fig.38. /41,42,51/ The valves can be switched via the time frame generator (Fig.8). This allows to establish a pressure in the mbar range in the previously evacuated cell in a few seconds. (Fig.39)

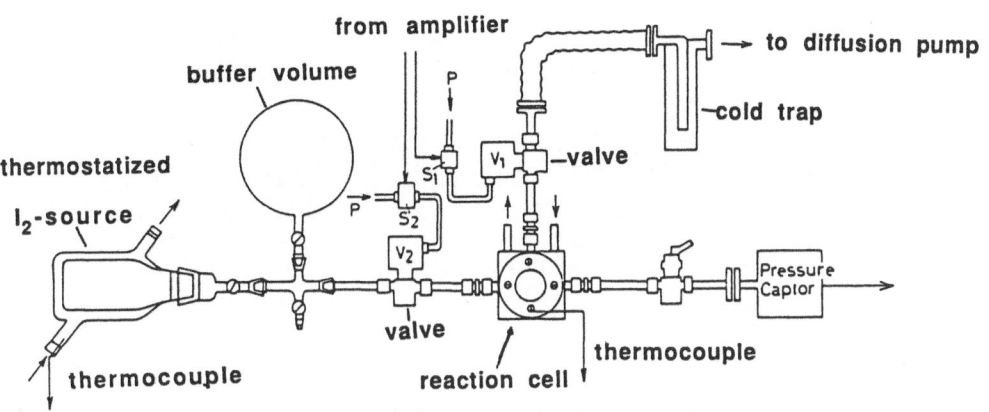

Fig.38 Reaction cell used to study the reaction of PA with I_2

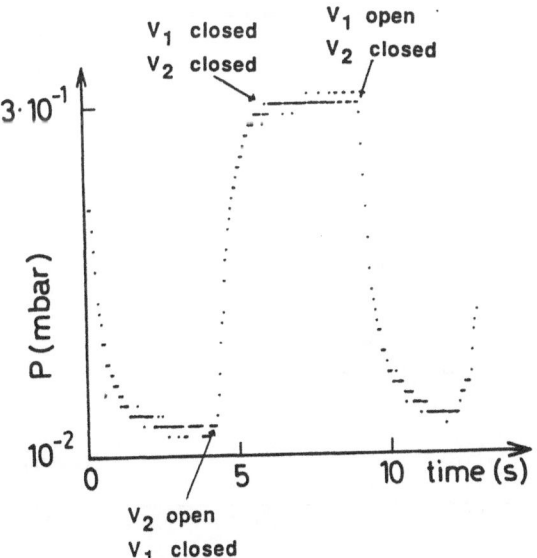

Fig.39 Development of I_2-pressure in the reaction cell upon switching valves V1 and V2. The development of pressure was followed with t_m=50 ms and t_w=10 μs, determined by the *Time Frame Generator*. (III.3)

472

A sequence of spectra taken during the reaction of Shirakawa type material for $P/P_0\approx1$ is shown in Fig.40./41/ The sample was exposed at r.t. to a constant I_2-pressure during the reaction.

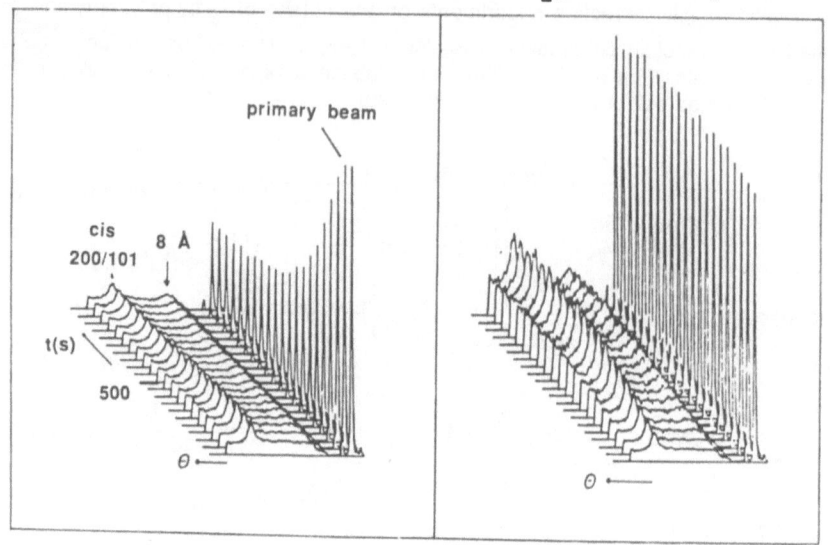

Fig.40 Change of WAXS-patterns upon reaction of Shirakawa-PA with I_2. The raw patterns are shown on the left side while the patterns on the right side have been corrected for the change in transmission./51/ (90 μm foil)

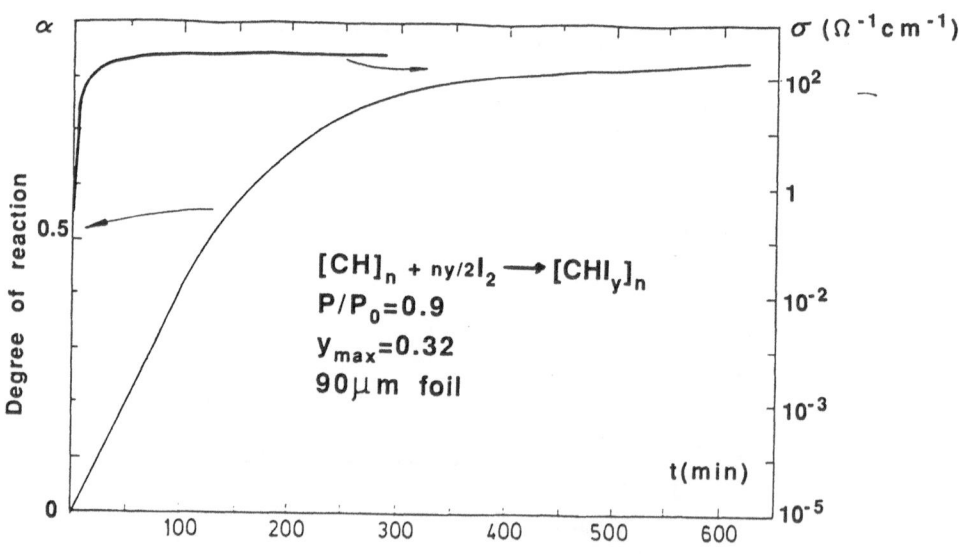

Fig.41 Change of specific electrical conductivity and volume reaction (determined

gravimetrically) for the reaction of I_2 with PA./52/

The formation of a new phase is evident from the appearance of a new peak. Note that at $\lambda=1.5$ Å a considerable change in transmission is taking place upon I_2-uptake. This can be seen by the decrease of the primary beam intensity, recorded also by the 1D-PSD. (Fig.40; sequence at left) Although this can be used for a correction of the transmission change/51/ (Fig.40; sequence at right) a systematic error is introduced by the expansion of the foil in the beam (see Fig.32), which allows such a correction only for the onset of the reaction. Shorter wavelength and thinner samples which are "bathed" in the beam would reduce such error sources.

It is of interest to compare the timescale of the appearance of this peak with that of the electrical conductivity./52/ (Fig.41) Obviously the formation of the peak and the increase of conductivity correlate well, while the bulk uptake (α was determined gravimetrically and scaled to 1 for $y_{max}=1$ /52/) is on a different timescale, i.e. much slower.

Shirakawa polyacetylene has a fibrillar structure (Fig.42) with chains aligned parallel to the fibre axes. The different kinetics may be explained by a rapid diffusion of I_2 into more open surface layers and a slower diffusion into the bulk. (Fig.42) According to this model, the electrical conductivity is determined by the surface layers of the polymer foil. The slow uptake has also been observed by real time neutron diffraction of the volume reaction of $[CD]_n$ with IBr, SbF_5 and I_2, which can be modelled by macroscopic diffusion law./53/

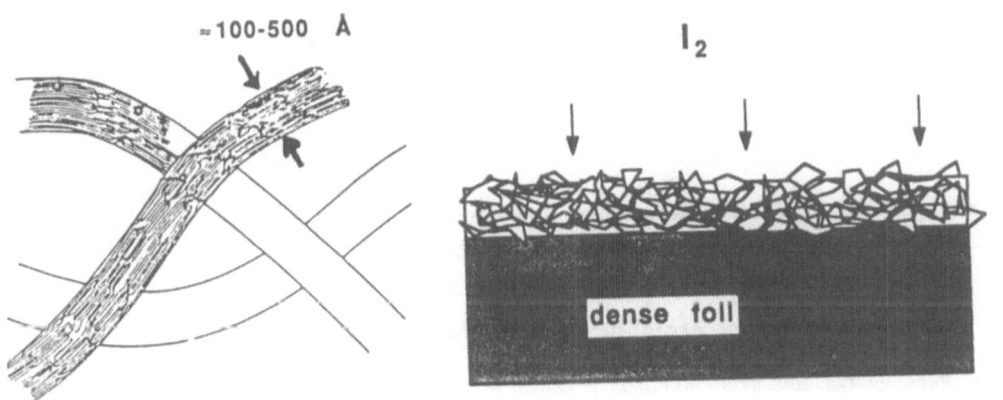

Fig.42   Model for a foil of Shirakawa-PA. The fibrils in the surface layers are only loosely packed. The fibrils are shown schematically on the left side

The reaction of PA with I_2 has also been studied for ≈30 μm foils of Durham-polymer by SAXS. /42/ The aim was to find out whether evidence for the formation of I_2-rich nuclei at the onset of the reaction -which have been discussed in the context of the mechanism of conduction /54/- can be obtained. A distribution of such nuclei with a sufficiently narrow size distribution in the matrix is expected to produce a SAXS peak due to the average distance of the nuclei. Such phenomena are common in crystallization processes of amorphous alloys.

The results of an experiment at $P/P_o\approx0.096$ show, however, no evidence for a discrete peak. (Fig.43) This suggests that the results of a high resolution electron microscopy study on single fibers of I_2-doped polyacetylene -where no nuclei were observed /55/ - do also apply to the onset of the reaction in Durham-PA.

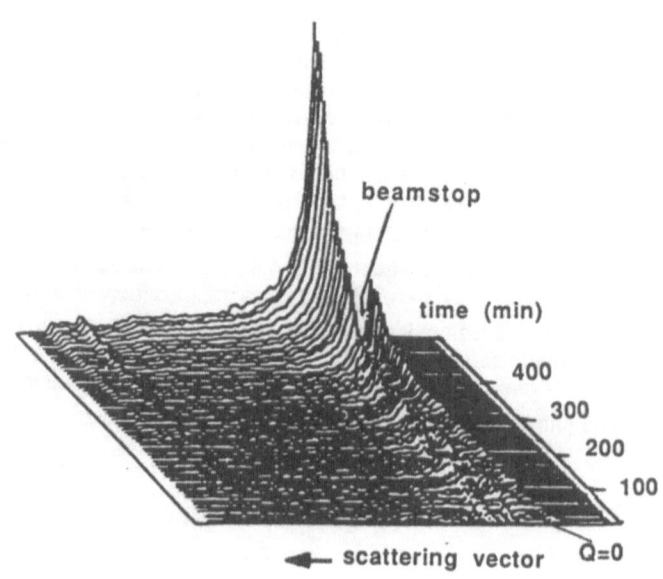

Fig.43   Change of SAXS-patterns of Durham-PA upon reaction with I_2 at 0.022 Torr. The pattern of unreacted PA has been subtracted from all other patterns./42/

V.   Future developments in instrumentation

The presently available instrumentation is limited in a number of areas such as:

(i)   <u>lower limit in Q</u>

Long periods >600 Å cannot be separated from the diffuse scattering background.(Fig.13) This has limited for example the study of the melting process in 1-octene copolymer of PE. (Fig.16b)
For crystallization studies, an overlap of the Q-range with that of light scattering methods will be of interest in order to study density fluctuations due to the onset of crystallization, the development of larger morphological units (e.g. spherulites) or spinodal decomposition /56,57/ up to the μm-range.

(ii)   <u>sample size</u>

Structural studies are ideally done on single crystal or highly oriented material. Polymer single crystals have often very small linear dimensions and phase transformations cannot be studied in real time with the flux of a bending magnet source. This holds also for single fibrils with < 10 μm diameter. Electron microscopy has been the principal research tool in structural research on small sample volumes. This method is, however, limited in real time applications and can introduce defects.

For a number of perturbation methods (e.g. heating, chemical reactions) the time necessary to establish a homogeneous perturbation across the sample depends on the sample size. The illuminated sample volume should therefore be as small as possible.

The **European Synchrotron Radiation Facility** -which is expected to come into operation ≈1993- will allow a considerable improvement in these areas ./58/ Table 2 lists a number of parameters of this source.

Table 2   Parameters of the ESRF storage ring./58/

Energy	(GeV)	6
Current	(mA)	>100
Bunch length	(ps)	80
Revolution time	(μs)	2.71
Maximum number of Insertion devices		29
Free-length of straight sections	(m)	6
Number of Bending Magnet Ports		26 at 20 keV
		16 at 10 keV

The improvements are principally due to the increase of the **brilliance** which is defined as: Photons/(s*mm^2*mrad^2*10% BW). (BW: bandwidth)

The brilliance is increased -as compared to existing sources- by two factors:

(i)   an optimized **emittance**
(ii)   the use of straight sections with periodic magnetic structures

The horizontal and vertical emittances are defined as:

$$\varepsilon_x = x * x' \qquad \varepsilon_z = z * z' \tag{3}$$

where x,x'; z,z' are the size and divergence of the electron beam.
The ESRF emittance $\varepsilon_x = 6.8 * 10^{-9}$ m*rad is significantly smaller than the DORIS value of $\varepsilon_x = 2.7 * 10^{-7}$ m*rad at 3.7 GeV. (vertically: $\varepsilon_z \approx 0.1 * \varepsilon_x$; upper limit, actual value can be smaller! /12/)
        Part of the magnetic lattice of the ESRF is shown schematically in Fig.44. Synchrotron radiation can be extracted both from **bending magnets** (≈ 6 mrad horizontally; source point close to the entrance of BM2), as well as from so called **insertion devices (wigglers** or **undulators)** which are installed in the straight sections.
Wigglers or **undulators** consists out of **N** periods of length λ_0./12,58/ While radiation is collected in a bending magnet over ≈ 15 cm, for a periodic magnetic structure the emitted photons add up in the observation direction over the full length. (≈ 6 m for the ESRF)

476

For a periodic magnetic structure, the emittances of the photon and the electron beam are practically the same in the hard X-ray part of the spectrum.

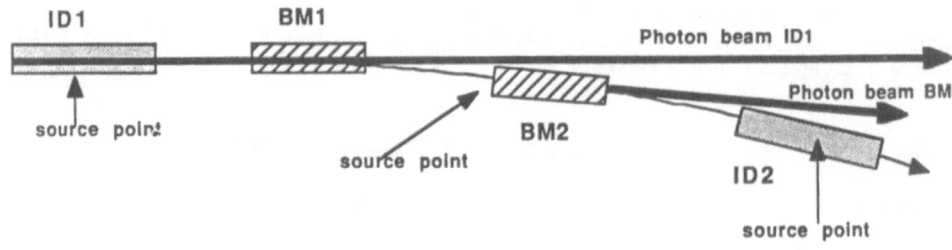

Fig.44 Part of the magnetic lattice of the ESRF-storage ring. The design shows two 6m long straight sections, where insertion devices like undulators can be incorporated (ID1 ,ID2) and two bending magnets (BM1,BM2). Synchrotron radiation can be extracted for experimental purposes from the insertion devices (shown for ID1; ≈1/γ opening angle for undulator; see Fig.45b) and from the bending magnets (≈ 6 mrad horizontally; source point close to the entrance of BM2)

Fig.45a shows schematically that every electron emits photons into a cone of opening angle 1/γ, where γ=1957*E (GeV). Here E is the energy of the ring. The maximum angle between the tangent to the electron beam and the on-axis observation direction is called α.

For α<1/γ the emitted spectrum is no longer continuous but shows discrete peaks due to an interference between the emitted photons. Such an insertion device is called an **undulator**. A smooth spectrum -similar to a bending magnet source- is observed in the X-ray part of the spectrum for α>1/γ . Such an insertion device is called a **wiggler**. In practice a **deflection parameter K** is defined as:

$$K=\alpha*\gamma \quad (4)$$

K values significantly above 1 imply a wiggler.

The distribution of the photon beam intensity in the plane of the insertion device is shown in Fig.45b. Note that an undulator has a nearly round beam spot of 1/γ diameter, in contrast to a wiggler. The figure shows also the emitted cone of an undulator at a large distance from the source point. The central -quasi monochromatic part- has a diameter of 1/($\gamma N^{0.5}$). In order to obtain the central part, a pinhole or slits have to be used.

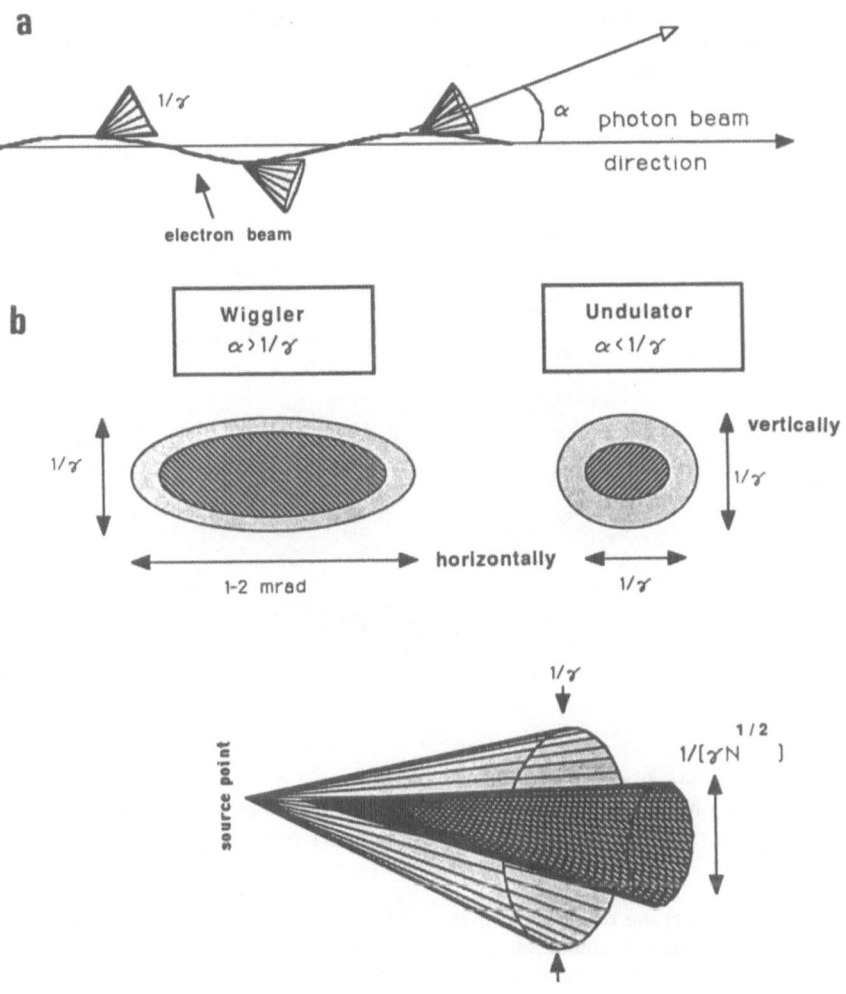

Fig.45a Model for a periodic magnetic structure. The electron beam shows oscillations in the plane vertical to the field of the magnets which have a wavelength λ_0. The photons are emitted by the electrons into cones of $1/\gamma$ opening angle.

Fig.45b Schematic representation of on-axis intensity projection of wiggler and undulator. The central part is surrounded by a zone of diffuse scattering. The undulator intensity distribution is shown in more detail below. Note that the horizontal intensity distribution of an undulator is increased by the in-plane amplitude of oscillation of the electron beam /12/ This amplitude increases with **K**.

478

The development of a specific undulator spectrum ,as a function of the size of a rectangular opening and at a large distance from the source point, is shown in Fig. 46. /59/ The K-value is 1, the vertical opening angle is 0.05 mrad and the horizontal opening angle is varied between 0 and 0.2 mrad.

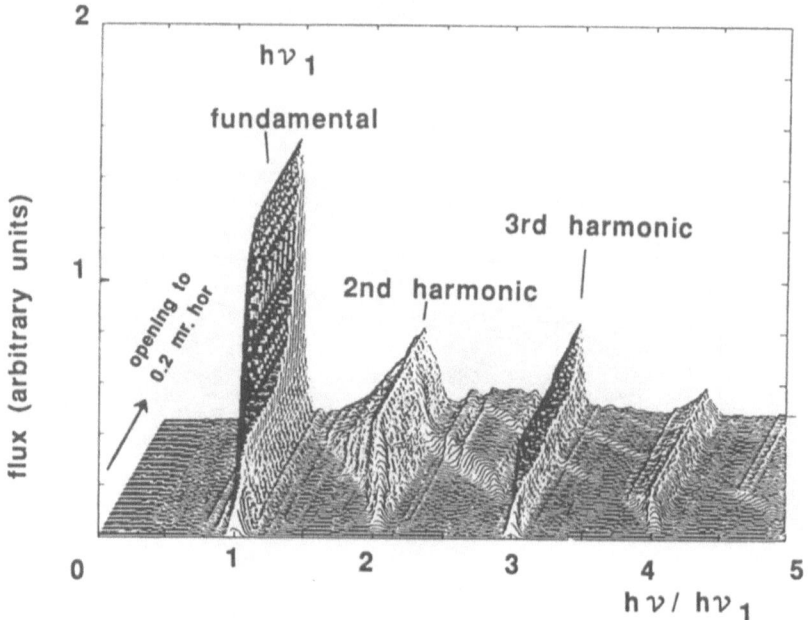

Fig.46 Development of the spectrum of an undulator with K=1. /59/ A possible wavelength of the fundamental would be 1.5 Å (≈8KeV). By changing the magnet gap, the peak can be displaced, i.e. the wavelength can be changed (tuning).

The small source point size and divergence of the radiation emitted by an undulator implies that the brilliance of such a radiation source is by several orders of magnitude larger than that of a **bending magnet.** In Table 3 the source parameters of DORIS (bending magnet) and the ESRF (undulator) are compared.

TABLE 3 Comparison of the source point parameters of an undulator with 6 m length and 150 periods at 0.1 A current. (central -quasi monochromatic- part)/58/ The T1-option is optimized for a small divergence. T2 for a small source point size. Relevant parameters from the A1-bending magnet at DORIS are added.(3.7 GeV)

	T1	T2	A1-DORIS
K	2.15	2.15	-
x (σ/mm)	0.437	0.062	1.70
z (σ/mm)	0.138	0.040	0.30
x' (σ/mrad)	0.016	0.103	0.31
z' (σ/mrad)	0.009	0.018	0.05

The nearly round, monochromatic central part and the high brilliance make the undulator highly interesting as insertion device for a SAXS pinhole camera.

The size of the beam at the slits of a pinhole camera and the length of the camera determine the limits of the diffuse scattering in the detector plane and therefore the size of he largest objects which can be resolved. /3/ Calculations for the limits in diffuse scattering, the size of the focus and the size of the beam at the sample position are shown in Fig.47 for a camera which is similar to the one shown in Fig.7./58/ Only the T1 option was considered, as it can be shown that a small divergence is more important in order to reduce the limits of diffuse scattering than a small source point. The max. resolution of the polymer beamline in Hamburg is shown for comparison.

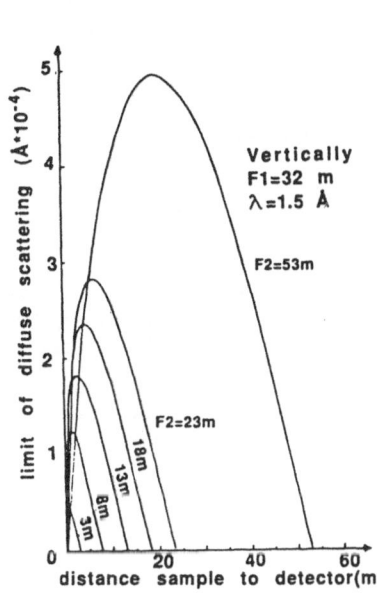

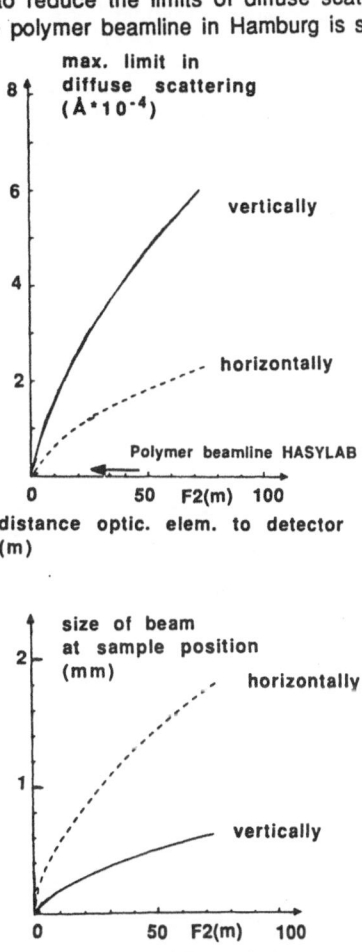

Fig.47 Calculation of the limit in diffuse scattering as a function of the distance optical element to detector , the maximum value of the limit in diffuse scattering and the size of the focus at the sample position . /58/ The deviation from a perfect round focal spot (Fig.45b) explains the differences in the horizontal and vertical directions.

480

The flux of such a camera is calculated to be ≈ 4*10^{13} as compared to ≈ 10^{11} for the polymer beamline at Hamburg. This should allow to record SAXS spectra of polymers like PE or PET in the ms-range provided that appropriate position sensitive detectors are developed. Alternatively the illuminated sample volume could be reduced by ≈ factor 100 while retaining a temporal resolution in the s-range.

A reasonable length of the camera is shown in Fig.48 Here the whole extension of the ESRF-experimental hall is utilized.

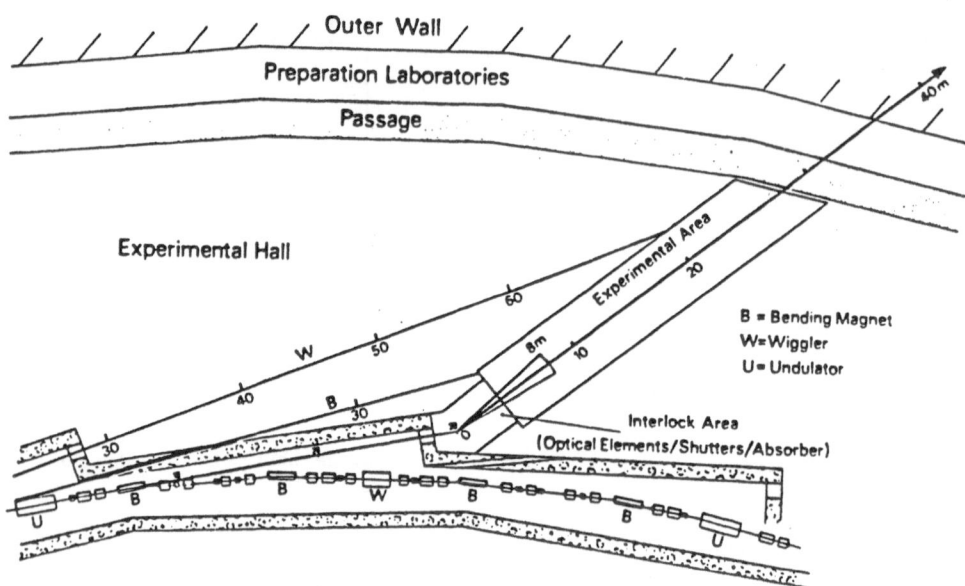

Fig.48 Possible implementation of a SAXS-camera within the experimental hall of the ESRF. /58/

Two modes of operation will be possible.

(i) For F2-values (Fig.6) of 8m (monochromator/detector distance), the optical bench can be turned around the monochromator axis which allows to vary the wavelength in the range 1<λ<2 Å. This will necessitate a tuning of the undulator spectrum by a change of the magnet gap./58/ A variable wavelength is of interest for the study of polymers with heavy atoms (e.g. [CHI$_y$]$_n$) or for crystallization studies of alloys or glasses.

(ii) For F2-values up to ≈25 m the wavelength will be fixed to 1.5 Å. The limit in diffuse scattering will be of the order of 3 μm (Q$_{min}$≈2*10^{-4} Å^{-1}) which suggests that the camera can overlap with the range of light scattering experiments. This would allow to study much larger morphological structures than presently accessible.

The range in applications of this **proposed** camera would therefore ly in the area of **real time studies** down to **small Q-values.** It should be noted, however, that real time light scattering experiments on the spinodal decomposition of mixtures of polystyrene and poly(vinyl-methyl-ether) suggest that Q_{min}-values of $\approx 10^{-5}$ Å^{-1} must be reached in order to see the interference peak due to the formation of domains. /56/ This is confirmed by a neutron small angle scattering experiment on metastable samples, using a double crystal diffractometer./57/ The advantages of the latter method and of X-ray scattering methods are the possibility to study opaque samples and to perform an absolute calibration.

Although it would in principle be possible to increase the length of the camera in order to decrease Q_{min}, it is felt that other methods like the **Bonse-Hart camera** /60/ -although not necessarily real time- might be more appropriate to obtain **ultrasmall Q-values** with short wavelength X-rays. /58/

A further camera with a **microfocus** has been **proposed** for SAXS-experiments on **ultrasmall samples.**/58/ In this case, focusing could be performed by a toroidal mirror. (Fig.49) The beam has to be lifted out of the plane of the storage ring in order to get rid of the hard background radiation (Bremsstrahlung). Assuming an undulator optimized for a small source point (T2 in Table 3) one calculates for a demagnification ratio of 2.7 (= F2/F1) a focus of 20*14 μm^2.

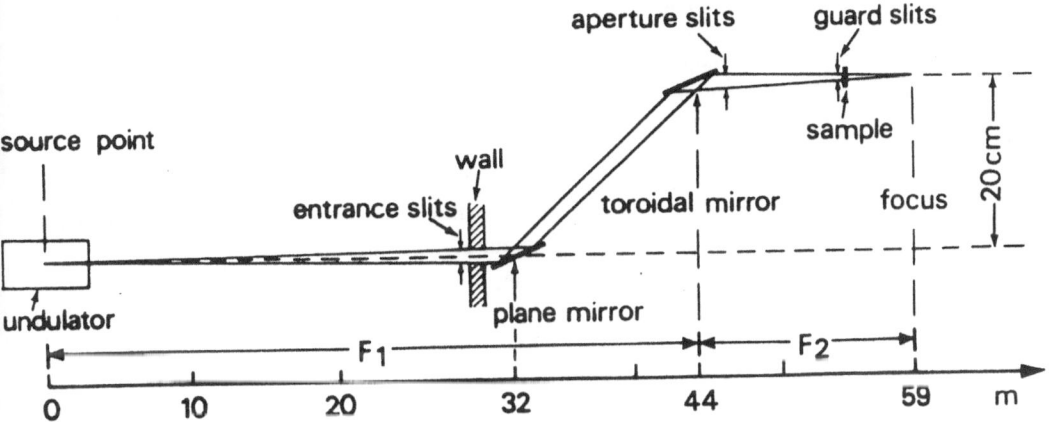

Fig.49 Schematical design of a point focusing camera using a toroidal mirror for focusing./58/

Such a focus appears to be feasible with the quality of existing optics. The mirror would be covered by gold in order to increase the angle of total reflectance. A double crystal monochromator with $\Delta E/E \approx 10^{-4}$ could be used to define the wavelength.

A more elegant design would be the use of multilayers in order to select the fundamental of the undulator spectrum.(Fig.46) An optics with cylindrical double mirrors and multilayers has been proposed for the ESRF./58/ A focus size of 40*10 μm^2 has recently been realized with W/C multilayers and this type of mirrors at the Stanford storage ring./61/

In case one would be able to use the full central cone of the fundamental of N =100 period undulator with $\Delta E/E \approx 1/N$ =10^{-2} /10,58/ instead of $\Delta E/E \approx 5*10^{-4}$ for a Ge-monochromator, the flux would be increased to $\approx 4*10^{15}$ Photons/s. Such a camera would be of interest to study dynamic processes in single fibrils of <1μm thickness, but also ultrasmall crystals with small unit cells. The broader wavelength band would allow to use the pseudo Laue technique. (stationary crystal)

In this context it is of interest to consider the minimum sample size one could examine. Thus the integrated intensity may be written as:/62,63/

$$I_H = I_o U \, \delta V \tag{5}$$

where I_o is the intensity impinging onto the sample, δV the crystal volume and U:

$$U = (e^2/mc^2V)^2 ((1+\cos^2 2\theta_B)/2)(| F_{hkl} |^2 \lambda^3/4\sin\theta_B) \tag{6}$$

where e is the charge and m the mass of the electron, θ_B is the Bragg angle and F_{hkl} the structure factor.

Taking the example of the Si(444) reflection (F_{444}=39.9), a wavelength of 1.5 Å and a crystal of 200 Å linear dimensions one calculates I_H=5800 counts/s ! This suggests that one could reach sample volumes which are at present only accessible by electron diffraction. In fact one could envisage an instrument where both methods would be coupled. Evidently sample handling and aligning will become a major problem for such experiments. One can expect ,however, that diffraction techniques for samples sizes < 1μm will at least partially replace experiments on unoriented samples (e.g. powders, foils).

In this respect it is also of interest to consider **sample heating problems.** For a sample in vacuum the temperature of the sample may be calculated from:/63/

$$W = \varepsilon\sigma(T_1^4 - T_2^4) \tag{7}$$

where W is the power absorbed per unit area, ε the emissivity, σ the Stefan-Boltzmann constant, T_1 the starting and T_2 the final temperature.

W may be calculated for a sample thickness t from:

$$W_t = W_o e^{-\mu t} \tag{8}$$

where

$$W = W_o - W_t \tag{9}$$

Assume values of W_o=30 W/mm^2 for the fundamental of an undulator (0.1 A electron current), t=200 Å, ε=0.1, σ=5.67*10^{-12} W/cm^2/°K, T_1=300K.

In the case of Si (μ=144 cm^{-1}) one calculates $T_2 \approx$1112°K, while for carbon (μ=4 cm-1) one calculates $T_2 \approx$304 °K ! This suggests that weakly absorbing organic, polymeric or biological material should encounter less heating problems than heavily absorbing inorganic compounds or metals. Note ,however, that crystals in a gas or in a liquid will behave quite differently and shorter wavelength would reduce the μ-values.

The possibility for such experiments will also depend critically on the **development cf detectors** with pixel sizes in the μm range and of **sample handling facilities,** which will be part of the **ESRF R&D** program./58/

References

/1/   For a review see: I.M. Ward, *Advances in Polymer Science*, Springer Verlag, Berlin (1985)

/2/   for a review see: J.C.W. Chien, *Polyacetylene*, Academic Press, New York (1984)

/3/   G. Elsner, C. Riekel, H.G. Zachmann, *Advances in Polym. Science,* Vol.67, H.H. Kausch and H.G. Zachmann, eds, Springer Verlag (1985)

/4/   C. Riekel in *Progress in Solid State Chemistry,* Vol.13,N°2 (1980), G.M. Rosenblatt, W.L. Worrell eds, Pergamon Press, Oxford

/5/   see for example: B. Wunderlich, *Macromolecular Physics*, Vol.1, Academic Press, New York (1973)

/6/   A. Guinier, *X-Ray Diffraction*, W.H. Freeman, San Francisco (1963)

/7/   B.K. Vainshtein, *Diffraction of X-Rays by Chain Molecules*, Elsevier, Amsterdam (1966)

/8/   *Small Angle X-ray Scattering*, O.Glatter and O. Kratky eds, Academic Press, London (1982)

/9/   G.R. Strobl and R. Eckel, *Progr. Coll. & Polym. Sci.* $\underline{62}$,9/15 (1977)

/10/   H. Meier, G.R. Strobl, *Macromolecules*, $\underline{20}$,649 (1987)

/11/   For a review see: G. Rosenbaum and K.C. Holmes in *Synchrotron Radiation Research*, E. Winick and S. Doniach eds, Plenum Press, New York (1980)

/12/   S. Krinsky, M.L. Perlman, R.E. Watson in *Handnook on Synchrotron Radiation*, Vol.1A, E.E. Koch, ed., North-Holland, Amsterdam (1983)

/13/   J. Hendrix, M. Koch, M.H.J. Bordas, *J. Appl. Cryst.*$\underline{12}$, 467(1979)

/14/   for a review see: J. Hendrix in *Advances in Polymer Science,* Vol.67, H.H. Kausch, H.G. Zachmann eds, Springer Verlag, Berlin (1985)

/15/   C. Boulin, D. Dainton, E. Dorrington, G. Elsner, A. Gabriel, J. Bordas, M.H.J. Koch, *Nucl. Instr. and Methods*, $\underline{201}$, 209 (1982)

/16/   P. Bösecke, D. Ercan, C. Riekel, *Journal de Physique*, Coll.C5, Tome 47, C5-175(1986)

/17/   E.M. Mandelkow, A. Harmsen, E. Mandelkow, J. Bordas, *Nature*, $\underline{287}$, 595 (1980)

/18/   Y. Inoko, H. Kihara, M.H.J. Koch, *Biophysical Chem.*,$\underline{17}$,171(1983)

/19/   H.E. Huxley, A.R. Faruqi, J. Bordas, M.H.J. Koch, J.R. Milch, *Nature*, $\underline{284}$, 140 (1980)

/20/   D.T. Grubb, J.J.H. Liu, M. Caffrey, D.H.. Bilderback, *J. of Pol. Sci., Polym. Phys.,* Ed.$\underline{22}$,367(1984)

/21/   R. Zietz, M. Dettenmaier, E.W. Fischer, *HASYLAB Jahresbericht* (1986)

/22/   Wen-li Wu, H.G. Zachmann, C. Riekel, *Polym. Comm.*, 25, 76 (1984)

/23/   B.C. Larson, C.W. White, T.S. Noggle, D.M. Mills, *Phys. Rev. Lett.*$\underline{48}$, 337 (1982)

/24/   A. Keller in *Structural Order in Polymers*, F. Ciardelli, P. Giusti, eds, Pergamon Press, Oxford (1980)

/25/   G.R. Strobl, M.J. Schneider, I.G. Voigt-Martin, *J. Polym.Sci.,Phys.Ed.*$\underline{18}$,1361 (1980)

/26/   J.H. Magill, C. Riekel, *Makrom. Chem., Rapid Comm.,*$\underline{7}$,287(1986)

/27/   P. Schouterden, C. Riekel, M.H.J. Koch, G. Groeninckx and H. Reynears, *Polym. Bull.*,13,533 (1985) and P. Schouterden, M. Vandermarliere , C. Riekel , M.H.J. Koch ,G. Groeninckx and H. Reynaers, in preparation

/28/   P. Schouterden, G. Groeninckx, B. Vanderheijden and F. Janssen, *Polymer*, accepted for publication

/29/   H.G. Zachmann, D. Wiswe, R. Gehrke, C. Riekel, *Makrom. Chem. Suppl.*, $\underline{12}$, 175 (1985)

/30/   R. Gehrke, *PhD-Thesis*, University of Hamburg (1986)

/31/   W. Prieske, *PhD-Thesis*, University of Hamburg (1985)

/32/ W. Fronk, B. Heise, B. Neppert, H.R. Schubach, W.Wilke, *Colloid & Polym. Sc.,* 262,99(1984)

/33/ T.H. Gilman, M.R. Resetarits, B. Christ, *Polym.Eng.Sci.,*18(6),477(1978)

/34/ T.P. Russell, *J. Pol.Sc. Phys.Ed.,*23,1109 (1985)

/35/ M. Bark, H.G. Zachmann, *HASYLAB Jahresbericht* (1986)

/36/ R.H. Baughman, S.L. Hsu, L.R. Anderson, G.P. Pez, A.J. Signorelli, *Molecular Crystals, Proc. NATO Adv. Res. Inst.,* Plenum Press, Les Arcs (1979)

/37/ H. Shirakawa, S. Ikeda, *Polym. J.,* 2,231(1971)

/38/ H.W. Hässlin, C. Riekel, K. Menke, S. Roth, *Makrom. Chem.,*185,397(1985)

/39/ T. Ito, H. Shirakawa, S. Ikeda, *J.of Polym. Sci.:Polym.Chem.Ed.,*13,1943(1975)

/40/ G. Perego, G. Lugli, U. Pedretti, *Mol.Cryst. Liq. Cryst.,* 117,59(1985)

/41/ C. Riekel, Makrom. *Chem. Rapid Comm.,* 4, 479 (1983)

/42/ K. Stegen, *Diplomarbeit,* University of Hamburg (1987)

/43/ G. Perego, G.Lugli, U. Pedretti, E. Cernia, *Journal de Physique, Colloque C3,* Tome 44, C3-93 (1983)

/44/ M. Rolland, P. Bernier, S. Lefrant, M. Aldissi, *Polymer,*21,1111 (1980)

/45/ C.I. Simionescu, C. Cascaval, V. Blascu, I.I. Negulescu, *Polym. Comm.,* 23, 1862 (1982)

/46/ J.C.W. Chien, F.E. Karasz, G.E. Wnek, *Nature,* 285, 390 (1980)

/47/ J.H. Edwards, W.J. Feast, *Polymer,* 21, 595 (1980)

/48/ J.H. Edwards, W.J. Feast, D.C. Bott, *Polymer,* 25, 359 (1984)

/49/ D. C. Bott, C.S. Brown, J.N. Winter, J. Baker, *Polymer,* 28,601 (1987)

/50/ D.C. Bott, C.S. Brown, C.K. Chai, N.S. Walker, W.J. Feast, P.J.S. Foot, P.D. Calvert, N.C. Billingham, R.H. Friend, *Synthetic Metals,* 14,245 (1986)

/51/ C. Riekel and K. Menke, *Mol. Cryst. Liq. Cryst.,* 105, 245 (1984)

/52/ T. Danno, K. Miyasaka, K. Ishikawa, *J. of Polym. Science,* 21, 1527 (1983)

/53/ C. Riekel, H.W. Hässlin, K. Menke, S. Roth, *Mol. Cryst. Liq. Cryst.* 117(1-4), 99 (1985)

/54/ K. Mortensen, M.L.W. Thewalt, Y. Tomkiewicz, T.C. Clarke, G.B. Street, *Phys. Rev. Lett.,* 45, 490 (1980)

/55/ E.K. Sichel, M. Knowles, M. Rubner, J. Georges, *Phys. Rev.*B25,5574 (1982)

/56/ T. Hashimoto, M. Itakura, N. Shimidzu, *J. Chem. Phys.*85(11), 6773(1986)

/57/ D. Schwahn, K. Mortensen, H. Yee-Madeira, *Proceedings of the Symposium F,* MRS Fall Meeting, Boston (1986)

/58/ Foundation Report (*The Red Book*) , European Synchrotron Radiation Facility, B.P. 220, F-38043 Grenoble Cedex (1987)

/59/ A. Luccio, unpublished

/60/ C. Nave, G.P. Diakun, J. Bordas, *Nucl. Instr. and Methods,* A246, 609 (1986)

/61/ A.C. Thompson, Y. Wu, J.H. Underwood, T.W. Barbee, *Nucl. Instr. and Meth. in Physics,* A255, 603 (1987)

/62/ W.H. Zachariasen, *Theory of X-Ray Diffraction in Crystals,* Dover Publ., N.Y.(1967)

/63/ J. Hastings, M. Lehmann and C. Riekel in Ref.58

Acknowledgement

Helpful discussions with J. Bordas, C. Williams, J. Hastings, M. Koch, A. Luccio and H.G. Zachmann are acknowledged.

DISCUSSION

POLYACETYLENE STRUCTURE

Time resolved X-ray diffraction shows only a transformation
from the cis-into the trans-modification. Further intermediate
steps, such as a change in packing of a given isomer, cannot,
however, be excluded. No evidence for such steps has been obtained
from other spectroscopic techniques, such as Raman and infrared.

In the Shirakawa polyacetylene foil samples the more loosely
packed surface layer was a few μm thick on a total sample thickness
of 100 μm.

SAMPLE HEATING

The heating problem noted above for small crystals could be
reduced by decreasing the X-ray wavelength from 1.5 Å to 0.5 Å,
but the difference this would make has not yet been calculated.

RADIATION DAMAGE

It is thought that polymers are more stable to radiation damage
than biological samples. It might be possible to raise the limit
of 10^{13} photons per second for biological samples to 10^{15} photons
per second for polymers, but it remains to be tested.

TIME-RESOLVED APPLICATIONS OF SYNCHROTRON RADIATION IN PROTEIN CRYSTALLOGRAPHY

H. D. Bartunik
Max-Planck Society
Research Unit for Structural Molecular Biology
2000 Hamburg 52
Federal Republic of Germany

ABSTRACT. Investigation of conformational changes in protein molecules during biological reactions by means of crystal structure analysis of intermediate states is feasible with time-resolved diffraction techniques using synchrotron radiation. Non-cyclic reactions may be studied on time scales of 100 msec to 100 sec; crystal cooling to subzero temperatures may be required for prolongating lifetimes and enhancing populations of intermediates. Submillisecond time resolution may be achieved in the case of cyclic reactions; utilization of the pulsed time structure of synchrotron radiation may eventually allow to reach even nanosecond time scales. The techniques and first applications are described.

1. INTRODUCTION

Many proteins, in particular most globular proteins, carry out their biological functions through dynamic changes in their three-dimensional structures. In enzymatic reactions, binding of the substrate may induce distortions in the structure of the substrate and conformational transitions in the enzyme molecule resulting in enhanced catalytic activity (see, e.g., Walsh, 1979). Another type of reactions like, e.g., ligand binding to myoglobin (Kuryian et al., 1986), involves diffusion of a small molecule into the interior of the protein which is accompanied by transient opening and closing of channels in the structure and reorientational motions of specific side chains. As a further example, light-energy transducing systems like bacterio-rhodopsin appear to undergo in their photocycles changes in their tertiary structures during transitions between short-lived excited states (Stoeckenius and Bogomolni, 1982).

  Study of such conformational changes requires determination of the three-dimensional structure with an accuracy of, at least, 0.1-0.2 Å in interatomic distances. A priori structural analysis of the entire folded conformation can at present only be obtained with crystallographic techniques; the required accuracy may be reached if diffraction

M. A. Carrondo and G. A. Jeffrey (eds.), Chemical Crystallography with Pulsed Neutrons and Synchrotron X-Rays, 487–507.
© *1988 by D. Reidel Publishing Company.*

data to high resolution are available. Certain local features of the structure may, on the basis of the known crystal structure, possibly be analyzed with higher accuracy by complementary techniques like NMR or EXAFS.

Conventional techniques, however, of macromolecular crystallography are not adequate for investigation of short-lived conformational states. New experimental techniques have been developed which are based on the use of synchrotron radiation (SR) from storage rings with high brilliancy in the hard X-ray range. Area detector (AD) systems which have recently become available reduce the total exposure time for high-resolution SR data collection (using monochromatic techniques) to a few minutes (Bartsch et al., 1986). Exposure times may be further reduced through application of real-time data handling and processing (Bartunik et al., 1986) and possibly of "white-beam" Laue diffraction techniques (Moffat et al., 1984). In combination with methods of cryoenzymology (Douzou, 1977), it is now feasible to investigate at atomic resolution "frozen-in" images of metastable states in conformational transitions.

Time-resolved determination of changes in the crystal structure, following the time course over a series of subsequent time channels on a μsec-to-msec time scale, is feasible in the case of reactions which may be cyclicly repeated and triggered by external stimulation (Bartunik, 1983). The time structure of SR may eventually be utilized in stroboscopic diffraction experiments with nanosecond time resolution. Such ultra-high time-resolution techniques have until now only been applied to small molecule structures (Larson et al., 1982; Pruss et al., 1984); their application to protein structures may become feasible on very-high intensity (wiggler or undulator) beamlines.

The present paper describes mainly the techniques. First applications which have been carried out demonstrate the feasibility of time-resolved studies. The expected further development of this new field in protein crystallography is discussed.

2. CONCEPTS AND TECHNIQUES

Crystal structure analysis of conformational changes is only feasible, if high population of intermediate states is achieved. For a priori structure determination, measurement of a more or less complete set of Bragg reflection intensities to high resolution is required. Data acquisition techniques have to be applied which allow to reach both short exposure times and high accuracy in the structure factors.

2.1. Population of intermediate states

The electron density distribution which may be derived from crystal structure analysis corresponds to an average over typically 10^{15} molecules in the crystal and over the total time period of data collection. Coexistence of structural substates corresponding to a discrete or continuous distribution in the conformation will lead to a superposition of electron density which may only interpreted unequivocally, if high occupancy of at least about 30% is achieved for a particular substate.

In addition to static disorder, dynamic (vibrational) disorder is present which arises from intra- and intermolecular motions occurring on time scales between picoseconds (for local modes) and some 100 nanoseconds (for reorientational motions of extended parts of the polypeptide chain). Collective intramolecular modes may be considered as transitions betwen energetically comparable substates (Frauenfelder et al., 1979). Such transitions will in the crystal, despite some coupling between intra- and intermolecular modes, in general occur out of phase. Static and dynamic contributions to the total disorder in the structural conformation will thus be indistinguishable in a coherent elastic diffraction experiment, except if they are investigated over a wide range in temperatures (Frauenfelder et al., 1979; Walter et al., 1984; Parak et al., 1987).

The population (or "occupancy") problem may be solved, if the system can be trapped in a discrete intermediate state which is energetically widely separated from neighbour states. This would reduce the total disorder to dynamic contributions arising from vibrations around the structural equilibrium of the intermediate. Relaxation to such an intermediate should occur on (nsec-) time scales characteristic for collective intramolecular modes. Trapping of intermediates in enzymatic reactions may possibly be realized by means of cryoenzymological techniques; this is discussed below.

2.1.1. External triggering of reactions. External stimulation, e.g., through heat jumps or photolysis may - in case of a unique pathway- provide a means of enhancing the population of intermediates.

Pumping into absorption bands in the optical and near-UV range, where dye lasers with high pulse energies are available, may be used to produce heat pulses provided that suitable labels can be bound to the protein. As an example, the acridin dye proflavin binds to serine proteases. Its absorption peak is located near 450 nm. A 1 Mol% concentration in a crystal with linear dimensions of 100 um corresponds to a peak absorption of about 0.5 OD; hence, homogeneous heating of the entire crystal volume should be feasible. Laser pulse heating may eventually be used in combination with cryotechniques for promoting a reaction in the protein crystal from one (immediately quenched) step to another in the reaction pathway.

Laser chemistry may be employed to initiate biological reactions in a number of proteins. One example is photodissociation of ligands bound to the prosthetic group in haem proteins which is followed by ligand rebinding; this has been applied in diffraction studies of carbonmonoxy myoglobin (see below).

In-situ activation of a reactant in the protein crystal may constitute a potentially powerful technique. ATP, Ca^{2+}, and a few other substrates or cofactors in enzymatic reactions of, e.g., kinases are available in "caged" form in which they are inactivated by complexing with photolabile groups. Such inactive precursors may be activated through laser pulse photolysis in the near UV (Kaplan et al.,1978; Walker et al., 1987).

2.1.2. Optical pumping of excited states.

Excited states in proteins which undergo a photocycle may be populated by optical pumping into broad pump bands followed by radiative and non-radiative transitions. An example is light-adapted bacteriorhodopsin (Stoeckenius and Bogomolni, 1982). Heterogeneities in population may result from branching of transitions.

2.2. Monitoring of reaction kinetics

Determination of the reaction kinetics in the protein crystal is important for defining experimental conditions under which high populations may be achieved during X-ray structure data collection. The reaction kinetics will in general be different in the crystal as compared to the protein in solution, due to differences in concentrations, solvent constitution, pH, and to effects of molecular packing in the crystal.

Reaction kinetics may be monitored by repeated remeasuring of structure factor amplitudes for a limited number of Bragg reflections. Since the structure of the protein molecule in its native or ground state conformation will in general be known in such studies, reflections may for this purpose be selected on the basis of model calculations.

A complimentary technique is in many cases provided by optical monitoring of absorption or emission spectra of prosthetic groups, reactants, or dye labels bound to the protein. Proflavin, as an example for a dye label, binds near the active site of serine proteases, and is removed upon binding of inhibitors or substrates to the enzyme in solution or in crystalline form (Tobias et al., 1987). A shift in the absorption peak wavelength allows to distinguish between free and bound proflavin.

2.3. Crystallographic concepts

Conformational transitions which are suitable for investigation by crystallographic techniques will in general involve small changes by not more than about 1% in the linear cell dimensions of the unit cell (otherwise, the lattice may get disordered or the crystal may even crack). Under such conditions, the phases of crystallographic structure factors corresponding to the native (ground state) structure of the protein will in general represent an acceptable approximation to the phases of the intermediate structure. Determination of structural changes then requires only measurement of structure factor amplitudes.

(Difference) Fourier techniques may be applied to obtain a priori structural information. At least about 20 % of all independent reflections which should be more or less evenly distributed in reciprocal space to a given resolution have to be included in the Fourier summation; an essentially complete data set, however, is desired in order to avoid artefacts and to reduce the noise level in the electron density maps. Model-dependent interpretations may be based on a much smaller number of reflections. For instance, the kinetics of substrate binding to an enzyme may possibly be followed in the crystal by recording the Bragg intensities of a few hundred reflections chosen on the basis of computer modelling.

An accuracy of 0.1-0.2 Å in interatomic distances can only be obtained if structure factors are determined to high Bragg resolution, i.e., to d-spacings below about 2 Å. For a number of small and medium-sized protein structures, crystals may be grown which diffract to at least 1.5 Å resolution. The use of SR has in many cases considerably extended the resolution range, to which diffraction data could be collected for a given structure; further, density maps calculated to the same nominal resolution show in general higher contrast with synchrotron data than with data from conventional X-ray sources. Such improvements are due to shorter exposure times and hence reduced effects of radiation damage, and to higher collimation leading to better signal-to-noise.

It is of particular interest in such studies to not only determine atomic coordinates but also mean squared amplitudes of fluctuations around these space (and time) averages. Information on local mobilities is contained in the crystallographic temperature factors (and in thermal diffuse scattering). Full description of time-averaged motions of atoms around their equilibrium positions through tensors is in general not feasible for proteins, because of the relatively limited number of observables as compared to the number of parameters. However, refinement of isotropic temperature factors for each atom may already provide valuable information (Frauenfelder et al., 1979; Artymiuk et al., 1979). Such refinement calculations again require diffraction data to high resolution and with a high degree of completeness.

2.3.1. <u>Monochromatic diffraction techniques</u>. Determination of structure factors requires measurement of integrated reflection intensities. Diffraction from a stationary protein crystal will in general only yield partially excited reflection intensities. The partiality depends on a number of parameters including divergences and wavelength spread in the incident beam and the mosaic distribution in the sample crystal. Standard data collection strategies involve stepwise screen-less rotation of the protein crystal in a monochromatic incident beam for integrating over the full rocking curve corresponding to the crystal's mosaic spread, and for exploring reciprocal space. Usually, wavelength bandwidths of ($\Delta\lambda / \lambda =$) 10^{-3} are employed at double-focussing SR beamlines (Bartunik et al., 1982); the angular full width of reflections is then typically about 0.1-0.2°. New monochromator crystals with broadened mosaic spread may increase the bandwidth - and hence both the angular width of reflections and the intensity incident on the sample - by one order of magnitude.

2.3.2. <u>Laue diffraction techniques</u>. The number of simultaneously excited reflections increases for a given protein crystal with increasing divergence or wavelength spread in the incident beam; this is the basis for Laue techniques of crystal data collection (Fig. 1).

492

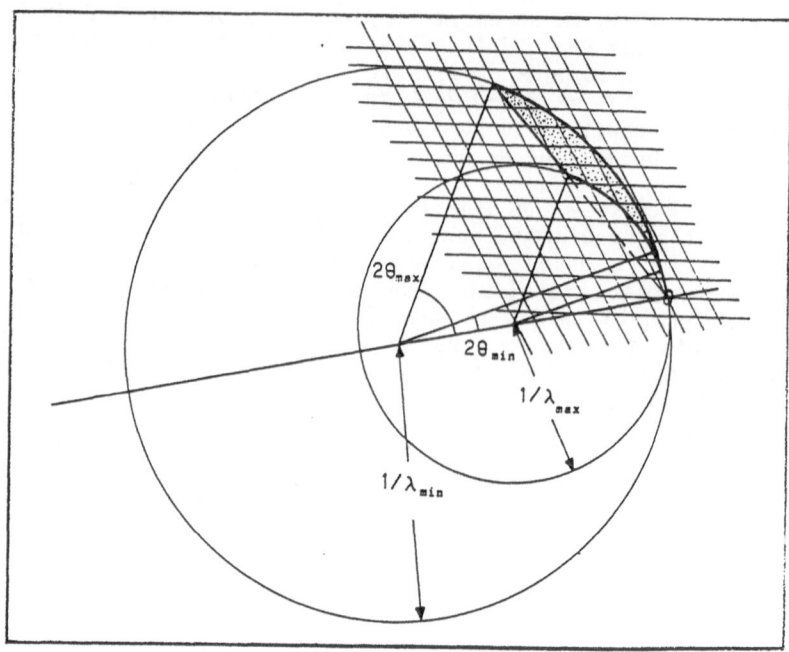

Fig. 1:   Ewald construction in two dimensions illustrating the prin-
          ciple of white-beam Laue diffraction techniques. Reciprocal
          lattice points contained in the shaded area between the Ewald
          spheres corresponding to the edges of the wavelength bandwidth
          give rise to simultaneously excited reflections.

The continuous spectrum of SR may be used in "white-beam" (broad-
bandpass) Laue photographs (Moffat et al., 1984; Bilderback et al.,
1984) which may yield in the order of 10% of all possible reflections
to high resolution in a single exposure taken from a stationary
crystal; even more simultaneous reflections may be (fully) excited in
the case of high crystal symmetry and choice of a favourable
orientation. Using integrating area detectors like conventional
photographic film or recently developed luminescent foils, such an
exposure can on a high-intensity beamline at a storage ring like CHESS
be made within less than 100 msec (Volz et al., 1986). Exposure times
could be further reduced on wiggler or undulator beamlines of new
storage rings which are in planning or under construction like the
ESRF, provided that the crystal may survive the heat input. It might
then eventually be possible to reach msec time resolution. Area
detector systems with on-line data handling capability could also be
applied in such Laue work; however, they are at present so strongly
limited in their maximum counting rates that the incident beam

intensity on existing bending magnet beamlines would have to be attenuated by at least two orders of magnitude.

Utilization of Laue techniques poses at the present moment still a number of problems due to the requirement to perform corrections for various wavelength-dependent effects in the observed reflection intensities; such systematic problems explain why Laue techniques, although known (mainly in neutron crystallography) since long time, have to date apparently not yet delivered accurate structure factors. Further problems may arise from wavelength or spatial overlaps which could affect 50% or more of all simultaneously observed reflection intensities (Hajdu et al., 1986), loss in signal-to-noise, and heat input into the crystal which may influence reaction rates. However, theoretical treatment of Laue diffraction geometry for the case of protein structures indicated that superposition of orders will mainly affect low and medium resolution reflections, whereas the bulk of outer shell reflections will be single Laue spots (J.Helliwell, personal communication). Development for processing Laue photographs is in progress. Radiation damage appears to be less harmful than expected (Moffat et al., 1986). Collection of a series of Laue exposures in order to obtain a complete data set will, of course, involve relatively long time scales defined by the time needed for rotating the crystal through (e.g., 20°) between exposures. In total, Laue techniques are potentially very attractive provided that systematic problems of data correction and scaling can be solved. They may be of particular interest in preliminary studies aiming to determine adequate experimental conditions, e.g., for trapping of intermediates, prior to data collection possibly employing other techniques.

The problems involved in quantitative evaluation of reflection intensities recorded with broad-bandpass Laue techniques may possibly be avoided with a "Pseudo-Laue" technique. In this technique, data are collected from a stationary sample crystal in a series of exposures which are taken as function of the incident wavelength varying through a broad wavelength range. Wavelength variation may e.g., be achieved with a rotating crystal monochromator. In this way, the total wavelength range is exactly defined, and monitoring of the momentary (wavelength-dependent) intensity incident on the crystal will provide a better basis for scaling. Optimum signal-to-noise is achieved, if (area detector) data are required in a series of short subsequent wavelength ranges. This will be feasible with real-time data handling systems like DACOM (see below). Such a procedure should allow to combine the advantages of both stationary-crystal Laue and monochromatic techniques. Fig. 2 shows a pseudo-Laue exposure which was taken from a stationary orthorhombic trypsin crystal (with cell dimensions between 55 and 68 Å) using a FAST area detector on a 1:1 imaging synchrotron beamline at HASYLAB/EMBL. During the exposure, the incident wavelength was continuously scanned through the range of 1.0 to 1.5 Å by rotating a Si(111) double crystal monochromator. The total exposure time was 10 seconds.

494

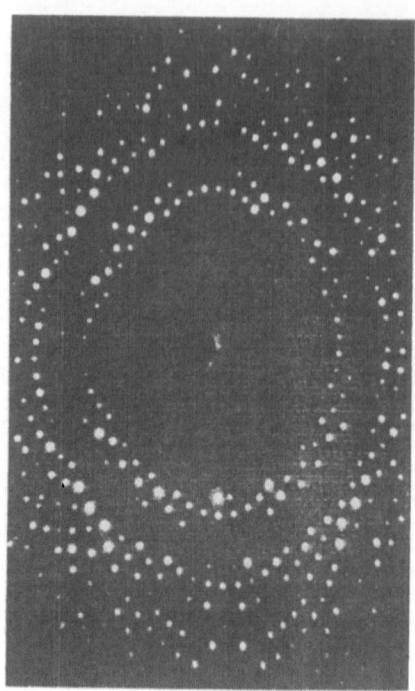

Fig. 2: Pseudo-Laue exposure taken with a FAST area detector from a stationary trypsin crystal during a wavelength scan from 1.0 to 1.5 Å. The total exposure time was 10 sec. The minimum d-spacing is 2.2 Å.

2.4. Low-temperature protein crystallography

Application of cryoenzymological techniques is of potential interest for slowing down reaction rates and for trapping the system in a transient state (Petsko et al., 1986). Assuming Arrhenius type behaviour, the rate constant, k, of an enzymatic reaction step with an activation energy of E will exponentially vary with temperature:

$$k = A \cdot \exp(-E_A / RT)$$

In the example of E = 15 kcal/mol and cooling from room temperature to -70°C, the rate constant will decrease by five orders of magnitude. The corresponding increase in lifetime, e.g., from a few milliseconds to a few minutes, may make crystallographic data collection to high resolution feasible, if SR and area detectors are applied (see below). Furthermore, the relative lifetime of intermediates states, corresponding to subsequent steps along the reaction pathway, will change as a function of their activation energies. This may lead to accumulation of the system in an intermediate state preceeding a rate-limiting step in the reaction. Cooling to subzero temperatures may in such cases, irrespective of the time scale of data collection, be necessary in order to obtain high enough population for crystallographic studies.

Low temperature studies which involve diffusion of reactants into the protein crystal require cryosolvents of low viscosity. Suitable cryosolvents have been developed and enzyme kinetics under cryo-conditions have been studied for a number of enzymes including serine proteases (Douzou et al., 1975; Petsko, 1975; Fink and Ahmed, 1976; Douzou, 1977; Fink and Petsko, 1981; Petsko et al., 1986). Attention has to be paid to possible changes in the reaction pathway and in enzymatic activity, e.g., due to competitive inhibition by an organic solvent. Instrumentation for crystal cooling to low temperatures is available (Bartunik and Schubert, 1982). Fig. 3 shows the scheme of a cold chamber which has been applied in SR protein data collection to very high resolution at temperatures down to 100°K.

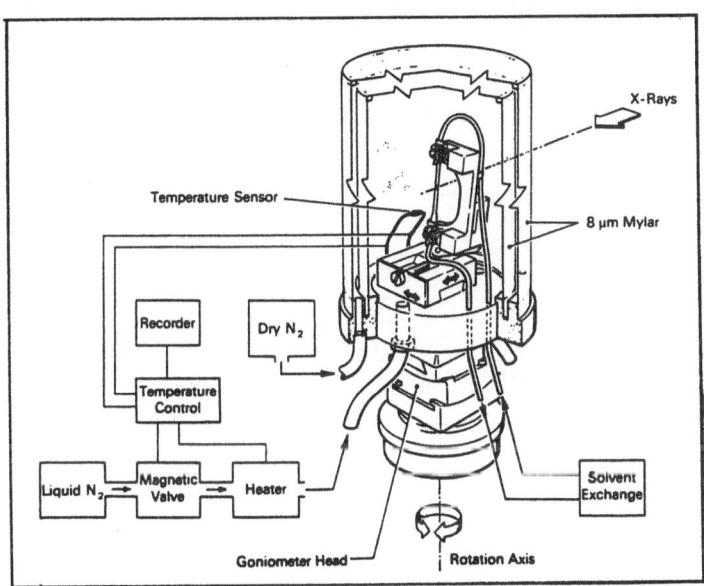

Fig. 3: Scheme of a cold chamber for high resolution data collection at temperatures T > 100°K. It incorporates a flow cell for solvent exchange.

In practice, the enzyme crystal is mounted in a Debye-Scherrer tube and fixed either through wedging (in a slightly conical tube) or with cotton linters. Silicon tubing is attached to form a flow cell. The solvent is replaced by a cryosolvent under stepwise increase in concentration and decrease in temperature. Then, diffusion of reactants into the crystal may start. If temperatures below about -50°C are employed, effects of radiation damage in the SR beam are strongly

496

reduced, and it will often be possible to collect complete data sets to high resolution from a single protein crystal.

As an alternative to the use of crysolvents in low-temperature work, shock-freezing of protein crystals in a liquid nitrogen or undercooled propane jet may be employed in order to quench the system in specific reaction steps. Rapid enough cooling will maintain the amorphous structure of the solvent, provided that the temperature is then not raised above ca. 180ºK. This is not easily achieved, if the crystal is mounted in a capillary tube. Too slow cooling will lead to formation of crystalline ice causing broadening of the mosaic spread and hence loss in Bragg resolution or even destruction of the protein crystal.

2.5. Time-resolved data acquisition using area detectors

The requirement to measure a great number of Bragg intensities within the short lifetime of an intermediate can only be met if simultaneously excited reflection intensities are recorded with an area detector (AD) system including facilities for rapid data handling and processing. Fig. 4 shows schematically the experimental set-up for such measurements on a double-focussing SR beamline.

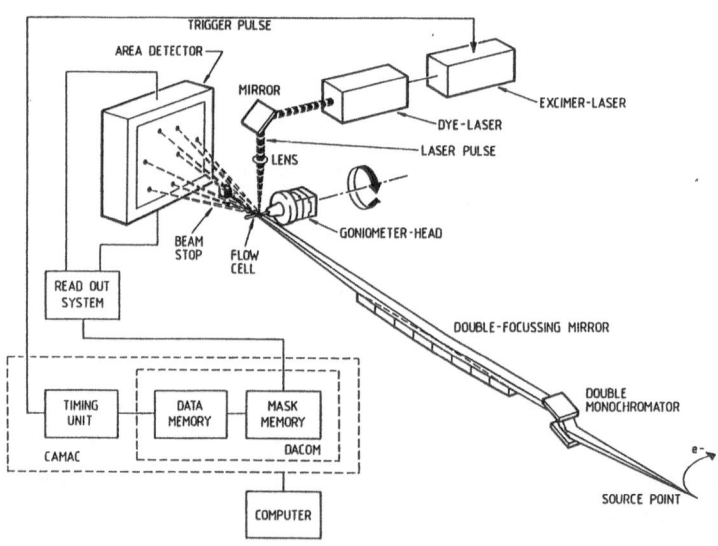

Fig. 4: Experimental set-up for time-resolved protein crystallography on a double-focussing SR beamline using an AD, time-paging data handling, and laser equipment for external stimulation of reactions in the crystal.

2.5.1. <u>Rapid data collection from proteins in metastable states</u>. With a scanning TV type AD (FAST), a full high-resolution data set can presently be recorded within a total exposure time of a few minutes. This has recently been demonstrated in trypsin data collection (in 1° rotation intervals covering in total 90°) on a bending magnet beamline at DORIS (Bartsch et al., 1986). The exposure time may, e.g., on a wiggler beamline at DORIS possibly be further reduced by one order of magnitude. The diffracted intensity incident on an AD will on the wiggler reach up to a few 10^5 photons/sec per (average) reflection and more than 10^7 in the entire pattern (assuming a wavelength near 1.0 Å and 0.1% bandwidth). Next-generation SR sources which are in planning will provide a further gain by a couple of orders of magnitudes in the primary intensity. The total measuring time is then more and more defined by the performance of the AD system.

At the present moment, fast-scanning TV camera systems (Arndt, 1984) offer the highest overall counting rate capability of all available ADs with on-line read-out of data. They are (with encoding into 8 bits) limited to maximum rates of about 10^5 a.u./sec per area of 5x5 pixels (corresponding to about 0.5x0.5 mm^2 which may be the size of a reflection spot; a.u.=amplitude units). Considering gas detector systems, multi-wire proportional chambers (MWPCs) can cope with maximum local incident rates of 10^4 photons/sec/mm^2; this limit might eventually be exceeded by one order of magnitude with parallel electrode devices which are being developed (Hendrix, 1986). The overall counting rate of a MWPC is limited to, at present, less than 1 MHz by the time needed for encoding of events.

Transfer of data onto disk or other backup media introduces severe limitations to the maximum speed of AD data collection. In the above-mentioned FAST data collection, the total time needed for dump of the original data onto disk (on a MicroVAX II) was about 10 minutes. However, data handling can be speeded up enormously by using hardware techniques for spot identification, integration, and background subtraction; a basis for such real-time techniques is provided by the precise predictability of (the positional parameters of) diffraction patterns (Bartunik and Boulin, 1986). A first system (DACOM) of this type has been developed for 256x256 resolution elements (pixels) and 1 MHz (stochastic) counting rates (Bartunik et al., 1986); it is at present further developed to be used with a scanning TV camera system (512x512 pixels, 10 MHz). DACOM compresses the total amount of data by a factor which corresponds - in the extreme case of hardware integration over entire mapped reflection spots - to the ratio between the number of pixels and the number of simultaneously excited reflections; this factor may amount to 10^3-10^4. As a consequence, it is feasible to store a full data set in a memory and to pre-process integrated data in real time, even in submillisecond time-resolved experiments. Mapping of reflection spots by hardware may eventually be combined with the use of high-speed array processors for more complex data reduction in pseudo-real time by software.

2.5.2. <u>Time-resolved data collection in subsequent time channels</u>. This is feasible in the case of cyclic reactions with a data handling system like DACOM (see above) providing a facility for time-paging. The minimum channel width is in such experiments defined by the intrinsic time resolution of the AD, and by the read-out and encoding times. The intrinsic time resolution of MWPCs is of the order of 100 nsec (corresponding to drifting of electrons over a distance of about 5 mm); events may at present be encoded within a few microseconds. Fast-scanning TV detectors have an intrinsic time resolution of 40 msec for read-out of one full image.

2.5.3. <u>Stroboscopic time-resolved data collection</u>. Time scales which are shorter than the intrinsic time resolution of the AD system may be reached in stroboscopic experiments with pulsed incident X-radiation. Use of a mechanical chopper will easily provide pulse lengths of 0.1-1 msec; the pulse repetition rate is defined by a second chopper. Such a device may, e.g., be applied in msec time-resolved data acquisition with an integrating AD like a TV camera system, if the intrinsic AD noise is tolerable despite an unfavourable duty cycle.

The pulsed time structure of synchrotron radiation provides a means for reaching subnanosecond time scales. The time width of an electron bunch at DORIS is of the order of 100 psec; the time interval between subsequent bunches may be varied from 2 nsec (in multi-bunch mode) to 960 nsec (in single-bunch mode). The master clock of the storage ring allows synchronization between stimulating (e.g., laser) pulse and probing X-ray pulse. A stroboscopic snapshot of transient conformational changes in proteins may be taken at a particular moment of a cyclic reaction by adjusting the time delay between both pulses. If the intrinsic time resolution of the detector is high enough, the system may in addition be probed by a series of SR pulses at constant time intervals corresponding to bunch-to-bunch distances.

3. APPLICATIONS

The techniques of time-resolved protein crystallography which have been described above can only be used up to their full power, if ADs are applied. More or less adequate AD systems have only very recently become available. Nevertheless, a number of experiments which have been carried out with photographic film or linear position-sensitive detectors demonstrated already the feasibility of time-resolved structural investigations. These experiments include studies of relatively long-lived metastable states in enzymatic reactions, and monitoring of the time course in ligand-protein interactions with sub-millisecond time resolution. Nanosecond time-resolved techniques have until now only been applied to small molecule structures.

3.1. Metastable states in enzyme-substrate interactions

The first, pioneering experiment applying cryoenzymological techniques for protein structure analysis at subzero temperatures was undertaken by Petsko and coworkers (Alber et al., 1976). In this study, the structure of an elastase-substrate complex forming an intermediate in the hydrolysis of N-carbobenyoxy-L-alanyl-p-nitrophenol ester was determined at a temperature of -55°C. However, the modest resolution of 3.5 Å which was reached in this experiment using a conventional source did not allow detailed analysis of the mechanism of serine protease catalysis in this reaction.

High resolution has been reached in recent SR experiments aiming to investigate enzyme-substrate bindingog. Using monochromatic techniques, a productive complex of trypsin with a slowly reacting substrate has been investigated at 1.8 Å resolution (Bartsch et al., 1987); adequate conditions for binding of the substrate were chosen on the basis of optical monitoring of the reaction in the single crystal (Tobias et al., 1987).

Louise Johnson and coworkers (Hadju et al., 1987) recently studied a complex of glycogen phoshorylase b with the substrate analogue heptenitol and its interconversion to heptulose-2-phosphate which inhibits continuation on the enzymatic reaction. The enzyme was activated by AMP. The turnover for heptenitol-arsenate is quite low. Since interconversion of the ternary enzyme-substrate complex is known to be the rate-limiting step in the kinetic mechanism, it was expected that transient accumulation of this complex could be observed before its conversion into product. In fact, difference density maps calculated at 2.5 to 3.0 Å resolution showed peaks in the vicinity of the catalytic site for the conversion which were attributed to either a ternary complex of enzyme-heptenitol-phosphate, or a mixture of two binary complexes enzyme-heptenitol and enzyme-heptulose-2-phosphate. These studies involved a number of experiments on time scales of one hour to one week for data collection using SR and monochromatic techniques. An application of broad-bandpass Laue techniques to a determination of the productive complex at 3 Å resolution is in progress (Hadju et al., 1986). Further SR studies of enzyme-substrate binding, involving cooling to subzero temperatures and the use of ADs, are underway in various laboratories.

3.2. Millisecond time-resolved studies

Laue techniques have found first test application in protein data collection using photographic film. As an example, a Laue diffraction pattern of lysozyme to high resolution has been taken at CHESS with an exposure time of 64 msec (Volz et al., 1986). Similar experiments have been carried out at the SRS Daresbury (Garner and Helliwell, 1986; Hajdu et al., 1986). Severe problems appear to still exist in quantitative evaluation of structure factor amplitudes from Laue exposures; the errors in scaling data and correcting for wavelength dependent effects appear to be substantial, in particular for broad-bandpass applications (see above).

Time resolution of 0.5 msec was reached in a study of ligand rebinding to carbonmonoxy-myoglobin (MbCO) following laser pulse photolysis of the CO ligand (Bartunik, 1983). The experimental set-up

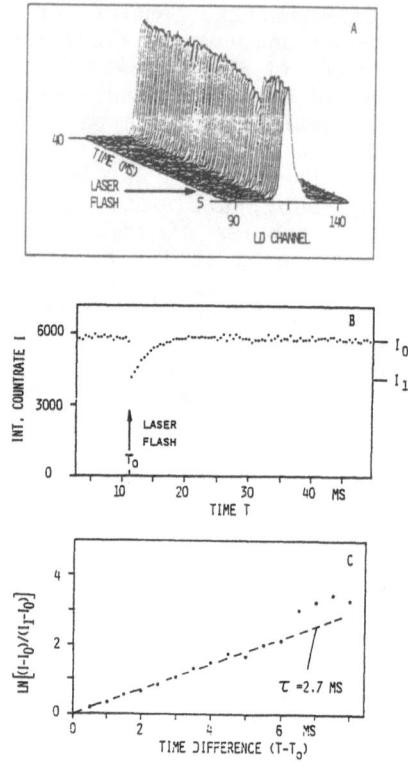

Fig. 5: Bragg reflection intensity from MbCO measured with 500 μsec time resolution before and after laser pulse photolysis of the ligand. Relaxation is exponential.

was very similar to the scheme depicted in Fig. 4, except for the use of a linear detector instead of an AD. The protein crystal was mounted in a flow cell. The time course in (up to a maximum of about five) simultaneously excited reflection intensities was followed in 100 time channels before photodissociation of the ligand by a 10 nsec laser pulse and during rebinding to the haem (Fig. 5). The kinetics of ligand rebinding were also investigated in separate time-resolved optical transmission experiments (Fig. 6). Both types of measurement yielded the same exponential time behaviour with a lifetime of 3-5 msec. This first msec time-resolved study demonstrated that homogeneous stimulation of a crystal volume (of about $20 \times 200 \times 200 \ \mu m^3$) which is suitable for SR experiments, and high population can be achieved for such a cyclic reaction under conditions of high-resolution X-ray data collection; it showed further that synchronization between the triggering laser pulse and the time paging data acquisition system can

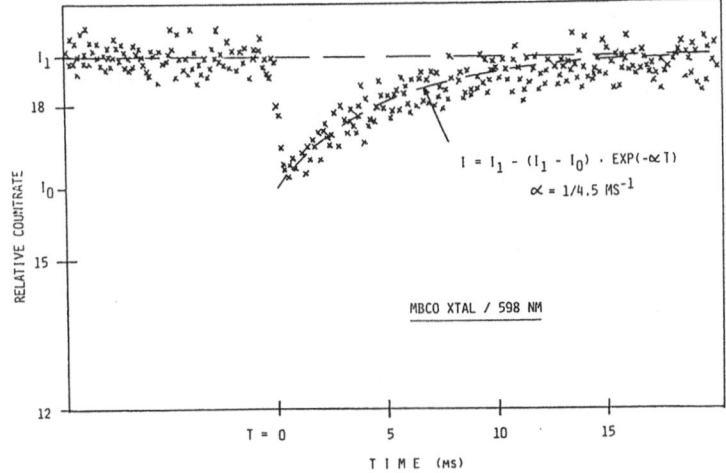

$$I = I_1 - (I_1 - I_0) \cdot EXP(-\alpha T)$$

$$\alpha = 1/4.5 \ MS^{-1}$$

MBCO XTAL / 598 NM

Fig. 6: Optical transmission at 598 nm through a single crystal of MbCO measured with 500 μsec time resolution before and after laser pulse photolysis.

be realized. The lifetime of the crystal was in these experiments essentially limited by effects of radiation damage in the SR beam, and not by effects of laser irradiation. It is planned to extend this study to the measurement of a full time-resolved data set with an AD at low temperatures (around 100°K) for possible investigation of the conformation during the lifetime of a transient state formed upon photodissociation.

3.3. Nanosecond time-resolved studies

The technique of stroboscopic diffraction data collection using the time structure of SR has until now only been applied to inorganic crystal structures for investigations of melting and recrystallization phenomena in silicon (Larson et al, 1982) and germanium (Larson et al, 1986) during pulsed laser annealing, and of structural changes in cerium pentaphosphate (CeP_5O_{14}) induced by laser pulse excitation of a short-lived ($T_{1/2}$ = 18.6 nsec) Ce^{3+} $4f^05d^1$ electronic state (Pruss et al., 1984; Bartunik, 1984). Fig. 7 shows the rocking curves of the (020) reflection of CeP_5O_{14} (monoclinic spacegroup $P2_1/c$) derived from stroboscopic measurements before the exciting laser pulse, at the moment of maximum population inversion during the lifetime of the excited state, and 80 nsec later. The cell dimension, b, contracts during the lifetime of the excited state, then expands due to the effect of heating, and finally relaxes (on a msec time scale).

502

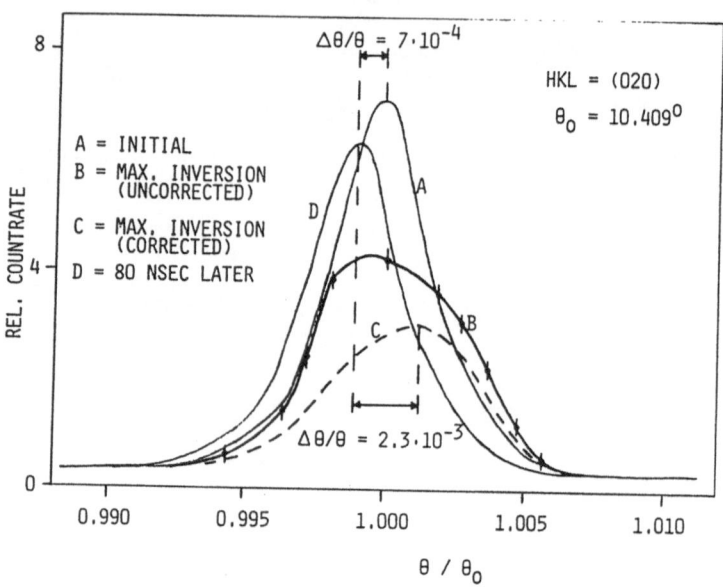

Fig. 7: Rocking curves of the (020) reflection of CeP_5O_{14} in the
ground state, in a short-lived excited state at the moment of
maximum inversion, and 80 nsec later.

In these studies, an overall time resolution of 15-20 nsec was
reached, corresponding to width and jitter of the laser pulse.
Stroboscopic measurements were made by varying, as described above, the
time delay between the laser and SR pulses. Stimulation of the
reactions was cyclicly repeated at frequencies of several Hz.
Diffracted or scattered X-ray photons were detected with plastic
scintillators and photo-multipliers with short rise times as compared
to interbunch distances.

4. CONCLUSIONS AND FUTURE DEVELOPMENTS

Investigation of conformational transitions in proteins during
biological reactions by means of crystal structure analysis of
intermediate states has become feasible through utilization of SR in
combination with ADs.
In the case of reactions which may not or not often enough be
cyclicly repeated, the time resolution which may be achieved with
presently available techniques is of the order of 100 sec. This refers
to collection of a complete high-resolution data set as a basis for a

priori structure determination, assuming that phases from a known native structure may be used. Partial data sets may possibly be measured with Laue techniques within less than 100 msec and used, at least, in preliminary studies for determining optimum experimental conditions; it is at the present moment not yet clear whether such methods will yield high enough accuracy in structure factors for a priori structural analysis. One has to take into account that the on average small size of changes in conformation which will occur makes high accuracy in the structure factors a crucial requirement. Independent of whether monochromatic or white-beam techniques are employed, it will in general be necessary to cool the protein crystal to subzero temperatures in order to increase the population of a specific intermediate; at the same time, the lifetime of such an intermediate will increase.

Cyclic reactions may be studied on much shorter time scales. It may be expected that the time range of microseconds to milliseconds will be of particular importance for investigating intermediate states which are populated, e.g., during the photocycle of proteins involved in photosynthesis. In this time range, crystallographic data may even be collected in a series of subsequent time channels. Eventually, nanosecond time resolution may be reached in stroboscopic measurements utilizing the pulsed time structure of SR. This may in principle allow to follow transient changes in the conformation on the time scale (of 10-100 nsec) of intramolecular collective modes, e.g., in protein-ligand interactions or redox reactions.

The feasibility of time-resolved protein crystallography has been demonstrated for time scales ranging from less than one millisecond to a few minutes. Nanosecond time-resolved techniques have been applied to small molecule structures.

Insertion devices on existing storage rings like DORIS deliver already high enough flux at the sample position for millisecond time-resolved work. SR sources with even higher brilliancy which are in planning or under construction will push the door open for sub-microsecond applications. In all time-resolved experiments, high stability in SR beam height and inclination will be of particular importance.

Rapid AD data acquisition systems have recently become available through the development of fast-scanning TV detectors. Within short time, also luminescent foil systems may be used as integrating ADs working in an essentially off-line mode. Due to the interest in on-line read-out and processing capabilities in time-resolved work, high-speed data handling systems for real-time integration and reduction will play an important role.

Despite the need for further development of methods and techniques, it may be expected that a number of applications which are presently underway in various laboratories will within short time, beyond mere feasibility tests, yield structural information on short-lived states in proteins.

REFERENCES

Alber, T., Petsko, G.A. and Tsernoglou, D. (1976). Nature (London) **263**, 297.
Arndt, U.W. (1984). Nucl. Instr. Meth. **222**, 252.
Artymiuk, P.J., Blake, C.C.F., Grace, D.E.P., Oatley, S.J., Phillips, D.C. and Sternberg, M.J.E. (1979). Nature (London) **280**, 563
Bartsch, H.H., Bartunik, H.D., Evers, U. and Tobias, J. (1987). Manuscript in preparation.
Bartsch, H.H., Bartunik, H.D., Evers, U., Prel-Rüffer, S. and Prigge, H.P. (1986). DESY-HASYLAB Ann. Rep., p. 341
Bartunik, H.D. and Schubert, P. (1982). J. Appl. Cryst. **15**, 227.
Bartunik, H.D., Fourme, R. and Phillips, J.C. (1982). In: Uses of Synchrotron Radiation in Biology, Ed. H.B. Stuhrmann, Acad. Press, London.
Bartunik, H.D. (1983). Nucl. Instr. Meth. **208**, 523.
Bartunik, H.D. (1984). Rev. Phys. Appl. **19**, 671.
Bartunik, H.D. and Boulin, C. (1986). In: Structural Biological Applications of X-Ray Absorption, Scattering and Diffraction, Scattering and Diffraction, Eds. H.D. Bartunik and B. Chance, Acad. Press, N.Y.
Bartunik, H.D., Boulin, C. and Schwab, H. (1986). J. de Physique **47**(8), C5-157.
Bilderback, D., Moffat, K. and Szebenyi, D. (1984). Nucl. Instr. Meth. **222**, 245.
Douzou, P., Hui Bon Hoa, G. and Petsko, G.A. (1975). J. Mol. Biol. **96**, 367.
Douzou, P. (1977). Cryobiochemistry, an Introduction, Acad. Press, NY.
Fink, A.L. and Ahmed, A.I. (1976). Nature (London) **263**, 297.
Frauenfelder, H., Petsko, G.A. and Tsernoglou, D. (1979). Nature (London) **280**, 558.
Garner, C.D. and Helliwell, J.R. (1986). Chem. in Britain **22**(9), 835.
Hajdu, J., Machin, P., Campbell, J.W., Clifton, I., Zurek, S., Gover, S. and Johnson, L.N. (1986a). Inform. Quarterly for Protein Cryst., Daresbury Lab., **17**, 17.
Hendrix, J. (1986). Same reference as Bartunik and Boulin (1986).
Kaplan, J.H., Forbush, B. and Hoffman, J.F. (1978). Biochem. **17**, 1929.
Kuryian, J., Wilz, S., Karplus, M. and Petsko, G.A. (1986). J. Mol. Biol. **192**, 133.
Larson, B.C., White, C.W., Noggle, T.S., Barhorst, J.F. and Mills, D.M. (1982). Phys. Rev. Lett. **48**, 337.
Larson, B.C., Tischler, J.Z. and Mills, D.M. (1986). J. Mat. Res. 1(1), 144.
Moffat, K., Szebeyi, D. and Bilderback, D. (1984). Science **223**, 1423.
Moffat, K., Bilderback, D., Schildkamp, W., Szebeyi, D. and Loane, R. (1986). Same reference as Bartunik and Boulin (1986).
Parak, F., Hartmann, H., Aumann, K.D., Reuscher, H., Rennekamp, G., Bartunik, H.D. and Steigemann, W. (1987). Eur. Biophys. J., in press.
Petsko, G.A. (1975). J. Mol. Biol. **96**, 381.

Petsko, G.A., Kuryian, J., Gilbert, W.A., Ringe, D. and Karplus, M. (1986). Same reference as Bartunik and Boulin (1986).

Pruss, D., Huber, G., Danielmeyer, H.G. and Bartunik, H.D. (1984). Acta Cryst. **A40**, Suppl., C-401.

Stoeckenius, W. and Bogomolni, R.A. (1982). Ann. Rev. Biochem. **52**, 587.

Tobias, J., Evers, U. and Bartunik, H.D. (1987). Manuscript in preparation.

Volz, K.W., Schildkamp, W. and Moffat, K. (1986). Engineering: Cornell Quarterly **20**(4), 39.

Walsh, C. (1979). Enzymatic Reaction Mechanisms, W.H. Freeman and Comp., San Francisco.

Walter, J., Steigemann, W., Singh, T.P., Bartunik, H.D., Bode, W. and Huber, R. (1982). Acta Cryst. **B38**, 1462.

DISCUSSION

CRYOPROTECTORS

Organic solvents such as methanol, which can act as a
competitive inhibitor, may well influence the enzyme system
to which they have been added as cryoprotectors. But these
drawbacks must be accepted if there is no other way of slowing
the reaction to within the time scale of the experiment.
There will always be a need to check the biological relevance
of reactions studied in this way, and in some cases it is
possible to show that the reaction pathways are essentially
the same as those at room temperature. Even at room temperature,
however, the enzyme may be less active in the crystal than in
solution where its environment is different. These problems are
important ones which should be addressed by further research.
When the technical difficulties of real-time synchrotron radiation
diffraction experiments are also remembered, it is clear that
much further work is needed in structural studies of enzyme
reactions.

DATA HANDLING

Larger and faster memory for data storage is not the answer,
because the synchrotron can only be used efficiently if the
experimenter can learn within hours or minutes, rather than weeks,
if the experiment has been successful or not and so make any
necessary adjustments before the next run. The important need is
thus for very rapid data reduction and evaluation with special
hardware and array processors. An objective now possible is the
production in minutes of a difference Fourier map of some critical
part of the structure. It will always be necessary to store data
for subsequent refinement, but the information kept will not be
the raw data as at present but intensities automatically integrated
and corrected for instrumental factors such as detector efficiency
and background subtraction.

In neutron diffraction, instrumental problems are similar, but with the complication that the collection of time of flight data requires a basic multi-channel analyser with about 10^7 channels. Neutron and synchrotron sites should be able to benefit from each others experience.

Flexibility in systems of data handling is also important at central sites where many different types of experiment have to be accommodated.

SMALL PROTEIN CRYSTALS

Dr Harding has found that protein crystals, which are small but the largest which could be grown, are rather poorly ordered. The diffracted intensities are less than expected, and the rocking curve may have a width of about 2 - 3 degrees compared with 0.1 degree for a good crystal. Such problems have not been prominent at Hamburg, so they may be specific to particular systems, and mostly affected by the molecular packing of the protein in the crystals.

ENZYME-SUBSTRATE INTERACTIONS AND MOLECULAR MODELLING

Theoretical molecular modelling of enzyme-substrate interaction by computer graphics is now an important adjunct to enzyme crystallography, but cannot replace it because the experimental results are by no means always predictable. Some substrates are found to be significantly distorted when bound to the enzyme. Care must be taken in data refinement to avoid constraints which would preclude the recognition of such distortions.

Molecular dynamics calculations should throw some light on how reactants and products make their way through the enzyme to and from the active site. Choice of system in which diffusion of products out of the enzyme is restricted may improve the chances of observation of reaction intermediates.

CARBON MONOXIDE-MYOGLOBIN BINDING

The photodissociation of carbon monoxide from the haem by the laser light had a quantum efficiency of unity. The thickness of the crystal was chosen to be 20 μm so that the laser light would penetrate the crystal. Comparison of the intensity change with that expected from the thickness of the crystal and separate experiments on the optical spectroscopy of the system showed that dissociation was uniform throughout the crystal.

At room temperature the carbon monoxide had sufficient energy to leave the haem pocket and travel some 10 Å into the solvent. The cycle of photodissociation, movement of the carbon monoxide in and out, and rebinding to the haem had a period of about 3 ms. At lower temperatures the carbon monoxide was unable to escape from the haem pocket and rebinding was fast.

CRYSTALLOGRAPHIC STUDIES OF BIOLOGICAL MACROMOLECULES USING
SYNCHROTRON RADIATION.

Peter F. Lindley.
Department of Crystallography,
Birkbeck College, (University of London),
Malet St.,
London, WC1E 7HX,
United Kingdom.

ABSTRACT.   A synchrotron provides a source of high intensity,
wavelength tuneable, highly collimated X-radiation which can be
utilised in structural studies of biological macromolecules in a
number of ways.   Recent progress in the following areas is discussed;
(a) high resolution data collection for weakly diffracting single
crystals of biological macromolecules, (b) time-resolved structural
studies on enzyme-substrate binding in the crystalline state, (c)
anomalous dispersion techniques to derive relative phase information
for structure analysis, (d) extended X-ray absorption fine structure
analysis, EXAFS, in order to examine the precise environment of metal
binding sites in metallo-proteins, and (e) structural and
transformational studies on DNA fibres.

INTRODUCTION.

For the protein crystallographer, a synchrotron provides a source of
high intensity, highly collimated, wavelength tuneable X-radiation,
(see *e.g.* Helliwell, 1984).   The routine use of such sources has
brought about significant technical advances with respect to
structural studies on large biological macromolecules.   These studies
are complicated by the inherently weak diffracting power of the
biological samples, the large volumes of diffraction data which need
to be recorded and processed, the sensitivity of the samples to
radiation damage, and the 'on-going' problem of deriving relative
phase information for each reflection in order to synthesise an image
of the macromolecule.   Recently crystallographers have been able to
exploit the properties of synchrotron radiation to surmount some of
these problems and advances have been made in many areas.   The
following list is not meant to be comprehensive, but gives some idea
of recent developements which have been used to obtain structural
information at high resolution;

1.   High resolution X-ray diffraction data collection for weakly
      diffracting crystals of large biological macromolecules.

2.   Time-resolved structural studies on enzyme-substrate binding in
      the crystalline state.

M. A. Carrondo and G. A. Jeffrey (eds.), Chemical Crystallography with Pulsed Neutrons and Synchrotron X-Rays, 509–536.
© 1988 by D. Reidel Publishing Company.

3. Anomalous dispersion measurements, utilising the wavelength
    tuneability, in order to derive phase information.

4. Extended X-ray absorption fine structure, EXAFS, measurements
    for studying the precise environment of metal atoms in
    metallo-proteins.

5. Diffraction studies on DNA fibres.

6. Ultra-fast diffraction data collection from single crystals
    using the 'white-beam' Laue technique.

7. Diffraction studies on very small single crystals.

Items 6 and 7 will not be discussed in the present paper, (see
Harding, M.J., *this volume*).

1. HIGH RESOLUTION DATA COLLECTION.

The total energy, E_{hkl}, diffracted by an ideally mosaic crystal
rotated at a constant velocity, ω, through the diffracting position is
given by the expression:

$$E_{hkl} = K \cdot \frac{1}{\omega} \cdot I_o \cdot \lambda^3 \cdot ALP \cdot \frac{V_{cry}}{V^2_{cell}} \cdot |F_{hkl}|^2$$

where : $K = [e^2/mc^2]^2$

: I_o is the intensity of an incident beam of wavelength, λ.

: A is an absorption factor for the crystal specimen, the
  capillary tube that contains it, and the mother liquor, (or
  equivalent), which keeps the crystal wet.

: L is the Lorentz factor, dependent on the time that the
  crystal spends in the diffracting position and inversely
  proportional to λ.

: P is a polarisation factor dependent on the state of
  polarisation of the incident beam.   For synchrotron
  radiation the beam is predominantly polarised in the
  plane of the synchrotron ring.

: V_{cell} is the unit cell volume in a crystal of volume, V_{cry}.

and : $|F_{hkl}|^2$ is the diffracting power of the set of planes, (*hkl*),
       in the crystal lattice.

The equation readily shows us that large unit cells, large V_{cell}, coupled with weak diffracting power, small $|F_{hkl}|$, typical of biological macromolecules, will reduce E_{hkl}, but that the high incident beam intensity, high I_o, available from a synchrotron source, will increase the total energy diffracted. We may also notice that the wavelength, λ, can be readily varied with a synchrotron source and it is often convenient to reduce λ, (which also reduces E_{hkl}), in order to reduce systematic errors in the corrections for absorption, (absorption corrections are normally measured empirically for biological specimens due to the presence of mother liquor around the crystal and the capillary tube in which the crystal is mounted).

Radiation damage is also an important factor. Such damage is thought to occur mainly by the free radicals, (solvated electrons), formed by irradiation, diffusing through the crystal lattice causing chemical denaturation and subsequent loss of crystallinity. Its effect on the diffraction pattern is not entirely predictable, although in many cases radiation damage initially causes loss of the high resolution data and eventually the entire crystal ceases to diffract. The higher the value of I_o, the more free radicals will be formed, but if data collection is very rapid, it is often possible to record the high resolution reflections before diffusion of the free radicals causes significant crystal damage. That is, the rate of free radical diffusion appears more important than the overall number of free radicals produced in this context. The wavelength used, particularly if it is close to an absorption edge, may also determine the number and rate of free radical production.

In addition, the lower beam divergence and improved collimation that can be attained with a synchrotron source leads to a significant improvement in the signal to noise ratio and this is particularly important for the relatively weak, high resolution diffraction data. In general it is possible to obtain more diffraction data per crystal specimen using a synchrotron source than with a conventional rotating anode source; sometimes this is purely in the number of diffraction patterns, (i.e. photographic films), that can be recorded for a given resolution, but often an increase in resolution per pattern is also achieved.

Example 1. Serum Transferrin, (from rabbit).

Serum transferrin, the central protein of iron metabolism, is involved in the redistribution of iron between sites of absorption, storage, and haem degradation and utilisation. It has a molecular weight of $\approx$ 80Kd, some 650 amino acid residues, and crystallises in space group $P4_32_12$, $Z = 8$, with cell dimensions $a = b = 127.4Å$, $c = 145.4Å$; the solvent content is $\approx$ 68% by volume. The protein is bilobal and contains two iron binding sites, one per lobe. Crystals of rabbit serum transferrin are weakly diffracting because of the high molecular weight and solvent content, but yield some 15° - 20° of data, (1° oscillation steps with exposure times of 100 sec. per step), on beam-line 7.2 on the SRS at Daresbury compared to only 6° - 8°, (approximately 5½ hours per step), on a rotating anode source, (wavelengths of 1.488Å and 1.542Å respectively). However, the maximum resolution practically obtainable is around 3.3Å for both cases, Figure 1.1(a).

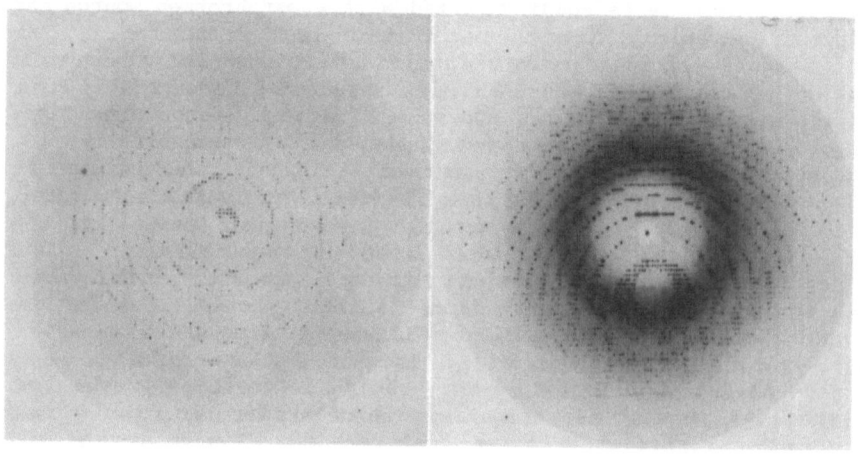

(a)   (b)

Figure 1.1. Typical oscillation patterns from protein crystals taken with synchrotron radiation.

(a) a gold chloride derivative of rabbit serum transferrin, taken on beam-line 7.2 at the SRS, Daresbury, λ = 1.488Å, film to crystal distance = 80 mm., osillation range = 1°, resolution ≃ 3.3Å.

(b) native bovine γ-II crystallin, taken on beam-line X11 at the EMBL Outstation, Hamburg, λ = 0.86Å, film to crystal distance = 80 mm., osillation range = 2°, resolution = 1.5Å. The film was 'over-exposed' to highlight the high resolution data; the black ring near the film centre is caused by scattering by the quartz glass capillary.

The collection of a complete data set involved some 7 crystals compared to about twice that number for a mercuric chloride derivative collected on a rotating anode generator. The use of the synchrotron source thus minimises problems of reproducing heavy atom soaking conditions and scaling of data sets from different crystals. A set of data for an uranyl acetate derivative was also collected at the SRS and these three derivatives have led to a structure analysis at 3.3Å resolution using the methods of multiple isomorphous replacement in conjunction with solvent flattening, (Wang, 1985). Figures 1.2 (a) and (b) show typical electron density in the N-terminal region and the overall topology of the complete molecule respectively.

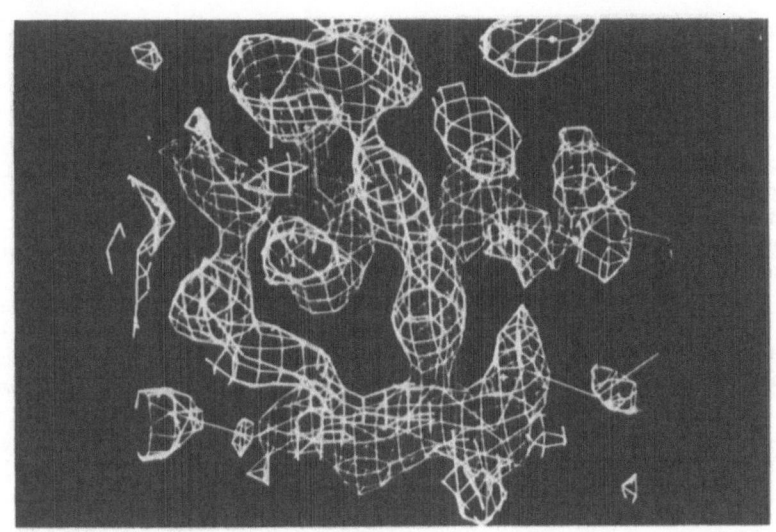

Figure 1.2(a). Rabbit serum transferrin. Electron density at 3.3Å resolution near the N-terminus showing a disulphide bridge, (centre of plate), between residue 20, part of an α-helix, and residue 37 on a β-strand running parallel to the helix; these elements of secondary structure form part of a βαβ subunit.

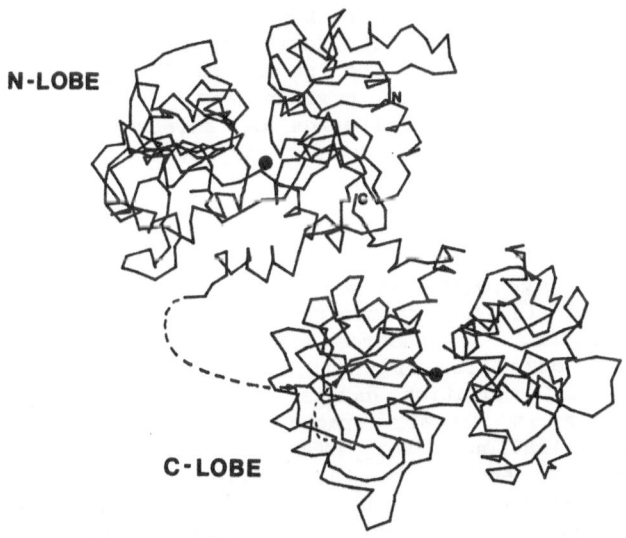

Figure 1.2(b). Rabbit serum transferrin; the C_{α} polypeptide chain through the bilobal molecule showing the two iron binding sites, one per lobe. The dotted line denotes part of the inter-lobe connecting peptide, which is not well-defined at the present stage of the structure analysis.

Example 2. γ-II crystallin, an eye lens protein, (from calf).

γ-II Crystallin, (Summers et al., 1984), crystallises in space group P4₁2₁2, Z = 8, with unit cell dimensions a = b = 57.8Å, c = 98.7Å. Using a conventional copper radiation source and a four-circle diffractometer intensity data were measured to 1.9Å resolution, although precession photographs suggested that the crystals diffracted to higher resolution. At the EMBL Outstation at DESY, Hamburg, using the synchrotron ring DORIS, it was possible to measure high quality data to a resolution of 1.5Å. The wavelength used was 0.86Å, minimising systematic errors due to absorption, and the low divergence and good collimation of the incident beam not only improved the signal to noise ratio for the relatively weak high resolution data but, in addition, because the crystals grow as elongated prisms, allowed the crystal specimen to be translated in the beam periodically to minimise the effects of radiation damage. Figure 1.1(b) is a 2° oscillation photograph, representative of the overall data set. In total some 120,000 reflections were measured of which ≈26,000 are unique; the internal consistency index for the symmetry related reflections was ≈8% on intensities.

Good quality high resolution diffraction data allow molecular models to be obtained at almost atomic resolution, even for proteins of the size of γ-II crystallin, (molecular weight ≈ 20Kd, 174 amino acid residues). Typical regions of electron density in the refined high resolution structure are given in Figure 1.3. Figure 1.4 indicates the overall topology of the γ-II crystallin molecule; a remarkably symmetrical structure consisting of four very similar 'Greek key' motifs organised into four anti-parallel β sheets. These detailed structural studies, which include the organisation of surface ion pairs, networks of aromatic, arginine and sulphur containing amino acid residues, and the distribution of sulphydryl residues, have enabled proposals to be formulated for the role of the protein in the healthy and cataractous lens.

Finally all the advantages of a synchrotron radiation source are especially needed in the recording of diffraction patterns from crystals of viruses. Exposures times can still be fairly long, several minutes, and often only one photograph can be obtained per crystal because of radiation damage. The very large unit cell parameters result in closely spaced diffraction maxima, so that incident beam collimation is very important. The natural collimation of the synchrotron source in the vertical direction is already provided; for example, for the SRS at Daresbury, the vertical divergence is about 0.25 mrad. Collimation in the horizontal direction can be achieved by placing suitable slits before the focussing monochromator.

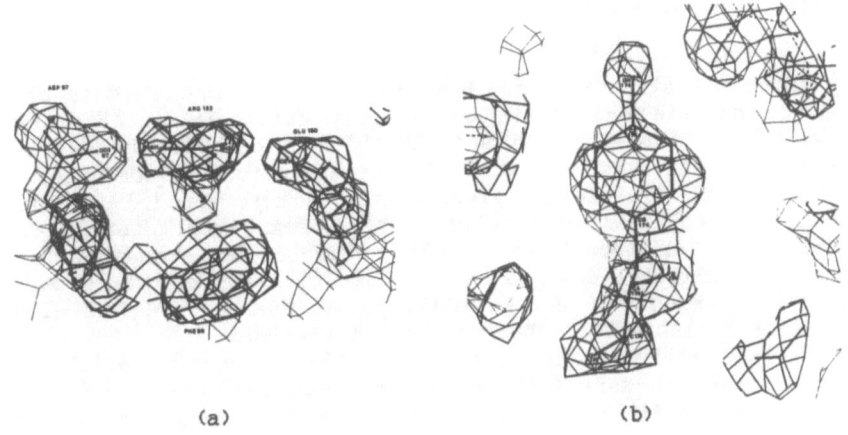

(a)   (b)

Figure 1.3.   Typical regions of electron density in the high
               resolution, 1.5Å, structure of bovine γ-II crystallin.

(a) Surface ion pairs involving Asp 97, Arg 152 and Glu 150; these
three residues enclose the internal hydrophobic Phe 98.

(b) The C-terminal Tyr 174 residue; the hydroxyl group is hydrogen
bonded to the main chain carbonyl group of residue 116 in a symmetry
related molecule.   In previous analyses electron density for Tyr 174
was very poorly defined.

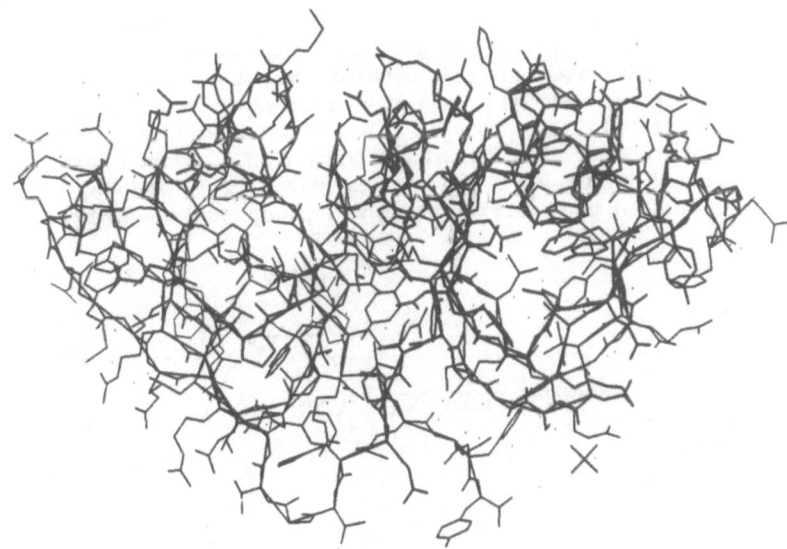

Figure 1.4.   The overall topology of the γ-II crystallin molecule; the
               dots represent the positions of water molecules.

2. TIME-RESOLVED CRYSTAL STRUCTURAL STUDIES.

X-ray data collection with conventional sources requires days or even weeks to complete and it is exceedingly difficult to obtain time-resolved three dimensional structural information about dynamic events in crystals.   Although cryogenic techniques allow, in some cases, the observation of transient intermediates, there are disadvantages, (apart from technical difficulties), and some uncertainties in the interpretation of then data.   For example, low temperatures may induce conformational changes and the use of cryoprotectants may also effect the geometry of active sites in enzymes as well as altering their accessibilities to substrates and inhibitors.

However, with the high intensity X-ray sources available at a synchrotron and in particular using the wiggler beam-line 9.6 at the SRS, Daresbury, data collection times can be dramatically reduced absolving the need for low temperatures.   Dr. Louise Johnson and her colleagues, (Hajdu et al., 1986 & 1987), have been able to carry out kinetic studies on the enzyme phosphorylase b using the SRS.

Glycogen phosphorylase catalyses the first step in the breakdown of glycogen and is the key enzyme through which hormonal, nervous and metabolite signals are relayed to meet energy requirements of the muscle cell.   The mechanism by whcih the enzyme uses its essential cofactor, pyridoxal-1-phosphate, in this reaction has long been the subject of debate.

Recent X-ray measurements, together with ^{31}P NMR measurements and reconstitution studies with the cofactor and substrate analogues, have indicated that the substrate phosphate and cofactor phosphate are in close proximity.   Typical of the crystallographic experiments is the inter-conversion of a substrate analogue, heptenitol, into heptulose-2-phosphate in the presence of inorganic phosphate.   The product is the terminating step of the reaction and heptulose-2-phosphate is a potent inhibitor of the enzyme; its structure may resemble the transition state.   The reaction, Figure 2.1, takes place in the crystal, where molecules of the enzyme are active, and can be studied by diffusion of the metabolites into the phosphorylase crystals.

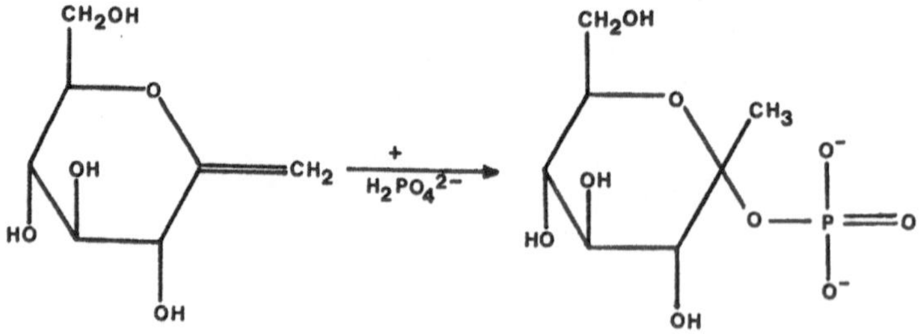

Figure 2.1. The conversion of heptenitol into heptulose-2-phosphate catalysed by the enzyme phosphorylase b.

In the kinetic experiments a crystal of the enzyme was mounted in a flow cell and the reaction started by allowing the substrates to diffuse into the crystal. Using fast data collection methods it was possible to trap the enzyme at an early stage in the reaction when there had been little turnover and to follow the accumulation of the product as the reaction proceded. X-ray diffraction data to 3.0Å or better were collected on an Arndt-Wonacott oscillation camera. In the best experiments some 100,000 reflections, (26,000 independent reflections), were recorded in 45 minutes and this compares with a time of about a week using a conventional laboratory rotating anode source. Interestingly, it was found that not only was the exposure time radically reduced, but significantly less radiation damage was observed which resulted in improved high resolution diffraction data. Table 2.1 shows typical soaking conditions used where (1) and (3) represent the initial and final reaction states and data were collected on a rotating anode generator and beam-line 7.2 at the SRS respectively, and (2) is one of several intermediate stages that were studied and where data were collected on the wiggler beam-line 9.6 at Daresbury.

Table 2.1. Phosphorylase crystal soaking experiments.

No.	Addition	Soaking Time	Temp.	Time for data collection	λ Å	Res. Å	N_{hkl} total
1.	100mM Heptenitol	18h.	25°C	1 week	1.542	3.0	69,500
2.	100mM Heptenitol 50mM P_i 2.5mM AMP	15min.	20°C	45min.	0.88	2.5	102,000
3.	As above	50h.	23°C	2.5h.	1.488	3.0	19,300

The reaction course was followed by computing difference Fourier syntheses using coefficients, $||F_{P+S}| - |F_P||$, where P+S represents the native protein plus substrate, and with native protein phases.

These electron density difference syntheses clearly indicated the sequence of events during the reaction and, for the first time, showed the incoming inorganic phosphate, P_i, above the double bond of the heptenitol molecule, Figure 2.2. The inorganic phosphate was also found to be in close contact with the phosphate moiety of the cofactor pyridoxal-1-phosphate on the enzyme as well as other key amino acid residues.

In the above series a different enzyme crystal was needed for each experiment whereas it would be extremely useful to follow the inter conversion from start to finish using the same crystal.

518

Indeed, the redevelopment of the white radiation Laue technique at the Daresbury Laboratory, (Machin, 1985), has made this possible for this particular enzyme.   Using the complete spectrum of the SRS wiggler beam-line, between 0.45Å and 2.5Å, a Laue photograph can be recorded in about ¼ sec, Figure 2.3.   Such a photograph contains some 21,000 measurable reflections, (neither spatially nor wavelength overlapped), which reduce to a unique set of 5000 reflections at a resolution of 3.0Å and corresponding to some 30% of a complete data set at this resolution.   Further analysis of this Laue data is in progress but, interestingly, the photographs show that on diffusion of either inorganic phosphate or glucose-1-phosphate, there is an immediate disorder of the crystal lattice which appears to heal after some 5-10 minutes.   It is likely that this transient disorder is caused by conformational changes in the enzyme in reponse to the substrates rather than osmotic shock, since no disorder is observed when the inhibitor glucose is diffused into the crystal or when phosphate is added to a crystal which has been incubated with glucose.

Clearly, for this particular enzyme system, the Laue method has great potential for allowing the recording of time-resolved structural information.

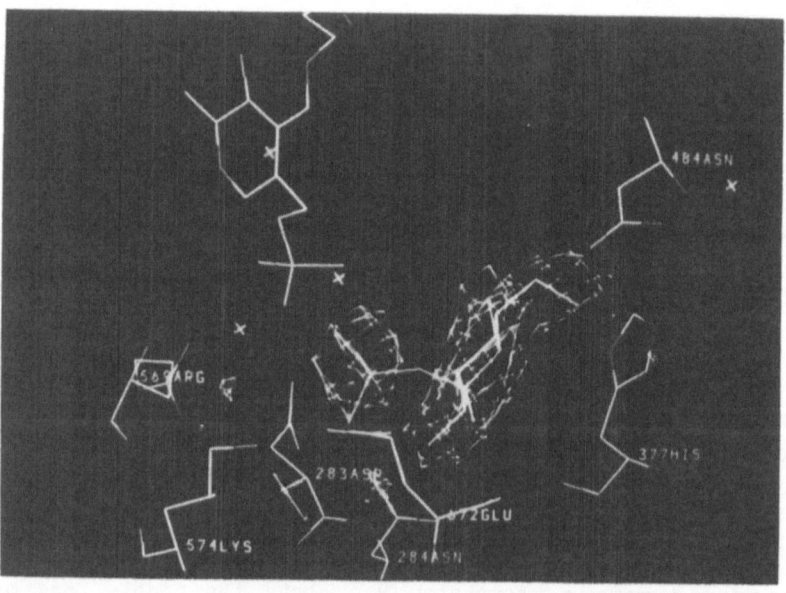

Figure 2.2(a).   The conversion of heptenitol to heptulose-2-phosphate corresponding to experiment (2) in Table 2.1.   The inorganic phosphate can be seen above the double bond of the heptenitol molecule and in close proximity to the phosphate moiety of the cofactor, pyridoxal-1-phosphate.

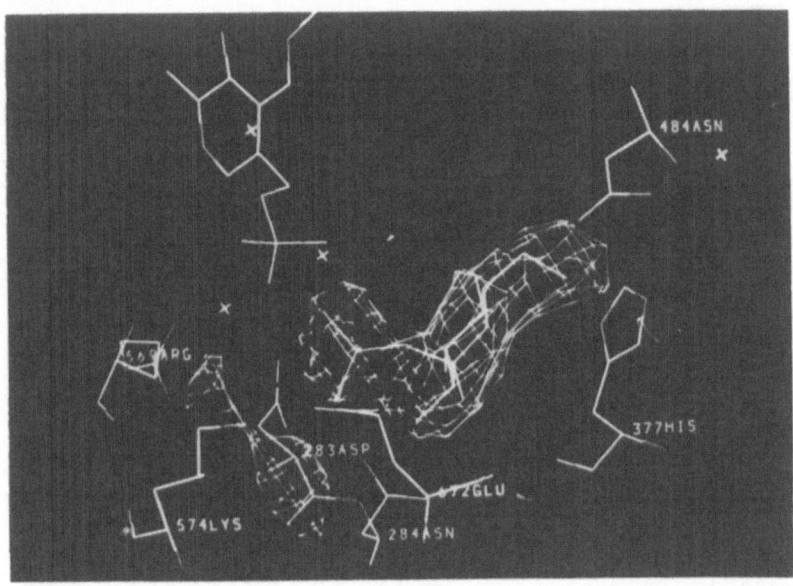

Figure 2.2(b). The conversion of heptenitol to heptulose-2-phosphate corresponding to experiment (3) in Table 2.1. Electron density for the product is clearly seen. In addition certain important amino acid residues near the active site appear to have changed position, for example, ASP 283 in the figure. However, there is no evidence of movement for the cofactor.

Figure 2.3. A typical Laue photograph of phosphorylase b. Exposure time approximately ¼ sec on beam-line 9.7 at the SRS, Daresbury.

3. THE TUNEABILITY OF THE SYNCHROTRON SOURCE : ANOMALOUS DISPERSION TECHNIQUES.

In order to construct an image of a diffracting object, not only are the amplitudes of the individual beams required, but also their relative phases.   In the X-ray case these latter quantities cannot normally be measured directly and this constitutes the classic PHASE PROBLEM in crystal structure analysis.

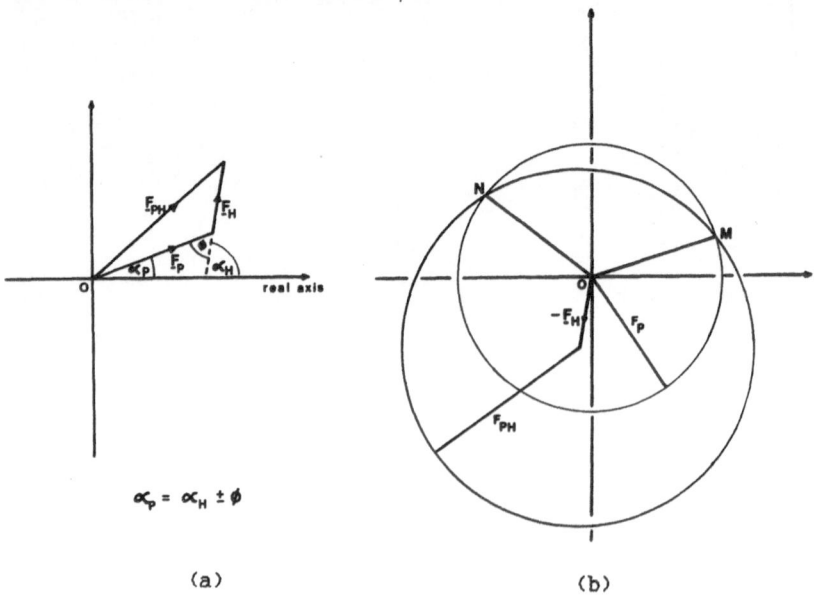

$$\alpha_P = \alpha_H \pm \phi$$

(a)   (b)

Figure 3.1   Isomorphous Replacement.

In protein crystallography, as indicated in Figure 3.1(a), the solution to this problem usually involves the preparation of heavy atom derivatives by soaking heavy atom reagents into single crystals of the native protein via the solvent channels.   The aim is to bind the heavy, (strongly diffracting), atoms to specific amino acid residues on or near the surface of the protein molecules in a regular manner throughout the crystal without disturbing the molecular conformation or the molecular packing.   In Figure 3.1(a), F_P, the native protein structure amplitude for the set of crystal planes, (hkl), and F_{PH}, the corresponding amplitude for the heavy atom derivative, are the quantities that are measured experimentally; α_P is the relative phase angle that is required to be determined.   The isomorphous difference, $||F_{PH}| - |F_P||$, can be used, for example via Patterson syntheses, to locate the heavy atom positions and from these $|F_H|$ and α_H can be calculated.   The angle, ϕ, may then be determined by the cosine rule on the vector triangle formed by F_P, F_{PH} and F_H. Then $\alpha_P = \alpha_H \pm \phi$.   Figure 3.1(b) indicates the nature of the phase ambiguity in the case of a single isomorphous derivative, that is, is E_H placed in phase advance or retardation with respect to E_P ?

This dilemma is normally solved by preparing a second heavy atom derivative which will also yield two solutions of α_P for each set of crystal planes, but one of which should coincide with one of the solutions from the first derivative. However, there are often problems such as lack of isomorphism for one or both derivatives, errors in the measurement of $|F_P|$ and $|F_{PH}|$, difficulties in estimating E_H, particularly for multi-sited or low occupancy heavy atom substitution, and often several derivatives are required for effective phasing. This means, of course, a considerable increase in the volume of diffraction data, in the time required for data collection and processing and, not least, in the supply of single crystals and protein required.

As a consequence, protein crystallographers have made prolonged attempts to use anomalous dispersion effects to yield phase information from only one heavy atom derivative or, for suitable metallo-proteins, from the native protein alone. The atomic scattering factor of an atom, f , is a measure of the amplitude of electromagnetic radiation scattered by that atom when radiation of a given amplitude falls upon it. This 'normal' scattering is disrupted if the frequency of the incident radiation is close to the natural resonance frequency of an absorption edge in that atom. In such circumstances the atom is said to scatter anomalously and the normal scattering factor, f_o, has to be modified as;

$$f = f_o + f' + if''$$

where f' is a dispersion component,

and f'' is a component which has a phase advance of $\pi/2$ with respect to f_o and is therefore out of phase with the incident radiation : f'' is therefore known as an absorption component.

The presence of a significant f'' component from an anomalously scattering atom leads to a breakdown of the Freidel law, i.e. the loss of a centre of symmetry from the overall diffraction pattern. The Argand diagram in Figure 3.2(a) shows the origin of the anomalous difference;

$$\Delta_{ano} = |F_{PH}+| - |F_{PH}-|$$

where the + and - signs refer to diffraction from the (hkl) and (-h-k-l) planes respectively. In the case of a metallo-protein, E_{PH}, E_H and E_P refer to the diffraction amplitiudes for the native protein, heavy atom and native protein minus heavy atom components respectively. In the absence of a significant f'' contribution,

522

the two amplitudes will differ in value giving rise to differences in the reflection intensities which can be determined experimentally.

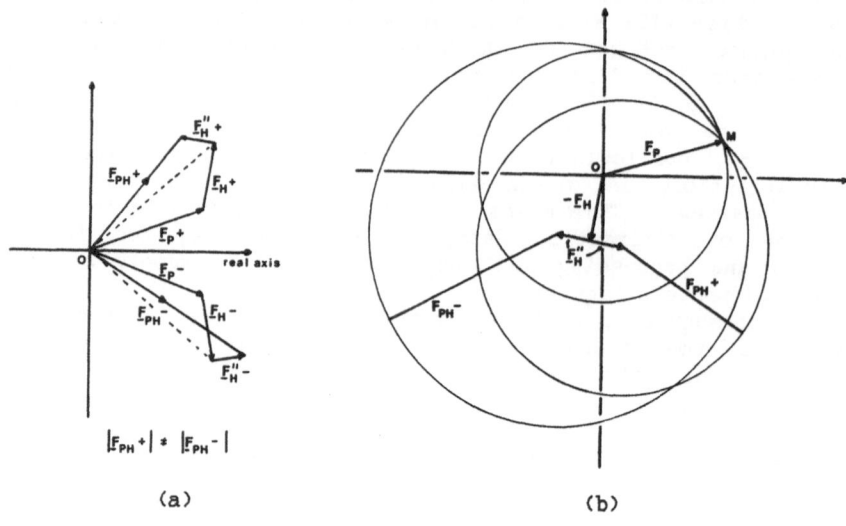

(a)   (b)

Figure 3.2.   Single Isomorphous Replacement with Anomalous Scattering.

The Harker construction in Figure 3.2(b) indicates, as implied by Peerdeman and Bijvoet, (1956), that the phase ambiguity can be resolved from a single isomorphous derivative.   Estimates of the accuracy of the intensity measurements required can be obtained using the following expressions, (Crick and Magdoff, 1956), *viz*:

$$\langle |F_{PH}|^2 - |F_P|^2 \rangle \simeq [2/N]^{1/2} . F_H / f_o \ \ldots \ldots \text{ Isomorphous differences.}$$

and $$\langle |F_{PH}+|^2 - |F_{PH}-|^2 \rangle \simeq 2[2/N]^{1/2} . F_H'' / f_o \ \ldots \text{ Anomalous differences.}$$

where : N is the number of light atoms in the structure,
     : f_o is the scattering power of a typical light atom at $\sin\theta/\lambda = 0$, *i.e.* $f_o = 7$ electrons,
     : F_H is the scattering power of the heavy atom at $\sin\theta/\lambda = 0$,
and   : F_H'' is the anomalous scattering power of the heavy atom.

For a heavy atom derivative of bovine γ-II crystallin containing one fully occupied mercury site, $N \simeq 1740$, $F_H = 80$ electrons, and $F_H'' \simeq 9$ electrons for Cu K_α radiation, the average isomorphous intensity difference calculates as 39% compared with an average anomalous intensity difference of only 9%.   The corresponding values for an equivalent heavy atom derivative of transferrin, $N \simeq 6500$, are 20% and 5% respectively.   These figures are maximum values and, in practice, the intensity differences will be considerably smaller, (for example, the occupancies may be less than unity).   However, they indicate that

accurate intensity measurements are required if significant use is to
be made of the anomalous scattering effect.

The coefficients, $|\Delta_{ano}|^2$, can be used in a Patterson synthesis
which, since Δ_{ano} contains contributions predominantly from the heavy
atoms, should show maxima corresponding to the 'heavy atom to heavy
atom' vectors.   Alternatively, if some estimates of the protein
phases are available, a Fourier synthesis computed using Δ_{ano} as
coefficients and phases, $[\alpha_P - \pi/2]$, should reveal the heavy atom
sites directly.   Figure 3.3 shows the confirmation of the iron atom
positions in rabbit serum transferrin using the latter technique.

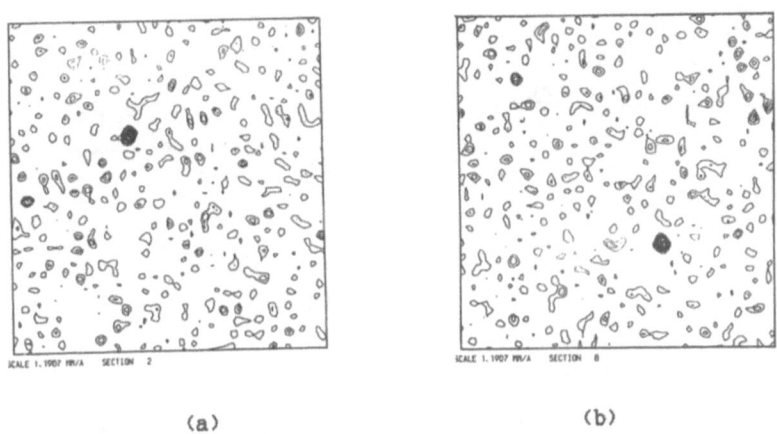

(a)   (b)

Figure 3.3. The location of the iron atoms in transferrin; (a) and (b)
are sections through the N- and C-terminal iron atom positions
respectively from an electron density synthesis computed using, Δ_{ano},
as coefficients and phases, $[\alpha_P - \pi/2]$.   The Δ_{ano} were derived from
intensity data collected photographically on an oscillation camera
with Cu K_α radiation.

The electrons giving rise to anomalous scattering are tightly
bound and close to the atom nucleus and hence the anomalous scattering
is much less dependent on the value of $\sin\theta/\lambda$ than the normal
scattering of the electrons as a whole.   However, it should be
stressed, and can easily be seen from Figure 3.4, that anomalous
dispersion information and isomorphous data are complementary. Δ_{ano}
is largest when F_P and F_H are almost at right angles to one another,
in this case the isomorphous difference, $|F_{PH}| - |F_P|$, is a minimum.
Alternatively, Δ_{ano} is zero if F_{PH} and F_P are collinear, and when Δ_{iso}
is a maximum.

The use of the anomalous effect with carefully selected
reflections is commonplace in conventional multiple isomorphous
replacement phasing, (North, 1965; Matthews, 1966; Hendrickson, 1979),
but the optimisation of any anomalous scattering must bring potential

benefits to a crystal structure analysis. Figure 3.5 indicates a typical variation of the anomalous scattering components f' and f'' near an absorption edge, in this case the Fe K edge in ferritin, (Sawyer, 1986). It can be seen that the value of f'', 7-8 electrons, at the edge is roughly twice that at the Cu K_α wavelength, $\lambda = 1.542\text{Å}$, $[(\lambda-\lambda_K)/\lambda_K = -0.115]$.

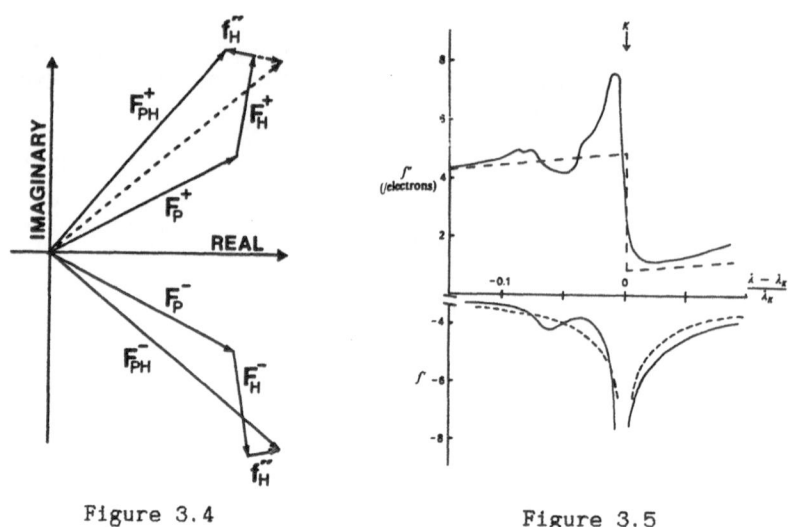

Figure 3.4   Figure 3.5

Figure 3.4.   The complementarity of the isomorphous and anomalous difference information.

Figure 3.5.   A typical variation of f' and f'' with λ, $[K = 1.743\text{Å}]$. The data have been derived, Sawyer (1986), from the work of Stuhrmann (1980) on the iron storage protein, ferritin; the dashed line is a theoretical estimation from Hoppe & Jakubowski (1975).

Table 3.1 lists metallic elements of interest in macromolecular crystal structure analysis and whose absorption edges can be readily accessed by the SRS at Daresbury, (Helliwell et al., 1984).
       Data collection in close proximity to an absorption edge where f'' is a maximum requires considerable technical expertise and care, (e.g. ensuring a suitable band-pass at the appropriate wavelength and measurement of the anomalous component at that wavelength), but will generally lead to optimised anomalous scattering information. Relatively little work has been published, as yet, in which full optimisation of anomalous scattering has played a significant role in the *ab initio* structure analyses of a biological macromolecules, (see Sawyer, 1986, *and references therein*; Karle, J., *this volume*).

Table 3.1   Absorption edges, Å, of interest with respect to proteins.

| | K Edge | | L Edges | | |
			L(I)	L(II)	L(III)
		U	0.570	0.592	0.722
Mo	0.620	Pb	0.782	0.816	0.951
Zn	1.283	Hg	0.835	0.878	1.009
Cu	1.380	Au	0.864	0.903	1.040
Ni	1.488	Pt	0.894	0.935	1.073
Co	1.608	Os	0.956	1.001	1.141
Fe	1.743	Re	0.990	1.037	1.177
Mn	1.896	Sm	1.601	1.694	1.845
		I	2.390	2.553	2.719

The metallic elements whose K edges are listed are those commonly found in metallo-proteins; those whose L edges are given are commonly used in the preparation of heavy atom derivatives.

However, a method of phasing, which has only become practicable with the advent of synchrotron radiation and which is increasingly being used, is the multiple wavelength technique.   If several data sets are collected at different wavelengths, chosen so that they yield optimal changes in both f' and f'', (Templeton et al., 1980; Narayan and Ramaseshan, 1981), the effect is similar to that of having several isomorphous derivatives, but without the associated hazards of lack of isomorphism; in suitable cases it is even possible to collect all the data from the same crystal specimen.   Figure 3.6 indicates the principal of the technique and full accounts of it are described by, for example, Phillips & Hodgson, (1980) and Hendrickson et al., (1985).   Because the absorption factor is wavelength dependent, it is important when using the method, that adequate corrections are made for absorption.
        A recent example of the application of the technique is by Harada et al., (1986), to the structure of cyctochrome c' from Rhodospirillum rubrum and which contains two haem iron moieties.   The X-ray intensity data were collected, to a resolution of 6.0Å,   on a four-circle diffractometer with three wavelengths, λ_1 = 1.077Å, λ_2 = 1.730Å and λ_3 = 1.757Å, using synchrotron radiation produced by the storage ring in the Photon Factory, National Laboratory for High Energy Physics, Japan. λ_2 and λ_3 straddle the Fe K absorption edge whereas λ_1 is far removed from the edge, (f' minimised); for λ_2 f'' is partially, but not fully, optimised.   The positions of the iron atoms could be determined from difference Patterson syntheses computed with coefficients, $||F\lambda_1| - |F\lambda_2||^2$ or $||F\lambda_1| - |F\lambda_3||^2$, (in each case simulating a heavy atom derivative), but a Patterson synthesis computed with the anomalous differences corresponding to λ_2 was unconvincing.   The best phase angles for the crystal were calculated from the X-ray intensity data measured with λ_1 and λ_2 and although these phase angles differed on the average by some 75° from those

obtained by multiple isomorphous replacement, the molecular boundary and several α-helices in the structure could be recognised in an electron density map, Figure 3.7.

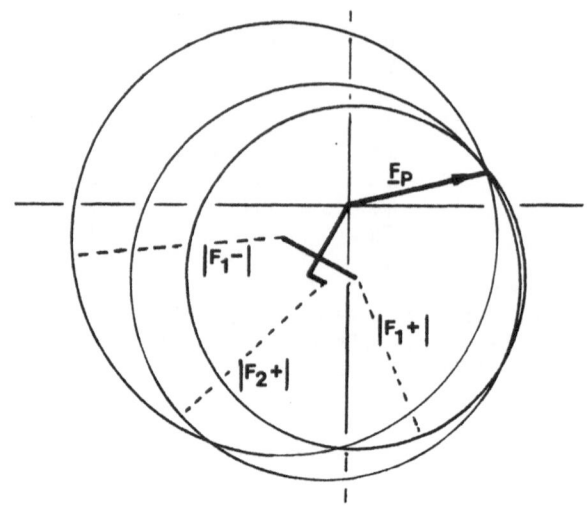

Figure 3.6. The simplest case of the multiple wavelength method. One Freidel pair of reflections is measured at the optimised wavelength for f'', λ(1), and another measurement is made at a different wavelength, λ(2), for which f' is small ; a unique phase solution is clearly possible.

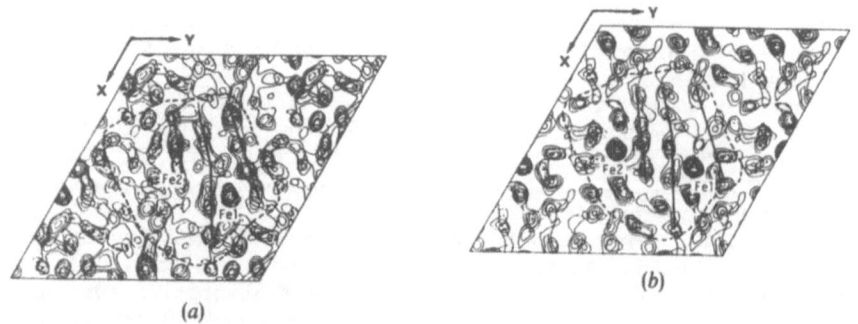

(a)   (b)

Figure 3.7. Composite electron density maps, [z = 0.10(0.02)0.16], around the two haem iron atoms in cytochrome c', (Harada *et al.*, 1986), calculated by (a) MIR, and (b) the multiple wavelength anomalous diffraction method. A dimeric molecule is surrounded by a dotted line in each map. Rods of electron density indicated by heavy lines correspond to α-helices.

4. EXTENDED X-RAY ABSORPTION FINE STRUCTURE ANALYSIS.

The structural detail which can be obtained using single crystal X-ray diffraction techniques is often limited for biological macromolecules by the resolution obtainable from the diffraction pattern. In the case of rabbit plasma transferrin, for example, the resolution is about 3.0Å, (Gorinsky et al., 1979). Although exposure times using the SRS at Daresbury are greatly reduced, (100 sec. against 5½ hr. for a conventional rotating anode source), and radiation damage also reduced, the high solvent content of the crystals, approximately 68% by volume, severely restricts the resolution obtainable. Other transferrins, human lactoferrin, (Anderson et al., 1987), and chicken ovotransferrin, (Abola et al.,1981), diffract to higher resolution, 2.3Å and 2.5Å respectively, but this is still rather poor compared to many small organic molecules where often the theoretical limit, $\lambda/2$, can be reached. In the transferrin molecule, there are two iron-binding sites which although similar are, on the basis of spectroscopic evidence, subtly different. Even with good refinement procedures it may not be possible to model small differences in the two iron environments from the X-ray diffraction data alone. What is required is a complementary technique and this is readily provided on a tuneable wavelength synchrotron source in the form of EXAFS.

X-ray absorption spectroscopy can be used to study the local environment of metal atoms, within a radius of 3 to 5Å, in biological molecules. It is not possible to obtain structural data for the whole protein, but the technique is relatively quick and easy to perform; it does not require the preparation of single crystals and most spectra are recorded from concentrated solutions of the metallo-protein. However, given favourable circumstances, metal-ligand distances can be determined with greater accuracy, often by an order of magnitude, compared to protein crystallographic studies. Eisenberger & Kincaid, (1978), Cramer & Hodgson, (1979), and Hasnain, (1986), have written useful reviews on the application of EXAFS to metallo-biological materials.

Figure 4.1(a) shows a typical absorption spectrum of a metallo-protein. The spectrum contains an absorption edge region consisting of discrete absorption bands originating from excitation of an electron from an inner shell of the metal to higher energy shells which are superimposed on another sharply rising band resulting from the excitation of an electron to the continuum. The intensities and positions of these transitions can provide information about the coordination of the metal and its oxidation state. The shape of the absorption edge and the spectrum extending to about 100eV, above the edge, termed the XANES region, (X-ray absorption near-edge structure), contain additional information about the metal site geometry and symmetry. The high energy side of the edge displays the oscillating modulations and constitutes the EXAFS region. The origin of these oscillations lies in the back scattering of some of the outgoing photoelectron wave by the atoms surrounding the metal. The interference between the outgoing wave and the several back scattered

528

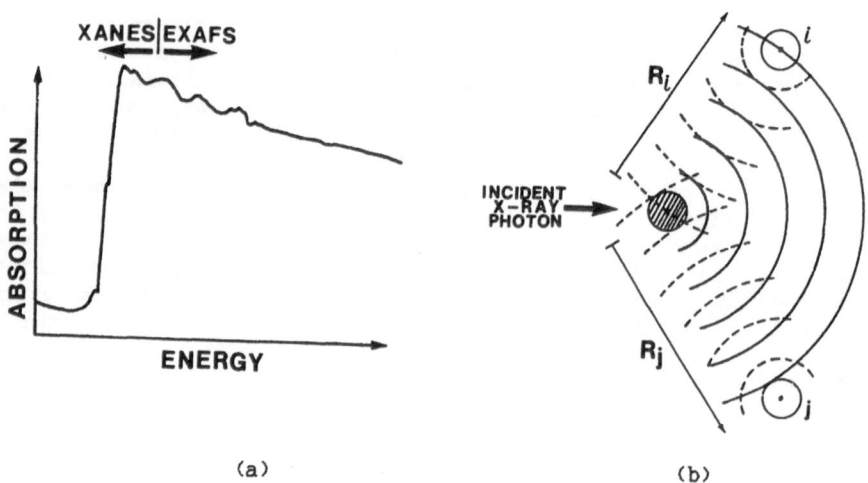

(a)   (b)

Figure 4.1.   The origin of EXAFS.

waves results in the EXAFS spectrum.   Figure 4.1(b) shows two atoms,
i and j, at distances R_i and R_j, from the absorbing metal atom.   For
a given photoelectron wavelength, (determined by the energy of the
incident X-ray photon), the back scattered wave from i interferes
constructively and that from j destructively with the outgoing wave.
The observed absorption is dependent on the sum of all the back
scattered waves from all the atoms which are situated in the vicinity
of the metal atom.   As the photoelectron wavelength is smoothly
varied by scanning the incident X-radiation, each back scattering atom
contributes, (to a first approximation), a damped sine wave to the
EXAFS spectrum.   Using this concept, Sayers *et al.*, (1971), were able
to develope the expression:

$$\chi(k) = \frac{1}{k} \cdot \Sigma_j \frac{N_j}{R_j^2} \cdot S_j(k) \cdot \sin[2kR_j + \delta_j(k)] \cdot \exp(-2\sigma_j^2 k^2) \cdot \exp(-\gamma R_j)$$

where : $\chi(k)$ is the EXAFS and k the photoelectron wave vector.

   : N_j is the number of atoms of type j at a distance R_j with
     back scattering amplitude $S_j(k)$.

   : the first exponential is a Debye-Waller term accounting for
     static and thermal disorder in R_j.

   : the second exponential accounts for the decay of the
     photoelectron.

: $\delta_j(k)$ is a phase shift correction term.

The expression assumes that the back scattering atoms are sufficiently small for the curvature of the spherical wave to be neglected, and that only single scattering events contribute to the EXAFS. If the phase shift correction term can be computed theoretically for all the back scattering atoms or derived from model compounds, then it is possible to extract from the experimental data the numbers, types and distances of the atoms surrounding the absorbing metal atom. This analysis is facilitated by performing a Fourier transform on the experimental data and this approximates to a radial density function about the metal atom after correction for the phase shift term.

Recent developments in the theoretical background to EXAFS have enabled the 'plane wave' approximation to be replaced by a curved wave treatment, (Lee & Pendry, 1975; Gurman et al., 1984), and allowance can now be made for multiple scattering events. This latter development is of particular importance with respect to the interpretation of the XANES region of the spectrum.

The crystal structures of at least two members of the transferrin family of iron metabolism proteins have recently been elucidated, viz. human lactoferrin at a resolution of 3.2Å, (Anderson et al., 1987), and rabbit serum transferrin at 3.3Å resolution, (Bailey et al. 1987). The two iron binding sites appear to have similar local geometry at this resolution and the amino acid ligands involved are two tyrosine, one histidine and one aspartic acid residue at each site; the current molecular model also allows for a (bi)carbonate anion, essential for iron binding, in a bridging position between the iron atom and a nearby arginine residue although this is not, as yet, completely resolved. These results compare with recent EXAFS measurements on chicken ovotransferrin, (Garratt et al., 1986) which indicate that the probable iron environment involves two low-Z ligands, (consistent with phenolate linkages), at 1.85(1)Å and four low-Z ligands at a slightly longer distance of 2.04(1)Å and that the two iron-binding sites have slightly different geometries. In this context a low-Z ligand implies oxygen or nitrogen between which EXAFS is not able to make a distinction. At first sight it would appear that the EXAFS could be predicting an extra ligand compared to the crystal structure analyses, but further EXAFS measurements on freeze-dried samples of the ovotransferrin, (Hasnain et al., 1987), indicate the presence of a labile water molecule at, or near, the iron-binding sites, Figure 4.2. EPR measurements provide confirmatory evidence of these EXAFS results; Figure 4.3 shows the changes in the EPR spectra of chicken ovotransferrin as a freeze-dried sample is exposed to a humid atmosphere. At the relatively low resolution of 3.2Å it is unlikely that such a water molecule would be revealed by the crystal structure analysis, nor indeed, small differences in the geometries of the two iron binding sites, but these studies usefully indicate the complementarity of the two techniques.

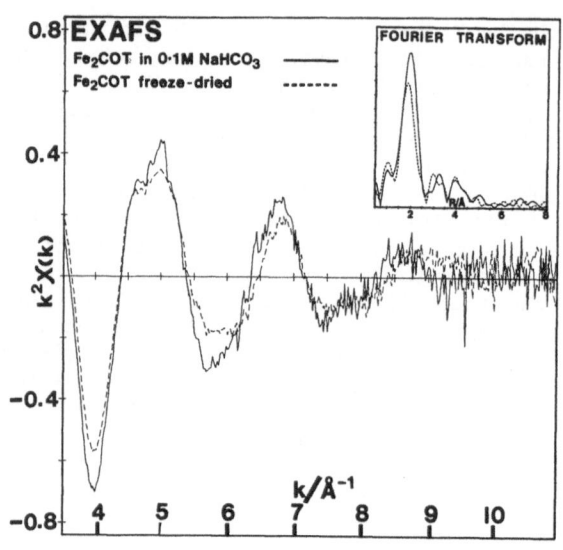

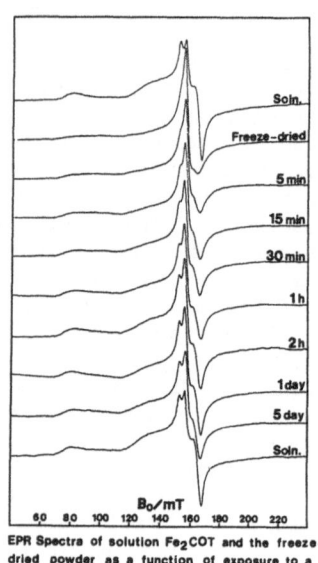

EPR Spectra of solution Fe₂COT and the freeze-dried powder as a function of exposure to a humid atmosphere

Figure 4.2   Figure 4.3

Figure 4.2. EXAFS spectra of a solution of diferric chicken ovotransferrin and a freeze-dried sample. Differences in the spectra can be interpreted in terms of a water molecule at, or near, each of the iron binding sites. Such molecules may play an important role in the physiological uptake and release of iron.

Figure 4.3. EPR spectra of chicken ovotransferrin showing the reversion from the freeze-dried to the solution spectrum on exposure of the freeze-dried sample to a humid atmosphere.

Future work with the transferrins will obviously involve detailed interpretation of the EXAFS spectra, including the XANES region, using the geometrical information available from the X-ray crystal structure studies, i.e. in a refinement mode.

Finally in this section, it should be pointed out that for biological problems such as those presented by the transferrins, the use of a synchrotron radiation source is a prerequisite for both the collection of good quality high resolution X-ray data and for the EXAFS studies.

5.   X-RAY DIFFRACTION STUDIES OF DNA STRUCTURE.

At Daresbury, Professor Watson Fuller and his colleagues, have carried out structural studies on DNA fibres which may be split into three related themes;

(a)   the explanation of observed diffraction patterns in terms of detailed molecular models,

(b)   the polymorphism of DNA and the factors that determine it, and

(c)   the dynamics and pathways of the transitions from one DNA conformation to another.

    Fibre X-ray diffraction patterns from natural DNA do not normally reveal significant data for Bragg spacings less than 3.0A, mainly because the molecular structure corresponding to the 'random' sequence of base pairs is not regular at such resolution.   Synthetic DNA however, such as poly d(A-T) with a regular alternating sequence of adenine and thymine bases along the strand, can be prepared as highly orientated fibres under the correct conditions of humidity and in the presence of appropriate counter ions such as Li^+.   Using synchrotron radiation these fibres give diffraction patterns showing exceptional crystallinity and exhibit data out to the 16th layer line, corresponding to a Bragg spacing of 1.5A.   With data of this resolution, highly detailed molecular models can be proposed for the DNA structure.
    Natural DNA extracted from a variety of sources, (calf thymus, salmon sperm), exhibits a remarkable degree of conformational polymorphism in its molecular structure.   Thus it can adopt double-helical conformations such as the A form, (11_1 helix, pitch = 28A), the B form, (10_1 helix, pitch = 34A) and the C form, (a family of helices with an approximate 9_1 symmetry and pitch around 31A), depending on the hydration and ionic content of the environment. Synthetic DNA fibres show even greater polymorphism.   Thus poly d(A-T).poly d(A-T) can adopt the D form, (an eightfold helix, pitch = 24A), whereas poly d(G-C).poly d(G-C), alternating guanine and cytosine bases, can adopt an S form.   The modification of conformation by base sequence has important implications for the biological function of DNA.   For example, natural DNA has stretches of highly repetitive base sequences which are potential sites for recognition and control processes.   Although it is generally agreed that the A and B forms are right-handed helices and that the S form is left-handed, both right and left handed models have been proposed for the D form.
    Transitions between conformations may be induced by varying the ambient humidity and Watson Fuller and his group, (Mahendrasingam et al., 1986), have used time-resolved X-ray fibre diffraction to follow the transitions occurring within a fibre of poly d(A-T).poly d(A-T) with the overall aim of determining the nature of the transition

532

mechanism.   During one set of these experiments, designed to study
the B to D transition, the SRS was operated at 2Gev and with beam
currents of between 210-70mA over periods of about 5 hours.   Using a
monochromatic beam, λ = 1.608Å, and a specimen to film distance of 5.6
cm, diffraction patterns were recorded in 4-10 minutes, a gain of
roughly <u>100</u> compared to a conventional rotating anode source.   During
a 17 hour period approximately 100 patterns were recorded as the fibre
was induced to undergo 4 D$\rightarrow$B$\rightarrow$D transitions in response to changes in
the relative humidity.   Transition times were typically 1 hour
allowing a sequence of 8-10 diffraction patterns to be recorded per
transition, Figure 5.1.

It was observed that the B$\rightarrow$D type transition was a smooth one,
implying that there is no change in hand during the transition.
Consequently, since the B form is right-handed, the D form must also
be right-handed and left-handed models proposed for D DNA must
therefore be incorrect.

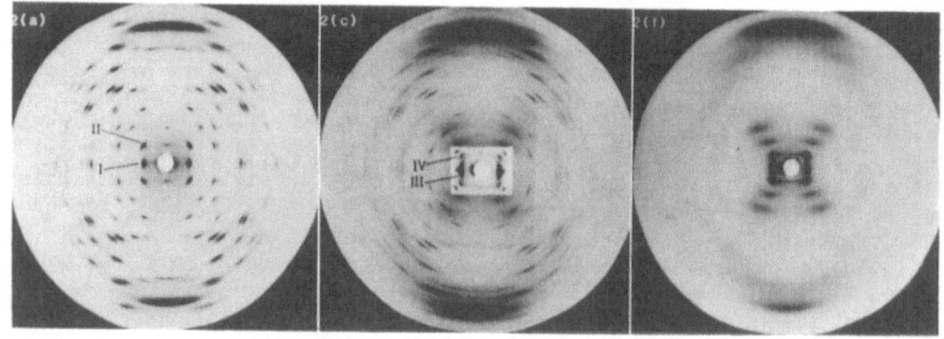

Figure   5.1.   The   D$\rightarrow$B   conformational   transition   for   poly
d(A-T).poly d(A-T) as the relative humidity is reduced from 98% to
55%.

The left hand photograph is typical of the crystalline D form;
reflections I and II are related to the lateral separation, 17.5Å, of
adjacent molecules in the fibre and the helical pitch, 24.3Å, of the D
form respectively.

In the centre photograph reflections III and IV are related to the
lateral separation and helical pitch of a semi-crystalline
intermediate conformation.

The right hand photograph shows the semi-crystalline B form with helix
pitch of 33.5Å.   The cross-like distribution of intensity is typical
of the B form.

CONCLUSION.

This article has attempted to describe some of the areas where synchrotron radiation has made a considerable impact on structural studies of macromolecules of biological interest. It has not attempted to be comprehensive, but rather to reflect the author's own interests and predjudices. Obviously the future is very exciting for the protein crystallographer. One can envisage the growth of areas such as the Laue technique to study reaction mechanisms in the crystalline state or for ultra-fast data collection, particularly from crystals which are very sensitive to radiation damage. The use of multiple wavelength measurements for surmounting the phase problem will also play an increasing role in macromolecular structure determination. However, it should be stated that although the actual data collection times may be much reduced, the experiments must be carried out with great care, particularly when measurements are made to maximise the absorption component of anomalous scattering; time spent in planning and preparing the experiment will be time well spent ! One considerable advantage of having a centralised facility such as a synchrotron is that it is possible to tackle a particular structural problem by a variety of complementary techniques, (crystal structure analysis and EXAFS), and this should be given consideration when planning experiments.

ACKNOWLEDGEMENTS.

The author would like to thank the many colleagues, at the SERC Laboratory, Daresbury, (J.Helliwell and S.S.Hasnain), the EMBL Out-station at DESY in Hamburg, (H.Bartunik and K.Bartels), and those at Birkbeck College, (T.L.Blundell, C.Slingsby, B.Gorinsky, S.Bailey, R.C.Garratt, H.Jhoti and S.Najmudin), for all the help that they have unstintingly given whilst he gained the user experience on which part of this article has been written. He is also very grateful to Dr. Louise Johnson , (University of Oxford), and Professor Watson Fuller, (University of Keele), and their respective colleagues for providing material and giving permission for it to be presented on their behalf. Acknowledgement is also due to the MRC and SERC for providing, in part, financial support.

The author is also grateful to the following publications for giving permission to reprint in part, or in whole, the following figures; Figure 2.2, (Embo J., in press, Copyright 1987, IRL Press Ltd., Oxford, UK), Figures 3.5 and 4.1, (Biochem. Soc. Trans., 14, 535-538, Copyright 1986, and 14, 542-545, Copyright 1986, Biochemical Society, London UK), Figure 3.7, (J. Appl. Cryst., 19 448-452, Copyright 1986, Munksgaard, Copenhagen, Denmark), and Figure 5.1, (Science, 233, 195-197. American Association for the Advancement of Science, Washington USA).

534

REFERENCES.

Abola, J.E., Wood, M.K., Chweh, A., Abraham, D. & Pulsinelli, P.D. (1981). In *The Chemistry and Physiology of Iron*, Eds. Saltmann, P. & Hegenauer, J., pp27-34, Elsevier, Amsterdam.

Anderson, B.F., Baker, H.M., Dodson, E.J., Norris, G.E., Rumball, S.V., Waters, J.M. & Baker, E.N. (1987). *Proc. Nat. Acad. Sci.*, in press.

Bailey, S., Gorinsky, B., Garratt, R.C., Jhoti, H., Lindley, P.F. & Sarra, R. (1987). *Unpublished*.

Cramer, S.P. & Hodgson, K.O. (1979). *Prog. Inorg. Chem.*, $\underline{25}$, 1-39.

Crick, F.H.C. & Magdoff, B.S. (1956). *Acta Cryst.*, $\underline{9}$, 901-908.

Eisenberger, P. & Kincaid, B.M. (1978). *Science*, $\underline{200}$, 1441-1447.

Garratt, R.C., Evans, R.W., Hasnain, S.S. & Lindley, P.F. (1986). *Biochem. J.*, $\underline{233}$, 479-484.

Gorinsky, B., Horsburgh, R.C., Lindley, P.F., Moss, D.S., Parkar, M. & Watson, J.L. (1979). *Nature*, $\underline{281}$, 157-158.

Gurman, S.J., Binsted, N. & Ross, I. (1984). *J. Phys.*, $\underline{C17}$, 143.

Hajdu, J., Acharya, K.R., Stuart, D.I., McLaughlin, P.J., Barford, D., Klein, H. & Johnson, L.N. (1986). *Biochem. Soc. Trans.*, $\underline{14}$, 538-541.

Hajdu, J., Acharya, K.R., Stuart, D.I., McLaughlin, P.J., Barford, D., Oikonomakos, N.G., Klein, H. & Johnson, L.N. (1987). *EMBO. J.*, in press.

Harada, S., Yasui, M., Murakawa, K. & Kasai, N. (1986). *J. Appl. Cryst.*, $\underline{19}$, 448-452.

Hasnain, S.S. (1986). *SERC Daresbury Laboratory Report DL/SCI/P523E.*, SERC Daresbury Laboratory, Warrington WA4 4AD, U.K.

Hasnain, S.S., Evans, R.W., Garratt, R.C. & Lindley, P.F. (1987). *SERC Daresbury Laboratory Report DL/SCI/P538E*. SERC Daresbury Laboratory, Warrington WA4 4AD, U.K. ; (1987). *Biochem. J.*, in press.

Helliwell, J.R. (1984). *Rep. Prog. Phys.*, $\underline{47}$, 1403-1497.

Helliwell, J.R., Cruickshank, D.W.J., Ellis, G.H. Habash, J. Papiz, M.Z., Rule, S. & Smith, J.M.A. (1984). In *Biological Systems: Structure and Analysis*. Eds. Diakun, G.P. & Garner, C.D., pp41-59, SERC Daresbury Laboratory Report DL/SCI/R22, SERC Daresbury Laboratory, Warrington WA4 4AD, U.K.

Hendrickson, W.A. (1979). *Acta Cryst.*, A35, 245-247.

Hendrickson, W.A., Smith, J.L. & Sheriff, S. (1985). In *Methods in Enzymology*, Eds. Wyckoff, H.W., Hirs, C.H.W. & Timasheff, S.N. 115, 41-55. Academic Press Inc., Orlando (USA) and London (UK).

Hoppe, W. & Jakubowski, U. (1975). In *Anomalous Scattering*, Eds. Ramaseshan, S. & Abrahams, S.C., pp 437-461. Munksgaard, Copenhagen, Denmark.

Lee, P.A. & Pendry, J.B. (1975). *Phys. Rev.*, B11, 2795.

Machin, P. (Ed.). (1985). In *Daresbury Laboratory Information Quarterly for Protein Crystallography*, No. 15., SERC Daresbury Laboratory, Warrington WA4 4AD, U.K.

Mahendrasingam, A., Forsyth, V.T., Hussain, R., Greenall, R.J., Pigram, W.J. & Fuller, W. (1986). *Science*, 233, 195-197.

Matthews, B.W. (1966). *Acta Cryst.*, 20, 82-86.

Narayan, R & Ramaseshan, S. (1981). *Acta Cryst.*, A37, 636-641.

North, A.C.T. (1965). *Acta Cryst.*, 18, 212-216.

Peerdeman, A.F. & Bijvoet, J.M. (1956). *Acta Cryst.*, 9, 1012-1015.

Phillips, J.C. & Hodgson, K.O. (1980). In *Synchrotron Radiation Research*. Eds. Winick, H. & Doniach, S., pp565-605, Plenum, New York.

Sawyer, L. (1986). *Biochem. Soc. Trans.*, 14, 535-538.

Sayers, D.E., Lytle, F.W. & Stern, E.A. (1971). *Phys. Rev. Lett.*, 27, 1204-1207.

Stuhrmann, H.B. (1980). *Acta Cryst.*, 36, 996-1001.

Summers, L.J., Wistow, G., Narebor, E.N., Moss, D.S., Lindley, P.F., Slingsby, C., Blundell, T.L., Bartunik, H. & Bartels, K. (1984). In *Pept. and Prot. Revs.* Ed. Hearn, M.T.W., 3, 147-168. Marcel Dekker Inc., New York (USA).

Templeton, L.K., Templeton, D.H., Phillips, J.C. & Hodgson, K.O. (1980). *Acta Cryst.*, B36, 436-442.

Wang, B.C. (1985). In *Methods in Enzymology*, Eds. Wyckoff, H.W., Hirs, C.H.W. & Timasheff, S.N. 115, 90-112. Academic Press Inc., Orlando (USA) and London (UK).

DISCUSSION

RADIATION DAMAGE

How radiation damage depends on the wavelength of X-rays is not well understood at present. Avoiding absorption edges might be expected to reduce damage, but little quantitative work has been reported, although a systematic study is currently in progress at the SRS, Daresbury.

Radiation damage at a particular wavelength, for example 1.488 Å on beam-line 7.2 at the SRS, varies from one protein to another, and even between crystals grown in the same batch from the same set of solutions.

Cryogenic techniques have been used in attempts to reduce radiation damage. An eye lens protein crystal coated with mineral oil and then cooled in a stream of nitrogen gas at approximately -150°C gave excellent diffraction data to 1.5 Å resolution using a rotating anode X-ray source with apparently no loss in intensity. Crystals of other proteins, however, fractured under those experimental conditions.

The Use of Synchrotron Radiation for Laue Diffraction and for the Study of Very Small Crystals

Marjorie M Harding
Department of Inorganic, Physical and Industrial Chemistry
Liverpool University,
P O Box 147,
Liverpool L69 3BX,
U K

Abstract

The lecture outlines the methods that have been used to record Xray diffraction data for structure determination for very small crystals; at Daresbury Laboratorys these employ either monochromatised synchrotron radiation and oscillation photographs or an area detector system, or the full white beam and Laue photographs; at other laboratories diffractometers have been used. The geometry of the Laue method and the procedure for deriving reflection intensities are described. Useful results have been obtained for crystals with dimensions in the range $10 - 30\mu m$. The scattering power of crystals is compared, taking into account both volume and composition. Many very small crystals are found to have a high mosaic spread or similar disorder and this results in additional difficulties. One structure, of a layer silicate, has been determined from area detector measurements and one, of an organometallic, from Laue photographs.

1 Introduction

In this lecture I shall look at two areas where synchrotron radiation (SR) has great potential — the study for structure determination of very small crystals, and the use of Laue diffraction patterns in structural studies.

The challenge with very small crystals is to record the diffraction pattern with a good enough signal to noise ratio to get reasonable intensity measurements — and often to determine the unit cell before that. I have been primarily interested in very small crystals or organic compounds of

M. A. Carrondo and G. A. Jeffrey (eds.), Chemical Crystallography with Pulsed Neutrons and Synchrotron X-Rays, 537–561.
© 1988 by D. Reidel Publishing Company.

biological significance, and organometallics and zeolites, but when good data on these can be achieved the next step will be to turn our attention to proteins, oligonucleotides, etc. The objective has been to get data good enough for structure determination.

At several sources 4-circle diffractometers have been used to record intensity (1)-(3). At Daresbury work on small crystals has been done in close association with protein crystallographic projects and the three available methods of intensity data collection are:

1. monochromatised SR, oscillation photographs, densitometry, processing of the digitised film image,

2. monochromatised SR, area detector (with geometry equivalent to oscillation photography), suitable software,

3. SR 'white beam', Laue photographs, densitometry and processing of the digitised film image.

Intensity data measurement for protein crystals (of fairly normal sizes) is being done by each of these strategies at Daresbury Laboratory so the development effort can be shared. Table 1 illustrates some achievements. Because many of the principles and experimental considerations are common to the first two approaches I shall describe the first in some detail and the second more briefly.

2   The Diffracted Intensity

The factors governing the intensity in a diffraction pattern are given by Woolfson (4).

$$I(hkl) = \frac{\lambda^3}{w} I_0 \left(\frac{e^2}{4\pi\epsilon_0 c^2 m} \right)^2 \frac{(1 + cos^2 2\theta)}{2 sin 2\theta} \frac{v_{crystal}}{v_{unitcell}^2} |F(hkl)|^2$$

where I_0 is the incident intensity at the wavelength of interest. The variation of the source intensity with wavelength is illustrated in fig 1. There are also many local factors that affect the intensity incident on the crystal — Be windows, monochromator characteristics, focussing mirrors. At

Table 1: Some small crystals studied with synchrotron radiation.

method	compound	crystal size /$(\mu m)^3$	result
at Daresbury $\lambda = 0.99$Å oscillation photos	trisaccharide disaccharide	$25 \times 20 \times 250$ $5 \times 25 \times 150$	photos that could probably be processed to give intensities
$\lambda = 0.90$Å area detector	piperazine silicate zeolite	$18 \times 175 \times 8$ $4 \times 125 \times 8$	intensity data, → structure intensity data, ?structure
white beam Laue photos	pip.silicate zeolite organometallic	$12 \times 4 \times 125$ $4 \times 125 \times 8$ $60 \times 50 \times 320$	film from wich intensities could probably be measured intensity data → structure
at Hamburg $\lambda = 0.91$Å 4 circle diffractometer	(reference 2) CaF_2	$6 \times 6 \times 6$	refinement, B's extinction study
at Stanford $\lambda = 1.74$Å 4 circle diffractometer	(reference 3) zeolite	$35 \times 25 \times 25$ $20 \times 10 \times 4$	a few reflections well measured as pilot study; more practicable

540

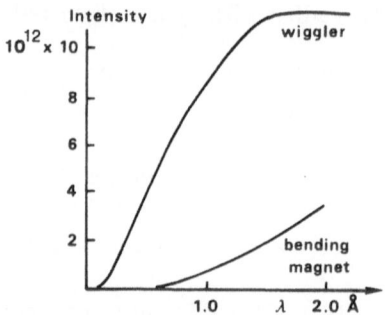

1. Intensity, for SRS at 2Gev, $250 mA$, in photons/horizontal mradian/sec in 0.1 % bandwidth.

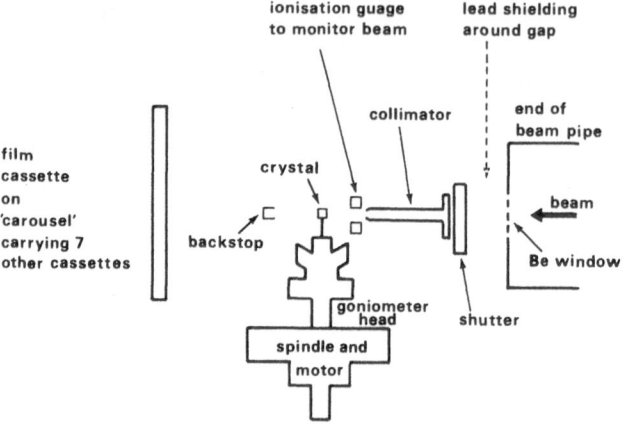

2. The essential parts of the oscillation camera.

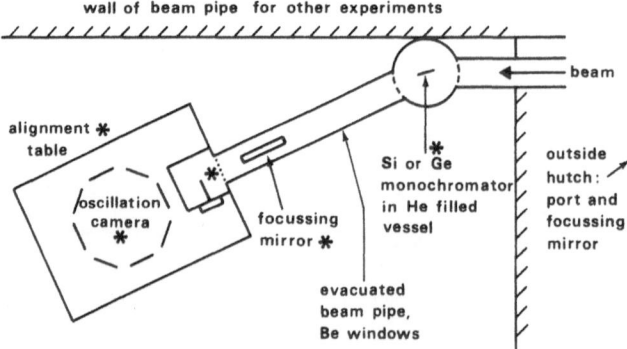

3. Sketch of arrangement in the 'hutch'. Each asterisk indicates movements controlled from outside the hutch, many by stepping motors. For major changes of wavelength the oscillation camera and associated beam pipe are moved manually.

any one source, and wavelength, though, we can look at the average reflection intensity to be expected from crystals of different sizes and of different chemical compositions.

Table 2: Comparison of estimated average intensity to be measured for crystals of different compositions and sizes

	I $\sum f_n^2/V_{cell}^2$	II volume of crystal used	I × II
	$/e^2 \text{Å}^{-6}$	$/(\mu m)^3$	$/e^2 \text{Å}^{-3}$
CaF_2	0.35	200	7×10^{13}
piperazine silicate	8×10^{-3}	25×10^3	20×10^{13}
trisaccharide	3.2×10^{-3}	125×10^3	40×10^{13}
protein, e.g. cytochrome c4	1.6×10^{-6}	45×10^6 'normal' size	7×10^{13}

Bachmann et al.(2) quote the estimated scattering power of a crystal, proportional to $(F(000))^2 \, v_{crystal}/v_{cell}^2$, and for crystals of very simple compounds this is indeed useful. $F(000)^2$ is $(\sum f_n)^2$, the square of the sum of atomic scattering factors in the primitive cell. For more complex crystals we are interested in the average value of $|F(hkl)|^2$, rather than the theoretical maximum, and therefore should replace $F(000)^2$ by $(\sum f_n^2)$. Table 2 compares these estimates of scattering power for four crystals — for more complex unit cells, or for those containing only light atoms, a larger crystal is required to give the same average intensity in the reflections to be measured.

3   Monochromatic Oscillation Photographs with SR

The Arndt-Wonacott camera is fully described (5), the experimental arrangements at SRS (6) and the film scanning procedures (7). Here I will give only a brief outline in the diagrams (fig 2, 3). At Daresbury Laboratory the whole system is designed and set up by J R Helliwell and coworkers. As 'users' we receive at the oscillation camera a very intense monochromatic Xray beam ($\Delta\lambda$ is controlled by slits), highly polarised. For safety the whole arrangement is enclosed in a 'hutch' — actually the size of a small room — and interlocked so that no person may be inside when the Xray Port is open. This means that all alignment operations and some camera operations must be done by remote control.

This route to intensity data for proteins will be discussed more fully in another lecture (P Lindley), but I think it is important to understand the diffraction geometry of the oscillation photograph well. The strategy of the measurement procedure is first to establish the cell and orientation well, then, with this knowledge to predict the position of all reflections expected on the film, then to integrate the intensity within specified boxes at these positions in the digitised film image. For a good protein crystal, when carefully done, this is a well established route to good intensity data (8). Can we use it for very small crystals?

If the crystal is too small for photography with conventional sources its unit cell and symmetry may not be known, or it may well not be praticable to set it up on the oscillation camera in a chosen orientation. A reasonable procedure is to take a series of oscillation photographs covering $180°$ around the spindle, then by the converse of the above geometry, the positions of the spots on each photograph can be mapped in reciprocal space and the reciprocal lattice looked for. The exact spindle angle, ϕ, at whitch each reflection occurred is not known, only that it occurred within the range of that photograph, and therefore each reflection is represented by a small arc, say $2°$ to $5°$ — this makes it a little more difficult to recognise the lattice and to determinate it accurately.

In principle this is an acceptable route to get the unit cell, and then the intensity data; in practice our programs to refine the cell and the orientation

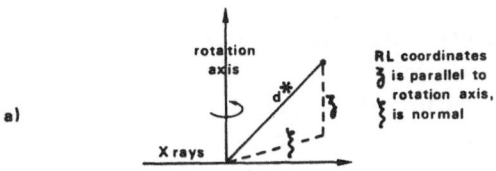

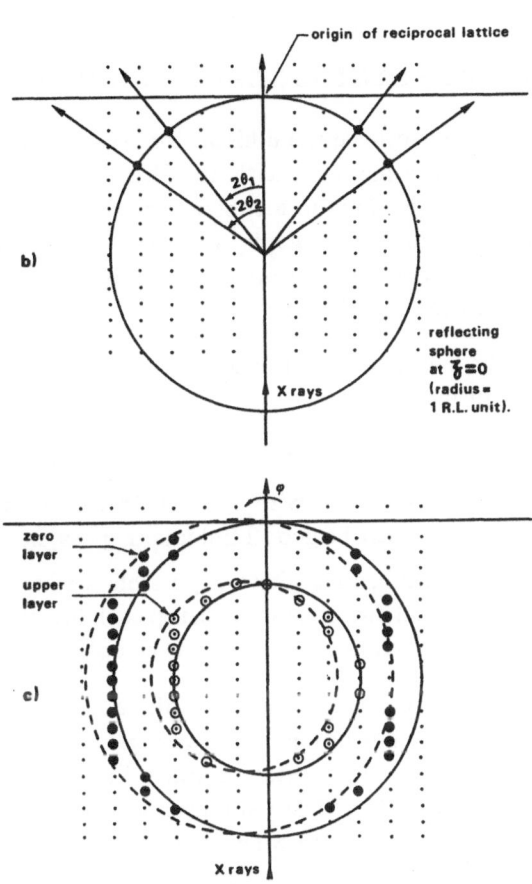

4. Geometry in oscillation photographs.

(a) Reciprocal Lattice coordinates ς and ξ.

(b) Reciprocal Lattice (Zero layer) with reflecting sphere. The rotation axis is through the R.L. origin and normal to the paper.

(c) If the crystal rotates through 10° all the points marked • in the zero layer will pass through the reflecting sphere and thus give rise to reflections. The radius of the reflecting sphere as it cuts an upper layer is $\sqrt{1-\varsigma^2}$. The points marked ⊙ will reflect.

are not yet in a satisfactory state, and the FAST area detector system is becoming available to do the job better. One other hazard needs attention with film: if the spots on the film are very small and sharp the measurement of transmission through the film needs to be made on a correspondingly small raster, otherwise the optical density is not properly integrated (the Wooster effect - (9)).

4 Some Experimental Considerations

1. Crystal mounting has not proved too difficult for crystals that can be studied in air; crystals are mounted with araldite on a $1mm$ strand of glass wool which itself is glued to a substantial glass fibre. Micromanipulators were available, but people working with me have not chosen to use them — a good microscope, a stable bench and a draft free environment are the top priorities.

2. Steps will need to be taken to keep the background intensity as low as possible; these can include

   (a) use of the smallest collimator praticable (we use 0.2 mm)

   (b) crystal mounting as described, with no thick fibre in the beam, or if in capillaries, they must be good Lindemann glass, not quartz,

   (c) keeping the beam path in air as short as possible; we have worked with collimator-crystal and crystal-backstop each less than 1 cm; (alternatively a He environment could be set up).

3. When studying very small crystals of fairly simple compounds we normally wish to collect higher resolution data (say $dmin = \lambda/2sin\theta$,max $= 1\text{Å}$) than is typical in protein crystallography. Various factors affect the choice of wavelength and are summarised in an appendix; $\lambda = 0.8\text{Å}$ was a popular choice at SRS last year.

4. At this stage a few comments on the pattern of work on the SRS are perhaps relevant; it is very different from working in a home laboratory with a conventional Xray source running continuously! Synchrotron radiation arises from the electron current in the synchrotron

ring. No experimental work is done while electrons are being accelerated and the ring is being 'filled'. Through much of 1986 the electron energy was 2 GeV and when first 'injected' into the ring the current 280-300 mA. Experimental work starts intensively at this stage. The current, and correspondingly the Xray intensity, decay rapidly at first and then more slowly; in good times it will have dropped to 100mA after 8-10 hours; then the beam is 'dumped' and a fresh filling started which may take around an hour. That is the time when experimenters relax, have a meal or catch up on scores of undeveloped films! If any part of the ring system has vacuum problems or has recently had them, beam lifetimes are much shorter and unpredictable. The facility is used intensively. A usual pattern on the crystallographic work-stations at Daresbury is for a team of 2, 3 or 4 people to run continuously for 24 hours, or sometimes for 2 or 3 days (and then sleep!); in routine data collection it is quite practicable to produce 50-100 films in one 8 hour shift. There must always be a device to monitor the incident Xray beam near the crystal, usually a small ionisation gauge.

5 Use of an Area Detector System

Arndt has given a very useful review of area detectors (10). At Daresbury we use the Enraf-Nonius FAST diffractometer. The following is a greatly simplified description. Consider initially the oscillation camera and its use for data collection:

1. The film is replaced with an area detector, a phosphor, sensitive to Xrays, which is fibre-optically coupled to a low-light-level television camera. The image on the area detector is also visible to the operator on a monitor.

2. The number of radiation quanta arriving at each pixel of the image in the course of the oscillation are summed and stored. In practice oscillation ranges such as 0.1° are chosen. Each reflection then appears on several successive images or frames.

3. Software is provided which will integrate the intensity of each reflection

   (a) over the appropriate area in each frame, and

   (b) over the successive frames in which it appears.

   At present software from Munich is being used (11).

4. In the Nonius system the crystal is mounted on a goniostat so that any direction on the crystal may be chosen as oscillation axis, and software allows manipulation of this.

The area detector should have major advantages for very small crystals. First compared with film, recording is over a much smaller oscillation range and so gives a much more favourable ratio of reflection intensity (signal) to background (noise); secondly it avoids all problems that might arise from small sharp spots and the Wooster effect or the grain size of the film; thirdly, since the spindle angle, ϕ, is determined for each reflection, unit cell recognition and refinement is easier and can be done by routines equivalent to those for autoindexing on conventional diffractometers. It is expected to be the method of choice for very small crystals, and quite recently two sets of measurements have been successfully made at Daresbury.

The data quality is at least satisfactory; the structure of the first compound was readily determined; refinement at present has given $R = 0.08$ for 511 reflections, but there is almost certainly scope for improving the processing of the data. The measurements have been fairly time-consuming — 2-3 shifts of beam-time each, a good part of wich was spent on the initial finding of the unit cell.

6   On the Quality of Very Small Crystals

Several years ago I undertook to try to collect SR intensity data on very small crystals because there were many interesting compounds for which growing good sized crystals had proved difficult or impossible. I naively assumed the main problem to be that the diffracted intensity is small because the crystal volume is small. But there is another problem:

Table 3: Structure determination for $C_4N_2H_{12}^{2+}\ Si_6O_{13}^{2-}$ (EU19)

unit cell and data collection on FAST diffractometer

crystal size $18 \times 75 \times 8\mu m$
wavelength 0.9Å
monoclinic $a = 13.57$, $b = 4.90$, $c = 22.46$Å, $\beta = 91.67°$
space group $C2/c$, $Z = 4$
crystal in arbitrary orientation,
   180° data set recorded in 0.5° frames,
   dumped on tape for latter processing
   second data set with tilted detector, similarly
dmin 1Å
processing by MADNES software, gave
   511 reflections with $F > 2\sigma(F)$
structure solved by MULTAN and refined by SHELX,
R=0.08 at present.

(S J Andrews, M Z Papiz, at Daresbury Laboratory,
crystals from B M Lowe, A S Blake, Edinburgh)

Table 4: Data collection on a zeolite (composition approximatly SiO_2)

crystal size $125 \times 4 \times 8(\mu m)^3$
wavelength 0.9Å

set 1 : 100° of data about one axis
set 2 : 100° of data about a second axis at 90° to first

| crystal rocking width | set 1 | 2° | |
| | set 2 | 10° | ! |

	no observations	independent reflections	Rmerg (on I)
set 1	785	331	0.06
set 1 & 2	1248	486	0.08

(work of M Z Papiz and S J Andrews)

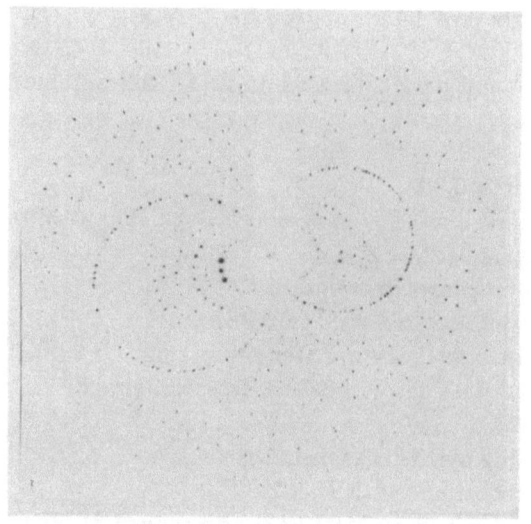

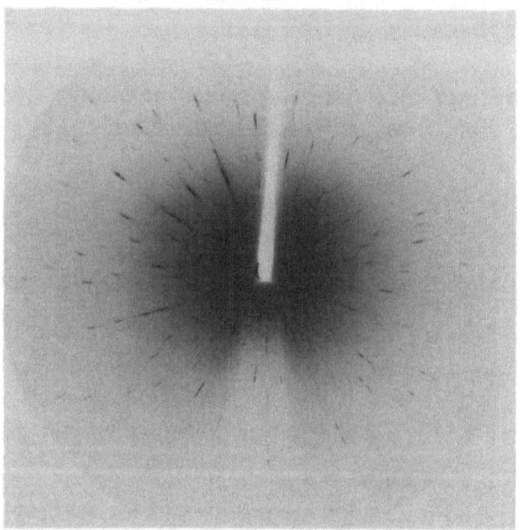

5. Laue diffraction patterns of

(a) proflavine hemisulphate, a good quality, normal size crystal, ($> 0.4mm$). Exposure time 1/8 sec (SRS at 2 GeV, 130 mA).

(b) a nucleotide crystal, a very thin plate $75 \times 100 \times 10 \mu m$. Exposure times 2 sec (SRS at 1.8 GeV, 200 mA).

(both crystals from Dr. S Neidle)

most of these crystals are not very well ordered, and this may often be the reason why they do not grow bigger.

The most striking evidence for poor ordering has come from Laue photographs (fig 5), but the evidence is also there in the area detector data collection, and to a more limited extent in the oscillation photographs. It is quantified for the example in Table 4 by the rocking width; this is extremely large compared with values for protein crystals, normally $0.1°$. Samples, and individual crystals, vary considerably, as may different directions in one crystal. In oscillation photographs it means that the spots are not so small and sharp and therefore the Wooster effect is not so serious. Some other examples, and the details of the geometry of the effect in Laue photographs are given (12).

7 Laue Diffraction Patterns

Laue diffraction patterns are recorded with a stationary crystal and a continuous range of wavelengths. Traditionally they have been used by metallurgists and mineralogists for checking crystal orientation or symmetry. They offer a way of using the continuous range of wavelengths available in SR. SR Laue photographs have been taken for proteins and simple organic and organometallic crystals and much progress made towards using these for intensity measurements. Exposure times are very short, fractions of a second for normal size crystals, so there are possibilities of 'kinetic crystallography'; for very small crystals they are still only a few seconds or minutes, hence my initial interest.

Experimentally, recording Laue diffraction patterns seems fairly simple. At Daresbury we have used the basic components of the Arndt-Wonacott camera. For a few experiments on the protein crystallography work station (9.6) the monochromator has been bypassed and the beampipe and camera swung round to the 'straight through' position (fig 2). For most experiments we have used a different work station (9.7) without monochromator; we have a simple stationary film cassette holder (there is not room for the carousel). Janos Hajdu has devised a fast shutter wich can give accurately timed exposures as short as 0.1 sec. For intensity measurements a filmpack is used; to cover the total range of intensity (and the different absorption

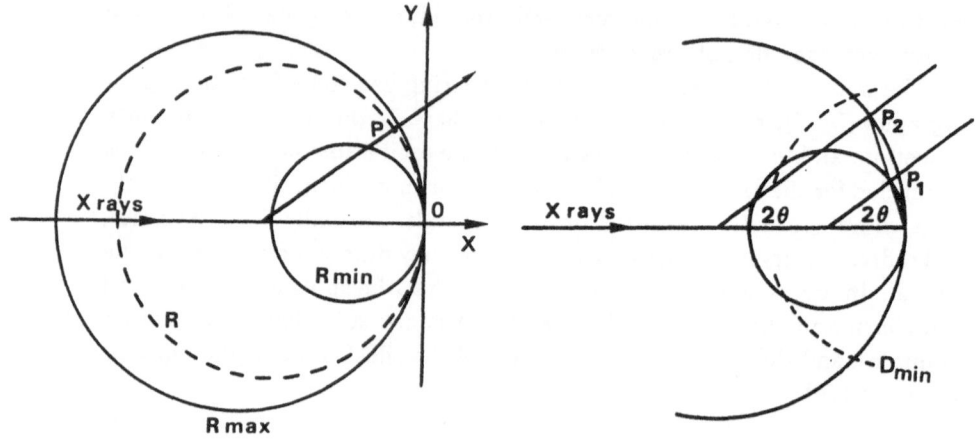

6. Geometry of Laue diffraction pattern; see text for explanation. (transmission mode)

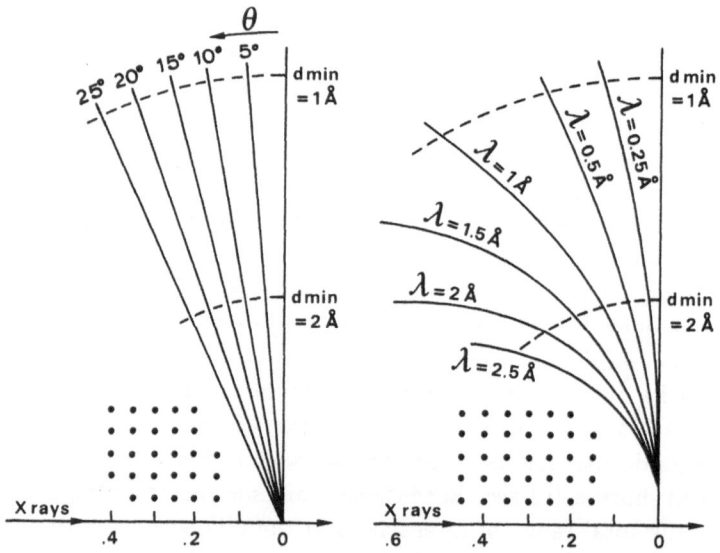

7. Properties of the Laue diffraction pattern are mapped onto the reciprocal lattice. Reciprocal distances shown are λ/d, etc for $\lambda = 1.0$Å. The Laue geometry has cylindrical symmetry about the beam direction. The volume of the reciprocal lattice recorded is bounded by λmin and λmax, dmin and θmax.

at different wavelengths) this needs to contain about 12 films, or 6 films interleaved with metal foils (aluminium and copper, 0.15 or 0.20 mm have been used). No currently available area detector has a high enough resolution combined with speed for use with these Laue diffraction patterns — those of proteins may contain 5000 or more spots on one film, and it will be desirable to measure spots as close as 0.25mm. The new technology of the 'imaging plate' (13) may well be suitable in the future, and a system for Xray diffraction work is being developed by Rigaku.

The diffraction geometry of Laue photographs is shown in fig 6. XYZ are orthogonal axes in reciprocal space, for a fixed wavelength, say 1.0Å; conventionally Z is parallel to the spindle of the camera. The circles of radius Rmax and Rmin represent the spheres of reflection for the wavelengths λmin and λmax (Rmax=1.0/λmin here). Any reciprocal lattice point, eg P at x,y,z between these two spheres may then give rise to a reflection; the centre of the relevant reflection sphere is at $-R,0,0$ where

$$R = \frac{(x^2 + y^2 + z^2)}{2x}$$

and the wavelength giving rise to the reflections is 1.0/R. Fig 6b shows the reflecting spheres for wavelengths $\lambda min = 0.6\text{Å}$ and $\lambda max = 1.6\text{Å}$ and shows the region of the reciprocal lattice covered if $dmin = 1.8\text{Å}$. It also shows one of the problems in using Laue patterns, that of superimposed reflections. The reciprocal lattice points P_2 and P_1 lie on a line which passes through the origin; these reflections will be superimposed in the Laue pattern since the angles of diffraction are the same for both; if the indices are h, k, l and nh, nk, nl then the wavelengths are in the ratio $n : 1$. It is also useful to look at the region of reciprocal space recorded on one Laue photograph and the variation of θ and λ within it, fig 7. It can be seen that the longer wavelength reflections will tend to occur near the edge of the film (large θ), the shorter wavelengths near the centre, and that high resolution reflections (large d^*) can occur at all θ values, not just at the edege of the film as in monochromatic work. The limits λmin, λmax and $dmin$ are all 'soft' limits, not sharp limits — and between λmin and λmax the incident intensity is not constant but varies continuously (see fig 1 and 9).

Some trial measurements of intensities in Laue diffraction patterns of simple compounds have been reported (14,15). Laue photographs of a very small crystal of an oligopeptide, gramicidin A, have been reported (16), and possible uses of Laue photographs with other biological materials extensively discussed (17). At Daresbury, intensity measurement procedures suitable for proteins as well as simpler compounds have been developed and tested and are still being further improved by M Elder, P Machin, J R Helliwell, J W Campbell, I Clifton and others (18, 19). In outline these procedures involve

1. Prediction of the Laue diffraction pattern to match the densitometered film image, including refinement of orientation angles, crystal-film distance, film centre position, cell dimensions (if necessary, and all but one). The refinement program, GENLAUE, includes simultaneous display on the computer terminal of the film image and the predicted pattern; the match is judged visually and then by the rms deviation of the calculated spot positions from the observed ones for a selection of spots (for example 100 spots with indices < 9); a match with rms deviation < $0.05mm$ is regarded as good enough to proceed to integration. The soft limits λmin, λmax and $dmin$ must also be carefully estimated at this stage, often by comparison of enlarged printouts of the prediction with the original photographs.

2. Integration of density in film image in a suitable box at all the predicted spot positions, for all the films in the pack.

3. Derivation of interfilm scalefactors, each of which is a function of wavelength (fig 8) and scaling together of all films in a pack (and Lp correction).

4. Unscrambling of the reflection intensities in which harmonics are superimposed; this requires the film factors determined in a stage 3 (20).

5. Wavelength normalisation, scaling and merging of the film packs: each reflection intensity must be multiplied by a wavelength-dependent factor $1/f(\lambda)$ and a pack scaling factor $c(i)$, both of which are determined empirically (21). For example, a set of 3700 intensity

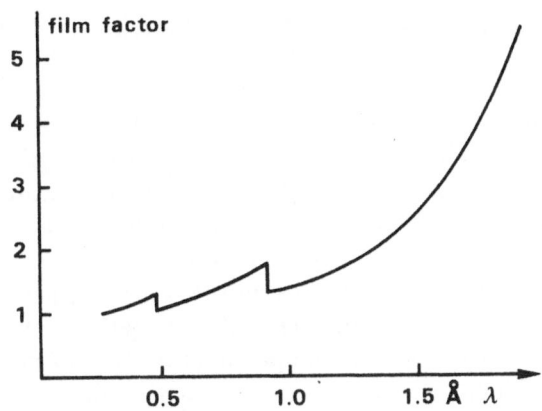

8. Variation of the film factor, empirically determined, with wavelength. Note the discontinuities at the *Ag* and *Br* absorption edges.

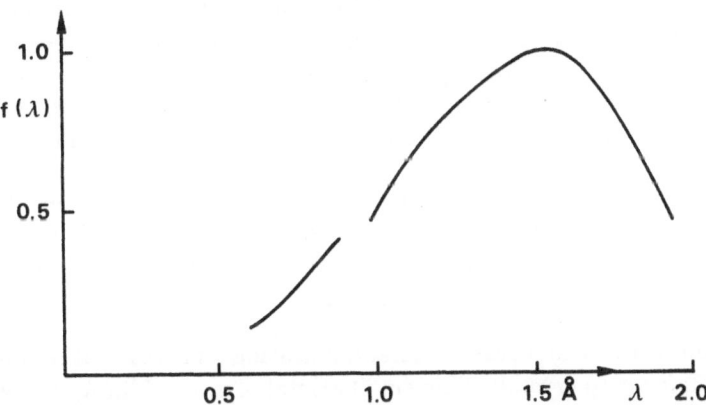

9. Example of $f(\lambda)$ in the wavelength normalisation. This combines the effects of the variation of the incident intensity at the crystal, and the film response.

measurements from 10 filmpacks represented 830 unique reflections and included sufficient overlap to determine $f(\lambda)$ and $c(1)$ to $c(10)$.

8 Applications of Laue Diffraction Patterns

Intensity measurements are likely to be particulary useful in two areas, time dependent studies and the study of very small crystals. Apart from this they have very considerable use for quick screening of crystals for quality and for assessing mosaic spread. Some of the most beautiful work has been done by Janos Hadju in association with L Johnson and others from Oxford on crystals of the enzyme phosphorylase.

This will be described in another lecture (P Lindley); here I shall describe a pilot study for the very small crystals application — the structure determination of an organometallic compound (12). Most of the details are given in Table 5, but improvements are still being made in the intensity measuring procedures and more satisfactory refinement is expected soon.

Laue photographs have some characteristics which should be born in mind:

1. Cell dimensions cannot be determined, only their ratios.

2. The superposition of harmonics means that intensity data cannot be measured for axial reflections ($h00$ etc) and others of simple order ($h00$, hhh); this is very serious if direct methods are to be used (or for determining space group), but not a great disadvantage for Fourier or difference Fourier methods. The proportion of reflections that can be measured as singlets approaches 87% for a large unit cell (23).

3. Since different reflections are measured at different wavelengths there are additional problems in measurement of data and scaling them together; if there are anomalous scatterers present their f', f'' values will be different in diferent reflections!

4. The full white beam causes substantial heating , and often other radiation damage to crystals. The small crystal of gramicidin A survived four or five exposures before the Laue pattern deteriorated too seriously (16). All protein crystals deteriorate after a small number of

exposures — two to ten, depending on the protein; the reflections become elongated, consistent with increased mosaic spread or similar disorder. The small organometallic crystal for which we took data was begining to deteriorate by the end of the series.

5. Laue geometry is very sensitive to mosaic spread or crystal quality; we have photographs of many crystals for which intensities could be measured on a diffractometer or by other monochromatic techniques, but whose Laue patterns could not have been well measured; they are too streaked. Conversely, though, Laue diffraction may be useful in assessing mosaic spread or crystal quality (12).

On the other hand the great advantage of the Laue method is the extremely short exposure time required to record a large proportion of the total diffraction pattern. The possibility of looking at the structure at intervals of 1 min or even less must completely outweigh these disadvantages; you will see in another lecture the sort of results that have been obtained for phosphorylase. For this reason it holds great promise for the future.

9 Notes added after the Study Institute

The structure determination of a fairly small crystal of piperazine silicate was described in section 5 (see also reference 24); during the data collection the SRS was running at 2 GeV, 10 mA (single bunch mode). Under normal running conditions (2 GeV, 250 mA multibunch mode) similar diffracted intensities would have been observed for crystal of 1/25 of this volume. After the 'high brightness lattice' installation at the SRS is complete, beam intensities are expected to be 7-8 times greater; therefore it should be possible to record useful diffraction data for a crystal of this complexity and quality, of dimensions $5 \times 5 \times 5 \mu m^3$. Hitherto powder diffraction has been the only method suitable for investigation of crystallites of this size.

An extensive study has been made of the regions of reciprocal space accessible in a single Laue diffraction photograph under different conditions, and of the proportion of reflections recorded singly and in multiplets (25).

Table 5: Structure determination of $RhFe(CO)_5Cl(dppee)$

dpppee is $Ph_2P.C(=CH_2).PPh_2$

monoclinic $a = 18.24$, $b = 19.81$, $c = 8.26$Å. $\beta = 86.5°$
   (known approximately from $CuK\alpha$ Weissenberg)
apparent space group $B2_1$, $Z = 4$
   (which is equivalent to $P2_1$, $Z = 2$)

crystal size $60 \times 50 \times 320\mu m$
SRS at $2GeV$, $158 - 146\,mA$

10 films packs at intervals of $18°$ around the spindle
$\theta max = 21°$
λ assumed to be 0.3-2.1 Å
$dmin$ assumed to be 1.2 Å
3707 reflection intensity measurements yielded
   754 unique reflections, $Rmerg = 0.14$

structure solution and refinement by SHELX
   R=0.17 at present (improvement expected)
space groups is $B2_1/m$ and structure is disordered.
(crystals from R Dawson and A K Smith, Liverpool)

The organometallic structure referred to in Table 5 has been refined to $R < 0.15$ (22, 26).

10   Acknowledgement

Many colleagues at Daresbury have contributed to this work and I am grateful to them all, but I wish to pay a special tribute to Pella Machin and Mike Elder; they played a central part in the software development for all the Laue work and they died tragically in a mountaineering accident in March 1987.

References

1. F S Nielsen, P Lee and P Coppens, Acta Cryst. (1986) **B42**, 359-364.

2. R Bachmann, H Kohler, H Schulz, H Weber Acta Cryst. (1985) **A41**, 35-40.

3. P Eisenberger, J M Newsam, M E Leonowicz and D E W Vaughan, Nature (1984) **309**, 45-47.

4. " Xray Crystallography", M M Wolfson, Cambridge University Press, Cambridge, 1970.

5. "The Rotation Method in Crystallography", ed U W Arndt and A J Wonacott, Elsevier/North-Holland Biomedical Press, 1977, Amsterdam.

6. J R Helliwell, Rep. Prog. Phys. (1984) **47**, 1403-1497.

7. A J Wonacott

8. T J Greenough and F L Suddath, J Appl. Cryst. (1986) **19**, 400-409.

9. W A Wooster, Acta Cryst. 1964, **17**, 878.

10. U W Ardnt, J Appl Cryst (1986) **19**, 145-163.

558

11. "MADNES", Munich Area Detector NE Software, J Plugrath, A Messerchmidt, 1986, Max-Planck-Institut fuer Biochemie, D8033 Martinsried, West Germany.

12. S J Andrews, J E Hails and M M Harding, and D W J Cruickshank, Acta Cryst. (1987) **A43**, 70-73.

13. Sonoda et al., Radiology (1983) **148**, 833. and J R Helliwell in "Information Quaterly for Protein Crystallography" **No 19**, Daresbury Laboratory, 1986.

14. I G Wood, P Thompson and J C Matthewman, Acta Cryst. (1983) **B39**, 543-547.

15. D Rabinovich and B Lurie

16. B Hedman, K O Hodgson, J R Helliwell, R Liddington and M Z Papiz, Proc. Nat. Acad. Sci. (USA) (1985) **82**, 7604-7.

17. K Moffat, D Bilderback, W Schildkamp and K Volz

18. M Elder, M M Harding, J R Helliwell, P A Machin et al., in preparation for Acta Cryst.

19. P A Machin and M M Harding in association with others, "Information Quaterly for Protein Crystallography", **No 15**, Daresbury Laboratory, 1985.

20. S Zurek et al., in "Information Quaterly for Protein Crystallography", **No 16**, 1985.

21. J Campbell, J Habash, J R Helliwell and K Moffat, in "Information Quaterly for Protein Crystallography", **No 18**, Daresbury Laboratory, 1986.

22. M M Harding, S J Maginn, J W Campbell, I Clifton, P Machin, submitted to Acta Cryst.

23. D W J Cruickshank, 1985, personal communication.

24. S J Andrews, M Z Papiz, R McMeeking,A J Blake, B M Lowe, K R Franklin, J R Helliiwell, M M Harding, submitted to Acta Crys. B, 1987.

25. D W J Cruickshank, J R Helliwell and K Moffat, Acta Cryst., 1987, in the press.

26. M M Harding, in "Computational Aspects of Protein Data Analysis", Proceedings of a Daresbury Study Weekend, (January 1987), S.E.R.C. Daresbury Laboratory.

Appendix

Considerations which may affect the choice of wavelength:

1. The maximun incident intensity could be selected from the synchrotron source characteristics, but the diffracted intensity is proportional to $Io.\lambda^3$, or, when an Lp factor has been take into account, to $Io.\lambda^2$ approximately (Arndt).

2. The film or detector response is likely to be wavelength dependent. A rough guide to how these factors combine at Daresbury Laboratory, on the Wiggler beam is given by Fig 9, taken from reference (21). Note the discontinuity at the bromine and silver absorption edeges.

3. Shorter wavelength will usually be desirable for getting high resolution data, and for reducing absorption errors, and perhaps radiation damage.

4. Changing wavelength may not be quick, so at the experimental station continuous working at one wavelength is likely to be preferred.

DISCUSSION

NORMALISATION OF LAUE INTENSITIES

An alternative to the internal method of normalisation described above is to obtain a pattern, immediately before the experiment, from a standard specimen of known intensities. However, one orientation of the standard might not give sufficient data. The internal method should be subject to fewer errors because of the large number of observations available to correlate different wavelengths; it is potentially as good as the use of a standard and involves much less work.

SELECTION OF CRYSTALS

Smaller crystals, provided they gave enough intensity to solve and refine the structure, might be better than larger ones if they had a smaller absolute mosaic spread. Variability of crystallisation suggested that choice of the individual crystal might be important; selection is easy for the largest crystals but difficult for the smallest ones.

Dr Harding expected that crystals which had grown larger were likely to be better, and thought that the process of crystallisation might be significant. Proteins and other organic crystals, growing slowly from aqueous solution and depending on van der Waals' contacts and hydrogen bonds for molecular packing, might more easily attain a configuration of minimum energy than zeolites, where the addition of smaller units by covalent bonding might offer more possibilities for misorientation. Variability within systems was again emphasised, however, by the excellent diffraction profile given by the 50 x 10 x 5 μm ZSM-5 (albeit twinned) microcrystal at Brookhaven (figure 6 of Dr A Kvick's paper). Systematic research is needed in this field.

LAUE INSTRUMENTATION

The beam used for Laue work at Daresbury had no facilities for selecting the wavelength range or limiting the background, but it was hoped that improvements would soon be made.

LAUE INTENSITIES AND DIRECT METHODS

Prof Beurskens recommended that multiple reflections should be treated as follows. When the individual reflection intensities could be estimated, as might be possible when 2 or 3 reflections overlapped, they should be included in the direct methods analysis even though their accuracy might be low. For refinement, however, it might be better to neglect them. Similar considerations applied to low wavelength reflections. Even if their normalisation was uncertain by 50%, those with a large normalised structure factor should be included in the direct methods analysis but rejected for refinement. Special reflections in which many orders overlapped were troublesome in direct methods so their absence was immaterial, but very low wavelength, high θ reflections were more interesting for direct methods.

ACCESS TO FACILITIES

Representatives of the major neutron and synchrotron radiation
sources in France, Germany, the United Kingdom and the United
States emphasised that their institutes warmly welcomed good
applications to use their facilities. With only minor variations
of detail, the procedures are broadly as follows.

(i)   Formulate a scientifically good experiment.

(ii)   Discuss it with a senior scientist who is working in the
       same or a closely related field as a member of the
       institute's staff or one of its project teams.

(iii) Complete the official application form so that the
       proposal can be considered by the institute's selection
       committee.

(iv)   Attend the institute to do the experiment, and at least
       the initial processing of the results, as a member of a
       team adequate in number (typically 2-4 people) and
       expertise.

Administrative sections will send information about closing dates
for applications (usually twice a year), the availability or
otherwise of grants for travel and accommodation, and in many cases
an Annual Report or Year Book which gives the flavour of recent
research and describes the infrastructure. Various Newsletters are
also informative: contact Dr M S Lehmann at ILL, Grenoble, for
neutrons, and NSLS, Brookhaven, for X-rays.

US institutes are well equipped and well funded for experiments,
but are short of manpower. Brookhaven, for example, has some 55
workstations covering every type of synchrotron research and each

M. A. Carrondo and G. A. Jeffrey (eds.), Chemical Crystallography with Pulsed Neutrons and Synchrotron X-Rays, 563–566.

564

operated by a project team (PRT) of 3-4 people. Arrangements for access to both Brookhaven and the Argonne National Laboratory are flexible, liberal and international. Applications from all nationalities are welcome for work at Brookhaven to be done in collaboration with the institute's PRTs which have 75% of the time, or independently through scientifically competitive bids for the remaining 25% of the time. The Intense Pulsed Neutron Source, IPNS, at the Argonne gives a higher proportion of time to external projects, and graduate students are encouraged to submit their own proposals even though the science may be slightly less advanced.

Access to ILL at Grenoble for neutrons is similar to that to the Argonne. UK sources, pulsed neutrons at the Rutherford Appleton Laboratory and synchrotron X-rays at Daresbury, are funded in collaboration with some of the larger European countries, which consequently have preferential access. But good proposals from elsewhere are very welcome and often lead to good science being done; a convenient route is collaboration with a senior scientist from one of the participating countries. At Hamburg access is flexible, and can often be arranged at short notice to EXAFS, small angle scattering and protein crystallography. In Europe graduate students are encouraged to gain experience of fairly advanced science by attachment to a team working on its leader's ideas.

Attendance at the institute is very strongly urged on account of the importance to a scientist of doing experimental work, the educational value, and the availability there of the special programs needed for data handling. Only in exceptional circumstances will some centres run samples for investigators who cannot attend. There is, however, an SERC EXAFS service at Daresbury for researchers in UK universities, the Netherlands and Sweden: it will mount samples, run the experiment and do preliminary processing of the data, but interpretation is the responsibility of the user.

POSSIBLE TRENDS IN FUTURE RESEARCH

This part of the discussion attempted to supplement the extensive information already given about new techniques with thoughts it might have generated, in lecturers and audience, about general trends in chemical crystallography over the next few years.

Instrumentation is a physical problem mainly tackled by physicists, but many of the applications are in chemistry. Progress may become increasingly dependent on bringing chemists and physicists together to work as teams or to transfer information. The International Union of Crystallography acts as a meeting place for the sciences, so a useful contribution might be to suggest the formation by the Union of a Commission on Synchrotron Radiation, to supplement the existing one on neutron diffraction.

Competition for limited resources in the new methods will require applications to be of substantial scientific merit.

The strikingly increased precision of instruments such as the High Resolution Powder Diffractometer (HRPD at ISIS) will lead to accurate structures of many materials which are too complex for normal powder methods and do not give suitable single crystals, and perhaps also to the revision of some structures previously studied by conventional X-ray techniques. Structures of interest may include those of molecular crystals containing H or D atoms and with not more than 100 positional parameters. High resolution will also greatly benefit work on phase transitions and chemical transformations, especially when combined with real time studies of kinetics and mechanism. Other improvements in precision can be expected as work on the measurement of neutron Bragg intensities without interference by thermal diffuse scattering spreads from simple inorganic to molecular crystals. The associated measurement of sound velocity and that by ultrasonics are distinguished by their time scales as 'zero sound' (non-interacting phonons) and 'first sound' (phonons interacting through anharmonic terms). Inelastic neutron scattering in general has considerable potential for spectroscopy, and the crystallographic approach may become a useful supplement to nmr studies of fluxional molecules.

The high intensity of synchrotron radiation will ensure its productive use in protein crystallography, as will its pulsed nature in time-resolved studies such as those on muscle action. Synchrotron radiation is also likely to be used in some of the better experiments on charge and deformation densities of smaller molecules. Thermal motion, absorption, and extinction which lacks a working model, will be problems in any such work; absorption and extinction may become critical if charge densities are to be determined round heavy metal atoms as in metal clusters, organometallics and catalysts. Good $(X)-(X)$ experiments will require high resolution data extending to large values of $(\sin \theta)/\lambda$, small crystals and short wavelengths to minimise absorption and extinction, and low temperatures to minimise thermal motion. Synchrotron sources will meet all these conditions, including Displex refrigeration to 10 K which may be helpful for molecular crystals with large proportions of non-zero-point vibrational motion.

Both neutron and synchrotron sources will contribute to the expanding field of surface science. The critical reflection neutron spectrometer CRISP at ISIS will soon become a scheduled instrument. It will provide high reflectivity at narrow glancing angles, time of flight wavelength scanning of $(\sin \theta)/\lambda$ without moving parts, sensitivity to magnetism, and a configuration compatible with the study of liquids. Highly collimated synchrotron radiation incident on surfaces at angles of 1-3 mrads has sufficient intensity for the study of monolayers. At Daresbury surface scientists are collaborating in the development of a glancing angle X-ray instrument which will be as flexible as a 4-circle diffractometer but contained within a vacuum system capable of maintaining clean surfaces. Further contributions can also be expected from surface EXAFS (SEXAFS).

CONSOLIDATION AND EXTENSION OF KNOWLEDGE

Prof Jeffrey urged everyone not previously involved with these disciplines to consolidate and extend what they had learned from the lectures, tutorials and informal discussions of this Advanced Study Institute. Those with teaching responsibilities should include the latest material as early as possible in their courses, introducing synchrotron sources with X-ray tubes, spallation sources with nuclear reactors, and time of flight expressions with diffraction theory. Researchers should at once plan experiments at the new facilities.

Now was a time of golden opportunity, especially for young people, to enter these important new fields at an exciting stage of rapid development.

Structure of Proteinase K Using Synchrotron Radiation, and Binding of Two Dipeptide Chloromethyl Ketone Inhibitors to the Active Site

Ch. Betzel ˙. G.P. Pal, J. Bajorath and W. Saenger
˙European Molecular Biology Laboratory (EMBL),
c/o DESY, Notkestrasse 85, D-2000 Hamburg 52, FRG
Institut fur Kristallographie, Fu-Berlin,
Takustrasse 6, D-1000 Berlin 33, FRG

The three-dimensional folding and the sequence of the 279 residue long polypeptide chain of proteinase K show a high degree of homology with the subtilisins. A high resolution refinement was carried out using restrained least-squares methods. The native enzyme and 180 water molecules were refined to R=16.5% for 31000 synchrotron diffraction data between 1.5 and 5.0 Å resolution. The binding mode of two synthetic carbobenzoxy-Ala-Ala-(Ala-Phe)-chloromethyl ketone inhibitors to the active site of proteinase K were determined by difference fourier methods and refined by restrained least-squares methods.

M. A. Carrondo and G. A. Jeffrey (eds.), Chemical Crystallography with Pulsed Neutrons and Synchrotron X-Rays, 567.
© 1988 by D. Reidel Publishing Company.

Protein Crystallography Using Synchrotron Radiation at the EMBL Outstation Hamburg

Ch. Betzel, K. Petratos, N. Pipon, H. Terry & K.S. Wilson
European Molecular Biology Laboratory (EMBL)
c/o DESY, Notkestrasse 85, D-2000
Hamburg 52 FRG

Intense, tunable synchrotron radiation (SR) plays today an important role in the determination of protein structure. The properties of synchrotron radiation can be summarized as high flux, high intensity, fine collimation, tunable over a wide λ -range, polarised and with defined time structure.

The high intensity is used routinely for the measurement of high resolution-data and of data that cannot be measured on a conventional source due to problems associated with weakly diffracting crystals. The tunability of SR is increasingly used to optimize anomalous scattering effects for the phase determination.

The fine collimation is useful for diffraction studies of crystals with very large unit cells (> 300 Å).

Finally, with high intensity beam lines, likes the X11 at the EMBL Outstation, and the use of electronic area detectors the field of kinetic crystallography is tractable.

The two synchrotron beam lines, X11 and X31, and the facilities for protein crystallography in Hamburg will be described.

M. A. Carrondo and G. A. Jeffrey (eds.), Chemical Crystallography with Pulsed Neutrons and Synchrotron X-Rays, 568.
© *1988 by D. Reidel Publishing Company.*

The Structure of $MnPO_4.H_2O$ by Synchrotron X-Ray Powder Diffraction

Philip Lightfoot* and Anthony K. Cheetham; Chemical Crystallography Laboratory, University of Oxford, U.K. and
Arthur W. Sleight; E.I. du Pont de Nemours and Co. Central Research and Development, Experimental Station, Wilmington, DE 19898, U.S.A.

The crystal structure of $MnPO_4.H_2O$ has been determined from high resolution synchrotron X-ray powder diffraction data collected on the X13A instrument(1) at the NSLS, Brookhaven. The powder pattern was indexed on the basis of 20 accurately measured reflections by the automatic indexing program of Visser(2). Integrated intensities were obtained for 61 unambiguously indexed reflections and used to generate a Patterson map from which the position of the manganese atom was determined. The remaining non-hydrogen atoms were located by Fourier methods, and the hydrogen atom placed geometrically and refined without constraints. Refinement of the entire diffraction profile, by the Rietveld method(3) and employing a "Voigt" peak-shape function(4), converged to final agreement factors $R_{wp} = 0.161$, $R_p = 0.122$, $R_{nuc} = 0.047$. The final observed, calculated and difference profiles are show in Figure 1. The compound crystallises in the monoclinic space group, $C2/c$, with lattice parameters a = 6.91169(1), b = 7.46958(1), c = 7.35664(1) Å, $\beta = 112.317(1)°$, Z = 4. The structure consists of axially distorted MnO_6 octahedra linked together, through the oxygen atom of the water molecule at a common vertex, to form zig-zag -Mn-O-Mn-chains running parallel to [101]. These chains are interconnected by PO_4 tetrahedra to form a continuous three-dimensional network.

This poster demonstrates three major uses of the ultra-high resolution and intensity available on the X13A instrument: (i) indexing of previously unknown phases, (ii) structure determination using integrated peak intensities and (iii) structure refinement by the whole-profile method, using a suitable description of the X-ray peak shape. The results confirm the recent conclusion by Attfield et al.(5) that *ab initio* structure determination from

M. A. Carrondo and G. A. Jeffrey (eds.), Chemical Crystallography with Pulsed Neutrons and Synchrotron X-Rays, 569–570.

synchrotron powder data is now a realistic alternative when single crystals are not available.

1. Cox, D.E.; Hastings, J.B.; Cardoso, L.P.; Finger, L.W. Mater. Sci. Forum. **9**, 1 (1986).

2. Visser, J.W. J. Appl. Crystallogr. **6**, 380 (1969).

3. Rietveld, H.M. J. Appl. Crystallogr. **2**, 65 (1969).

4. Young, R.A.; Wiles, D.B. J. Appl. Crystallogr. **15**, 430 (1982).

5. Attfield, J.P.; Sleight, A.W.; Cheetham, A.K. Nature. **322**, 620 (1986).

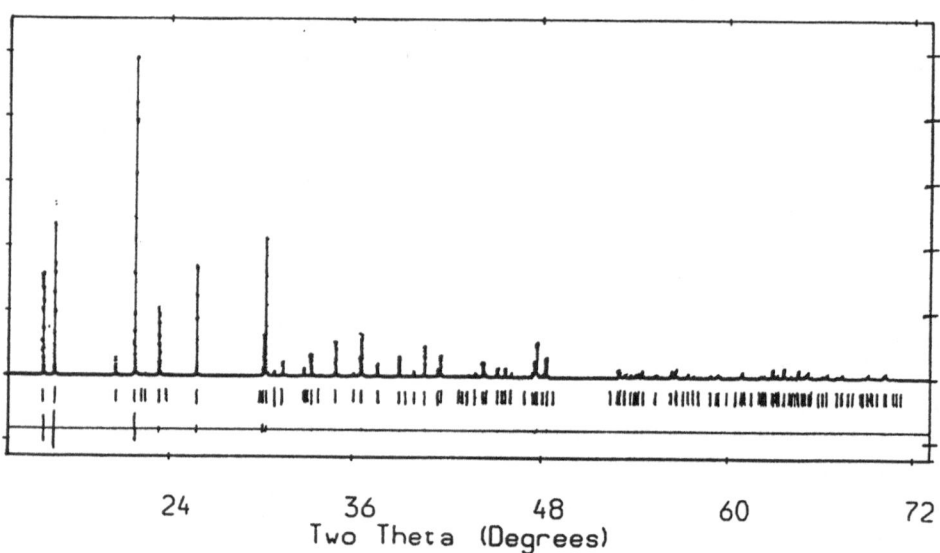

Figure 1: Final observed (points), calculated (solid line) and difference (below) plots for the Rietveld refinement.

ENDIX — a computer program to simulate energy dispersive X-ray and synchrotron powder diffraction diagram

E. Hovestreydt
Institut fur Kristallographie der Universitat
Karlsruhe (TH), Kaiserstr. 12, D-7500 Karlsruhe
Federal Republic of Germany

E. Parthé
Laboratoire de Cristallographie aux Rayons X,
Université de Genève, 24 Quai Ernest Ansermet,
CH-1211 Geneva 4, Switzerland

and

U. Benedict
Europaisches Institut fur Transurane,
Postfach 2266, D-7500 Karlsruhe,
Federal Republic of Germany

Summary

A FORTRAN 77 computer program is described which allows to simulate energy dispersive X-ray and synchrotron powder diffraction diagrams. The input consists of structural data (space group, unit cell dimensions, positional atom coordinates) and information on the experimental conditions (fixed Bragg angle, type of X-ray tube and applied voltage or operating power of synchrotron radiation source). The output consists of a list of normalized intensities of the diffraction lines ordered with increasing energy (in keV) and optionally also of an intensity-energy plot. The intensities are calculated with due consideration of the wavelength dependence of both the anomalous dispersion and the absorption coefficients. For a better agreement between the observed and the calculated spectra a provision is

M. A. Carrondo and G. A. Jeffrey (eds.), Chemical Crystallography with Pulsed Neutrons and Synchrotron X-Rays, 571–572.
© *1988 by D. Reidel Publishing Company.*

made to superimpose optionally on the calculated diffraction line spectrum all additionally observed lines such as fluorescence –, emission lines and escape peaks.

The different effects which have been considered in the simulation are discussed to some detail. The program is demonstrated with a calculation of the energy dispersive powder diffraction pattern of UPt_3 whith Ni_3Sn structure type.

An Investigation of the Structure and Molecular Ordering of Liquid Crystal Phases

R.W. Date, G.R. Luckhurst and J.M. Seddon,
Department of Chemistry, The University, Southampton, SO9 5NH, U.K.

In recent years interest has grown in liquid crystals which contain two similar mesogenic cores linked by a flexible alkyl chain. Such compounds are often called siamese twins in contrast to the conventional or monomeric liquid crystals which only possess one mesogenic unit per molecule.

In order to further our understanding of these novel materials we are investigating the α,β bis (4-n-alkylanilinebenzilidine-4'-oxy) alkanes (m.OnO.m's), the siamese twin form of the monomer 4-n-alkyloxybenzilidine-4'-alkylanilines (nO.m's):

$$H_{2n+1}C_nO\text{-}\bigcirc\text{-}CH{=}N\text{-}\bigcirc\text{-}C_mH_{2m+1} \qquad nO.m$$

$$H_{2m+1}C_m\text{-}\bigcirc\text{-}N{=}CH\text{-}\bigcirc\text{-}O(CH_2)_nO\text{-}\bigcirc\text{-}CH{=}N\text{-}\bigcirc\text{-}C_mH_{2m+1} \qquad m.OnO.m$$

These series are particularly attractive as the nO.m's possess a high degree of liquid crystalline polymorphism and have been extensively studied by X-ray and NMR techniques.

We have synthesised the siamese twin compounds over the range of terminal chain lengths m = 0 to 10 for middle spacer lengths n = 1 to 12 and investigated their liquid crystalline behaviour using polarising microscopy, X-ray diffraction and differential scanning calorimetry. We have found members of this series with m = 0,1 or 2 do not form liquid crystal phases. For higher values of m, nematic phases are observed except when m is approximately equal to or greater than n where smectic phases occur. This switchover to smectic behaviour can be explained by the preference of

573

M. A. Carrondo and G. A. Jeffrey (eds.), Chemical Crystallography with Pulsed Neutrons and Synchrotron X-Rays, 573–574.
© *1988 by D. Reidel Publishing Company.*

the alkyl chains in a particular molecule to be adjacent to the alkyl chains of neighbouring molecules, resulting in the formation of layers.

We are particularly interested in the relationship between the middle spacer chain length and ordering in the mesophases. Initially this was investigated by differential scanning calorimetry which produced a dramatic odd-even variation for the mesophase to isotropic transition entropy. A higher degree of mesophase ordering arises when the central chain has an even number of carbon atoms than when odd. We are now studying the orientational and translational ordering of mesophases by monodomain X-ray diffraction to see whether these properties also show odd-even effects, as implied by the calorimetry results.

Relation Between the Crystalline Structure and the S_E- Phase of VINYL-P-(P'-N-HEXYLOXYPHENYL)BENZOATE

R.Maertens[a], L. Van Meervelt[a], G.S.D. King[a] and G. Germain[b]
[a] Katholieke Universiteit Leuven, Celestijnenlaan 200,
B-3030 Heverlee, [b] Université Catholique de Louvain,
Place Louis Pasteur 1, B-1348 Louvain-la-Neuve, Belgium

Liquid crystalline polymers can be divided in two groups; either the mesophase inducing group is part of the main chain or the liquid crystalline behaviour is induced by mesogenic side-chains. As part of a more elaborate work in the latter area the following compound was obtained by synthesis: vinyl-p-(p'-n-hexyloxyphenyl)benzoate. Liquid crystalline transformations for this compound were detected by DSC and optical microscopy. An arced focal-conic form texture was observed at T=61.91°C and another transformation occurred at T=100.03°C. The latter structure was identified to be a smectic A-phase and the former was assigned a smectic E-phase on the basis of X-ray observations. This low molecular weight compound was selected for studies of polymerisation in organized phases. There seems to be little literature information on the relation between crystal structure and the structure on the highly organized S_E- phases. Hence the present work was undertaken to explore these structure features as well as to broaden knowledge in the field.

The compound ($C_? H_{24}O_3$) crystallizes from n-hexane in the space group P/1 with a=9.415(1), b=9.822(2), c=21.122(8) Å, α=89.03(3), β=89.82(2), γ=72.25(2), $D_c(Z = 4) = 1.159 g.cm^{-3}$, D_m not measured (air inclusions in crystals).

Intensities from a first crystal were measured at 290 K with a Syntex P2$_1$ diffractometer using Nb-filtered MoKα - radiation, ω-scan, (sin θ/λ)max=0.55. The structure was solved by direct methods: 27 of the 48 heavy atoms were found by MULTAN80 after elimination of the 10 highest

M. A. Carrondo and G. A. Jeffrey (eds.), Chemical Crystallography with Pulsed Neutrons and Synchrotron X-Rays, 575–576.

E-values, the rest structure was solved by using DIRDIF81. The refinement was less good, probably due to a less quality of the measurement and the crystal.

Intensities from a second crystal were measured using graphite monochromatized Cu Kα -radiation, $2\theta - \theta$-scan, $(\sin \theta / \lambda)$max=0.5, and a semi-empirical absorption correction based on a ψ-scan was applied. The known structure obtained from the Mo-radiation measurement was used to calculate an electron density map with the Cu-radiation intensities.

Full-matrix refinement (using the X-RAY76 System) with anisotropic temperature factors for the heavy atoms, and isotropic temperature factors for the calculated hydrogen atoms led to a final discrepancy factor R of 0.123 for the observed reflections. The maximum and minimum heights in the difference Fourier are respectively 0.48 and -0.39 e.A^{-3}, giving no hints of disorder.

The molecules are arranged in the unit cell with their long axes in the [011] direction. The two crystallographically independent molecules are in trans configuration. The angle between the calculated best planes through the planar biphenyl parts is 53°. The vinyl system makes an angle of 11° (molecule 1) or 20° (molecule 2) with the biphenyl part. The average C-C length in both phenyl rings is 1.38 Å in both molecules. The mobility of the hexylchain is reflected by the higher temperature factors (U_{eq} C221 =0.0210 A^2, Ueq C222 = 0.0184 A^2) and the decreased C-C bond lengths. No intermolecular atomic distances shorter than the sum of the Van der Waals radii are found in the crystal structure. The molecular packing confirms the herringbone structure of the smectic E-phase.

We are attempting to measure the diffraction pattern of the single crystal above the Cr-S_E transition temperature.

PLUTO drawing

On the Influence of the Non-Bonded Pair of $Pb(II)$ in the Novel Ordered Perovskite $Pb[Sc_{0.5}(Ti_{0.25}Te_{0.25})]O_3$

J. A. Alonso, I. Rasines `
Instituto " Elhúyar".C.S.I.C. Serrano, 113. 28006 Madrid. (Spain).

A large number of ABO_3 perovskites are known in which the B positions are occupied by two kinds of cations (1). When their radii or charges differences are small they distribute at random, but when those differences are big enough both cations adopt a long-range ordered arrangement, giving rise to a superstructure with unit-cell parameters related to that of the ideal perovskite cell, a_0. In the perovskite $Pb[Sc_{0.5}(Ti_{0.25}Te_{0.25})]O_3$ both features are present and, at the same time, the stereochemical influence of the lone pair of $Pb(II)$ is shown.

Experimental and results

$Pb[Sc_{0.5}(Ti_{0.25}Te_{0.25})]O_3$ was obtained as a yellowish-white powder from mixtures of PbO, Sc_2O_3, TiO_2 and TeO_2, heated in air at $1223K$. The X-ray diffraction pattern was characteristic of a cubic ordered perovskite, $a = 2a_0 = 8.0609(2)$ Å, S.G. Fm$\bar{3}$m (n° 225), $Z = 8$. The best discrepancy R factor between observed, I_0, and calculated, I_c, intensities, $R = 0.027$, was obtained for Sc at $4(a)$, 1/2 1/2 1/2 ; Ti and Te at random (1 : 1) at $4(b), 0\,0\,0$; Pb at a quarter of 32(f), x x x (x=0.229); and oxygen at 24(e) u 0 0 (u=0.240).

Discussion

The acceptable agreement found between I_o and I_c for superlattice reflections (111, 311, 331...) confirms that Ti and Te are randomly distributed, whilst Sc is ordered with regard to them giving a 1:1 long-range ordered arrangement. This could be expected from the ionic radii and charges of Sc, Ti and Te.

M. A. Carrondo and G. A. Jeffrey (eds.), Chemical Crystallography with Pulsed Neutrons and Synchrotron X-Rays, 577–578.

On the other hand, the occupancy of 32(f) positions by Pb atoms, at 0.29 Å from those usually adopted, 8(a), 1/4 1/4 1/4 (1), can be explained as a consequence of the electrical repulsion between the non-bonded $6s^2$ pair of *Pb(II)* and the PB-O bonds of its own coordination polyhedron, supposed the lone pair directed towards one of the vertexes of the cube: the shift of the Pb from 1/4 1/4 1/4 to the x x x (x=0.229) along the [111] direction increases slightly the angles between the lone pair and all the Pb-O bonds, minimizing the repulsion effects.

References

1. Nomura, S.; "Crystallographic and magnetic properties of perovskite and perovskite related compounds", in Landolt-Bornstein Zahlenwerte und Funktionen aus Naturwissenshaften und Technik. Neue Series, edited by Hellwege, H.G. and Hellwege, A.M. Berlin, Springer 1978, Gruppe III, Band 12 a, pp. 368-520.

Syntesis and Crystal Structure of $Ta_2Te_2O_9$

J. A. Alonso, A. Castro, E. Gutiérrez Puebla, M. A. Monge, I. Rasines *
and C. Ruiz Valero
Instituto " Elhúyar".C.S.I.C. Serrano, 113. 28006 Madrid. (Spain).
Dpto. Q. Inorgánica. Facultad de Ciencias Químicas.
28040 Madrid. (Spain).

During the last years some systematic X-ray diffraction studies on the ternary oxides of Te(IV) and M(V) (M = V, Nb, Ta) have been performed (1-5) but only the crystal structures of $V_2Te_2O_9$ and $Nb_2Te_3O_{11}$ (4,5) have been described. This work aims to report the preparation and crystal structures of $Ta_2Te_2O_9$, and to compare it to the known structures in the $M_2O_5 - TeO_2$ (M = V, Nb) systems.

Experimental

Single crystals of $Ta_2Te_2O_9$ were grown by slow cooling of $Ta_2O_5 - TeO_2$ melts from 1173 K to 773 K at 2 Kh $^{-1}$. A flat crystal of $0.1 \times 0.1 \times 0.025$mm was mounted in a Nonius CAD-4 diffractometer. The intensities of 1327 unique reflections with $1 < \theta < 25^{\circ}$ were measured with monochromatic MoK_α radiation and a $\omega/2\theta$ scan technique. The intensities were corrected for Lorentz and polarization effects and 1184 of them were considered as observed. The heavy atoms were located from a three-dimensional Patterson map, and those of oxygen from Fourier synthesis. An empirical absorption correction was applied at the end of the isotropic refinement. Mixed full matrix least-squares refinement with anisotropic thermal parameters for the Ta and Te atoms, and isotropic for the oxygen atoms led to R=0.061.

Results and discussion

$Ta_2Te_2O_9$ is monoclinic, S.G. $P2_1/c$ (No. 14), Z = 4, a = 7.100(1), b = 7.486(2), c=14.625(5) Å, $\beta=102.98(2)^0$. The structure consists of infinite puckered layers of composition $(Te_4O_{12})_n$ parallel to the ab plane,

M. A. Carrondo and G. A. Jeffrey (eds.), Chemical Crystallography with Pulsed Neutrons and Synchrotron X-Rays, 579–581.

alternating along the c axis with infinite sheets constituted by nearly regular TaO_6 octahedra sharing corners. The two kinds of tellurium atoms are four- and five-fold oxygen coordinated, the non-bonded pair being directed towards the vacant position of a trigonal bipyramid and a strongly deformed octahedron, respectively.

The known oxides in the $M_2O_5 - TeO_2$ (M = V, Nb, Ta) systems show a gradual increase of the tellurium coordination from M=V to M = Ta. In $V_2Te_2O_9$ Te atoms are threefold coordinated, the trigonal pyramids sharing corners to give isolated Te_2O_5 units; the three- and four-fold coordinated Te atoms of $Nb_2Te_3O_{11}$ form Te_3O_8 groups, whereas in $Ta_2Te_2O_9$ the two kinds of Te polyhedra constitute bidimensional nets, $(Te_4O_{12})_n$. On the other hand, V atoms in $V_2Te_2O_9$ occupy the center of trigonal bypiramids, wich share corners to give $(VO_4)_n$ chains; Nb in $Nb_2Te_3O_{11}$ and Ta in $Ta_2Te_2O_9$ are both six-fold coordinated, but NbO_6 octahedra share corners to give double chains while those of Ta form a bidimensional network. The different stacking of the polyhedra along the sequence chains-double chains-layers can be explained as a consequence of the gradual increase of the ionic character of M-O bonds from M=V to M = Ta.

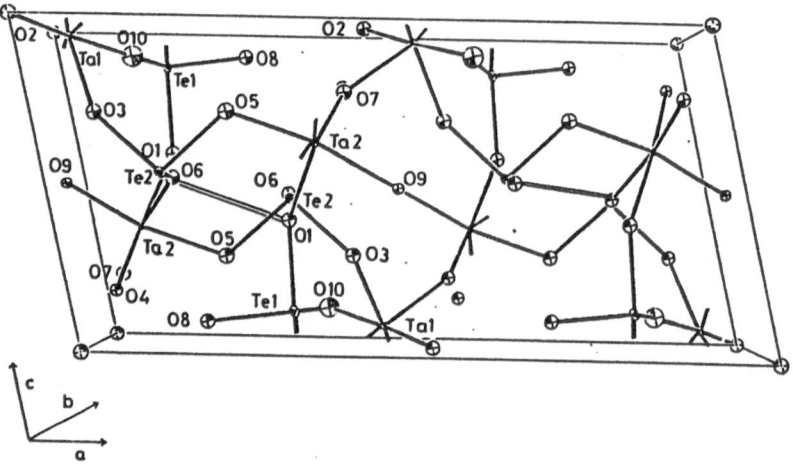

Figure 1: A view of the unit cell.

References

1. G.A. Chase, C.J. Phillips, J. Amer. Ceram. Soc. **47**, 467 (1964).

2. J.C.J. Bart, G. Petrini, Z. Anorg. Allg. Chem. **466**, 81 (1980).

3. J.C.J. Bart, G. Petrini, Z. Anorg. Allg. Chem. **465**, 51 (1980).

4. J. Darriet, J. Galy, Cryst. Struct. Comm. **2**, 237 (1973).

5. J. Galy, O. Lindquist, J. Solid State Chem. **27**, 279 (1979).

Remarks on the Structures and Phase Transitions in Aluminate Sodalites

W. Depmeier, Institut fur Kristallographie der Universitat Karlsruhe (TH)
D-7500 Karlsruhe

Aluminate sodalites of general composition $M_8[Al_{12}O_{24}](XO_4)_2$, with M=Ca,Sr ... and X=S, Cr, Mo, W ... constitute a structural family exhibiting ferroic phase transitions. In the course of our investigations concerning this family we encountered several problems, some of which might be of general interest, viz.:

1. **"Strain"**. The mismatch between the latent cubic symmetry of the framework and that of the orientation of the cage anions results in marked repulsive interaction between XO_4 O atoms and framework O atoms; thereby strong angular and bond length distortions are produced.

2. **"Phase transitions"**. The deviation from cubic symmetry can be cancelled by order-disorder processes; these might be brought about by raising the temperature or by changing the composition of mixed crystals. The phase transitions are of the ferroic type; ferroelastic and ferroelectric species have been found so far. The transitions occur at the zone boundary, fact which accounts for the formation of superstructures and of pronounced pseudomerohedral twinning.

3. **"Theory"**. First steps towards a phenomenological theory have been done. However, the progress suffers from a lack of information; e.g., it is still questionable whether there is only **one** theory for all aluminate sodalites or more than one according to the various degrees of interaction XO_4-framework.

4. **"Multicritical points"**. The investigation of a mixed crystal series indicated that for certain compositions the enthalpy of transition approaches zero; however, the existence of a tricritical point could not been proved, yet. Its existence would be highly interesting because of the six-component order parameter.

582

M. A. Carrondo and G. A. Jeffrey (eds.), Chemical Crystallography with Pulsed Neutrons and Synchrotron X-Rays, 582.
© 1988 by D. Reidel Publishing Company.

RELATIONSHIP BETWEEN CHEMICAL REACTIVITY AND CRYSTAL STRUCTURE IN THE ORGANIC SOLID STATE.

by C. BAVOUX, A. THOZET, M. PERRIN,
Laboratoire de Cristallographie
and R. LAMARTINE, R. PERRIN, J. VICENS,
Laboratoire de Chimie Industrielle
U.A. 0805 : Cristallographie et Chimie des Matériaux
Université Claude Bernard LYON I, France.

During the last twenty years there has been a number of detailed studies devoted to the chemistry of organic molecular solids and made by a number of groups involving chemists and crystallographers.

These workers were interested in the reactions of the organic solids subjected either to a physical agent like light or heat or to a chemical one which can be a solid, a liquid or a gas. Our studies concerne especially phenol molecules and gas (chlorine, hydrogen) ; they can be divided in three groups :

1. Chemical reactivity on different polymorphs of the same compound.
2. Chemical reactivity on different faces of a single crystal.
3. Chemical reactivity on a definite molecule variously surrounded.

In the first group, the poster will give the example of chlorination of 4-chlorophenol which crystallizes in metastable or stable forms. Single crystals of metastable variety react more rapidly than those of the stable form.

About reactivity on different faces of a crystal w'll give two examples, one about 4-chlorophenol, the other about 4-isopropyl-3-methyl phenol (parathymol). In each case faces parallel or perpendicular to a given direction do not react in the same way. In addition, for the case of thymol which crystallizes in the chiral group $P4_1$ w'll give results showing an asymetric synthesis.

At last about the third group, results of chlorination of 3,5-dichlorophenol molecule in pure crystals, in a binary compound with 2,6-dimethyl-phenol and in a complex with ether-18-crown-6 will be given.

M. A. Carrondo and G. A. Jeffrey (eds.), Chemical Crystallography with Pulsed Neutrons and Synchrotron X-Rays, 583–584.
© 1988 by D. Reidel Publishing Company.

For all examples it is necessary to envisage accessibility to the reactive centers of the molecules. So the arrangements of molecules in the crystal and the way in which the molecules appear on the faces are necessary to explain the different results.

For these reasons chemists have to work with crystallographers and pluridisciplinary groups are necessary.

In the future, studies during reactions will be very interesting.

The last example on the poster (transformation of orthocyclohexadienone into paracyclohexadienone by light) shows arrangements of molecules at the beginning and at the end of the reaction. But what happens during the reaction ? perhaps the answer could be done if we study the dynamic of reaction with pulsed neutrons and synchrotron X-rays.

Study by EXAFS of $Pt\text{-}Ru/Al_2O_3$ Catalysts

P. Esteban†, F. Garin, P. Bernhardt and G. Maire‡

† Universitat Autonoma de Barcelona. Dept. Química.
Divisio Quimica Inorganica. Bellaterra. Barcelona.
‡ Laboratoire de Catalyse et Chimie des Surfaces, U.A. 432 du CNRS,
Universite Louis Pasteur, 4, Rue Blaise Pascal 67000 Strasbourg (France).

Two Pt-Ru catalysts (8%Pt-2%Ru and 6%Pt-4%Ru, in weight) have been studied by EXAFS during a "in situ" treatment under H_2 flow. Spectra was collected at room temperature, 473 K, 673 K and again at 473 K.

Analysis of the EXAFS on the L III edge of Pt and the K edge of Ru shows:

1. The reduction of Pt and Ru is complete at 473 K (no such atoms as Cl or O are detected in the first coordination sphere of Pt or Ru).

2. Separate phases of Pt and Pt-Ru are formed:

   (a) two independent Pt-Pt distances (2.78 Å and 2.65 Å), and one Pt-Ru distance (2.65 Å) are detected on the Pt edge.

   (b) two equal Ru-Ru and Ru-Pt distances (2.65 Å) are detected on the Ru edge.

This analysis confirme the formation of bimetallic aggregates as suggested previously by electron-microscopy and catalytic tests.

M. A. Carrondo and G. A. Jeffrey (eds.), Chemical Crystallography with Pulsed Neutrons and Synchrotron X-Rays, 585.
© *1988 by D. Reidel Publishing Company.*

Electron Distribution in the Laves Phase $TiFe_2$

M.J.M. de Almeida, M. Margarida R. Costa, Etelvina M. Gomes
Centro FC1 INIC - Departamento de Física - Universidade de Coimbra
3000 COIMBRA - PORTUGAL

The structure of binary Laves phase compounds, which have in general a stoichiometic composition AB_2, can be described in terms of triangular nets of A and B atoms stacked between a Kagomé net of B atoms. Different stacking possibilities of these nets give rise to three distinct crystallographic structures, namely C 14 $(MgZn_2)$, C 15 $(MgCu_2)$ and C 36 $(MgNi_2)$, which have been investigated by Friauf (1) and Laves et al. (2).

The closest packing of these alloys — which have been considered size-factor compounds — occurs when the radius ratio is $R_A/R_B = 1.225$. Deviations from this ideal value have, however, been observed; this suggests that changes in the valences of the constituent elements take place, causing the necessary adjustments of atomic volumes in order to preserve the closest possible packing. A careful determination of the electron density distribuition in these compounds is therefore of great interest in order to test the above possibility.

Thus, a complete set of structure factors for the hexagonal Laves phase $TiFe_2$ (C 14 type structure) have been measured by X-ray diffraction, using a CAD 4 four-circle diffractometer and $Mo - K\alpha$ radiation. The observed structure factors have been correct for Lorentz, polarization, absorption and extinction effects, and placed on an absolute scale, as described in a previous work (3).

The refinement of structural and temperature parameters using a full-matrix least-squares technique was carried out, based on a postulated spherical model for the distribution of atomic electrons in the solid. The main features of the electron density will be presented and visualized in terms of difference Fourier maps for several sections of the unit cell.

1. Friauf, J.B. - J. Am. Chem. Soc., **49**, 3107 (1927)

2. Laves, F. and Witte, H. - Metallwirtsch. Metallwiss. Metalltech. **14** 645 (1935)

3. M. J. M. de Almeida, M. Margarida R. Costa, Etelvina M. Gomes - Actas da Conferência "Física 86", 1986.

M. A. Carrondo and G. A. Jeffrey (eds.), Chemical Crystallography with Pulsed Neutrons and Synchrotron X-Rays, 586.
© 1988 by D. Reidel Publishing Company.

The Structure of Squaric Acid by Powder Neutron Diffraction

R.J. Nelmes and Z. Tun, Physics Dept., University of Edinburgh
W.I.F. David and W.T.A. Harrison, Rutherford Appleton Laboratory

The location of hydrogen using powder diffraction techniques has hitherto been extremely difficult using traditional X-ray and neutron diffractometers. With X-rays, hydrogen scatters so weakly in the presence of other nuclei as to be almost invisible, whereas with neutrons, naturally occurring hydrogen produces very large backgrounds, resulting from a large incoherent scattering cross-section, which masks all but the strongest Bragg peaks. Deuteration obviates the above problems in neutron diffraction studies but is an expensive and often complicating option. The considerably higher resolution of the high resolution powder diffractometer, HRPD, at ISIS over similar machines results in a substantial increase in the peak to background ratio thus allowing problems of significant complexity to be tackled.

Room-temperature single-crystal neutron and X-ray structure determinations of squaric acid $(H_2C_4O_4)$ indicate that the material has a planar structure that possesses a small monoclinic distortion from tetragonal symmetry. The small pseudosymmetry ($\Delta d/d < 0.001$) and hydrogenous content of squaric acid make it an ideal test candidate for high resolution studies. The data recorded in 12 hours from a sample 2cm. x 1.5cm. in cross-section and 0.5cm. deep were refined straightforwardly to give an excellent fit (weighted profile R-factor $=4.60\%$; expected R-factor $= 4.55\%$; $\chi^2 = 1.02$). Atomic positions agree closely with the single-crystal neutron diffraction study; the locations of the hydrogens are substantially different and less well-determined in the single-crystal X-ray study. Lattice parameters obtained from profile refinement indicate the smallness of the monoclinic distortion and the high lattice precision available using the time-of-flight method (a= 6.12890(5)Å; b=5.26781(5)Å; c=6.14025(6)Å; β=89.9632(5)°: spontaneous ferroelastic strains are $\epsilon_{11} = 9.25(6) \times 10^{-4}$; $\epsilon_{13} = 3.21(4) \times 10^{-4}$).

M. A. Carrondo and G. A. Jeffrey (eds.), Chemical Crystallography with Pulsed Neutrons and Synchrotron X-Rays, 587.
© *1988 by D. Reidel Publishing Company.*

Data Analysis in Time-Resolved Powder Diffractometry

J. Rodriguez(1), M. Anne(2) and J. Pannetier(3)

(1) Instituto de Ciencia de Materiales de Barcelona (ICMAB-CSIC). Marti i Franques s/n, 08028 Barcelona, Spain.

(2) Laboratoire de Cristallographie du CNRS associe a l'USTMG, 166 X, 38042 Grenoble Cedex, France.

(3) Institut Laue-Langevin, 156 X, 38042 Grenoble Cedex, France

Thermodiffractometric or time-resolved experiments in powder diffraction give us a great amount of data, characterizing the surfaces $I(\theta,T)$ or $I(\theta,t)$.

It is not unusual to have 1000 diffraction patterns with 400 point each one. These diffraction patterns content all the information about phase transitions, chemical reactions, kinetic of transformations and so on.

In order to obtain the relevant physical parameters, it is necessary to perform a numerical treatment of the data before any quantitative analysis of the physico-chemical problem involved. It is very hard to do this with conventional computer programs conceived to handle an unique diffraction pattern.

We have developed a program system to analyse these data in an automatic form. For the regions of unique phase, two kind of situations may occur:

1. The structure is not known, but we know aproximately the unit cell. In this case a modified version of the Pawley program (1), together with a command file, can be used to obtain the evolution of the cell parameters, profile parameters characterizing the shape of peaks and integrated intensities as a function of temperature or time.

2. The structure is known and it is not too complex. In this case a modified version of the Wiles and Young program (2) can be used to perform an structure refinement as a function of temperature. This program uses the Rietveld Method (for nuclear structure only) and must be used together with a command file.

M. A. Carrondo and G. A. Jeffrey (eds.), Chemical Crystallography with Pulsed Neutrons and Synchrotron X-Rays, 588–589.
© *1988 by D. Reidel Publishing Company.*

In both cases the command file controls the execution of the program in a loop. In every step of the loop, the input files are actualized to the current diffraction pattern. The fitted parameters of a patern are the input for the next one.

In the zones of coexistence of two phases of known structure the case 2) could be used, provide that some constraints are stablished between parameters. From the fitted scale factors one can deduce the relative amount of the two phases.

If the structure of the existing phases is unknown, a selected zone of the diffraction paterns may be used to obtain the evolution of position, integrated intensity and breadth of some independent peaks. This can be done with a more conventional program used to fit line profiles. At present, we are trying to modify the Pawley program to handle at least three different phases at the same time.

1. Pawley G.S. (1981). J. Appl. Cryst. 14, 357.

2. Wiles, D.B. & Young, R.A. (1981). J. Appl. Cryst. 14, 149.

Phase Transitions In $Sr_2CO_2O_5$: a Neutron Diffraction Study

J. Rodriguez(1), J.M.G. Calbet(2), J.C. Grenier(3)
J. Pannetier(4) and M. Anne(5)

(1) Instituto de Ciencia de Materiales de Barcelona (ICMAB-CSIC). Marti i Franques s/n, 08028 Barcelona, Spain
(2) Dpto. Quimica Inorganica, Facultad de Quimicas,
Universidad Complutense, 28040 Madrid, Spain.
(3) Laboratoire de Chimie du Solide du CNRS,
33045 Talence Cedex, France
(4) Laboratoire de Cristallographie du CNRS associe a l'USTMG, 166 X, 38042 Grenoble Cedex, France.
(5) Institut Laue-Langevin, 156 X, 38042 Grenoble Cedex, France.

$Sr_2Co_2O_5$, when quenched to room temperature from 910°C, adopts a metastable brownmillerite structure (B). This structure as a result of vacancy ordering in the parent perovskite $SrCoO_3$-x substructure, is characterized by an alternating sequence of octahedral CoO_6 and tetrahedral CoO_4 layers along the b axis. This phase is antiferromagnetic and the Co+3 cations are in a high spin state (1).

By heating $Sr_2Co_2O_5$ (B) in inert atmosphere at 750°C, a rhombohedral phase (R) is obtained. The Co^{+3} cations are in an intermediate state between high and low spin (1,2).

We have carried out an experiment at the Institut Laue-Langevin in Grenoble, using a position sensitive detector (D1B) in order to study this phase transition. Starting with $Sr_2Co_2O_5$ (B) at room temperature, the sample was heated in a vacuum up to 920°C at a controlled rate (0.6°C/min), then cooled again to room temperature. Every 2 minutes, a diffraction pattern was stored.

In the surfaces representing the intensity versus the Bragg angle and temperature, five transitions can be detected:

1) Bm → Bp, which correspond to the transition from antiferromagnetic ordered B-phase to the paramagnetic state (278°C).

M. A. Carrondo and G. A. Jeffrey (eds.), Chemical Crystallography with Pulsed Neutrons and Synchrotron X-Rays, 590–591.
© 1988 by D. Reidel Publishing Company.

2) Bp → R, which has a coexistence range of temperatures (530-588"C). This is a reconstructive first order phase transition.

3) R → R', it is an unknown transition which takes place at about 750°C and it is characterized by an anisotropic displacement of the peaks which can not be indexed in a simple way. The R' phase is probably an inconmensurate state of the R phase.

4) R' → P, it was expected to be the R → Bp transition. This transition to a cubic perovskite structure is also reconstructive and takes place progressively at 882°C. The transformation to the P phase is completed at 915°C.

5) P → R', on cooling, the perovskite phase is transformed again in the R' phase. R' begins to appear at 840°C and the P phase disappears at 772°C.

This study shows that the anionic vacancy ordering, from the disordered cubic perovskite at high temperature to the brownmillerite structure at room temperature, takes place in a few seconds during the quenching process.

1. J.C. Grenier et al., Mat. Res. Bull., 14, 831 (1979).

2. J. Rodriguez & J.M.G. Calbet, Mat. Res. Bull., 21, 429 (1986).

A Rietveld Profile Analysis Treatment of Line-Broadening in $KAlF_4$ layered compound

A. Gibaud
Université du Maine
Faculté des Sciences
72017 Le Mans Cedex, France

In the $KAlF_4$ layered compound, some of the lines present in the X-ray and neutron powder diffraction patterns are found to be abnormally enlarged. In such a case, the classical Rietveld method is not satisfactory to take into account the enlargement. We have therefore used a multi-pattern version of the Rietveld program modified by Le Bail in order to take into account the line broadening due to size and strain effects. It is shown that the broadening can be interpreted as coming from antiphase domains resulting from opposite rotations of AlF_6 octahedra.

M. A. Carrondo and G. A. Jeffrey (eds.), Chemical Crystallography with Pulsed Neutrons and Synchrotron X-Rays, 592.
© 1988 by D. Reidel Publishing Company.

600

601

FORMULA INDEX